Frontier Topics in Nuclear Physics

NATO ASI Series

Advanced Science Institutes Series

A series presenting the results of activities sponsored by the NATO Science Committee, which aims at the dissemination of advanced scientific and technological knowledge, with a view to strengthening links between scientific communities.

The series is published by an international board of publishers in conjunction with the NATO Scientific Affairs Division

A	Life Sciences	Plenum Publishing Corporation
B	Physics	New York and London
C	Mathematical and Physical Sciences	Kluwer Academic Publishers
D	Behavioral and Social Sciences	Dordrecht, Boston, and London
E	Applied Sciences	
F	Computer and Systems Sciences	Springer-Verlag
G	Ecological Sciences	Berlin, Heidelberg, New York, London,
H	Cell Biology	Paris, Tokyo, Hong Kong, and Barcelona
I	Global Environmental Change	

Recent Volumes in this Series

Volume 328 — Quantum Field Theory and String Theory
edited by Laurent Baulieu, Vladimir Dotsenko, Vladimir Kazakov, and Paul Windey

Volume 329 — Nonlinear Coherent Structures in Physics and Biology
edited by K. H. Spatschek and F. G. Mertens

Volume 330 — Coherent Optical Interactions in Semiconductors
edited by R. T. Phillips

Volume 331 — Hamiltonian Mechanics: Integrability and Chaotic Behavior
edited by John Seimenis

Volume 332 — Deterministic Chaos in General Relativity
edited by David Hobill, Adrian Burd, and Alan Coley

Volume 333 — Perspectives in the Structure of Hadronic Systems
edited by M. N. Harakeh, J. H. Koch, and O. Scholten

Volume 334 — Frontier Topics in Nuclear Physics
edited by Werner Scheid and Aurel Sandulescu

Volume 335 — Hot and Dense Nuclear Matter
edited by Walter Greiner, Horst Stöcker, and André Gallman

Series B: Physics

Frontier Topics in Nuclear Physics

Edited by

Werner Scheid

Justus-Liebig-University
Giessen, Germany

and

Aurel Sandulescu

Institute of Atomic Physics
Bucharest, Romania

Springer Science+Business Media, LLC

Proceedings of a NATO Advanced Study Institute on
Frontier Topics in Nuclear Physics,
held August 24–September 4, 1993,
in Predeal, Romania

NATO-PCO-DATA BASE

The electronic index to the NATO ASI Series provides full bibliographical references (with keywords and/or abstracts) to more than 30,000 contributions from international scientists published in all sections of the NATO ASI Series. Access to the NATO-PCO-DATA BASE is possible in two ways:

—via online FILE 128 (NATO-PCO-DATA BASE) hosted by ESRIN, Via Galileo Galilei, I-00044 Frascati, Italy

—via CD-ROM "NATO Science and Technology Disk" with user-friendly retrieval software in English, French, and German (©WTV GmbH and DATAWARE Technologies, Inc. 1989). The CD-ROM also contains the AGARD Aerospace Database.

The CD-ROM can be ordered through any member of the Board of Publishers or through NATO-PCO, Overijse, Belgium.

Library of Congress Cataloging-in-Publication Data

On file

ISBN 978-0-306-44831-7 ISBN 978-1-4615-2568-4 (eBook)
DOI 10.1007/978-1-4615-2568-4

© 1994 Springer Science+Business Media New York
Originally published by Plenum Press New York in 1994

All rights reserved

No part of this book may be reproduced, stored in a retrieval system, or transmitted in any form or by any means, electronic, mechanical, photocopying, microfilming, recording, or otherwise, without written permission from the Publisher

PREFACE

This volume contains the lectures and contributions presented at the NATO Advanced Study Institute (ASI) on "Frontier Topics in Nuclear Physics", held at Predeal in Romania from 24 August to 4 September 1993. The ASI stands in a row of 23 Predeal Summer Schools organized by the Institute of Atomic Physics (Bucharest) in Predeal or Poiana-Brasov during the last 25 years.

The main topics of the ASI were cluster radioactivity, fission and fusion. the production of very heavy elements, nuclear structure described with microscopic and collective models, weak interaction and double beta decay, nuclear astrophysics, and heavy ion reactions from low to ultrarelativistic energies. The content of this book is ordered according to these topics.

The ASI started with a lecture by Professor Greiner on the "Present and future of nuclear physics", showing the most important new directions of research and the interdisciplinary relations of nuclear physics with other fields of physics. This lecture is printed in the first chapter of the book.

Cluster radioactivity means the spontaneous emission of carbon, fluorine or other light nuclei out of very heavy nuclei. This new field of nuclear physics began its development in the Institute of Atomic Physics in Bucharest 15 years ago when lifetimes for cluster radioactivity were theoretically predicted. Today cluster radioactivities are measured at various laboratories in the world. The recent theoretical and experimental aspects of cluster radioactivity were thoroughly presented in lectures and contributions at the ASI and are collected together with those of cold fission and fusion in the second chapter of this volume.

Another major point of the ASI were the lectures on nuclear structure presented in chapters III (Heavy elements) and IV (Nuclear structure). This important field is presently experiencing a renaissance which is supported by new experimental techniques such as the crystal-ball spectrometers and by refined theoretical models including cluster and continuum states. The search for new elements with charge numbers larger than 109 is now based on new estimates of the nuclear stability against alpha-decay and fission.

The double beta decay allows to fix an upper bound of the neutrino mass. The newest results of double beta decay of ^{76}Ge, obtained in the Gran Sasso experiment, are given in chapter V on weak interaction and double beta decay.

The last three chapters of the book are mainly devoted to nuclear rections and heavy ion collisions in connection with astrophysics (chapter VI), heavy ion physics (chapter VII) and miscellaneous physics (chapter VIII). Out of the topics of these lectures we like to point to investigations of resonance structures in cross sections seen in light and medium heavy ion collisions, which are essential for our understanding of the formation of nuclear molecules and of the cosmological (astrophysical) synthesis of light nuclei.

The atmosphere of the ASI was very exciting and stimulating. We were very satisfied with the excellent and pedagogically presented lectures and the very interesting contributions of the ASI-students. The lectures were very lively, especially because of the interesting questions, new suggestions and ideas raised by the participants.

The ASI had a very high educational, scientific and political rank in Romania. In the opening session on 24 August 1993 Professor Sandulescu welcomed the President of the Romanian Senate, Professor Dr. O. Gherman, the Romanian Minister of Science and Technology, Professor Dr. D. Palade, the President of the Romanian Academy, Professor Dr. M. Draganescu, the Vice-President of the Romanian Academy, Professor Dr. R. Grigorovici, the Director of the Unesco-Office in Bucharest, Professor Dr. I. Vaideanu and the Director of the Institute of Atomic Physics in Bucharest, Professor Dr. G. Pascovici. We are grateful to them for their interest in the ASI and their addresses to the participants.

The NATO-ASI was awarded and generously supported by the Scientific and Environmental Affairs Division of the NATO. It was also supported by the UNESCO, the Romanian Ministry of Science and Technology, the Romanian Academy, the Institute of Atomic Physics in Bucharest, the Romanian Physical Society and, last but not least, by the German Ministry of Research and Technology through the Romanian-German collaboration contract. We express our deep gratitude to these institutions for their financial help.

The Romanian scientific secretaries, Dr. M.I. Cristu and Dr. S. Stoica, and the technical secretaries have done an extremely valuable and very difficult job with great success during these days. We are obliged and thankful to them for their very efficient work.

Finally, we thank Dipl.-Phys. Stefan Hofstetter very much for his very excellent collaboration in editing this volume.

Giessen, January 1994

Aurel Sandulescu
Werner Scheid

CONTENTS

I. Present and Future of Nuclear Physics (Introductory Lecture)   1

1. **Present and future of nuclear physics**
   W. Greiner   3

II. Cluster Decay, Fission and Fusion   37

2. **Cluster radioactivity**
   P. B. Price   39

3. **Recent advances in cluster radioactivities**
   D. N. Poenaru and W. Greiner   45

4. **Cluster radioactivity. Two-soliton solutions on sphere**
   A. Ludu, A. Sandulescu and W. Greiner   57

5. **Alpha and ^{12}C radioactivities of ^{114}Ba**
   G. Popa, A. Sandulescu, I. Silisteanu and W. Scheid   71

6. **Fine structure in the cluster decays**
   O. Dumitrescu   73

7. **Microscopic calculation for alpha and ^{12}C emissions from proton-rich Ba and Ce isotopes**
   A. Florescu and A. Insolia   79

8. **Clustering and the continuum in nuclei**
   R. J. Liotta   81

9. **Cluster formation and decay in a microscopic model**
   R. G. Lovas, K. Varga and R. J. Liotta   87

10. **The alpha-particle mean field and consistent pre-equilibrium and statistical emission**
    M. Avrigeanu, V. Avrigeanu and P. E. Hodgson   99

11. **Neutron multiplicities in spontaneous fission and nuclear structure studies**
    J. H. Hamilton, J. Kormicki, Q. Lu, D. Shi, K. Butler-Moore, A. V. Ramayya, W.-C. Ma, B. R. S. Babu, G. M. Ter-Akopian, Yu. Ts. Oganessian, G. S. Popeko, A. V. Daniel, S. Zhu, M. G. Wang, J. Kliman, V. Polhorsky, M. Morhac, J. D. Cole, R. Aryaeinejad, R. C. Greenwood, N. R. Johnson, I. Y. Lee and F. K. McGowan   101

12 Cold fission
F. GÖNNENWEIN 113

13 Theory of super-asymmetric cold fission and cluster-decay
R. K. GUPTA 129

14 Superasymmetric fission trajectories in a tridimensional configuration space
M. MIREA 141

15 Quantum tunneling spectrum and application to cold fission
E. STEFANESCU and A. SANDULESCU 143

16 Fusion and quasi-elastic reactions at near-barrier energies
L. CORRADI, D. ACKERMANN, S. BEGHINI, G. MONTAGNOLI, L.MUELLER, D. R. NAPOLI, C. PETRACHE, G. POLLAROLO, N. ROWLEY, F. SCARLASSARA, G.F.SEGATO, C.SIGNORINI, P.SPOLAORE, F.SORAMEL, A.M.STEFANINI 145

17 A pocket in cold fusion potential barrier
R. GHERGHESCU 147

III. Heavy Elements 149

18 Properties of heaviest nuclei
R. SMOLAŃCZUK, J. SKALSKI and A. SOBICZEWSKI 151

19 The prospects of heavy element research
G. MÜNZENBERG 157

20 Multinucleon transfer reactions – an alternative path to heavy element synthesis
M. T. MAGDA 169

21 Microscopic and semi-microscopic approach to the properties of transactinide nuclei
L. BITAUD, J.F. BERGER, J. DECHARGÉ and M. GIROD 181

IV. Nuclear Structure 187

22 On the origin of rotations and vibrations in atomic nuclei
J. P. DRAAYER, C. BAHRI and D. TROLTENIER 189

23 Particle-rotor model description of deformed nuclei
A. COVELLO, A. GARGANO and N. ITACO 207

24 Spin-dependent generalized collective model
M. GREINER, D. HEUMANN, W. SCHEID, G. BRAUNSS and P. HESS 217

25 Two phonon excitations in heavy nuclei studied in photon scattering experiments
P. VON BRENTANO, A. ZILGES, R.-D. HERZBERG, U. KNEISSL and H. H. PITZ 221

Contents ix

26 Toward a complete understanding of the scissors mode
N. LO IUDICE 229

27 Electric excitations in backward electron scattering
M. DINGFELDER, R. NOJAROV and A. FAESSLER 243

28 Observation of non yrast states in Barium nuclei following isomeric β-decay with the OSIRIS cube detector array
P. VON BRENTANO, K. KIRCH, U. NEUNEYER, G. SIEMS and I. WIEDENHÖVER 245

29 Identification of the two pairing-vibration phonon$\otimes$particle mode in ^{145}Sm and neighbours
A. M. OROS, L. TRACHE, G. CATA-DANIL, P. VON BRENTANO, K. O. ZELL, G. GRAW, D. HOFER and E. MÜLLER-ZANOTTI 253

30 Triaxial deformation in proton-odd Cs nuclei
O. VOGEL, A. GELBERG and P. VON BRENTANO 255

31 Shape coexistence in neutron-deficient ^{111}Sb
V. E. IACOB, C. STAN-SION, M. PARLOG, A. BERINDE, N. SCINTEI, N. NICA and L. TRACHE 257

32 $\mathcal{J}^{(2)}$ anomalies in the yrast superdeformed band of ^{149}Gd
G. DUCHÊNE and the EUROGAM collaboration 259

33 Lifetime measurements with the gamma ray induced Doppler (grid) broadening method
A. JUNGCLAUS, H. G. BÖRNER, J. JOLIE, K. P. LIEB and S. ULBIG 261

34 On line study of neutron deficient Hafnium isotopes using fast radiochemical separations
D. TRUBERT, M. HUSSONNOIS, J. F. LE DU, L. BRILLARD, V. BARCI, G. ARDISSON, Z. SZEGLOWSKI, O. CONSTANTINESCU and YU.TS. OGANESSIAN 263

35 Toroidal multipole transitions in collective excitations of atomic nuclei
Ş. MIŞICU 267

36 How important is the proper treatment of translational invariance in the analysis of electron scattering from nuclei?
K. W. SCHMID 269

37 Variational approach to complex nuclear structure problems
K. W. SCHMID and A. PETROVICI 279

38 Description of the continuum in calculating partial decay widths of giant resonances
T. VERTSE, P. LIND, R. J. LIOTTA and E. MAGLIONE 281

39 High-quasiparticle calculation in spherical nuclei
J. BLOMQVIST, A. INSOLIA, R.J. LIOTTA and N. SANDULESCU   283

40 Cluster approach to atomic nuclei
G. S. ANAGNOSTATOS   285

41 Deformation, clusterization, fission and SU(3)
J. CSEH, G. LÉVAI, K. VARGA, R. K. GUPTA and W. SCHEID   295

42 Nuclear shapes at very high angular momenta
R. K. GUPTA, J. S. BATRA, S. S. MALIK, P. O. HESS and W. SCHEID   307

V. Weak Interaction and Double Beta Decay   309

43 Double beta decay and neutrino mass. The Heidelberg–Moscow experiment
H. V. KLAPDOR-KLEINGROTHAUS   311

44 Anharmonic and deformation effects in $2\nu\beta\beta$ decay
A. A. RADUTA   331

45 Nuclear aspects of double beta decay
S. STOICA   343

46 Progress report on an experiment to measure parity mixing of the $J^\pi T = 0^+1; 0^-1$ doublet in ^{14}N
M. PREISS and G. CLAUSNITZER   353

47 A new look to the nuclear structure calculations of the PNC cases in A=18-21 nuclei
M. HOROI, G. CLAUSNITZER, B. A. BROWN and E. K. WARBURTON   355

VI. Nuclear Astrophysics   357

48 Various problems of nuclear astrophysics approached by Coulomb dissociation experiments
H. REBEL   359

49 Weakly interacting massive particles as the dark matter of the universe
P. B. PRICE, D. P. SNOWDEN-IFFT and E. S. FREEMAN   369

50 Recent topics from nuclear reactions in the energies ranging from keV to GeV
K. KUBO   379

51 An electromagnetic calorimeter for spectroscopy of TeV cosmic rays muons
I.M. BRÂNCUŞ, H.J. MATHES, J. WENTZ, H. BOZDOG, M. DUMA, M. KRETSCHMER, G. PASCOVICI, M. PETCU, H. REBEL and K. W. ZIMMER   395

52 Thermonuclear reaction rate uncertainties from nuclear model calculations
V. AVRIGEANU and A. HARANGOZA   403

VII. Heavy Ion Collisions   405

53 Nuclear molecular phenomena in heavy ion collisions
J. Y. PARK, A. THIEL, W. SCHEID and W. GREINER   407

54 Energy dependence of the inverted scattering potentials of the $^{12}C + ^{12}C$ system in the range $E_{cm} = 8 - 12$ MeV
B. APAGYI and W. SCHEID   419

55 The Ni+Ni puzzle: Resonances in the scattering of medium heavy ions?
N. CINDRO, U. ABBONDANNO, Z. BASRAK, M. BETTIOLO, M. BRUNO, M. D'AGOSTINO, P.M. MILAZZO, R.A. RICCI, W. SCHEID, J. SCHMIDT, G. VANNINI and L. VANNUCCI   421

56 Quasimolecular states - a particular case of the new-class resonant states
C. GRAMA, N. GRAMA and I. ZAMFIRESCU   433

57 Classical cluster formation, application to nuclear multifragmentation
J. B. GARCIA, C. CERRUTI, S. GAZAIX   435

58 Azimuthal anisotropies of pions in heavy-ion collisions: a new chance of probing the hot and dense reaction phase?
ST. A. BASS, C. HARTNACK, H. STÖCKER and W. GREINER   439

59 The ratio of the Φ and J/Ψ meson yields as a possible signature of the quark-gluon plasma formation
I. LOVAS   441

60 Processes in peripheral ultrarelativistic heavy ion collisions
M. GREINER, M. VIDOVIĆ and G. SOFF   447

61 Electron-positron pair creation in relativistic atomic heavy ion collisions
J. THIEL, J. HOFFSTADT, N. GRÜN and W. SCHEID   453

VIII. Miscellaneous Topics   465

62 Gamma and meson production by Cherenkov-like effects in nuclear media
W. STOCKER and D. B. ION   467

63 Instantons in QCD
C. ALEXA   479

64 Wigner distribution for the harmonic oscillator within the theory of open quantum systems
AURELIAN ISAR   481

65 The quantum deformation of su(2) into su(1,1) and the potential picture
   A. LUDU   483

66 Potentials in the algebraic scattering theory
   A. ZIELKE and W. SCHEID   485

67 Jahn-Teller distorted excited states of the C_{60} cluster
   P. SURJÁN, L. UDVARDI and K. NÉMETH   487

68 New classes of poles and resonant states
   C. GRAMA, N. GRAMA and I. ZAMFIRESCU   489

69 Classical phase space structure induced by spontaneous symmetry breaking
   M. GRIGORESCU   491

   Author Index   493

   Subject Index   495

I.

Present and Future of Nuclear Physics

(Introductory Lecture)

PRESENT AND FUTURE OF NUCLEAR PHYSICS

Walter Greiner

Institut für Theoretische Physik, Johann-Wolfgang-Goethe Universität, Frankfurt am Main, Germany

The future of a science depends on its open fundamental problems. Therefore we have to ask: are there open fundamental tasks in nuclear physics? My answer is yes, and I will proof it by discussing several such fundamental issues in nuclear physics.

Be reminded that there are certain "Ur-Fragen" as e.g. how did our world begin, how were the present day elementary particles formed (lepton and baryon synthesis) and how the elements in nature built up (nuclear synthesis). Another "Ur-question" is how life has been formed, or how our brain functions.

It is presently believed that our world began with a big bang, a giant phase transition from a true vacuum into a Higgs minimum. This inflationary phase went extremely fast and was followed by the "ordinary" expansion of spacetime. During these very early stages our universe consisted of a phase of matter which we believe is similar to or identical with what we call a quark gluon plasma (QGP). It was originally visualized as hot and dense matter consisting of free quarks and gluons. But during the last couple of years D. Rischke showed by analysing the results of lattice gauge calculations that the QGP seems to be, actually, a cluster plasma, i.e. quarks and gluons are still dominantly coupled to colour singlets. Figure 1 illustrates these pictures. When such a plasma cools down ordinary particles like baryons, mesons etc. are formed. This is then the stage of baryon synthesis.

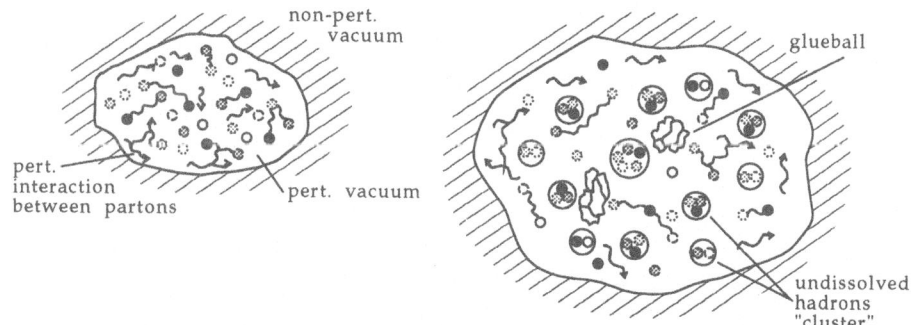

Figure 1. *The so-called gluon plasma seems to be a cluster plasma (second picture) where quarks and gluons still form color singlets. According to D. Rischke only 15% free quarks and gluons are present at the critical temperature T_c.*

Clearly, theoretical concepts like these have to be verified or falsified. Fortunately, through nuclear shock waves, proposed by Scheid and Greiner in 1969 and 1973 and studied theoretically in great detail by Stöcker, Maruhn and others [1], one has a mechanism at hand to compress and heat nuclear matter. It is the key to investigate nuclear matter, its phase transition and its equation of state. When it is compressed and heated, nucleons are excited into Δ's and other resonances, Δ-matter is created (Stöcker, Greiner, Scheid 1978 [2]). Some people renamed it recently "resonance matter" (Metag 1992). This, as well as meson condensates could cause a density isomer of the form shown in figure 2.

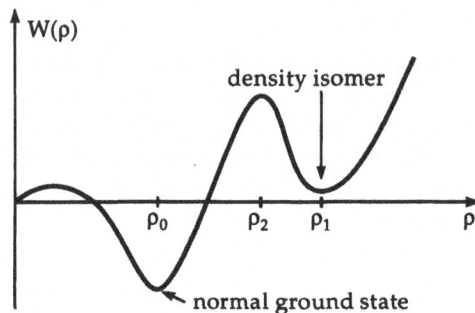

Figure 2. *Illustration of a density isomer in the equation of state (schematic). If the shock wave density reaches the instability window between ρ_1 and ρ_2 the shock will collapse (disperse due to negative pressure) and reappear again for $\rho_{shock} > \rho_1$*

How can it be discovered? Well, when reaching the negative pressure domain of such a density isomeric configuration, shock waves ("flow") should disappear (smear out) and reappear again after densities on the positive pressure side of the second minimum are reached. Such a possibility has already been discussed in the 70'ties (Stöcker,....Greiner 1976 [3]) and E. Schopper's early shock wave experiments (suffering of low statistics) seem to indicate such physics (fig. 3). It must be confirmed or disproved! Using medium heavy projectiles at the SIS at GSI and utilizing the Fopi-detector should enable us to comply.

According to Relativistic Quantum Molecular Dynamics (RQMD) high baryonic densities and temperatures are reached in Pb+Pb collisions at AGS-, CERN- and even RHIC energies. The so-called Bjorken-scenario, according to which the colliding nuclei should fly through each other more or less untouched (like two swarms of insects), is not valid. Instead nuclear matter is stopped to large extend at RHIC energies. Hence it will be the baryon rich quark gluon plasma which is formed. In Pb-Pb collisions at $\sqrt{s}$=200GeV/A about 8000 pions, 30 Λ's and 500 K^+K^- and $K_0\bar{K}_0$ pairs are produced at midrapidity (see fig. 4). This opens the door to generate completely new forms of nuclear matter:

a) In such a micro bang there is a good chance to form multi-pionic atoms, in which the **bosonic character** of the pions enters essentially. They would constitute new atoms, different form fermionic atoms. The π-π interaction will finally be responsible to limit the number of pions in an orbit. The problem is to obtain clusters (i.e. small nuclei) which can capture the pions and to find the location of these clusters in phase space together with the appropriate number of pions in their vicinity. The capture rate is then essentially given by folding the product of this quantity together with the pion capture cross section. b) Multi-Λ-hypernuclei and other Memos (Multiply Strange Exotic Mesonic Objects) are likely to be created. These

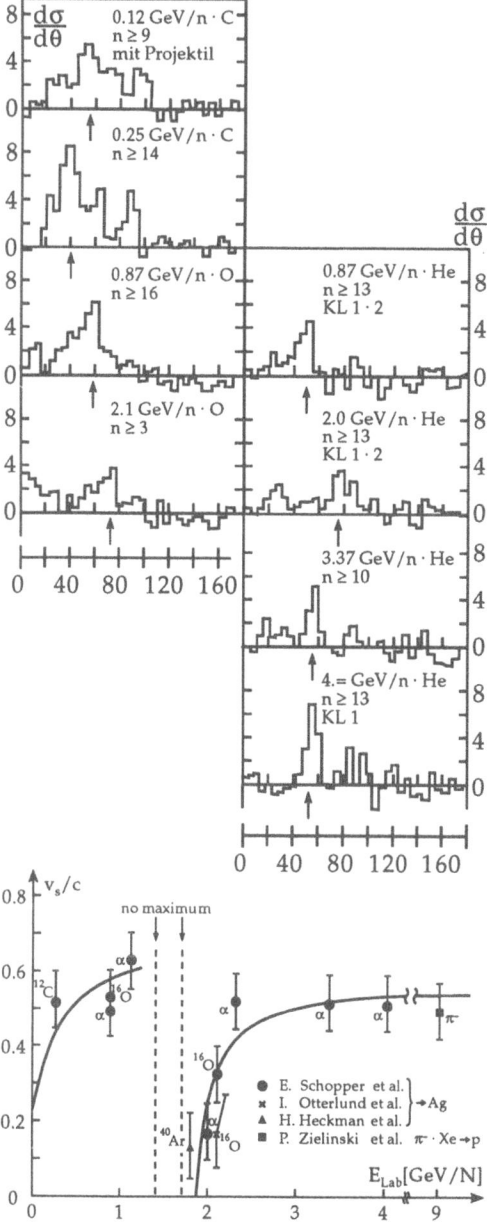

Figure 3. *Early data by E. Schopper show as a function of energy a flow-peak with irregular behaviour between 1200-1800 Mev/A. The shock velocity v_s/c deduced from these data (second figure) magnifies the irregularity. This could be the signature of a density isomere (phase transition to Δ-matter or some sort of meson condensate or both).*

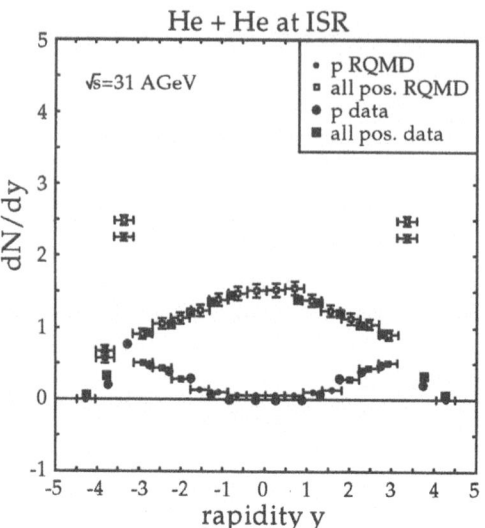

Figure 4. 4a. *Transparency and stopping. Figure 4a) shows He+ He at $s^{1/2} = 31$ AGev (ISR), which is the up to now highest energy achieved for nucleus-nucleus collisions. The RQMD-calculations of T. Schönfeld yield a qualitatively very similar rapidity distribution as $p + p$ (respectively $p + \bar{p}$). This is not unexpected. Nevertheless the agreement of RQMD with data is impressive.*

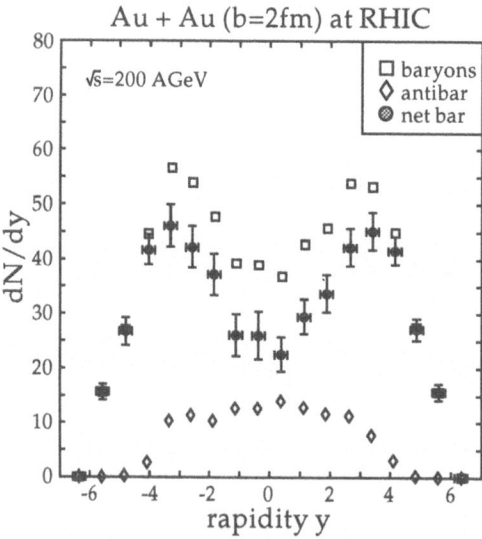

Figure 4. 4b: *Figure 4b) shows the rapidity distribution for Au+Au at RHIC-energies $s^{1/2} = 200$ AGeV. For nearly central collisions the impact parameter b=2 fm has been chosen. Obviously the nuclei do not stop, but there is also no baryon free region at y=0, as many people had expected. The analysis of these results show that mesons contribute about 1/4 to the stopping.*

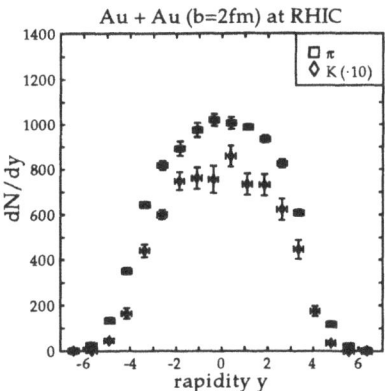

Figure 4. 4c: Figure 4c) shows the π's produced (about 7400) and their rapidity distribution. There is now a plateau at midrapidity and therefore also no Bjorken-scenario. However, the maximum is broad; hence one is probably not too far from a Bjorken-scenario. The π's peak at y=0; the baryons at $y \neq 0$. Hence baryons collide preferentially with slow mesons. This accounts for the contribution $\Delta y_z \approx 0.6$ to baryon stopping. If there would be a meson plateau, one would have equal probability for a baryon to collide with slow and fast meson. In this case there would be no contribution to baryon stopping.

exotic objects have been proposed and intensively studied by Carsten Greiner, J. Schaffner and Horst Stöcker. The dissolved form of the Memos is what has been called strangelet (Carsten Greiner and Horst Stöcker are the pioneers in these possibilities [4]). Tremendous research oppourtunity lies in this domain. For example, the periodic system of elements is generalized by not only having proton and neutron numbers as "degrees of freedom" but also the strangeness number. One might even generalize to negative p-n-axis, i. e. $\bar{p}$, $\bar{n}$, $\bar{\lambda}$-axis and enter into the sector of antimatter (anti-deuterons, anti-carbons, anti-hypernuclei etc.). Fig. 5 illustrates the general ideas of what might happen in a micro-bang. The periodic system of elements has suddenly reached new dimensions, which are breath-taking. The Memos's may exhibit exceptional properties: bound neutral (e. g. $^4M_{2\Lambda}^{2n}$, $^{10}M_{2\Lambda}^{8n}$, pure Λ-droplets like $^8\Lambda$) and negatively charged composite objects with positive baryon numbers (e. g. $^4M_{2\Sigma^-}^{2n}$, $^6M_{2\Lambda,2\Xi^-}^{2n}$) could be formed in rare events. Such negatively charged nuclei can easily be identified in a magnetic spectrometer. They could be considerably more abundant than antinuclei with the same A. The properties of these objects were studied by J. Schaffner, Carsten Greiner and H. Stöcker within the relativistic meson-baryon-field theory, which gives an excellent description of normal nuclear and single-Λ hypernuclear properties. They calculated the rich spectrum of such exotic objects, their stability and structure (J. Schaffner, C. Greiner, H. Stöcker, Metastable Exotic Multihypernuclear Objects...[5]). Also solutions for a large variety of bound short-lived nuclei (e. g. $^8M_{2\Lambda2\Sigma^-}^{2p2u}$) were found, which may decay strongly via the formation of cascade (Ξ) particles. Multi-Ξ-hpernuclei are also possible. It turns out that the properties of such exotic multihypernuclear objects reveal quite similar features as the strangelets, which were propsed as a (unique) signal for quark-gluon-plasma formation in heavy ion collision.

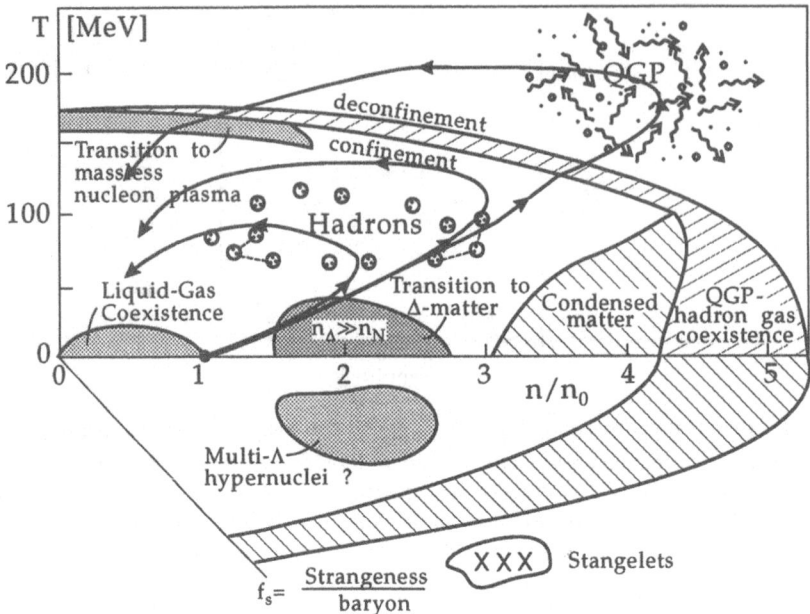

Figure 5. *Phase diagram of hadronic matter, including strangeness. In principle one could supplement this figure by a "negative density" axis and an anti-strangeness axis, which would then allow to include antimatter and its cluster effects in the diagram. Normal nuclear matter occurs at $T = 0$, $n/n_0 = 1$. Deconfinement into a so-called quark-gluon-plasma is expected at high T, n, f_s. Multi strange objects like multi-Λ-hypernuclei and strangelets (which are more or less the dissolved form of multi-Λ-hypernuclei) are also expected.*

Multi-Λ-hypernulei and Memos

The Lagrangian of the Relativistic Mean Field Theory (RMFT), which has been proven to give a very good description of normal nuclei [6], has the following structure:

$$\mathcal{L} = \mathcal{L}_{Dirac} + \mathcal{L}_\phi + \mathcal{L}_V + \mathcal{L}_R + \mathcal{L}_A$$

with additional terms for the hyperons Λ, Σ, and Ξ:

$$\mathcal{L}^{Hyp}_{Dirac} = \bar{\psi}_\Lambda(i\gamma^\nu\partial_\nu - m_\Lambda)\psi_\Lambda + \bar{\psi}_\Sigma(i\gamma^\nu\partial_\nu - m_\Sigma)\psi_\Sigma + \bar{\psi}_\Xi(i\gamma^\nu\partial_\nu - m_\Xi)\psi_\Xi$$

$$\mathcal{L}^{Hyp}_\phi = -g_{\sigma\Lambda}\phi\bar{\psi}_\Lambda\psi_\Lambda - g_{\sigma\Sigma}\phi\bar{\psi}_\Sigma\psi_\Sigma - g_{\sigma\Xi}\phi\bar{\psi}_\Xi\psi_\Xi$$

$$\mathcal{L}^{Hyp}_V = -g_{\omega\Lambda}V^\mu\bar{\psi}_\Lambda\gamma_\mu\psi_\Lambda - g_{\omega\Sigma}V^\mu\bar{\psi}_\Sigma\gamma_\mu\psi_\Sigma - g_{\omega\Xi}V^\mu\bar{\psi}_\Xi\gamma_\mu\psi_\Xi$$

$$\mathcal{L}^{Hyp}_R = -\frac{1}{2}g_{\rho\Sigma}\vec{R}^\mu \cdot \bar{\psi}_\Sigma\vec{\tau}\gamma_\mu\psi_\Sigma - \frac{1}{2}g_{\rho\Xi}\vec{R}^\mu \cdot \bar{\psi}_\Xi\vec{\tau}\gamma_\mu\psi_\Xi$$

$$\mathcal{L}^{Hyp}_A = -\frac{1}{2}eA^\mu\bar{\psi}_\Sigma(1+\tau_0)\gamma_\mu\psi_\Sigma - \frac{1}{2}eA^\mu\bar{\psi}_\Xi(1+\tau_0)\gamma_\mu\psi_\Xi$$

Variation of the fields yields within the mean field approach and in the no sea approximation, i.e. the sum of the densities runs only over the occupied states, four boson fields and three baryon field equations which can be solved numerically for the nuclear ground states.

The parameters of the meson fields and of the nucleon coupling are fitted to the properties of eight spherical nuclei. Calculations done with lower effective masses show instabilities

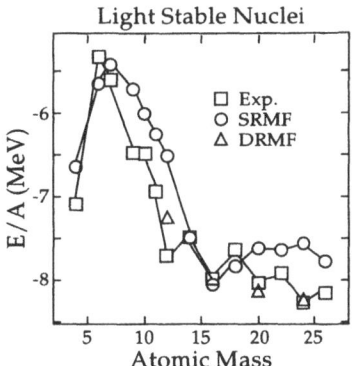

Figure 6. *The binding energy of light stable normal nuclei in spherical relativistic mean field model (SRMF) and in the deformed RMF (DRMF) compared to experimental data.* [5].

for highly dense nuclear systems as ^{12}C and unphysical behaviour in nuclear matter for a negative coefficient c in the standard nonlinear self-interaction of the effective scalar field. The determination of the various parameters is discussed in [5]. An appreciation of the accuracy of these calculations can be obtained from fig. 6, which shows the binding energies for normal light nuclei. In fig. 7 the single particle energies of single-Λ-hypernuclei in the Relativistic Mean Field Theory (RMFT) are compared to recent data. The binding energies of such nuclei are shown in fig. 8. Obviously the RMFT describes existing data rather well. The density distribution of the hyperons is shifted outwards relative to the nucleons according to their smaller binding energy (fig. 9). Thus an extended neutral Λ-halo appears which results in an enhanced interaction radius. Here "anomalons" [7] come in mind.

Early discussions about double-Λ-hypernuclei can be found in [8]. The multi-Λ-hypernuclei, the exotic objects as well as the possibility of their creation in heavy ion collisions came in [5], [6]. In figures 10 and 11 the binding energies of various spherical multi-Λ-hypernuclei are depicted as a function of the number of Λ's. Note that the binding energy increases when hyperons are added to normal nuclei [6], because a new degree of freedom is opened for which the Fermi energy is small for small number of Λ's. The argument is analogous to the discussion of the stability of strange quark matter [4]. All possible pairs of nucleons and hyperons can be sorted according to their strangeness number and their charge (both conserved in strong interactions). The pair with lowest mass is shown in the following tables 1 and 2. Three different baryons can be combined to six metastable configurations (e.g. $\Lambda\Xi^-n$ - see table 2). It is not possible to have more than three baryon species in a metastable configuration, because they will react immediately to form a configuration with two or three different baryons, if the mass difference cannot overcome by binding energy differences inside the bound system.

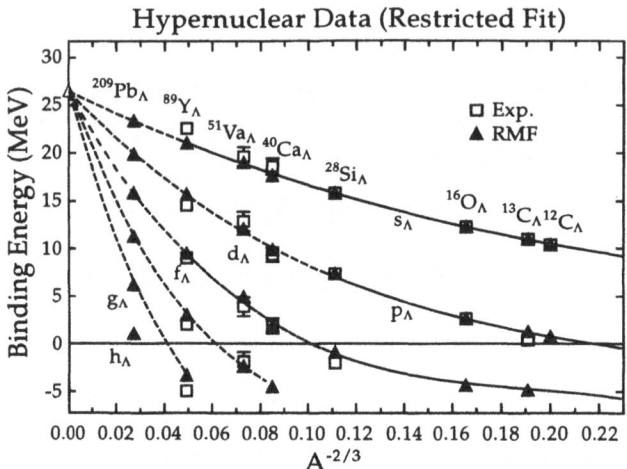

Figure 7. *Single particle energies of single-Λ-hypernuclei in RMFT compared to recent data from* [5].

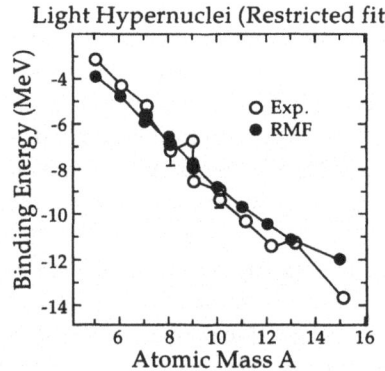

Figure 8. *Binding energies of light hypernuclei calculated in RMFT and compared with experimental data from* [5]

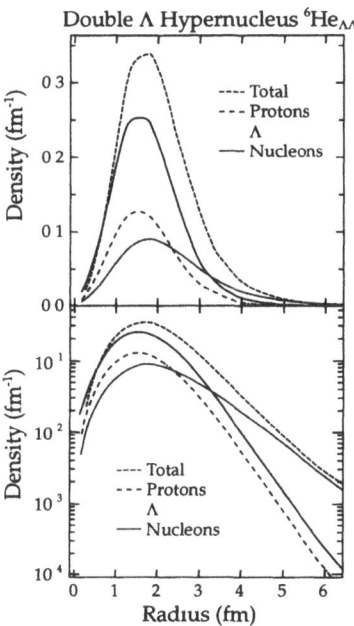

Figure 9. The density distribution times r^2 of the double-Λ-hpernucleus $^6He_{\Lambda\Lambda}$. The Λ-distribution creates a halo around the nucleons and enhances therefore the interaction radius (from [5]).

It is important to note that the combinations listed in tables 1 and 2 cannot decay via meson (pion or kaon) emission into another state. The channels with the smallest Q-value are given by the reactions

$$\Sigma^- + \Sigma^- \longrightarrow \pi^- + \Xi^- + n \quad (\Delta E = -6 \text{MeV})$$
$$\Sigma^+ + \Sigma^+ \longrightarrow \pi^+ + \Xi^0 + p \quad (\Delta E = -14 \text{MeV}) \quad ,$$

where ΔE denotes the mass difference of the reaction. Binding energy differences in a strange composite can cancel the mass differences and these strong interactions would be energetically feasible. Therefore metastable combinations which include Σ's might not be bound. Mesonic reaction channels can only occur for the two reactions considered, because all other mesonic channels need more than 50 MeV binding energy difference in nuclei to overcome the mass difference (i.e. to be exothermic). In view of the maximum binding energy (the potential depth in nuclear matter is ≈ 30 MeV for hyperons, ≈ 50 MeV for nucleons) this seems rather unlikely.

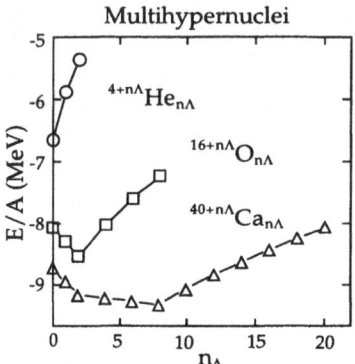

Figure 10. *The binding energy of various multi-Λ-hypernuclei in dependence of the number of Λ's in the nucleus. Multi-Λ-hypernuclei are more stable than normal nuclei.*

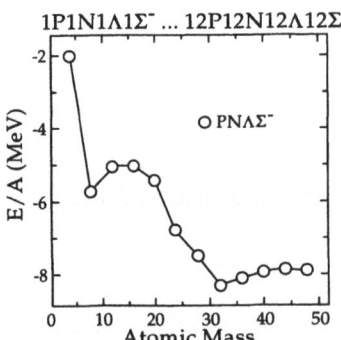

Figure 11. *Binding energy of exotic multi-Λ-hypernuclei consisting of equal number of neutrons, protons, Λ's and Σ's. Note the strong shell effects!*

Table 1. *The configuration of metastable duplets of nucleons and the baryons sorted according to strangeness s and charge q. The numbers given stand for the sum of the masses (in MeV).*

$-s\backslash q$	-2	-1	0	$+1$	$+2$
0			nn 1879	np 1878	pp 1877
1		Σ^-n 2137	Λn 2055	Λp 2054	Σ^+n 2128
2	$\Sigma^-\Sigma^-$ 2395	Ξ^-n 2261	$\Lambda\Lambda$ 2231	$\Xi^0 p$ 2253	$\Sigma^+\Sigma^+$ 2379
3	$\Xi^-\Sigma^-$ 2519	$\Xi^-\Lambda$ 2437	$\Xi\Lambda$ 2431	$\Xi^0\Sigma^+$ 2504	
4	$\Xi^-\Xi^-$ 2643	$\Xi^0\Xi^-$ 2636	$\Xi^0\Xi^0$ 2630		
5	$\Xi^-\Omega^-$ 2994	$\Xi^0\Omega^-$ 2987			
6	$\Omega^-\Omega^-$ 3345				

Table 2. *The configuration of metastable triplets of the nucleons and the hyperons sorted according to strangeness s and charge q.*

$-s\backslash q$	-2	-1	0	$+1$	$+2$
1				Λnp	
2					
3	$\Xi^-\Sigma^-n$	$\Xi^-\Lambda n$		$\Xi^0\Lambda p$	$\Xi^0\Sigma^+ p$
4					
5		$\Xi^-\Xi^0\Lambda$			
6					
7	$\Omega^-\Xi^-\Xi^0$				

Potential metastable configurations, which might be formed in relativistic heavy ion collision, are states consisting of 1) several Λ's, 2) several Λ's (or Σ^-'s) with neutrons or 3) several Ξ's with Λ's and neutrons.

J. Schaffner, Carsten Greiner, and H. Stöcker calculated the total binding energy for a large variety of Memo's. The critical line for zero binding energy can be compared in the two dimensional parameter plane to the region of parameter values compatible with hypernuclear data. They found that droplets of lambdas with or without neutrons are less likely to be bound within our model for parameters consistent with hypernuclear data. As mentioned before, the introduction of correlations for the lambdas may change this conclusion due to the magnification of the $\Lambda\Lambda$ interaction.

Let us summerize the discussion so far: J. Schaffner, C. Greiner and H. Stöcker have demonstrated that Λ-hypernuclei can be described rather well in the relativistic mean field model. Multi-Λ-hypernuclei are more strongly bound than normal nuclei, which is of

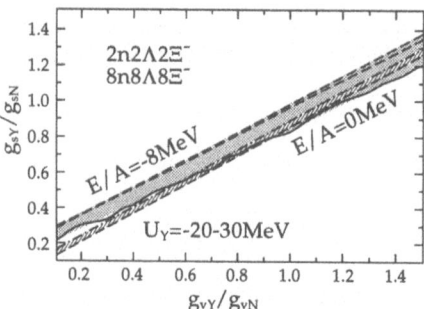

Figure 12. *The binding energy of spherical Memos consisting of neutrons, Λ 's and Ξ 's are compared to the potential depth of the hyperons in nuclear matter. These negatively charged objects cannot be ruled out at the present stage of knowledge of the hypernuclear properties.*

great interest for the formation of multi-hypernuclei in heavy ion reactions. Exotic multi-hypernuclei consisting of Λ's and Σ^-'s may be very stable for certain magic numbers. However, their constituents react strongly and these objects are not stable. Metastable exotic multi-hypernuclear objects (MEMO) can not be excluded within the mean field theory if the coupling constants extracted from hypernuclear data are employed. They are equal to their baryon number ($f_s = 1$), similar as the strangelets. If formed, they could be observed easily in relativistic heavy ion collisions because of their unusual Z/A-ratio.

These hadronic objects compete with the stranglets, their quark counterpart. It is of course viable that Memos form a doorway state to the strangelet production in heavy ion collisions or vice versa: Memo's may first coalesce in high multiplicity region of the reaction. If strangelets are more bound than our "conventional" multihypernuclear clusters, the Memo may then be transformed into a strangelet. Naturally, I cannot and will not describe all the challenges in nuclear and heavy ion sciences: instead I shall concentrate on a few more issues which I consider of particular interest and fundamentality.

The decay of the neutral to the charged vacuum

Let me begin with the overcritical QED-vacuum [9], which can be studied in the collision of very heavy nuclei. You all remember our famous figure showing the supercritical shells during the course of a collision (see figs. 14 and 15). The dynamical production, in itself a high-order QED-process (the cross section being proportional to $\sigma \sim (Z_1 + Z_2)^{20}$) [10], has been calculated by the Frankfurt school and is quantitatively confirmed by experiments (see fig 18). The "diving" of the electronic K-shell in giant atoms has also been experimentally verified up to the diving point (see fig. 16). These binding energies have been deduced from the measurements of the K-ionisation probabilities as a function of the impact parameter (fig. 17). In future it is important to demonstrate the diving, i.e. to follow the K-shell binding energy beyond the diving point. For that it is necessary to obtain precision measurements of K-shell ionisation probabilties at small impact-parameters. Since – in the course of the collision – the created K-hole may be transferred to higher shells, and holes in higher

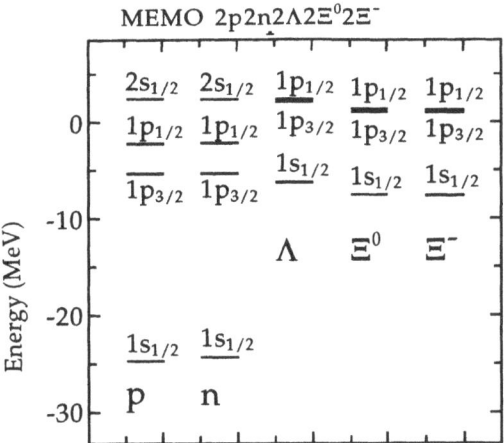

Figure 13. *The single particle energy of a Memo consisting of two of each of the baryons in the baryon octet, except the Σ's. The binding energy difference cancels the mass difference of the strong reaction channels so that the whole "nucleus" is unstable.*

shells may be transferred down to the final atomic K-shell, these transfer-processes need to be calculated with high accuracy in order to ensure the precise determination of K-hole-creation at small impact parameters. Eventually such measurements must be supplemented with hole creation measurements in the L- and M-shells as well.

Great attention has been focused on the e^+e^--coincidence line structures observed in various heavy ion-ion collisions. Such structures showed up recently (T. Cowan) also in

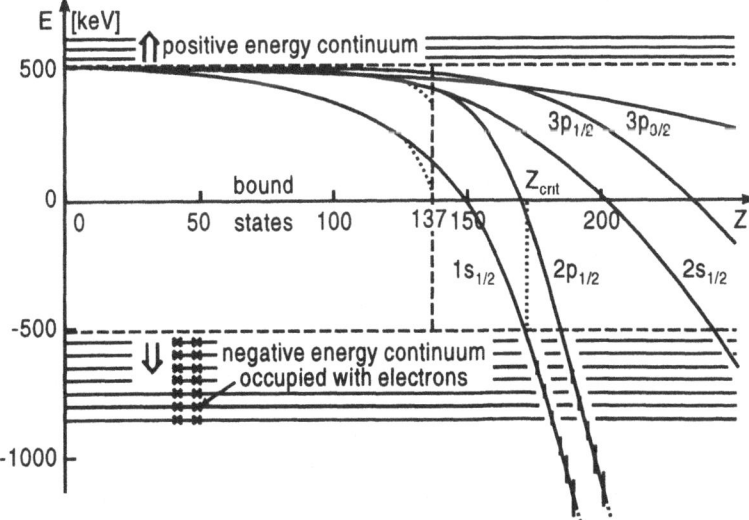

Figure 14. *Energy levels of electrons in a Coulomb potential of central charge Z. At Z_{crit} the 1s level "dives" into the negative energy continuum. This causes the instability of the neutral vacuum which decays under spontaneous positron emission into a charged vacuum.*

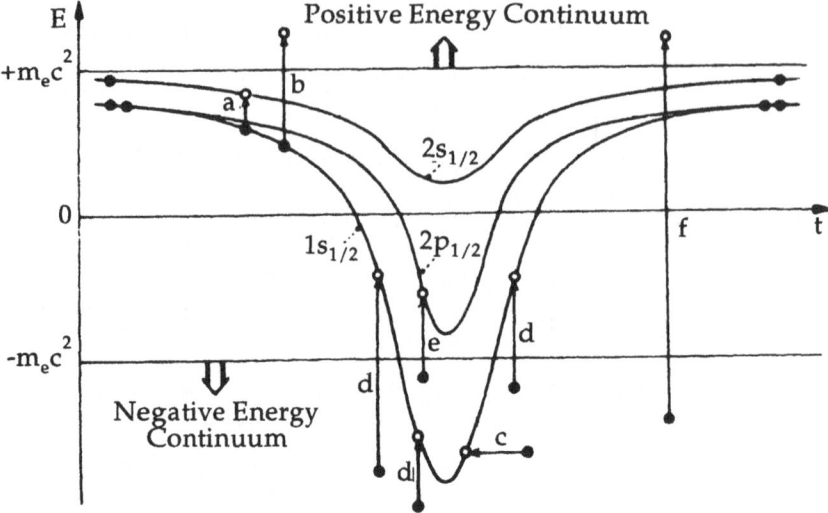

Figure 15. *Electron levels during the course of a heavy ion collision, e.g. U+Cm. Dynamically induced processes are indicated by vertical arrows while spontaneous positron emission is represented by the horizontal arrow.*

medium heavy systems like Xe-Au. I wonder, in fact I will not be surprised if similar observations will also be made in very light systems like ^{16}O+^{24}Mg, etc. Let me explain why: It was believed that new particles were discoverd, but it could quickly be shown that new point (elementary) particles would violate precision measurements (Lamb shift in electronic and myonic atoms, (g-2) for electrons and myons, ...) [11][12]. Complex particles were constructed, bound in the Coulomb field of nuclei, but their decay would always yield dominantly forward-backward e^+e^--correlations [13]. For five coinidence lines observed such forward-backward correlations were originally reported. Meanwhile for four of the e^+e^- sum energy lines the forward backward peaking seems to be invalid, and e^+e^- pairs seem to be emitted into more or less the same direction. In view of the changes in the experimental observations and particularly in view of the extremely complex models to be invented from the theory side, I dare to return to my suggestion made first in my opening address of the German Physical Society Nuclear Physics meeting in spring 1988 in Berlin, where I suggested that the most natural explanation of the observed e^+e^--structures is nuclear pair conversion. The conversion could take place in either the target-, projectile-remnant, in an escaping light nuclear cluster or in all of these. Most likely it will be a $0^+ \rightarrow 0^+$ monopole conversion, but also $E1$-, $M1$- and $E2$-scenarios are thinkable. We must go "back to simplicity" and not believe changing and partly conflicting observations which force us to consider unbelievably complex models for their theoretical interpretation. Meanwhile the observation by Wollersheim, that the e^+e^--line intensity has – as a function of the closest approach between the colliding nuclei – a very similar slope as the neutron transfer (see fig. 19) strengthens my conviction. The excitation and conversion of single nuclear states is, in fact, expected to have steeper slopes as the inclusive single neutron transfer. I consider it a formidable task for the experimentalists to confirm this conjecture and to proceed then to the most important task, i.e. the search for the decay of the neutral vacuum into a charged one.

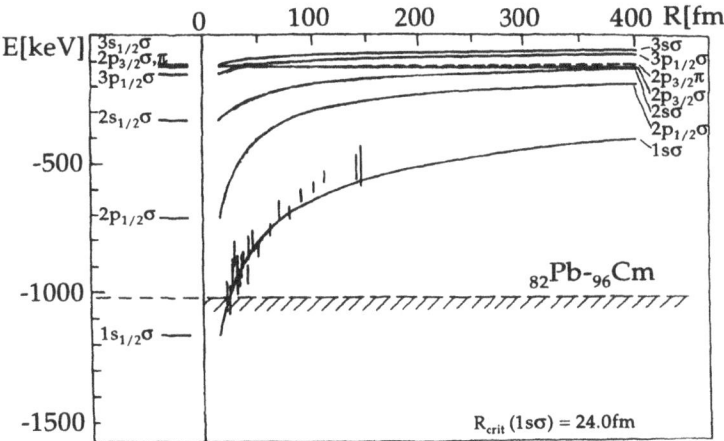

Figure 16. *If one uses the $P(b)$ measurement of fig. 17, one may deduce the binding energy of the electronic K-shell as a function of the distance between the two nuclei. The full curves show the Two-Center-Coulomb calculation of the various levels. Obviously the electronic K-shell acquires a binding energy up to $2m_0c^2$, i.e. up to the diving point on the K-shell.*

For that to be possible, i.e. to see spontaneous single positron emission lines, it is necessary that nuclei stick together (form giant nuclear molecules) for 10^{-20} sec and longer. This can happen if nuclear pockets develop, as were predicted with simple folding potentials [14] - see figs. 20, 21, 22, and 23. Better theoretical studies of this nuclear physics questions are necessary, but equally important is the experimental investigation of this question: Do giant nuclear systems exist; how pronounced are nuclear potential pockets; in which systems do they occur, are third (smaller) clusters emitted; what is the mass transfer?

It seems that such questions can only be solved through measurements of elastic and inelastic excitation functions and mass transfer similary as carried out by D. A. Bromley and collaborators in the study of nuclear molecular phenomena in light ion systems. Nuclear physics at the Coulomb barrier of heavy nuclei is a fascinating topic itself. The possible existence of e.g. U-U giant nuclear system, even for only 10^{-20} sec, would be the most super-hyper-deformed nuclear complex we can imagine! Do such giant systems exist generally or only for specific combinations of projectile and target?

Let me also mention an old idea of mine [15]: When I asked Christoph Schmelzer in our weekly discussions back in 1975, whether it would be technically feasible **to cross beams at very small angles**, he said: "This needs development, but I think it should work".

What I meant is to have Uranium (or other) beams stripped off all electrons (i.e. naked nuclei), to split the beam and to have them crossed at such small angles that their relative energy is less than or equal to their Coulomb energy (see fig. 24). One would then have naked nuclei colliding with naked nuclei and especially superb conditions for studying QED of strong fields. In particular the vacuum decay in strong field can be studied with such systems [16]. Also, of course, single electron-high Z-atoms and all the directly measurable

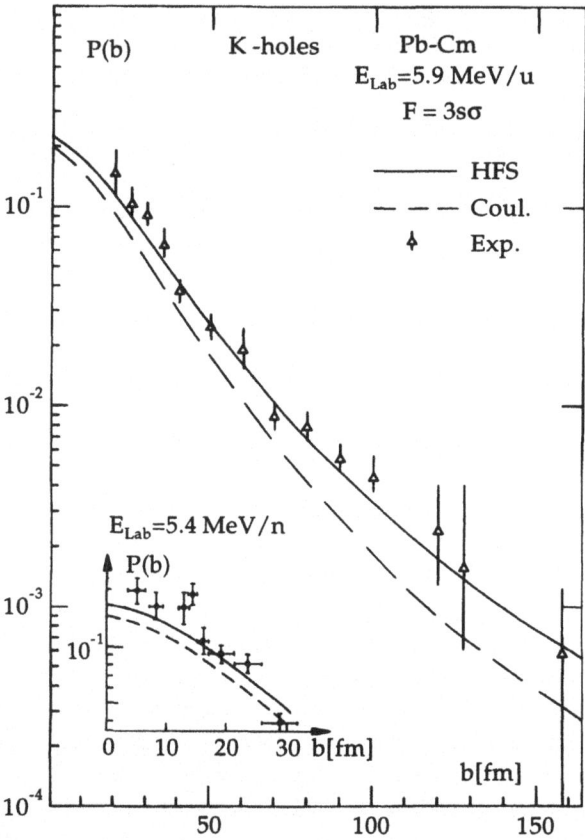

Figure 17. K-hole creation probability $P(b)$ for a Pb-Cm collision as a function of the impact parameter. The insert shows the same for rather small impact parameters. The full curve represents the theory with screening due to other electrons within Dirac-Hatree-Fock calculations. The dashed curves are for pure two-center Coulomb potentials.

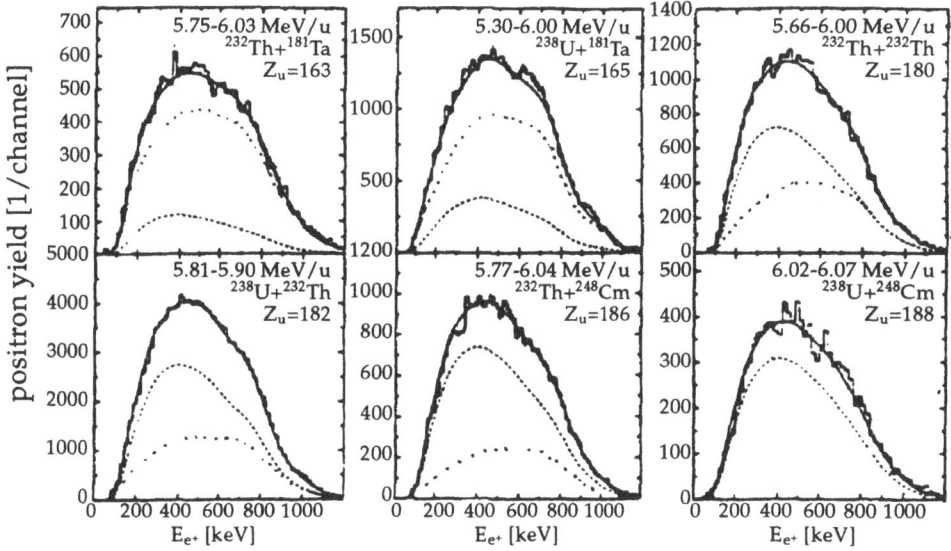

Figure 18. *Comparison of measured positron spectrum (EPOS-group) with the calculation of the dynamical positron production (dashed) and conversion (dotted), the full curve represents the sum of both. In general one finds quantitative agreement between theory and experiment.*

QED-effects are then becoming accessible for studies. The ESR at GSI emerged out of these initial ideas and I look forward that it will indeed be used for such fundamental experimental investigations.

The experimental verification of the idea of decay of the e^+e^--vacuum in strong electric field is of utmost importance, not only because it is one of the most fundamental (if not **the** most fundamental) effects in Quantum Electrodynamics, but also because it has its analogues in strong gravity: The so-called Hawking radiation from black holes is nothing but a vacuum decay process [9]. Also the sparking of cosmic strings belongs into this domain of problems. Thus, the implications of strong field QED in astrophysics are enormous.

Cluster decay, cold fission, cold fusion

The experimental discovery in 1984 of the spontaneous emission of carbon 14 from ^{223}Ra by H. J. Rose and G. A. Jones has confirmed theoretical predictions by A. Sandulescu, D. Poenaru and W. Greiner [17] that a new decay mode intermediate between alpha decay and fission must exist. It is recognized worldwide that this has opened up the new field of cluster radioactivities. In our 1980 paper, besides others, the ^{14}C radioactivity of ^{222,224}Ra had been clearly stated, four years before the first successful experiment, in which ^{14}C emission from ^{223}Ra was detected at Oxford. These predictions were based on the idea of cold rearrangements of a large number of nucleons in decay and fusion, i.e. decays from a compound nucleus (A_i, Z_i) to a daughter nucleus (A_d, Z_d) and an emitted cluster (A_{cl}, Z_{cl}) with $A_i = A_d + A_{cl}$ and $Z_i = Z_d + Z_{cl}$, and vice versa in fusion, in which one of the partners is the double magic nucleus ^{208}Pb or a nucleus close to it.

An alpha-decay-like description gives minimum values for the difference between heights of the barriers and the Q values at these fragmentations, i.e. maximum barrier penetrabilities.

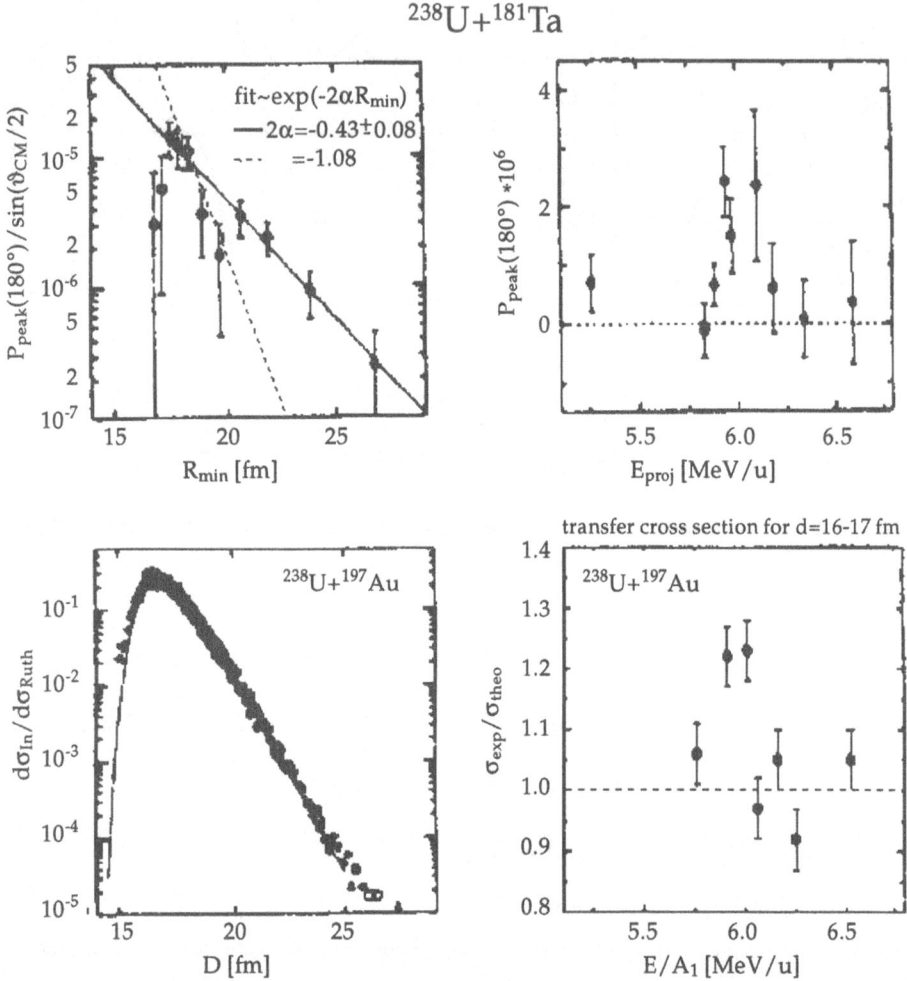

Figure 19. Positron peak probability as a function of R_{min} (upper figure) and one-neutron-transfer cross section as a function of closest approach D (according to H.-J. Wollersheim.) The excitation function of both are quite similar (right column). The neutron transfer falls steeper. Neutron transfer and Coulumb excitation of specific 0^+ levels might be involved, which subsequently decay by pair conversion.

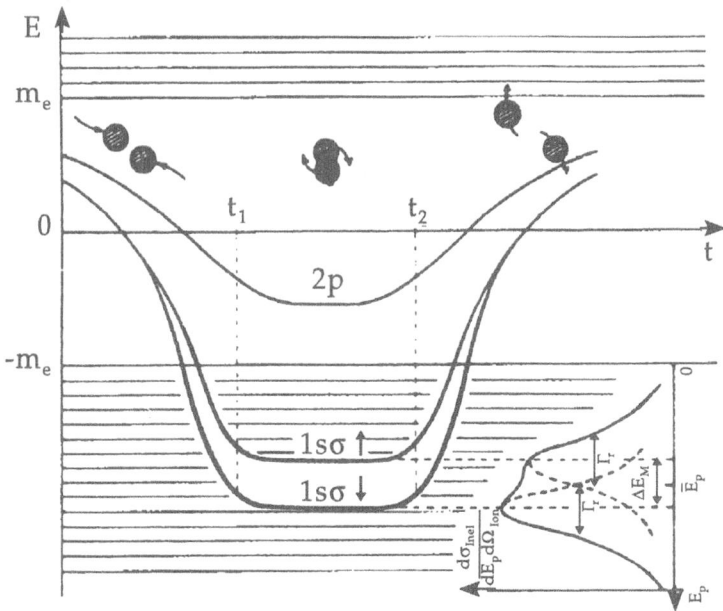

Figure 20. Time delay due to nuclear contact prolongs the overcriticality for the electronic K-shell. This leads to line-structure for the spontaneously produced positrons.

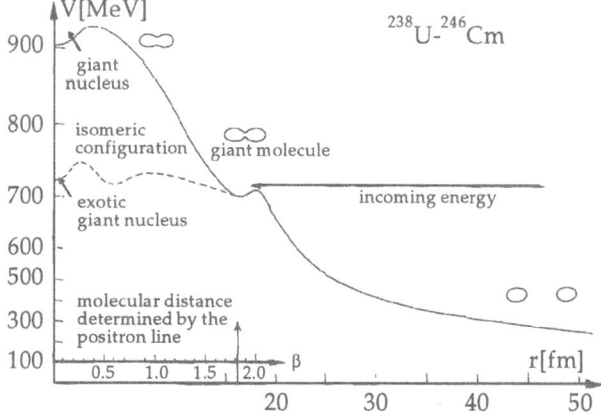

Figure 21. The potential between U and Cm shows a pocket at touching configurations. The potential was calculated by folding the densities with a $Y3M$-interaction. If such potential pockets do really exist, giant nuclear molecules would be formed, eventually causing very long time delays.

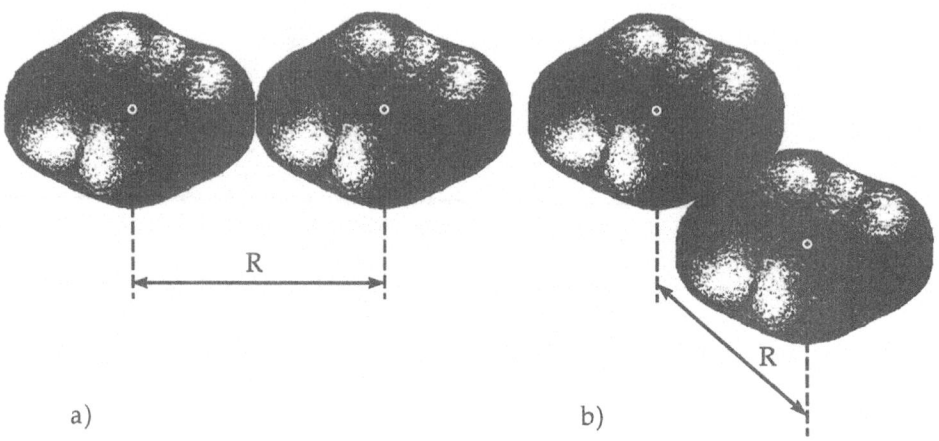

Figure 22. *Typical U-U configurations upon touching. a) nose-nose-, b) side by side-collision.*

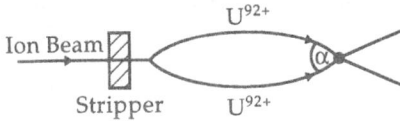

Figure 23. *Schematic figure of molecular levels in a giant molecular potential pocket.*

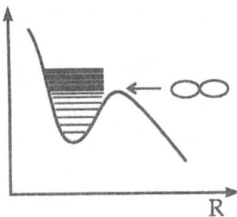

Figure 24. *Scheme of producing the collision of naked Uranium with naked Uranium. After stripping and storing the naked ions in a storage ring they might be brought into a collision under a small angle α, thus controlling the relative energy of the nuclei.*

It is worth mentioning that, due to the fact that both fragments are in the ground state, the touching configurations are situated inside the barriers. Consequently, since the nucleus-nucleus interaction is known from scattering data, the main part of the barrier can be calculated accurately, i.e. quite accurate half-lives are predictable if the preformation probabilities or spectroscopic factors can be evaluated correctly. We have developed a new semiclassical method [21] allowing to calculate easily these factors within our analytical superasymmetric fission model

A fission-like description gives **cold valleys** in the corresponding energy surface at these fragmentations. Outside the touching configurations the barriers are identical with the previous barriers. The difference consists in the inner barrier which is evaluated in the macroscopic-microscopic method by minimizing the potential energy surfaces or maximizing the WKB penetrabilities in a static description respectively. Quite different procedures are used for the evaluation of the corresponding inertia (collective masses).

Table 3. *The present experimentally measured cluster decay half lives.*

$Parent$		$Emitted$		$Daughter$		$Experiment$
Z	A	Z_e	A_e	Z_d	A_d	$\log T(s)$
88	222	6	14	82	208	11.02 ± 0.06
88	223	6	14	82	209	15.20 ± 0.05
88	224	6	14	82	210	15.90 ± 0.12
88	226	6	14	82	212	21.33 ± 0.20
89	225	6	14	83	211	17.34 ± 0.30
90	228	8	20	82	208	20.86 ± 0.30
90	230	10	24	80	206	24.64 ± 0.07
91	231	10	24	81	207	23.38 ± 0.08
92	232	10	24	82	208	21.06 ± 0.10
92	233	10	24	82	209	24.82 ± 0.15
92	233	10	25	82	208	
92	234	10	24	82	210	25.25 ± 0.05
92	234	12	28	80	206	25.75 ± 0.06
94	236	12	28	82	208	21.68 ± 0.15
94	238	12	28	82	210	25.70 ± 0.25
94	238	12	30	82	208	
94	238	14	32	80	208	25.30 ± 0.16
96	242	14	34	82	208	23.24

In the years after 1984 new natural decays with the emission of oxygen, fluorine, neon, magnesium and silicon were discovered, in which the daughter nuclei are ^{208}Pb or nuclei close to it. An illustrative example of the new decay modes in the ^{235}U natural chain is given in fig. 25. The current experimental cluster decay half-life data are presented in fig. 26 as a function of mass of the emitted cluster A_{cl}. The corresponding values are given in table 3. The experiments have been guided by the predicted half-lives and branching ratios relative to alpha decay within our superasymmetric fission model [19], and the measurements are in good agreement with the theory. Due to the great experimental activity in the field we expect other natural decays to be discovered soon. Recently, detailed surveys on these new radioactivities have been published [20]. In particular I point out the new island of cluster radioactivity between ^{114}Ba and ^{132}Sm which has been calculated by Poenaru and myself [21]. We are predicting ^{12}C, ^{16}O and ^{28}Si cluster decays (figs. 27 and 28). Preliminary reports on the observation of ^{114}Ba $\rightarrow$ ^{12}C+^{102}Sn by Oganessian, Mikheev and Tretyakova have been given. Presently Roeckl, Price and Bonetti are setting up an experiment at GSI searching for this **new island of cluster radioactivity**.

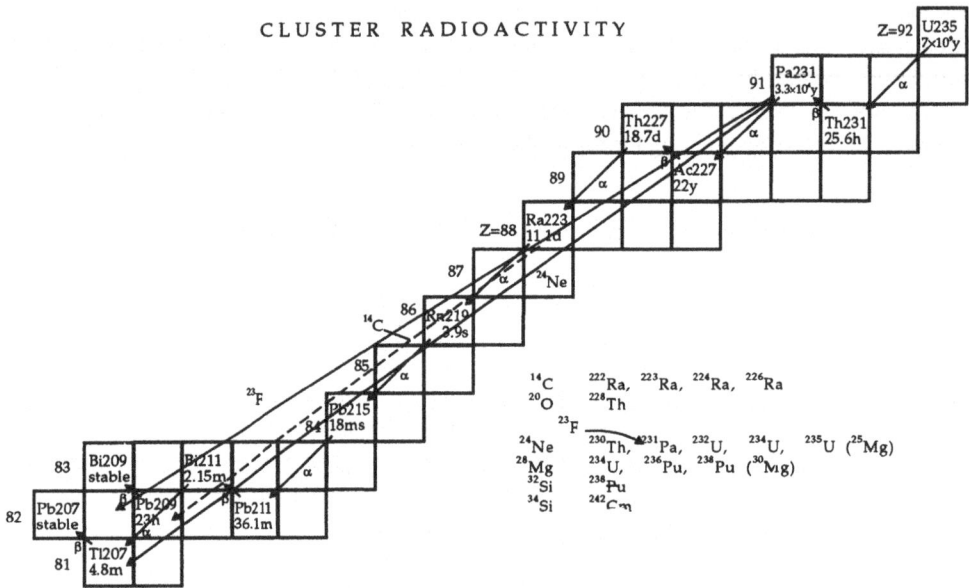

Figure 25. *Cluster radioactivity of various actinide nuclei. This figure demonstrates impressively the new radioactive process, in particular as compared to α-decay.*

Another interesting topic is that of the fine structure experimentally discovered [22] in ^{14}C radioactivity of ^{223}Ra, few years after the first theoretical consideration, by Martin Greiner and Werner Scheid [23], of transitions toward excited states of the final fragments.

There are many phenomenological and microscopic alpha-decay-like theories. The phenomenological ones reproduce excellently the experimental data. They are very useful for predicting new decay modes but give little information on the nuclear structure. The microscopic ones are limited only to very simple cases.

There are also many phenomenological and microscopic fission-like theories. The potential energy surface as a function of the distance between the fragments and the mass asymmetry shows an additional cold valley, corresponding to ^{208}Pb as one of the fragments (see e.g. fig. 29). Many-dimensional descriptions are necessary. Such calculations have been successfully carried out by H. Klein, J. Maruhn, and myself [24]. The collective mass parameters play a dominant role in the dynamics. We can find [25] the optimum fission path in a multidimensional space of deformation parameters by solving a nonlinear differential equation.

The most interesting interpretation of the cluster decays is based on the fact that the surface of a nucleus in its ground state fluctuates. By expanding the surface in multipoles

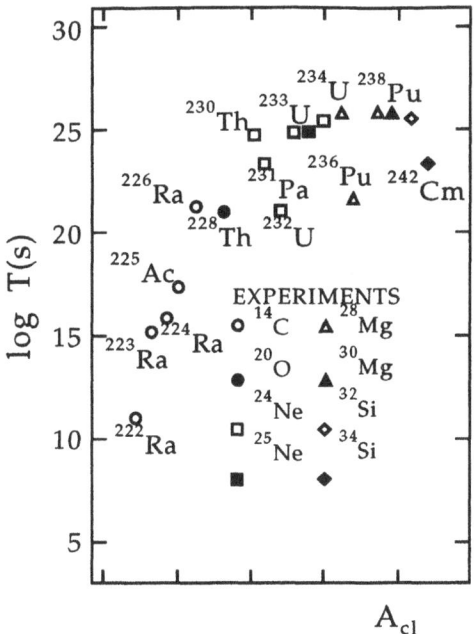

Figure 26. *Various daughter nuclei and their cluster decay, as presently experimentally verified.*

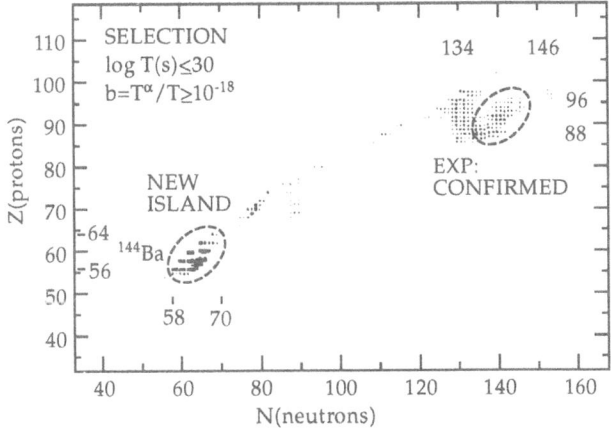

Figure 27. *The predicted new island of cluster radioactivity around ^{114}Ba. Also shown is the experimentally verified island of cluster radioactivity in the actinide nuclei.*

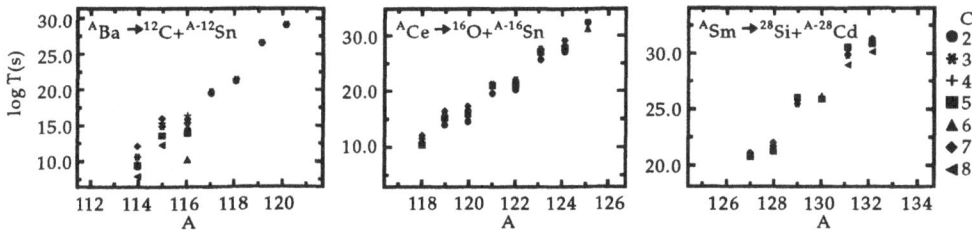

Figure 28. *Some typical decays in the predicted new island of cluster radioactivity.*

it is well known that the fluctuations associated with the low-lying quadrupole and octupole vibrations are quite large. They are smaller for higher multipoles.

Recently, by including nonlinear terms in the hydrodynamical equations we showed that these fluctuations lead to stable solitons on the surface of a circle with the shape $R = R_0 + R(\theta, t)$, or on a sphere if the perturbation has an axial symmetry [26]. These solitons were interpreted as clusters on the surface of the most rigid nucleus ^{208}Pb. This is a new large amplitude collective motion in nuclei different from the collective motions in the standard collective models. An illustrative example of a soliton located on a sphere is given in fig. 30. With that model a good description for the spectroscopic factors in cluster decays is obtaind.

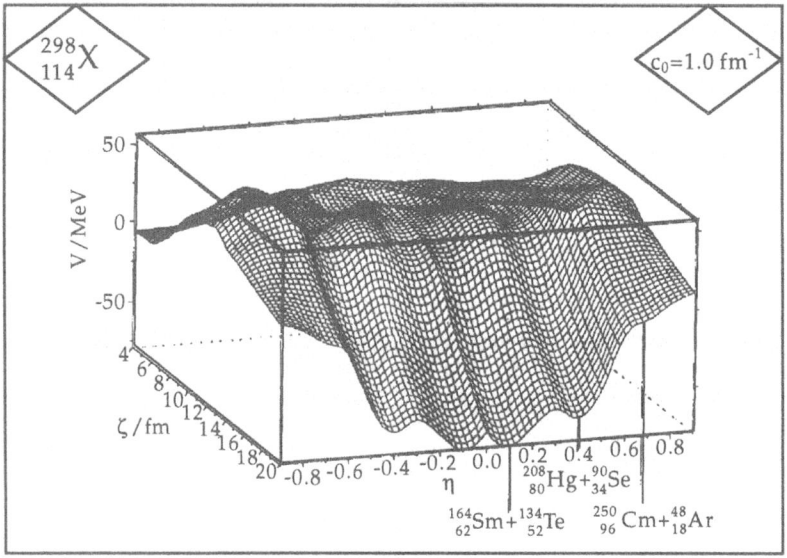

Figure 29. *Potential energy surfaces of the superheavy nucleus $^{298}X_{114}$. The abscissa shows the mass asymmetry coordinate $= (A_1 - A_2)/(A_1 + A_2)$, the ordinate the two center distance in fm. The "cold valleys" can be clearly recognized.*

Cold fission or neutronless fission is defined as the cold fragmentaion in spontaneous or thermal-neutron induced fission in which virtually all the available energy goes into the total kinetic energy of the fragments. It is completely analogous, even identical, with cluster radioactivity [20].

The main argument against such an interpretation is the fact that thermal-neutron induced fission is a very fast process of the order 10^{-22} s and consequently the fragments have no time to penetrate the barrier. Due to the fact that the yields for cold fission are $\approx 10^{-7}$ per fission, the half-lives become 10^{-15} s. This value can be easily obtained if we consider first that, due to the large Z_1, Z_2 values, the cold fission barriers are much thinner than for cluster barriers and, second, due to the deformations of both fragments the barrier heights are much smaller and close to the corresponding Q-value plus the binding energy B_n of the last neutron. Consequently the barrier penetrabilities are very large in such a way that the product with frequency ν of the collision with the barrier can explain the above value.

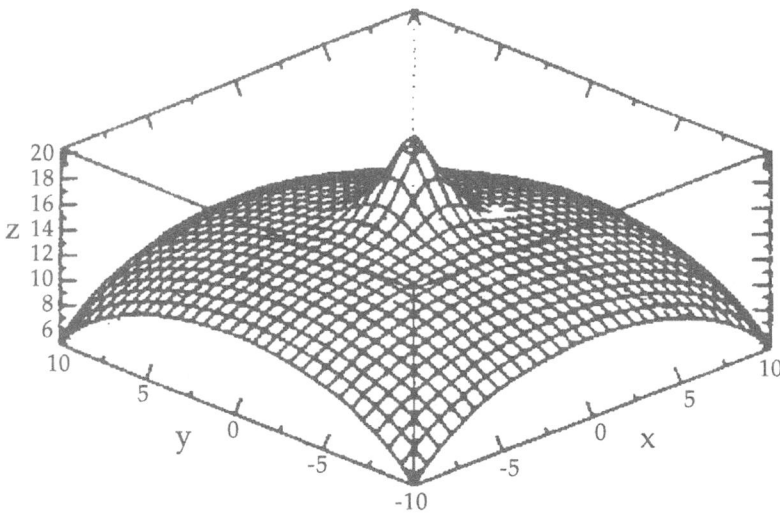

Figure 30. The α particle as a soliton wave on the surface of a big (e.g. Pb) nucleus.

From an experimental point of view the main difference is the fact that the cold fragments must be detected on a large background of fission fragments and not on a large background of alpha particles as in cluster radioactivity. Only recently it was possible to resolve the masses and the charges one by one at very high total kinetic energies of the fragments by using a twin-ionization-chamber (two back-to-back ionization chambers). This set-up which is practically a 4π detector allowed to measure the tails of the mass and charge yields at very high total kinetic energies of the fragments. No bumps, which could be related to definite half-lives of this process have been observed [27].

From a theoretical point of view, for real cold fragmentations for which total kinetic energy of the fragments equals the Q-value, we have to use the ground state deformations. Already earlier we have shown [28] that with **a single set of fragment deformations**, slightly modified relative to those suggested by the few existing experimental deformation values, it is possible to reproduce the experimental charge distribution for all mass fragmentations and for all currently measured nuclei. At lower kinetic energies we assume that the excited deformed fragments are additionally β-deformed (see fig. 31). This additional deformation can be easily evaluated by assuming an associated harmonic Hamiltonian for β-vibrations with the collective mass B given by the liquid drop model - and the stiffness parameter C given by the extrapolation of the energies $E_\beta = \sqrt{C/B}$ of the β-band heads.

For a given mass and charge split $(A, Z) \rightarrow (A_H, Z_H) + (A_L, Z_L)$, the absolute yields for the fragments in cold fission are taken proportional to the decay constant $(A_L, Z_L) = P(A_L, Z_L)$ for that particular desintegration. The frequency ν of the collisions with the barrier before penetration is considerd as very slowly varying with A and Z of the fragments. Hence for a given mass split A_H, A_L of the initial nucleus, the relative yield for a charge Z_L can be taken proportional with its penetrability factor $Y(A_L, Z_L) = P(A_L, Z_L)/\sum_{Z_L} P(A_L, Z_L) = P(A_L, Z_L)/P(A_L)$, and the total relative yield for a light mass A_L fragment is $Y(A_L) = P(A_L)/\sum_{A_L} P(A_L) = P(A_L)/P_{total}$. Similary the total relative yield for a light charge Z_L fragments is $Y(Z_L) = \sum_{A_L} P(A_L, Z_L)/P_{total} = P(Z_L)/P_{total}$.

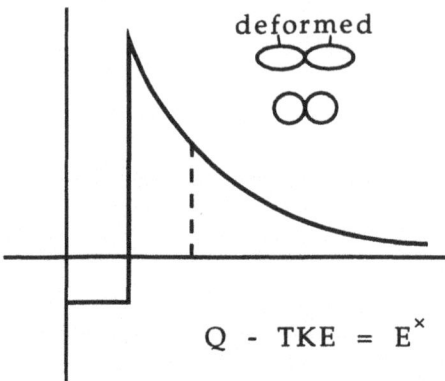

Figure 31. *Illustration of the effect if deformation energy E^* is increased while the total kinetic energy is decreased.*

The penetrability P through the potential barrier between the fragments is evaluated in the WKB approximation. The nuclear attractive potential acting between two coaxial deformed fragments is approximated with the proximity potential and the Coulomb potential is taken as that of two coaxial ellipsoids.

The comparison of the theoretical (open circles) and the experimental values (full circles) of the mass yields $Y(A_L)$ at fixed values of the light fragment kinetic energy E_{cl} for ^{234}U* and ^{236}U* and for the charge yields $Y(Z_L)$ of the same nuclei are given in fig. 32. We recognize a more or less perfect agreement with the experimental data. More than that, the individual light fragment yields $Y(A_L, Z_L)$ normalized to 100% at each energy are also well reproduced. Also the variation of $Y(A_L)$ and $Y(Z_L)$ with E_{cl} are perfectly explained. We therefore conclude that more than **150 new cluster decays** have been already observed in the present cold fission data.

Long ago, in 1977, based on the existence of the steep ^{208}Pb valley in the potential energy surfaces of very heavy elements, we predicted the best projectil-target combinations for producing superheavy elements [29]. As an example we give in fig. 33 the potential energy at the touching configuration as a function of the light fragment A_L for a few isotopes of elements $Z = 108$ and $Z = 110$. This potential is, in fact, the fragmentation potential $V(R, \eta)$ for $R = R_{touching}$. In these calculations the deformations of the colliding partners have been neglected. The element $Z = 108$ has already been observed in the fusion products of 208Pb with ^{58}Fe after emission of one neutron. Presently new experiments are planned at GSI to produce $Z = 110$ with ^{62}Ni. Already during the last couple of years, using ^{208}Pb and ^{209}Bi targets a new region of transactinide nuclei with $Z = 104 - 109$ has been discovered. These new nuclei are deformed and have relatively high stability against spontaneous fission due to a gap in the Nilsson levels coupled to a strong hexadecupole deformation [30]. They differ from earlier predicted superheavy nuclei around ^{186}110 ^{298}114 which are spherical and stabilized by a double shell closure.

The cross section for fusion-evaporating products is limited by the small survival probability of the excited compound nucleus on its way to the ground state, as well as by relative stability to fast fission. We should like to stress that this method leads not only to low excitation energy of the compound nuclei (as experimentalists usually mention) but also to a relative stability against fast fission which is due to the valley produced by ^{208}Pb in the

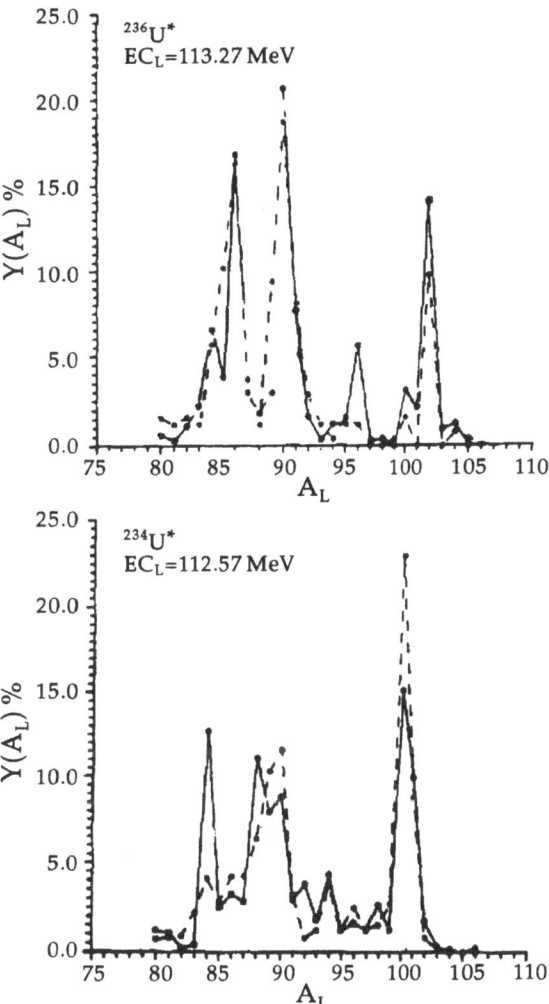

Figure 32. Cold fission as cluster radioactivity: For various daughter nuclei the yields for the emission of the light fragment A_L and Z_L respectively are given on the ordinates. The full curves show the calculations while the dashed curves connect the experimental values. E_{cl} denotes the kinetic energy of the light fragment.

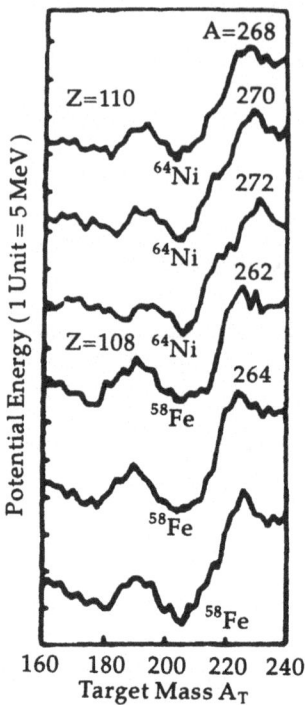

Figure 33. *Excerpt from the 1976 predictions of Gupta, Sandulescu and Greiner for the production of very heavy and super heavy elements. These predicted combinations were used in the GSI experiments for the production of the very heavy element 106...109.*

potential energy surface as a function of mass asymmetry and relative distance. For such reactions with large values of the fissility parameter the main decay process is fast fission.

On the other hand, the possible explanation of cold fission data as an emission of cold fragments suggests us the new possibility for the **synthesis of the superheavy elements**: not only spherical nuclei like Pb or Bi, but also **target-projectile combinations with two magic deformed clusters** can and should be used! For spherical magic partners one has a large Q value but also larger barrier heights. On the other hand, for deformed magic partners one has smaller Q values but also much smaller barrier heights. If, for some target-projectile combinations, the decrease in the barrier height due to the deformation is larger than the decrease in the Q values, we expect new fragmentation valleys which can be used for the synthesis of superheavy elements [20].

From a systematic study of the potential energy surface at the potential barriers computed with the same unique set of deformations deduced from present cold fission data (which includes also experimentally known deformations), we obtain new valleys which exist only for elements with $Z > 110$. As an example we give in fig. 34 the target projectile combinations for the synthesis of superheavy elements with $Z = 114$ and 116.

Finally we conclude this section by stating that the most favourable target-projectile combinations for the synthesis of the traditional spherical superheavy nuclei which are stabilized

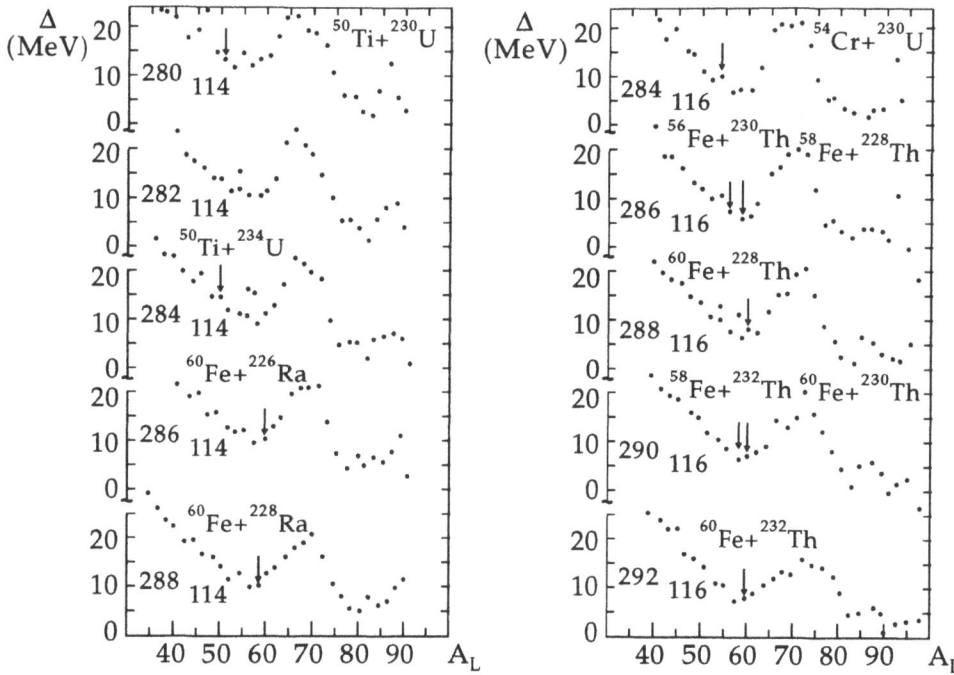

Figure 34. New fragmentation valleys for the synthesis of superheavy elements produced by the deformation of the colliding partners. The new predicted target-projectile combinations are indicated by arrows.

by high barriers to the mass symmetry (i.e. combinations which are relatively stable to fast fission), are $^{60}\text{Fe}+^{226}\text{Ra} \to {}^{286}114$; $^{58}\text{Fe}+^{232}\text{Th} \to {}^{290}116$ and $^{60}\text{Fe}+^{232}\text{Th} \to {}^{292}116$.

On the cluster structure of the quark gluon plasma

We mentioned at the beginning that the analysis of lattic gauge calculations by D. Rischke shows that the quark gluon plasma is most likely not a free gas of quarks and gluons, but still strongly correlated to local color singlets, i.e. baryons and mesons are still intact, only "swollen". This picture is supported by the much earlier observation of Theis et al. [31], that hot and dense baryonic-mesonic matter (no quarks and gluons present) shows a phase transition to vanishing baryon-masses, resembling the quark gluon plasma in detail, even quantitatively. In modern language one would call this a phase transition with restoration of chiral symmetry (see the effective mass m^* plotted in figure 35, but also at the detailed figures of ref. [31]). In other words the strong phase transition of nuclear matter called "quark gluon plasma" can also be described in the baryon-meson picture. The signatures for such a phase transition have to be newly considered. Antimatter clusters and strangelets may still be valid signals, but essentially excitation functions of various observables are needed, which are expected to show irregularities when passing over the phase transition energy.

This has several consequences: For example, such experiments seem only possible at a dedicated facility, because long beam times will be necessary! This seems to be impossible

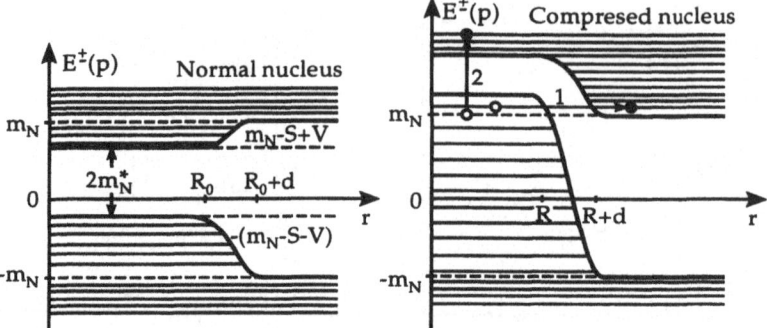

Figure 35. *The level spectrum for nucleons in an anormal and compressed nucleus within meson field theory. For strong enough compression the gap between negative energy states (antibaryons) and positive energy states (baryons) ceases and spontaneous antibaryon emission sets in. The Fourier frequencies available in the course of a violent nucleus-nucleus collision cause induced baryon-antibaryon production. Several bound antibaryons may therefore exist in the compression zone and be emitted as antibaryon clusters ($\bar{d}$, $\bar{He}$, etc.) The great analogy to the spontaneous vacuum decay in strong QED should be noticed!*

at CERN, but possible at RHIC. I plead for a heavy ion collider at GSI. Rudolf Bock and his associates (including me as a theoretical discussion partner) saw the need for such a facility already 15 years ago. The present SIS facility at GSI should have been the first step (injection) for such a collider!

On the production of antibaryons and anti-nuclei

Up to 1991 I believed that the formation of anti-nuclei like anti-deuterons, anti-helium, etc. would constitute a real signal for having had a quark gluon plasma during the course of the collision. The basic argument was that in a very hot plasma, besides quarks, also a lot of anti-quarks are present; and hence the chance that a larger number of anti-quarks join to form anti-nuclei is larger than under ordinary circumstances [32] [33]. However, also in the baryon-meson picture, i.e. in meson field theory a mechanism to produce enhancement for the production of anti-nuclei has been recently discovered [34]. The basic idea is that baryons are interacting via a scalar (S) and via a vector field (V). The energy spectrum of the Dirac equation is given by

$$E^+ = V + \sqrt{p^2 + m^{*2}}$$

where $m^* = m - S$ is the effective mass. Obviously for small momenta p one gets for nucleons N:

$$E_N(p) = E^+(p)_{p \to 0} \longrightarrow m_N \underbrace{-S + V}_{\text{shallow potential}}$$

and for antinucleons $\bar{N}$:

$$E_{\bar{N}}(p) = -E^-(p)_{p\to 0} \longrightarrow m_N \underbrace{-S-V}_{\text{deep potential}}$$

Now, selfconsistent meson field theory [35][36][31] yields $S \sim \rho_S$ (scalar density) and $V \sim \rho_V$ (vector density).

At normal nuclear cluster density $\rho_V = \rho_0 = 0.15 fm^{-3}$ one has $S \sim -250 MeV$, $V \sim 200 MeV$, depending on the coupling constants chosen. At high density and temperature one expects a decrease of the effective nucleon mass m_N^* and an increase of the vector potential V.

The thermal production of baryon-antibaryon ($B\bar{B}$) is then enhanced [37] according to

$$Y_{\bar{B}} \sim \exp(-2m_{B^*}/T), \qquad \bar{B} = \bar{N}\bar{\Delta}\bar{\Lambda}\cdots$$

Also spontanous and dynamical production of $N\bar{N}$ pairs becomes possible, quite analogous to the QED processes discused above. This happens if

$$(E_{\bar{N}} + E_N) = m_N^* - V + m_N = 0$$

at $\rho_V \sim (3-7)\rho_0$. While, as discussed above, the scalar field is attractive for baryons and antibaryons, the vector field is repulsive for baryons but attractive for antibaryons. With the dynamical and thermal production of a great number of N-holes (i.e. $\bar{N}$) in bound nucleon

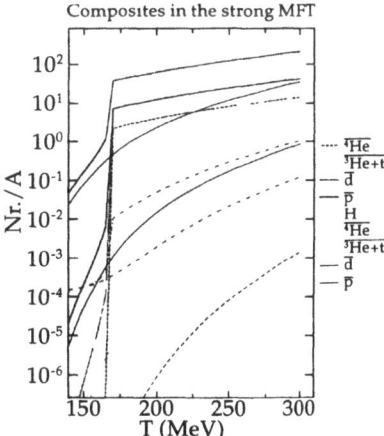

Figure 36. *Lacking still a complete dynamical calculation one may get estimates for antibaryon-cluster production from a thermal model. For $\bar{\alpha}$-production we indicated the enhancement over the free production ($\bar{\alpha}$-free) by more than 4 orders of magnitude. Similar enhancements are obtained for other antimatter clusters.*

orbits, one has a similar great number of $\bar{N}$ sitting closely together like nucleons in nuclei. This leads to enhanced production of anti-nuclei. For example $\bar{\alpha}/\bar{p} = 10^{-9} - 10^{-6}$.

A thermal model estimate for the enhanced production of anti-nuclei is illustrated in fig. 36 and shows 1.5 to 4 orders of magnitude enhancement as compared to the free gas estimates. Indeed a challenge for nuclear physics in the near future!

Acknowledgements

I am grateful to Dr. Carsten Greiner, Prof. J. Maruhn, Dr. D. Rischke, Dr. J. Reinhardt, Prof. A. Schäfer, Dr. Th. Schönfeld, Prof. H. Stöcker and Prof. G. Soff for valuable discussions.

References

[1] H. Stöcker and W. Greiner; *Phys. Rep.* **137**,279(1986)
    W. Scheid, R. Ligensa, and W. Greiner; *Phys. Rev. Lett.* **21**,1479(1968)
    W. Scheid, H. Müller, and W. Greiner; *Phys. Rev. Lett.* **32**,741(1974)
[2] J. Hofmann, H. Stöcker, W. Scheid, and W. Greiner; Workshop on GeV/A Coll. of Heavy Ions - How and Why; Bear Mountain, New York 1974; BNL Rep. 50445 ed L. Lederman and J. Weneser, 1975 p.39
    G. F. Chapline et.al. *Phys. Rev.* **D8**,4302(1973)
    H. G. Baumgardt et. al. *Z. Phys.* **A237**,359(1975)
    J. Hofmann, W. Scheid, and W. Greiner; *Nuovo Cimento* **33A**,343(1976)
    H. Stöcker, W. Greiner, and W. Scheid; *Z. Phys.* **A286**,121(1978)
[3] J. Hofmann, H. Stöcker, U. Heinz, and W. Greiner; *Phys. Rev. Lett.* **36**,88(1976)
[4] C. Greiner, P. Koch, and H. Stïcker; *Phys. Rev. Lett.* **58**,1825(1987)
    C. Greiner, D. Rischke, P. Koch, and H. Stöcker; *Phys. Rev.* **D38**,2797(1988)
    C. Greiner and H. Stöcker; *Nucl. Phys.* **B24**,239(1991)
[5] J. Schaffner, C. Greiner, and H. Stöcker; *Phys. Rev.* **C46**;323(1992)
[6] P. G. Reinhardt, M. Rufa, J. Maruhn, W. Greiner, J. Friedrich; *Z. Phys.* **A323**,13(1986)
    M. Rufa, H. Stöcker, J. Maruhn, P. G. Reinhardt, and W. Greiner; *J. Phys.* **G13**,143(1978)
    M. Rufa, P. G. Reinhardt, J. Maruhn, W. Greiner, and M. R. Strayer *Phys. Rev.* **C5**,390(1988)
    M. Rufa, J. Schaffner, J. Maruhn, H. Stöcker, W. Greiner, and P. G. Reinhardt *Phys. Rev.* **C42**,2469(1990)
    B. Waldhauser, J. Maruhn, H. Stöcker, W. Greiner; *Phys. Rev.* **C38**,1003(1987)
    J. Fink, V. Blum, P. G. Reinhardt, J. Maruhn, W. Greiner; *Phys. Lett.* **B218**,277(1989)
    P. G. Reinhardt; *Rep. Prog. Phys.* **52**,439(1989)
[7] B. F. Bayman and Y. C. Tang; *Phys. Rep.* **147**,155(1987)
[8] A. Kerman and M. S. Weiss; *Phys. Rev.* **C8**,408(1974)
[9] W. Greiner, B. Müller, J. Rafelski; QED of strong fields, Springer 1985
[10] J. Reinhardt et.al. *Phys. Rev.* **A24**,103(1981)
[11] K. Geiger et.al. in "Tests of Fundamental Laws in Physics" p.107, eds. U. Fackler and J. van Trank, Gif sur Yvette 1989
[12] A. Schäfer; *J. Phys.* **G15**,371(1989)
[13] E. Stein, S. Graf, J. Reinhardt, A. Schäfer, and W. Greiner *Z. Phys.* **A340**,377(1991)
[14] M. Seiwert, J. Maruhn, W. Greiner, and J. Friedrich *J. Phys.* **G11**,L21(1985)
[15] W. Betz, G. Heiligenthal, J Reinhardt, R. K. Smith, and W. Greiner
    Important Problems in future Heavy Ion Atomic Physics
    in: The physics of electronic and atomic collisions; eds. J. S. Risley, R. Geballe; Univ. of Washington Press, Seattle, 1976 p.531
[16] U. Müller, T. de Reus, J. Reinhardt, B. Müller, and W. Greiner; *Phys. Rev.* **A37**,1449(1988)

[17] A. Sandulesecu, D. N. Poenaru, and W. Greiner; *Sov. Journ. Part. Nucl.* **11**,528(1980)

[18] D. N Poenaru, W. Greiner; *Physica Scripta* **44**,427(1991)

[19] D. N. Poenaru, W. Greiner, K. Depta, M. Ivascu, D. Mazilu, and A. Sandulescu; *At. Data Nucl. Data Tables*, **34**,423(1986)
D. N. Poenaru, M. Ivascu, A. Sandulescu, and W. Greiner; *J. Phys* **G10**,L183(1984)
D. N. Poenaru, D. Schnabel, W. Greiner, D. Mazilu, and R. Ghergescu; *At. Data Nucl. Data Tables* **48**,231(1991)

[20] A. Sandulescu, W. Greiner; *Rep. Prog. Phys.* **55**,1423(1992)
D. N. Poenaru and W. Greiner in Nuclear Decy Modes, eds. D. N. Poenaru and W. Greiner; CRC Press, Boca Raton, in print

[21] D. N. Poenaru and W. Greiner; Proc. Int. Conf. on Radioactive Nucl. Beams, Louvain la Neuve, 1991, Adam Hilger IOP Pub., Bristol, p.203
D. N. Poenaru, W. Greiner, and R.Ghergescu; *Phys. Rev* **C47**,2030(1993)

[22] L. Brillard, E. G. Elayi, E. Hourrani, M. Hussonnois, J. F. Le Du, L. H. Rosier, L. Stab; *C. R. Acad. Sci. (Paris)* **309**,1105(1989)

[23] M. Greiner and W. Scheid; *J. Phys.* **G12**,L229(1986)

[24] H. Klein; P.Hd.-Theses (1991)

[25] M. Mirea, D. N. Poenaru, and W. Greiner; *Nuovo Cimento* bf 105A,571(1992)

[26] A. Ludu, A. Sandulescu, and W. Greiner; *Int. J. Modern Phys.* **E1**,169(1992)

[27] G. Simon, thése pour le grade de docteur d'état, université Paris-sud, Orsay, 1990

[28] A. Sandulescu, A. Florescu, W. Greiner; *J. Phys.* **G15**,1815(1989)

[29] R. K. Gupta, A. Sandulescu, and W. Greiner; *Phys. Let.* **B67**,254(1977); *Z. Phys.* **A283**,217(1977); *Z. Naturf.* **32**,704(1977)

[30] S. Cwiok and A. Sobiczevski; *Z. Phys.* **A342**,203(1992)

[31] J. Theis, G. Graebner, G. Buchwald, J. Maruhn, W. Greiner, H.Stöcker, J. Polonji; *Phys. Rev.* **D28**,2286(1983)

[32] U. Heinz, P. R. Subramanian, H. Stöcker, and W. Greiner; *J. Phys.* **G12**,1237(1986)

[33] U. Heinz, P. R. Subramanian, and W. Greiner; *Z. Phys.* **A318**,247(1984)

[34] I. N. Mishustin, L. M. Satarov, J. Schaffner, H. Stöcker, and W. Greiner; *Z. Phys.* **A341**,47(1991)
I. N. Mishustin, L. M. Satarov, J. Schaffner, H. Stöcker, and W. Greiner; Proc. Int. Conf. on High Energy Physics Problems (Dubna, 1990), eds. A. S. Baldin, V. V. Burov, and L. P. Kaptari, World Scientific (9191) p.614

[35] H.-P. Duerr and E. Teller, *Phys. Rev.* **101**,494(1956)

[36] J. D. Walecka; *Ann. Phys. (NY)* **83**,491(1974)

[37] J. Schaffner, I. N. Mishustin, L. M. Satarov, H. Stöcker, and W. Greiner; *Z. Phys.* **A341**,47(1991)

II.

Cluster Decay,

Fission and Fusion

CLUSTER RADIOACTIVITY

P. B. Price

Physics Department, University of California, Berkeley, CA 94720, USA

Introduction

Atomic, molecular, and nuclear clusters can be produced with surprising ease in various kinds of ovens. Clusters of sodium atoms are known to show sharp abundance peaks at numbers ranging from 8 atoms up to tens of thousands of atoms. Peaks at atom numbers up to $\sim 10^3$ are related to the closing of spherical shells and elliptical subshells of valence electrons in a positive jellium; peaks at much larger numbers are related to the stacking of atoms in three-dimensional solids. Under optimal oven conditions, clusters of carbon atoms show a sharp peak at 60 carbon atoms (the now-famous buckyball). In stellar ovens, cosmic abundances of nuclei lighter than Fe show peaks at even-A and at multiples of the alpha particle; there is a strong abundance peak at ^{56}Fe, the nucleus with the largest binding energy per nucleon; and the heavier nuclei show abundance peaks at closed shells such as N = 50 and 126 and at Z = 50 and 82. Cluster radioactivity of heavy nuclei in the ground state (not in a nuclear oven!) was predicted mainly by Romanian theorists[1]. Since the work of Rose and Jones[2] in 1984 nearly 20 nuclides have been found to emit clusters of intermediate mass, ranging from ^{14}C up to ^{32}Si. In this mode the nucleus undergoes two-body decay by barrier penetration, the rate of which is governed mainly by the Q value,

$$Q = M(A, Z) - M(A_1, Z_1) - M(A_2, Z_2) \tag{1}$$

which enters into the expression for the Gamow penetrability,

$$P = e^{-2G}; \quad G = \int [2\mu(r)(V(r) - Q)]^{\frac{1}{2}} dr. \tag{2}$$

The subscripts 1 and 2 refer to the light and heavy daughter, and $\mu(r)$ is the reduced mass. The decay rate, which is proportional to P, is maximized when the barrier height, $V(r) - Q$, evaluated at the scission point, is as small as possible. For given values of Z_1 and Z_2 this occurs when both of the emitted fragments are as tightly bound as possible, subject to the constraint that the parent be close enough to stability for the experiment to be practicable. Herein a brief review is given of experimental progress in this field. Ref. [3] is one of numerous recent reviews, with references to experimental and theoretical work.

Systematics of Decay Rates and Branching Ratios

The most elegant method, and the only one that gives accurate spectroscopic information, uses an electronic detector at the focal plane of a magnetic spectrometer to identify the cluster and its kinetic energy. Knowing the Q value, and assuming that the cluster is emitted in its ground state, one can infer the level of the heavy daughter nucleus. Hussonnois[4] has surveyed the results obtained with this method. The most exciting result is the discovery that the ^{14}C decay of ^{223}Ra, first reported in ref. [2], most often leaves the ^{209}Pb daughter in the first excited state rather than the ground state. Unfortunately, the spectrometric method is limited to decay modes with branching ratio $B(cl/\alpha) > \sim 10^{-11}$ and lifetime $< \sim 10^{17}$ sec. The overall systematics of cluster radioactivity has been established by using plastic or glass track detectors [4] to measure the range and ionization rate of the light cluster, typically with energy 2 to 2.5 MeV/nucleon. From these measurements the energy and charge of the cluster can be determined. Some advantages of track-recording solids are:

1. One can choose a track-recorder that is insensitive to all ions with $Z < Z_1$. In their studies of ^{14}C emitters the Berkeley group used polycarbonate film[6, 7]. To study ^{14}C and ^{20}O emitters the Milano group used barium phosphate glass[8, 9, 10]. Both of these detectors are insensitive to ions lighter than carbon. To study Ne and Mg emission, the detectors of choice have been polyethylene terephthalate[11] or PSK-50 phosphate glass[12]. To study Si emission, which has been seen only at very low branching ratios, one must discriminate against the accumulation of short tracks due to atoms struck by alpha particles impinging on the detector. LG-750 phosphate glass has been used because it is insensitive to Ne and lighter ions but can record Mg and Si[13].

2. Track-recording solids collect events with an efficiency that is independent of rate, and they are immune to backgrounds other than heavy charged particles such as fission fragments. There is thus no problem with pulse pileup during a narrow time interval.

3. They can be tailored in size and shape to fit into the collector of a facility such as ISOLDE, where the ^{14}C radioactivities of ^{221}Fr, ^{221,222,224}Ra, and ^{225}Ac were discovered, and into UNILAC, where a search for ^{12}C radioactivity of ^{114}Ba is planned.

To identify a charged particle with a track-recording solid one immerses the solid in a selective reagent that etches the surface at a uniform rate v_G and etches along the trajectory of the particle at a higher rate v_T, which is a strongly increasing function of ionization rate. The competition between the two rates leads to a conical etchpit of half-angle $q = \arcsin(v_G/v_T)$. Both Z and range (and thus energy) can be determined from microscopic measurement of the size and shape of the etchpit. Image processing techniques reduce the tedium of scanning, but track measurements are still done mostly manually.

Table 1 summarizes the results of searches for cluster radioactivities. Column 2 gives the kinetic energies for the light cluster, assuming ground state to ground state decays. The last two columns give the measured values of partial half-life and branching ratio relative to alpha decay. Where limits are given, they refer to 90% confidence level. The other six columns give half-lives predicted (or postdicted) by various authors. Beginning with the column labeled "Poe.", the calculations are discussed in refs. [14], [6], [15], [16], [17], and [18]. Most of the models have been fine-tuned with two or more adjustable parameters after seeing results of some of the measurements. Usually one parameter takes into account the fact that most odd-A nuclei emit clusters at a rate hindered by a factor as much as 10^2 relative to the rate for even-even nuclides. Figure 1 presents the results in the form of Geiger-Nuttall plots, for which the abscissa is defined in eq. 2. The data for even-even parents are shown in solid symbols; the data for odd-A parents are shown in open symbols. No cases have yet been found of cluster decay of odd-odd parents. One sees that the hindrance factor for an odd-A parent, defined as the factor by which its half-life deviates from the best straight line through

Table 1. *Calculated and Measured Values of log T (half-life, sec)*

Decay Mode	E_k (MeV)	Poe.	SqW	S-S	B-M	B-W	Measured log T (sec)	Measured $-\log B$
^{221}Fr $\to ^{14}$C	29.28	14.3	15.2	16.0	14.0	15.	14.5 ± .12	12. ± .12
^{221}Ra $\to ^{14}$C	30.34	14.2	14.1	14.8	> 12.4	14.2	13. ± .2	11.7 ± .2
^{222}Ra $\to ^{14}$C	30.97	11.1	11.2	11.6	11.4	11.8	11.0 ± .06	9.4 ± .06
^{223}Ra $\to ^{14}$C	29.85	15.1	15.0	15.7	15.3	15.1	15.2 ± .05	9.2 ± .05
^{224}Ra $\to ^{14}$C	28.63	15.9	16.0	16.8	16.1	16.2	15.8 ± .12	10.3 ± .12
^{225}Ac $\to ^{14}$C	28.57	17.8	18.7	19.7	18.5	18.6	17.16 ± .06	11.2 ± .06
^{226}Ra $\to ^{14}$C	26.46	20.9	21.0	22.2	21.0	21.1	21.3 ± .2	10.6 ± .2
^{228}Th $\to ^{20}$O	44.72	21.8	21.8	-.-	21.0	21.8	20.7 ± .08	12.9 ± .08
^{231}Pa $\to ^{23}$F	46.68	25.9	26.0	25.5	-.-	26.8	26.	14.
^{230}Th $\to ^{24}$Ne	51.75	25.2	24.8	24.9	24.7	24.8	24.6 ± .07	12.3 ± .07
^{232}Th $\to ^{26}$Ne	49.70	30.2	29.1	28.4	28.7	29.3	> 27.9	> 10.3
^{231}Pa $\to ^{24}$Ne	54.14	23.3	23.7	23.5	21.6	23.4	22.9 ± .05	10.9 ± .05
^{232}U $\to ^{24}$Ne	55.86	20.8	20.7	20.0	20.9	20.8	20.5 ± .03	11.1 ± .03
^{233}U $\to ^{24}$Ne	54.27	[25.2	24.9	24.8	23.7	25.4]	24.8 ± .03	12.1 ± .03
$\to ^{25}$Ne	54.32	[25.7	25.1	24.4	-.-	26.0]		
^{234}U $\to ^{24}$Ne	52.81	[26.1	25.8	25.7	25.5	25.6]	25.9 ± .2	13.0 ± .2
$\to ^{26}$Ne	52.87	[27.0	26.2	25.0	25.9	26.4]		
^{235}U $\to ^{24}$Ne	51.50	[29.9	29.7	30.1	-.-	29.9]	> 27.4	> 11.1
$\to ^{25}$Ne	51.68	[30.6	29.7	29.6	-.-	28.0]		
^{233}U $\to ^{28}$Mg	65.32	27.4	26.9	27.5	-.-	28.0	> 27.8	> 15.1
^{234}U $\to ^{28}$Mg	65.26	25.9	25.4	25.7	25.4	25.4	25.7 ± .2	12.8 ± .2
^{237}Np $\to ^{30}$Mg	65.52	28.3	28.3	27.7	> 27.3	29.9	> 27.4	> 13.6
^{236}Pu $\to ^{28}$Mg	70.22	21.1	21.2	20.5	21.5	22.0	21.7 ± .3	13.7 ± .3
^{238}Pu $\to ^{30}$Mg	67.00	[26.2	25.9	24.3	25.6	25.8]	25.7 ± .25	16.3 ± .25
$\to ^{28}$Mg	67.32	[26.2	25.5	-.-	25.7	26.]		
^{238}Pu $\to ^{32}$Si	78.95	26.1	25.7	-.-	25.8	25.7	25.3 ± .16	15.9 ± .16
^{241}Am $\to ^{34}$Si	80.60	25.8	26.5	26.2	25.3	28.8	> 25.3	> 15.1
^{242}Cm $\to ^{34}$Si	82.88	23.5	23.4	22.6	-.-	24.1	> 21.5	> 14.4

the data for the even-even parents for the same type of cluster emission, is typically one or two orders of magnitude.

Implications for Nuclear Structure

As Fig. 1 shows, the data for even-even parents rather accurately define a set of parallel lines on a Geiger-Nuttall plot of log half-life *vs* penetration factor, with a standard deviation no worse than a factor ~ 2. In a cluster model the displacement of the lines for heavy clusters from the ^{4}He line can be interpreted in terms of a preformation probability. In almost every case the half-lives for odd-A parents are hindered with respect to the lines for even-even

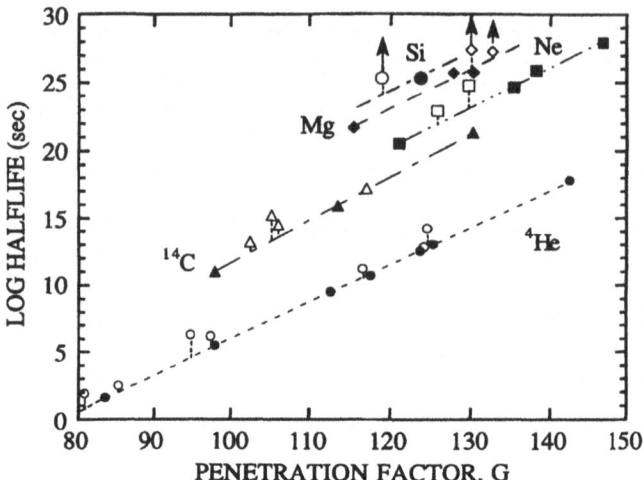

Figure 1. *Geiger-Nuttall plot of measured half-lives for cluster decays of even-even (filled symbols) and odd- A (open symbols) nuclides.*

parents. It has been known for decades that alpha decays of odd-A nuclides are strongly hindered when the odd nucleon has to change its quantum state[19]. In cluster radioactivity we use analogous reasoning to attribute the hindrance to the necessity, in almost every case of odd-A decays, of changing the quantum state of the odd nucleon in the ground state of parent and of heavy daughter. The increase in half-life is then due either to the poor overlap of initial and final wave function or to the smaller Q value for decay to an excited state with an improved overlap. We now discuss the experimental observations to date, indicating the nominal states of the unpaired nucleon before and after cluster emission:

(1). $\quad\quad ^{223}\text{Ra}(g\frac{9}{2}^+) \rightarrow\, ^{14}\text{C} +\, ^{209}\text{Pb}^*(i\frac{11}{2}^+)$ \quad\quad excited state

The preferred final state is the first excited state of ^{209}Pb. See ref. [4] for a detailed discussion of the spectroscopic data on branching ratios to various levels of ^{209}Pb. An explanation for the preference for the $i\frac{11}{2}^+$ state of ^{209}Pb is that the Nilsson wave function describing the ^{223}Ra ground state contains large components arising from the $i\frac{11}{2}$ neutron shell-model orbit but none arising from the $g\frac{9}{2}$ state.

(2). $\quad\quad ^{233}\text{U}(g\frac{9}{2}^+) \quad\rightarrow\quad ^{24}\text{Ne} +\, ^{209}\text{Pb}(g\frac{9}{2}^+)$ \quad\quad ground state
$\quad\quad\quad\quad\quad\quad or \quad\rightarrow\quad ^{28}\text{Mg} +\, ^{205}\text{Hg}(p\frac{1}{2}^-)$ \quad\quad parity change

A distortion of the spherical ground state of ^{209}Pb into a spheroid would break the fivefold degeneracy of the $g\frac{9}{2}$ state; the $\frac{5}{2}$ [633] Nilsson configuration of one of the resulting states is the same as the ground state of the deformed ^{233}U parent; thus the ground state to ground state transition in ^{24}Ne emission is favored. In contrast, for ^{28}Mg emission the ground state of the daughter ^{205}Hg is the neutron $p\frac{1}{2}$ hole state, which is unrelated (and has opposite parity) to the ground state of ^{233}U. Only by decaying to the fifth excited level of ^{205}Hg at 1.85 MeV would there be a good match, but the Q value is so small that the penetrability to this state would be negligibly small. The authors of ref. [20] used this explanation to account for the smaller branching ratio $B(\text{Mg/Ne})$ than predicted by the theoretical models.

(3). $\quad$ $^{231}\text{Pa} \rightarrow {}^{24}\text{Ne} + {}^{207}\text{Tl}$ $\quad$ even-even cluster

$\quad\quad or \rightarrow {}^{23}\text{F} + {}^{208}\text{Pb}$ $\quad$ odd-A cluster

The branching ratio $B(\text{F/Ne})$ was measured[21] to be about an order of magnitude smaller than calculated in the various models. This is the first example of emission of an odd-A cluster. The unpaired particle, here a proton, must go from the heavy parent, ^{231}Pa, to a very light nucleus, ^{23}F. Not surprisingly, this is a strongly hindered process.

(4). $\quad$ $^{225}\text{Ac}(\text{h}\frac{9}{2}^{+} \text{ or } \text{f}\frac{7}{2}^{-} ?) \rightarrow {}^{14}C + {}^{211}\text{Bi}(\text{h}\frac{9}{2}^{+} \text{ or } \text{f}\frac{7}{2}^{-} ?)$

Bonetti et al.[8] point out that one can account for the experimental hindrance factor of ~ 1 if the transition is from the Nilsson configuration $\frac{3}{2}$ [532], which originates in the $\text{h}\frac{9}{2}$ ground state of the deformed ^{225}Ac parent, to the $\text{h}\frac{9}{2}$ ground state of the nearly spherical ^{211}Bi. They also give an alternative explanation which takes into account reflection asymmetry terms in the deformation of Ac isotopes[22]. According to ref. [22], the odd proton of ^{225}Ac belongs to the $\Omega = \frac{3}{2}$, $n_\Omega = 13$ orbit which originates mainly from the $\text{f}\frac{7}{2}^{-}$ subshell; therefore the transition probably goes to the first excited state of ^{211}Bi ($E^* = 0.4$ MeV, $J^\pi = \frac{7}{2}^{-}$). Bonetti et al. estimate a hindrance of a factor 6 relative to the first explanation, which is a less good fit to the observed rate. With magnetic spectroscopy, Hussonnois hopes to be able to measure the kinetic energy of the ^{14}C accurately enough to infer the level of the ^{211}Bi daughter.

(5). $\quad$ $^{221}\text{Fr}(\text{h}\frac{9}{2}^{-}) \rightarrow {}^{14}C + {}^{207}\text{Tl}(\text{s}\frac{1}{2}^{+})$ $\quad$ parity change $\quad\quad$ (3)

(6). $\quad$ $^{221}\text{Ra}(\text{g}\frac{9}{2}^{+}) \rightarrow {}^{14}C + {}^{207}\text{Pb}(\text{p}\frac{1}{2}^{-})$ $\quad$ parity change $\quad\quad$ (4)

A recent study of ^{221}Fr and ^{221}Ra by Bonetti et al.[9] has shown that branching ratios for ^{14}C emission are hindered by one order of magnitude relative to the rate for even-even parents. In their brief communication they did not discuss how the hindrance could be accounted for in terms of nuclear structure.

For all but the first of these six examples of cluster emission by odd-A parents, one does not yet know experimentally the levels of the final states of the heavy daughters, and in several cases the Nilsson states of the parents are not definitively known. It seems clear that measured branching ratios provide useful information that can guide theorists in determining nuclear structure. An indication of the uncertainty in structure is that it has not been possible to extract predictions from theorists as to branching ratios for various cluster emission modes by odd-A parents in advance of experiments. Reasoning from the large hindrance factors found for alpha-decay of odd-odd parents, it would be interesting to try to detect cluster radioactivity of odd-odd nuclides. Candidates include ^{222}Fr ($\log t_\alpha (\text{sec}) = 5.3$; predicted $\log B(^{14}\text{C}/\alpha) = -12.8$) and ^{224}Ac ($\log t_\alpha$ (sec) = 5.6; predicted $\log B(^{14}\text{C}/a) = -10.7$). Bonetti et al.[9] have succeeded in observing a branching ratio as small as $\log B = -12.0$ for ^{14}C emission from ^{221}Fr.

References

[1] A. Sandulescu, D. N. Poenaru and W. Greiner, *Sov. J. Part. Nucl.* **11**, 528 (1980)
[2] H. J. Rose and G. A. Jones, *Nature* **307**, 245 (1984)
[3] P. B. Price, in: *Clustering Phenomena in Atoms and Nuclei*, p. 273, edited by M. Brenner, T. Lönnroth and F. B. Malik, (Springer-Verlag, Berlin, 1992)
[4] M. Hussonnois, *C14 radioactivities of radium isotopes*, in: proceedings of this conference
[5] P. B. Price, *Ann. Rev. Nucl. Part. Sci.* **39**, 19 (1989)
[6] P. B. Price et al., *Phys. Rev. Lett.* **54**, 297 (1985)
[7] S. W. Barwick et al., *Phys. Rev. C* **34**, 362 (1986)
[8] R. Bonetti et al., *Nucl. Phys. A* **562**, 32 (1993)
[9] R. Bonetti, C. Chiesa, A. Guglielmetti, C. Migliorino and P. Monti in: *Proceedings of Second International Conference on Clustering Phenomena in Atoms and Nuclei*, (Santorini, Greece, 1993)
[10] R. Bonetti et al., *Nucl. Phys. A* **556**, 115 (1993)
[11] See, for example, S. P. Tretyakova et al., *Z. Phys. A* **333**, 349 (1989)
[12] See, for example, Shicheng Wang et al., *Phys. Rev. C* **36**, 2717 (1987)
[13] See, for example, Shicheng Wang et al., *Phys. Rev. C* **39**, 1647 (1989)
[14] D. N. Poenaru et al., *At. Data Nucl. Data Tables* **34**, 423 (1986) and unpublished revisions
[15] Y.-J. Shi and W. J. Swiatecki, *Nucl. Phys. A* **438**, 450 (1985); *Nucl. Phys. A* **464**, 205 (1987); and unpublished tables
[16] B. Buck and A. C. Merchant, *Phys. Rev. C* **39**, 2097 (1989); *J. Phys. G: Nucl. Part. Phys.* **16**, L85 (1990); and unpublished tables
[17] R. Blendowske and H. Walliser, *Phys. Rev. Lett.* **61**, 1930 (1988); and unpublished tables
[18] M. Ivascu and I. Silisteanu, *Nucl. Phys. A* **485**, 93 (1988); *J. Phys. G: Nucl. Part. Phys.* **15**, 1405 (1989); and unpublished tables
[19] J. O. Rasmussen, *Ark. Fys.* **7**, 185 (1953)
[20] P. B. Price, K. J. Moody, E. K. Hulet, R. Bonetti and C. Migliorino, *Phys. Rev. C* **43**, 1781 (1991)
[21] P. B. Price, R. Bonetti, A. Guglielmetti, C. Chiesa, R. Matheoud, C. Migliorino and K. J. Moody, *Phys. Rev. C* **46**, 1939 (1992)
[22] S. Cwiok and W. Nazarewicz, *Phys. Rev. C* **39**, 2097 (1989)

RECENT ADVANCES IN CLUSTER RADIOACTIVITIES

Dorin N. Poenaru[1] and Walter Greiner[2]

[1] Institute of Atomic Physics, Box MG-6, RO-76900 Bucharest, Romania
[2] Institut für Theoretische Physik der J. W. Goethe Universität, Postfach 111932, D-60054 Frankfurt am Main, Germany

Introduction

The physics of nuclear decay modes by spontaneous or beta-delayed emission of charged or neutral particles from nuclei has been developed intensively.[1,2] A real progress has been achieved in the understanding of newly discovered phenomena on the basis of our unified approach of cluster radioactivities, cold fission, and α-decay.[3-5] A system exhibiting such kind of processes may be viewed as a nuclear molecule.[6]

The first reliable predictions[7] of cluster radioactivities have been made by using both fission models (fragmentation theory and the asymmetric two-center shell model, three variants of numerical- and one analytical- (ASAFM)[8] superasymmetric fission model) and the Gamow penetrabilities. Particularly our ASAFM proved to be very useful in getting the main informations to guide the experiments. A semiempirical relationship based on fission theory gave the best agreement with experimental partial half-lives of almost 400 α-emitters. In this way, cluster radioactivities are making a bridge between extremely asymmetric α decay and almost symmetric cold fission. Wrong conclusions have been drawn in other papers published before 1980, including speculations since 1924 concerning the abundance of some gases (like N and Ne) in uranium ores.

The experiment performed by Rose and Jones[9] on ^{14}C spontaneous emission from ^{223}Ra, confirming our predictions (based on penetrability calculations) that ^{14}C should be the most probable emitted cluster from such a nucleus, triggered a great excitement among the experimentalists and theorists. Solid state track detectors and the superconducting solenoidal spectrometer SOLENO, played a key role in obtaining further experimental results (see the review papers[10-14] and the references therein). The fine structure[15] in ^{14}C radioactivity of ^{223}Ra has also been discovered[16] with SOLENO.

In a new comprehensive table we have published the results of a systematic study within ASAFM, extended in the region of heavier emitted clusters and of parent nuclei far from stability and superheavies.[17] We predicted a new island of proton-rich cluster emitters[18] awaiting for future experimental confirmation, as well as the cluster radioactivity of some α-stable neutron-rich nuclei.[19]

Many other theoretical models have been presented elsewhere.[2,20-22] Clusters have been interpreted as solitons on the nuclear surface.[5,23] We had introduced[24] a new semiclassical method of calculating cluster preformation probability within a fission model, as the penetrability of the prescission part of the barrier.[25] On this basis we have shown that, in a way, preformation cluster models are equivalent with fission models and we got one universal curve for each kind of cluster radioactivity, pointing out both the advantages and the drawbacks of such a "linearization".

Extensive studies of two-dimensional fission dynamics over a wide range of mass asymmetry have been performed.[26] A study of the Werner-Wheeler dynamics of different fission paths[40] has clearly shown that the cluster-like shapes (intersecting spheres with constant radius of the light fragment) are more suitable (the action integral takes lower values) than the more compact ones (intersecting spheres with constant volumes of both fragments) for $A_e < 34$, and the smoothed-neck influence is stronger for lower mass asymmetry. We performed calculations for α-decay, ^{28}Mg radioactivity and cold fission (with the light fragment ^{100}Zr) of ^{234}U - the first nucleus for which all three groups of decay modes have been experimentally detected.

The essential difference between a model (FM) derived from fission theory and a model (PCM) based on traditional many-body theory of α decay is the shape of the potential barrier: it has both an inner (prescission) and outer (postscission) part in a fission model, but only an outer one in a PCM. This difference is also reflected in the expression of the decay constant:

$$\lambda_f = \nu_f P_f \; ; \; \lambda_p = \nu_p S P_p \tag{1}$$

where the subscripts f and p denote fission and preformation, respectively, ν is the frequency of assaults on the barrier, P is a potential barrier penetrability, and S is the preformation probability. By expressing the penetration probability P_f as a product of the two terms $P_{ov}P_s$ corresponding to the inner (overlapping fragments) and the outer part (separated fragments) of the barrier, respectively, we got P_{ov} as the preformation probability within a fission model. In this manner we can estimate without technical difficulties the spectroscopic factor for any binary reaction.

Thanks to the organizers, cluster radioactivity, cold fission phenomena,[27] and superheavy nuclei,[28,29] have been discussed in many talks at the present school (W. Greiner, P. B. Price, M. Hussonnois, R. K. Gupta, R. J. Liotta, R. G. Lovas, A. Sandulescu, I. Silisteanu, F. Gönnenwein, G. Münzenberg, A. Sobiczewski and others). Consequently in the following, after briefly updating the comparison of our predictions with experimental results, we shall be mainly concerned with cluster preformation in closed- and mid-shell nuclei,[30] as well as with nuclear structure effects in cluster radioactivity.[31]

Experimental Confirmations of Predicted Half-lives

In order to be able to take into consideration the large number of combinations parent - emitted cluster (at least of the order of 10^5), we developed since 1980, the analytical superasymmetric fission model (ASAFM) with which we made the first predictions of nuclear lifetimes. The half-life

$$T = [(h \ln 2)/(2E_v)] exp(K_{ov} + K_s) \tag{2}$$

of a parent nucleus AZ against the split into a cluster $A_e Z_e$ and a daughter $A_d Z_d$ is calculated by using the WKB approximation, according to which the action integral is given by

$$K = \frac{2}{\hbar} \int_{R_a}^{R_b} \sqrt{2B(R)E(R)} dR \tag{3}$$

with $B = \mu$ – the reduced mass, $K = K_{ov} + K_s$, and $E(R)$ replaced by $[E(R) - E_{cor}] - Q$. E_{cor} is a correction energy similar to the Strutinsky shell correction, also taking into account the fact that Myers-Swiatecki's liquid drop model overestimates fission barrier heights, and the effective inertia in the overlapping region is different from the reduced mass. R_a and R_b are the turning points of the WKB integral, $R_i = R_0 - R_e$ is the initial separation distance, $R_t = R_e + R_d$ is the touching point separation distance, $R_j = r_0 A_j^{1/3}$ ($j = 0, e, d$; $r_0 = 1.2249 fm$) are the radii of parent, emitted and daughter nuclei. The two terms of the action integral K, corresponding to the overlapping (K_{ov}) and separated (K_s) fragments, are calculated by analytical formulas (approximated for K_{ov} and exact for K_s in case of separated spherical shapes within LDM).

By choosing $E_v = E_{cor}$ we have reduced the number of fitting parameters. Both shell and pairing effects are included in $E_{cor} = a_i(A_e)Q$ ($i = 1, 2, 3, 4$ for even-even, odd-even,

Table 1. *Comparison between measured and calculated half-lives.*

Parent		Emitted		Daughter		log T(s)			
							ASAFM		Experiment
Z	A	Z_e	A_e	Z_d	N_d	Early	Ref. 32	Ref. 17	
87	221	6	14	81	126	15.0	14.4	14.3	14.47 ± 0.10
88	221	6	14	82	125	13.8	14.3	14.2	13.17^a
88	222	6	14	82	126	12.6	11.2	11.1	11.01 ± 0.06
88	223	6	14	82	127	14.8	15.2	15.1	15.15 ± 0.05
88	224	6	14	82	128	17.4	15.9	15.9	15.69 ± 0.12
88	226	6	14	82	130	22.4	21.0	20.9	21.22 ± 0.20
89	225	6	14	83	128	18.5	17.8	17.8	17.16 ± 0.10
90	228	8	20	82	126	22.4	21.9	21.9	$20.72 + 0.08$
91	231	9	23	82	126	24.8	25.9	25.9	$26.02^{+0.76}_{-0.51}$
90	230	10	24	80	126	24.9	25.3	25.2	24.61 ± 0.09
91	231	10	24	81	126	22.0	23.4	23.3	22.89 ± 0.05
92	232	10	24	82	126	20.4	20.8	20.8	20.40 ± 0.05
92	233	10	24	82	127	23.1	24.8	25.2	24.84 ± 0.06
92	234	10	24	82	128	25.7	26.1	26.1	25.88 ± 0.30
92	233	10	25	82	126	23.3	25.0	25.7	24.84 ± 0.06
92	234	12	28	80	126	24.6	25.8	25.9	25.75 ± 0.06
94	236	12	28	82	126	19.8	21.0	21.1	21.68 ± 0.15
94	238	12	28	82	128	24.8	26.0	26.2	25.70 ± 0.25
94	238	12	30	82	126	24.4	25.7	26.2	25.70 ± 0.25
94	238	14	32	80	126	23.7	25.1	26.1	25.28 ± 0.16
96	242	14	34	82	126	20.9	22.3	23.5	23.24^b

[a] preliminary (R. Bonetti et al); [b] preliminary (S. P. Tretyakova et al).

even-odd, and odd-odd parent nuclei) due to its proportionality with the Q value, which is maximum when the daughter nucleus has a magic number of neutrons and protons.

With only few exceptions, in the region of nuclei far from stability, measured α-decay partial half-lives are not available. In principle we can use ASAFM to estimate these quantities. Nevertheless, slightly better results can be obtained by using our semiempirical formula.[33]

The cluster emitters experimentally confirmed up to now (see Table 1) are heavy trans-francium nuclei: ^{221}Fr, $^{221-224,226}$Ra, ^{225}Ac, ^{228,230}Th, ^{231}Pa, $^{232-234}$U, ^{236,238}Pu, and (preliminary) ^{242}Cm. The corresponding daughter nucleus is the doubly magic ^{208}Pb in eight of these cases, and it is another nucleus from its neighbourhood for the other nuclides. The measurements have been recently reviewed.[13,14]

In 1986, after taking properly into account the even-odd effect, we have improved our earlier predictions. Generally, as one can see from the table, the measurements have confirmed, within one order of magnitude, the calculated values of ASAFM. This result looks not bad if we take into account that in spontaneous fission there are still disagreements of many orders of magnitude (3 or even 5 orders).

Shell and pairing effects are clearly seen. The shorter half-life, of 10^{11} s, corresponds to the ^{14}C radioactivity of ^{222}Ra parent nucleus, leading to a double magic daughter ^{208}Pb. The largest branching ratio with respect to α-decay is about $10^{-9.2}$ and has been observed in the first experiment on ^{14}C radioactivity of ^{223}Ra. With the high sensitivity of solid state track detectors, it was possible to measure a branching ratio as low as $10^{-16.25}$ for Mg emission from ^{238}Pu. The longer partial half-life determined up to now, of 10^{26} s, is that of ^{231}Pa nucleus against ^{23}F emission.

Usually only one kind of emitted cluster (the most probable) could be experimentally observed, for the next one the emission rate being smaller by some oders of magnitude. Nevertheless, one can see few exceptions in Table 1: both F and Ne emitted by ^{231}Pa; Ne and Mg from ^{234}U, and Mg + Si from ^{238}Pu.

For ^{14}C radioactivity of ^{223}Ra we presented the global result; details about the fine structure of different transitions observed in this case are discussed below.

Besides the main region of trans-lead parent nuclides, already experimentally explored, a new island of proton rich nuclides[18] looks promising. Light clusters are preferentially emitted from neutron-deficient nuclei and the heavy ones from neutron-rich parents. The particularly strong shell effects of the 126 neutron closure and of 82 magic number of protons combine in ^{208}Pb nucleus with a very favorable position from the liquid-drop model point of view, namely almost the bottom of the valley of beta-stabilty. None of the other double magic neighbours, ^{100}Sn and ^{132}Sn, are as privileged as ^{208}Pb. They are both far from stability, and the shell effects of Z = 50, N = 50, 82 magic numbers are weaker. Consequently the mass distributions in cold fission is very broad compared to cluster radioactivity and α-decay.

The above mentioned role of ^{208}Pb is the main reason why all up to now experimentally discovered cluster emitters are located in this region of nuclei. As we have already mentioned, eight of the confirmed cluster radioactivities (^{14}C, ^{20}O, ^{23}F, ^{24,26}Ne, ^{28,30}Mg and ^{34}Si) correspond to the ^{208}Pb daughter, and the other to some neighbouring nuclei.

We expect that the search for new radioactivities will continue in this region. Particularly interesting are the transitions in which at least one of the three partners possesses an odd number of nucleons, because of the spectroscopic informations one can get. Besides the even-odd effects, one should not forget about the ^{8}Be emission in competition with double-α-decay, and cold fission as cluster radioactivity.

Our estimations of half-lives are reliable enough, provided the masses M, M_e, M_d determining the released energy $Q = (M - M_e - M_d)c^2$, are accurately known. Any erroneous increasement of the Q-value is reflected in a lower K, producing a drastic decrease of the half-life, because of the exponential dependence shown in Equation (2). Consequently we need mass values as close as possible to the true ones.[34,35]

Whenever available, we use the measured masses or those determined from systematics by Wapstra et al.[36] Otherwise we try the mass estimations[37] one after the other, and then we compare the results. We have performed a large number of systematic calculations (25685 input masses times about 200 possible emitted clusters). From the results obtained in this way, we usually selected those which have $T < 10^{25} y$ and branching ratio relative to α-decay $b = T_\alpha/T > 10^{-18}$, because with presently available experimental techniques, the phenomena of cluster decays can only be detected when T is low enough and b is sufficiently high. Besides the main region of trans-radium parents, where experiments have been done, another one with $Z = 56 - 64$, $N = 58 - 70$ should be carefully investigated.

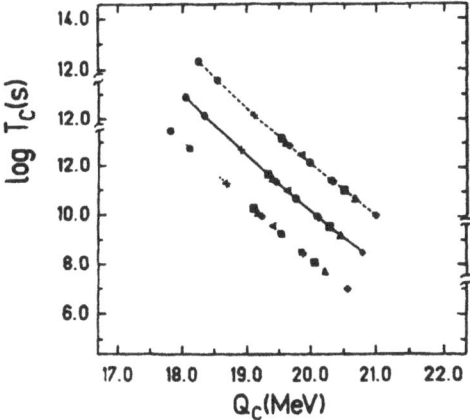

Figure 1. *The sensitivity of the ^{114}Ba half-life against ^{12}C radioactivity to the mass values of the parent nucleus (different points) and of the daughter ^{102}Sn (full line – 65.02 MeV; dashed – 65.23; dotted – 64.78).*

Let us consider, as an example, ^{12}C radioactivity of ^{114}Ba, we have mentioned as interesting few years ago. The sensitivity of the estimated half-life and branching ratio to an uncertainty in the mass value of parent nucleus and that of the daughter is relatively high, as can be seen on Figure 1. If we take into account the estimated β-decay half-life of 0.3 - 0.5 s, the branching ratio with respect to this decay mode could be in the range 10^{-8} - 10^{-12} s, which is measurable in an experiment on-line.

In a recently performed measurement, the exotic nucleus ^{114}Ba was produced by bombarding a Ni target with 280 MeV ^{58}Ni heavy ions. According to these preliminary results from Dubna[38] about 10 events (of ^{12}C) have been observed. There are difficulties connected with the high background produced by neutrons. In a new experiment to be done at GSI Darmstadt[39] one can improve considerably the experimental conditions, by transporting the reaction products far from the region with strong neutron background. Reaction cross section, β-, α-, and C-decays are planned to be determined. Radioactive beams could be alternatively used to produce nuclei far from stability which are cluster emitters. Even if the C-emission will not be detected at the beginning, from the measured kinetic energy of the α- and β particles we can get improved values of the implied masses, which in turn could be used to obtain more reliable estimates of half-lives and branching-ratios.

On the basis of a large number of systematic calculations, using different input mass tables to find the released energy, we expect that an island of cluster emission should exist in the neutron-deficient α-stable region, where the neutron-rich neutron ($N_e = 50$) and proton ($Z_e = 28$) magic numbers of the emitted cluster, as well as (in a less extent) the proton magic number of the daughter ($Z_d = 50$) are playing an important role.[19] However, the up to now estimated half-lives seem to be too long to be measured in competition with β-decay.

Cluster Preformation

It is extremely difficult to find within a microscopic theory a convenient method to calculate the preformation probability S for clusters heavier than α particle. In order to study a broader range of emitted clusters[21] a phenomenological law for the variation of S with A_e has been introduced. Any dependence on Z_e, Z_d and the neutron number N_d has been neglected. For cluster emission this choice is justified, from practical point of view, in what concerns the daughter nucleus. We have shown, indeed, on the basis of a large amount of systematic calculations within ASAFM, that in the most interesting region of parent nuclei, the heavy fragment is almost always the doubly magic nucleus ^{208}Pb or one of its neighbours.

The above mentioned technical difficulty is not present in a fission theory, where the preformation probability is obtained from an integral, analytically approximated within ASAFM. We got a formula according to which $-\log P_{ov}$ is proportional to the potential barrier width $(R_t - R_i)$ and to $\sqrt{\mu_A E_b}$. When A_e increases, the fission barrier height is also increasing up to a certain maximum value, beginning with which it starts to decrease. Consequently, a bending of the function $-\log P_{ov}$ should occur toward the large mass numbers of the emitted clusters. By studying fission dynamics we have also found[40] a shape transition around $A_e = 34$, from cluster-like shapes to more compact ones.

One may assume a linear dependence of the preformation probability in the range of emitted clusters already measured :

$$\log P_{ov} = \frac{(A_e - 1)}{3} \log P_{ov}^\alpha \qquad (4)$$

where the preformation probability of α particle P_{ov}^α may be determined by fit with experimental data.

On this basis we found a single universal curve $\log T = f(\log P_s)$, for each kind of cluster radioactivity of even-even parent nuclei. This has been done by making a further assumption that the frequency ν is independent on the emitted cluster and the daughter nucleus $\nu(A_e, Z_e, A_d, Z_d)$ = constant. From a fit of experimental data we obtained: $P_{ov}^\alpha = 0.0160694$ and $\nu = 10^{22.01}$ s^{-1}, leading to

$$\log T = -\log P_s - 22.169 + 0.598(A_e - 1) \qquad (5)$$

which is a straight line for a given A_e, with a slope equal to unity. For any combination of fragments $A_e Z_e$, $A_d Z_d$ one can calculate easily

$$-\log P_s = 0.22873(\mu_A Z_d Z_e R_b)^{1/2}\left[\arccos\sqrt{r} - \sqrt{r(1-r)}\right] \quad (6)$$

where $r = R_t/R_b$, $R_t = 1.2249(A_d^{1/3} + A_e^{1/3})$, $R_b = 1.43998 Z_d Z_e/Q$, and $\mu_A = A_d A_e/A$ is the reduced mass number.

The advantage of such a handy relationship is evident. The up to now 14 even-even half-life measurements are well reproduced (within a ratio 3.86, or rms=0.587 orders of magnitude). A similar result is obtained with the "hand-pocket" formula presented in Ref. 21. Another argument of its universality is illustrated in Fig. 2. Instead of having different lines for various parent nuclei, like in the "classical" systematics $\log T = f(Q^{-1/2})$, one gets practically only one line when we plot $\log T = f(\log P_s)$. Nevertheless one should be aware of its limitations. If we have a closer look, by plotting the deviations of $\log T$ values of 124 α emitters from the experimental ones, versus the neutron number of the daughter nucleus, one can see a structure due to the shell effects. Unlike the simple assumption of $S = f(A_e)$, given by eq. (5), we shall examine below a more realistic approach of the preformation probability.

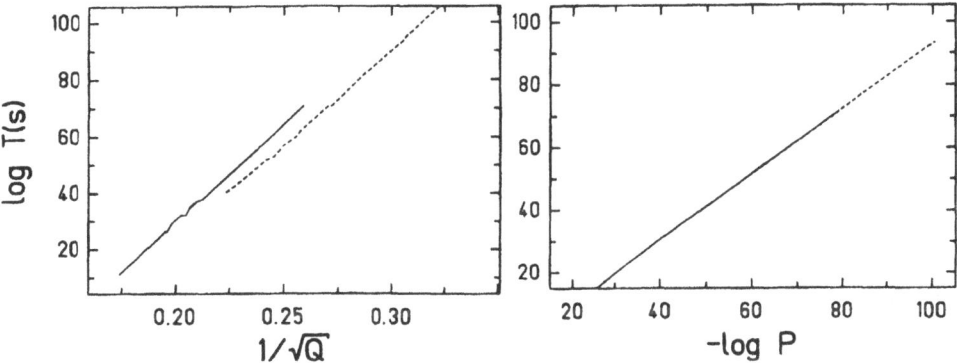

Figure 2. *Comparison of a "classical" (Geiger-Nuttall-like) systematics (left hand side) with universal curve-like systematics (right hand side), calculated within ASAFM for ^{14}C radioactivity of even-even Ra isotopes (full line) and of Pb isotopes (dashed line).*

Typically in a fission model the potential barrier with its two parts (inner or prescission and outer or postscission) is thicker in comparison with that of a PCM, where only the latter is present. This inner part, which is missing in a PCM, comes from the early stages of the process by considering the finite spatial extensions of the two fragments when the deformation energy is calculated. Strong interaction and electrostatic forces are present. Its relative importance is growing for heavier emitted clusters. In a two-center parametrization it extends from the initial distance between centers $R_i = R_0 - R_e$ to the touching point $R_t = R_d + R_e$, where R_0, R_d and R_e are the radii of the parent, daughter, and emitted nuclei, respectively. From the touching point configuration to the exit (or the classical turning point R_b) one has the outer barrier, practically dominated by the Coulomb forces if the angular momentum effects are negligible small.

The potential barrier in a PCM also extends from the "channel radius" R_t to the outer turning point, expressed as $R_b = Z_e Z_d e^2 / Q$ if spherical shapes of nuclei are assumed. Usually the finite cluster radius is not taken into account, but this barrier can be identical with the outer one from fission model, as long as the same nuclear radii constants are used, because the Coulomb interaction of two spheres is identical with that of point charges.

The quantum penetrabilities are usually calculated in the semiclassical WKB approximation. The relationship for K_p of the PCM is identical to that of K_s and in the Coulomb field $E(R) = e^2 Z_e Z_d / R - Q$ we got a well known analytical result (see eq. (6)). The two decay constants in Equation (1) look now very similar. Both ν_f and P_s from a fission model, have their counterparts ν_p and P_p in a PCM. In our fission model the preformation probability (in fact the probability of a touching point configuration to be formed), $S_{sc} = P_{ov}$, is the penetrability of the inner part of the barrier, given by

$$\log S_{sc} = -0.09537 (E_b^0 A_e A_d / A)^{1/2} (R_t - R_i) \left[\sqrt{1 - b^2} - b^2 \ln \frac{1 + \sqrt{1 - b^2}}{b} \right]. \quad (7)$$

We denote $b^2 = (E_{cor} + E^*)/E_b^0$, where E_{cor} has been defined above, E^* is the excitation energy concentrated in the separation degree of freedom, $E_b^0 = E_i - Q$ is the barrier height before correction. The interaction energy at the top of the barrier, in the presence of a nonnegligible angular momentum, $l\hbar$, is given by:

$$E_i = e^2 Z_e Z_d / R_t + \hbar^2 l(l+1)/(2\mu R_t^2) \quad (8)$$

where μ is the reduced mass.

Unlike within approximation leading to the "universal curves" of even-even nuclei, where $\log S = -0.598(A_e - 1)$ depends only on the mass number of the emitted cluster, the nuclear structure effects could be taken into consideration in the above relationship. The heavier the cluster, the larger is the difference between the probability to be formed in a nucleus leaving a daughter ^{208}Pb and the preformation in this doubly magic nucleus itself. An "experimental" value is defined as the ratio $S_{exp} = \lambda_{exp}/(\nu P_s)$.

The fact that a given cluster is preformed with a larger probability in a parent nucleus corresponding to ^{208}Pb daughter than in the doubly magic nucleus ^{208}Pb is illustrated in Fig. 3, in which we consider few clusters: α-particle; ^{14}C and ^{24}Ne. Instead of only one value of S at a given mass number A_e (a horizontal line) independent on Z_e, Z_d, A_d, one can see a whole range, increasing at a constant N_d from $Z = 82$ (dashed line) to $Z = 82 + Z_e$ (full line).

The maximum value is reached at the magic neutron number of the daughter $N_d = 126$. If we consider only the measured up to now emissions, they are lying quite well on a straight line, justifying the above considerations leading to universal curves. Nevertheless, the whole picture shows us once more that the simplifying assumptions are not generally confirmed. Figure 3 also illustrates very clearly that a given cluster has a larger probability to be formed in a parent nucleus corresponding to ^{208}Pb daughter than in a double magic nucleus ^{208}Pb. The difference between the two values increases for heavier clusters.

We plan to extend our calculations of spectroscopic factors to pickup and stripping reactions.

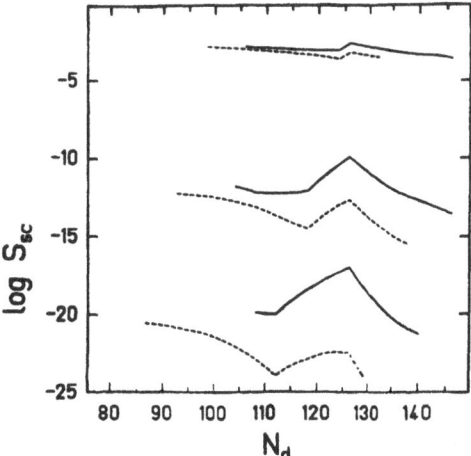

Figure 3. *In descending order from the top to the bottom we present the preformation probability of ^{4}He, ^{14}C and ^{24}Ne versus neutron number of the daughter. For each cluster the parent nuclei are either Pb isotopes (dashed line) or heavier nuclides leading to Pb isotopes as daughter nuclei (full line).*

Nuclear Structure Effects

Cluster emission leading to excited states of the final fragments have been considered for the first time by Martin Greiner and Werner Scheid.[15] In the first experiment[16] on the fine structure of ^{14}C radioactivity of ^{223}Ra it was shown that the transition toward the first excited state of the daughter nucleus is stronger than that to the ground state. In other words, like in spontaneous fission of odd-mass nuclei, or in fine structure of α-decay, one has a hindered and a favoured transition, respectively. Moreover, preliminary results with better statistics and energy resolution[41,42] do not exclude the possibility that the transition toward the second excited state of ^{209}Pb is favored too. Up to now theoretical considerations[43,44] have only been partly successful in explaining the experimental result. New experiments and calculations are in progress.

Like in α-decay, or in spontaneous fission of odd-mass or odd-odd nuclei, one can define a hindrance factor of a given transition

$$H = T^{exp}/T_{e-e} \tag{9}$$

where T^{exp} is the measured half-life, and T_{e-e} is the corresponding quantity for a hypothetical even-even equivalent, estimated either from a systematics ($\log T$ versus $Q^{-1/2}$ for example[14]) or from a model. A transition is favoured if $H \simeq 1$, and it is hindered if $H \gg 1$. Spontaneous fission rates of odd-A nuclei are slower than that of e-e nuclei, indicating a higher fission barrier, due to shell effects.

Within ASAFM we can easily simulate the even-even assumption. Unlike in the above mentioned systematics, we also can take into account the angular momentum, l, determined from the following conditions:

$$|J_f - J_i| \leq l \leq J_f + J_i \ ; \ \pi_i \pi_f (-1)^l = 1 \tag{10}$$

The spin and parity of the (initial) ground state of ^{223}Ra is $J_i^{\pi_i} = 3/2^+$. In the final states of ^{209}Pb one has $J_f^{\pi_f} = 9/2^+$ (gs), $11/2^+$ (first excited state at 0.779 MeV), and $15/2^-$ (second excited state at 1.423 MeV). Consequently the allowed angular momenta are 4 and 6 units of $\hbar$ in the first two cases, and 7 and 9 in the last one. The Q-values are 31.85, 31.07 and 30.43 MeV, respectively. If the intensities of the above mentioned ^{14}C transitions are 15 %, 81 % and 4 %, and the total half-life for cluster radioactivity is $10^{15.20}$ s, the corresponding partial lifetimes are $10^{16.02}$, $10^{15.29}$, and $10^{16.60}$ s. By ignoring the angular momentum, we get within ASAFM hindrance factors of 417; 2.4, and 2.5, respectively. They became 263; 1.5; 0.5 for $l = 4$; 4; 7.

We have performed similar calculations for other odd-mass parent nuclei. We got a large hindrance factor (H = 44) for ^{24}Ne decay of ^{233}U (gs $5/2^+$) to the gs ($9/2^+$) of ^{209}Pb; the best fit with $T^{exp} = 10^{24.84}$ s is obtained for the transition to the first excited state ($11/2^+$ at 778 keV) with $l = 4$, $T^{ASAFM} = 10^{24.89}$. Other hindered transitions could be: ^{23}F decay of ^{231}Pa (gs $3/2^-$) to the 0^+ gs of ^{208}Pb (H = 12); ^{14}C decay of ^{221}Fr (gs $5/2^-$) to the $1/2^+$ gs of ^{207}Tl (H = 7.5); ^{14}C decay of ^{221}Ra to the $1/2^-$ gs of ^{207}Pb (H = 5), and ^{24}Ne decay of ^{231}Pa to the gs of ^{207}Tl. The ^{14}C transition from ^{225}Ac (gs ($3/2^-$)) to the ($9/2^-$) gs of ^{211}Bi seems to be not hindered, unless the future fine-structure measurement will show a possible contribution from ^{221}Fr (alpha daughter) ^{14}C decay.

The favoured transitions are explained within present version of ASAFM (one should allow an accuracy of ± 1 order of magnitude of any theoretical calculations), but the hindered one needs a larger action integral, K. Such an increase can be obtained with a larger potential barrier, due to the so-called[45] *"specialization energy"*, in a similar way to what has been done for spontaneous fission of odd-mass nuclei.[46]

The specialization energy arises from the conservation of spin and parity of the odd particle during the fission process. With an increase of deformation (distance between fragment centers in the TCSM) the odd nucleon may not be transferred on a low energy at a level crossing, if in this way it cannot conserve the spin and parity. The corresponding barrier becomes higher and wider, compared to that of the even-even neighbour and the fission half-life becomes longer.

Large hindrance factors have been measured not only in the fine structure of α-decay, but also in spontaneous fission. Spontaneous fission rates of odd-A nuclei are slower than that of e-e nuclei, indicating a higher fission barrier, due to shell effects. In both kinds of theories the physical explanation relies on the single-particle spectra of neutrons or/and protons. If the uncoupled nucleon is left in the same state both in parent and heavy fragment, the transition is favoured. Otherwise the difference in structure leads to a large hindrance.

Technically in a fission theory the calculations are performed as mentioned above, and in a many-body theory of α-decay the numerical value of the square of the overlap integral of the initial and final wave-functions is taken as a measure of the hindrance factor, because it reflects the difference in the cluster preformation amplitude.

Unlike in α-decay, where the initial and final states of the parent and daughter are not so far from each other, in ^{14}C radioactivity of ^{223}Ra, one has a unique possibility to study a transition from a well deformed parent nucleus with complex parity-mixed odd-neutron orbitals to a spherical nucleus with pure shell model wave function.

In this way, one can get *direct spectroscopic information* on spherical components of deformed states. For example ^{223}Ra is known to be pear-shaped with a possible set of deformation parameters $\beta_2 = 0.129$, $\beta_3 = 0.10$, $\beta_4 = 0.075$, $\beta_5 = 0.01$, and $\beta_6 = 0.004$. According to the reflection-asymmetric rotor model the ground-state of ^{223}Ra is built from $3/2^+$ state emerging from $i_{11/2}$ orbital with small admixtures from $g_{7/2}$, $j_{15/2}$, and $g_{9/2}$. Three sets of opposite parity bands: $3/2[761] \otimes 3/2[631]$, $5/2[633] \otimes 5/2[752]$, and $1/2[770] \otimes 1/2[640]$ explain the low-lying states in many radium isotopes. The overlap integral was obtained[44] low (4%), and high (64%), for the transitions toward gs and first excited state, in agreement with the experiment, but also low (7%) for the transition to the second excited state of ^{209}Pb – a challenging result.

Acknowledgments

This paper was partly supported by Bundesministerium für Forschung und Technologie, by Gesellschaft für Schwerionenforschung Darmstadt, by Stabsabteilung Internationale Beziehungen des Kernforschungszentrums Karlsruhe, and by Institute of Atomic Physics. One of us (DNP) received a donation of computer equipment from Soros Foundation for an Open Society.

References

[1] D. N. Poenaru and W. Greiner, eds.: *Handbook of Nuclear Decay Modes* (CRC Press, Boca Raton, Florida, to be published)

[2] D. N. Poenaru, M. Ivaşcu and W. Greiner in: *Particle Emission from Nuclei*, Vol. III, p. 203, edited by D. N. Poenaru and M. Ivaşcu, (CRC Press, Boca Raton, 1989)

[3] D. N. Poenaru and W. Greiner, Chapter 7 in Ref. 1

[4] W. Greiner, M. Ivaşcu, D. N. Poenaru and A. Sandulescu in: *Treatise on Heavy Ion Science*, Vol. 8, p. 641, edited by D. A. Bromley, (Plenum Press, New York, 1989)

[5] A. Sandulescu and W. Greiner, *Rep. Prog. Phys.* **55**, 1423 (1992)

[6] W. Greiner in: *Clustering Phenomena in Atoms and Nuclei*, p. 213, edited by M. Brenner, T. Lönnroth and F. B. Malik, (Springer, Heidelberg, 1992)

[7] A. Săndulescu, D. N. Poenaru and W. Greiner, *Sov. J. Part. Nucl.* **11**, 528 (1980)

[8] D. N. Poenaru, M. Ivaşcu, A. Săndulescu and W. Greiner, *J. Phys. G.* **10**, L183 (1984)

[9] H. J. Rose and G. A. Jones, *Nature* **307**, 245 (1984)

[10] P. B. Price, *Annu. Rev. Nucl. Part. Sci.* **39**, 19 (1989)

[11] E. Hourani, M. Hussonnois and D.N. Poenaru, *Ann. Phys. (Paris)* **14**, 311 (1989)

[12] S. P. Tretyakova, V. L. Mikheev, Yu. S. Zamyatnin and A. A. Ogloblin in: *Proc. International Conference on Exotic Nuclei*, Foros, Ukraina, edited by Yu. E. Penionzhkevich and R. Kalpakchieva, (World Scientific, Singapore, to be published)

[13] R. Bonetti and A. Guglielmetti, Chapter 10 in Ref. 1.

[14] E. Hourani, Chapter 9 in Ref. 1.

[15] M. Greiner and W. Scheid, *J. Phys. G.* **12**, L285 (1986)

[16] L. Brillard, A. G. Elayi, E. Hourani, M. Hussonnois, J. F. Le Du, L. H. Rosier and L. Stab, *C. R. Acad. Sci.* **309**, 1105 (1989)

[17] D. N. Poenaru, D. Schnabel, W. Greiner, D. Mazilu and R. Gherghescu, *Atomic Data Nucl. Data Tables* **48**, 231 (1991)

[18] D. N. Poenaru, W. Greiner and R. Gherghescu, *Phys. Rev. C* **47**, 2030 (1993)

[19] D. N. Poenaru, W. Greiner, R. Gherghescu and D. Mazilu, *Fizika B* **1**, 221 (1992)

[20] R. A. Broglia, *Z. Phys. D*, to be published

[21] R. Blendowske, T. Fliessbach and H. Walliser, Chapter 8 in Ref. 1

[22] R. K. Gupta, W. Scheid and W. Greiner in: *Clustering Phenomena in Atoms and Nuclei*, p. 296, edited by M. Brenner, T. Lönnroth and F. B. Malik, (Springer, Heidelberg, 1992)

[23] A. Ludu, A. Sandulescu and W. Greiner, *Int. J. Mod. Phys. E* **1**, 169 (1992)

[24] D. N. Poenaru and W. Greiner, *Phys. Scripta* **44**, 427 (1991)
[25] D. N. Poenaru, invited talk, in: *Proc. 6th Internat. Conf. on Nuclei Far from Stability* and *9th Internat. Conf. on Atomic Masses and Fundamental Constants*, Bernkastel-Kues, Germany, p. 443, edited by R. Neugart and A. Wöhr, (IOP Publishing, Bristol, 1993)
[26] M. Mirea, D. N. Poenaru and W. Greiner, *Nuovo Cimento* **105A**, 571 (1992)
[27] F. Gönnenwein in: *The Nuclear Fission Process*, edited by C. Wagemans, (CRC Press, Boca Raton, 1991)
[28] G. Münzenberg, Chapter in Ref. 1.
[29] A. Sobiczewski, invited talk, in: *Proc. 6th Internat. Conf. on Nuclei Far from Stability* and *9th Internat. Conf. on Atomic Masses and Fundamental Constants*, Bernkastel-Kues, Germany, p. 403, edited by R. Neugart and A. Wöhr, (IOP Publishing, Bristol, 1993)
[30] D. N. Poenaru and W. Greiner, *Z. Phys. D*, in print
[31] D. N. Poenaru, W. Greiner, E. Hourani and M. Hussonnois, *Z. Phys. D*, in print
[32] D. N. Poenaru, M. Ivascu, D. Mazilu, R. Gherghescu, K. Depta and W. Greiner, *Central Inst. of Phys., Bucharest, Report* NP-54 (1986)
[33] D. N. Poenaru and W. Greiner, Chapter 6 in Ref. 1
[34] A. H. Wapstra, Chapter 1 in Ref. 1
[35] N. Zeldes, Chapter 2 in Ref. 1
[36] A. H. Wapstra, G. Audi and R. Hoekstra, *Atomic Data Nucl. Data Tables* **39**, 281 (1988)
[37] P. E. Haustein (special editor), *Atomic Data Nucl. Data Tables* **39**, 185 (1988)
[38] Yu. Ts. Oganessian, V. L. Mikheev and S. P. Tretyakova, *private communication*, 1992
[39] P. B. Price, E. Roeckl and R. Bonetti, *private communication*, 1992
[40] D. N. Poenaru, J. A. Maruhn, W. Greiner, M. Ivascu, D. Mazilu and I. Ivascu, *Z. Phys. A* **333**, 291 (1989)
[41] M. Hussonnois, J. F. Le Du, L. Brillard, J. Dalmasso and G. Ardisson, *Phys. Rev. C* **43**, 2599 (1991)
[42] E. Hourani et al., *Phys. Rev. C* **44**, 1424 (1991)
[43] M. Hussonnois, J. F. Le Du, L. Brillard and G. Ardisson, *Phys. Rev. C* **44**, 2884 (1991)
[44] R. K. Sheline and I. Ragnarsson, *Phys. Rev. C* **44**, 2886 (1991)
[45] J. A. Wheeler in: *Niels Bohr and the Development of Physics*, p. 163, edited by W. Pauli, L. Rosenfeld and V. Weisskopf, (Pergamon Press, London, 1955)
[46] J. Randrup, S. E. Larsson, P. Möller, S. G. Nilsson, K. Pomorski and A. Sobiczewski, *Phys. Rev. C* **13**, 229 (1976)

CLUSTER RADIOACTIVITY. TWO-SOLITON SOLUTIONS ON SPHERE

A. Ludu[1], A. Sandulescu[2] and W. Greiner[3]

[1] *Department of Theoretical Physics, University of Bucharest, Bucharest-Magurele PO Box MG-5211, Romania*
[2] *Department of Theoretical Physics, Institute of Atomic Physics, Bucharest-Magurele PO Box MG-6, Romania*
[3] *Institut für Theoretische Physik, Universität Frankfurt/Main, Robert-Mayer-Str. 8-10 Frankfurt/Main, Germany*

1. Introduction

The cluster radioactivity represents a new form of large amplitude collective motion in the nucleus. Cluster radioactivity was theoretically suggested in 1980 [1] and first experimentally discovered in 1984, ^{14}C emission from ^{223}Ra [2]. Since then, new experimental discoveries [3] of the new natural decays with emission of carbon, oxygen, neon, magnesium and silicon from heavy nuclei indicate a large enhancement of the preformation of clusters on the nuclear surface of the double magic nucleus ^{208}Pb or close to it. This suggests that the external nucleons join together to form the emitted cluster leaving the residual nucleus unpolarized, i.e. to new shapes which could not be described by multipole expansion of the nuclear surface. We consider that only a collective description is the most appropriate formulation to these exotic decays. In a phenomenological description, we have to find some arguments of collective type which may justify the existence of new exotic shapes. Probably such shapes have not been considered due to the fact that many body correlations of this type have not been included in the microscopic calculations. Another example of cluster enhancement is the alpha decay, whose preformation factors are underestimated by the shell model by at least two orders of magnitude. This is most likely due to the absence of the four-body correlations in the evaluations of the preformation factors. Other possible explanations are already discussed in the topical review on new radioactivities [4].

In the present paper, by introducing nonlinear terms of higher order in the hydrodynamical equations, we show that not only stable one-soliton solutions, but also two-soliton solutions can exist on the surface of the sphere. Contrary to the Bohr-Mottelson model, we assume that the outside nucleons do not polarize the double magic core ^{208}Pb. The soliton itself contains the polarization effect. Consequently taking into account the nonlinear equations up to the third order in the expansion of the nuclear surface we obtain a two-soliton solution which provides the conservation of the center of mass in the geometric center of the initial sphere. Such a model could describe many other natural phenomena, at different scales, like: the vortices in a shallow rotating atmosphere and in particular the Great Red Spot of Jupiter, pairs of double stars, drift waves in cylindrical plasmas (Z-pinch) etc. The total potential energy consists of four terms: surface, centrifugal, Coulomb and shell energies. The last term was introduced phenomenologically by the overlaps between the daughter and cluster nuclei (considered to be spheres) and the parent nucleus having quite different shapes than the shapes described by the usual multipole expansion of the nuclear surface in spherical harmonics. This term causes a new minimum in the total potential energy. Due to the fact that alpha and cluster decays are spontaneous decays, we choose this minimum to be degenerate in energy with the ground state minimum. The calculation of the total energy of the soliton shows that the energy density in soliton increases with increase of energy. This behaviour realizes the stability of the soliton on the spherical surface [5].

This description leads to a new coexistence model consisting of the usual shell model and a cluster-like model described by a soliton (or a pair of solitons) moving on a sphere. Due to the large barrier between the two minima, the amplitude of the cluster-like state is much smaller than the usual ground state. The ratio of the square of the two amplitudes gives the preformation probability of the corresponding cluster at the nuclear surface. Recently [6-10] we have shown that by introducing nonlinear terms in the hydrodynamical equations stable solitons exist on the surface of a circle with the shape $r = R + \eta(\theta, t)$, respectively on a sphere if the perturbation has an axial symmetry. The shell effects related to the emitted clusters lead to a new minimum in the total energy choosen, due to the fact that the decays are spontaneous decays, to be degenerated with the ground state minimum. In this way a new coexistence model consisting of the usual shell model and a cluster-like model describing a soliton moving on the nuclear surface was introduced. It was shown that the corresponding amplitudes describe excellently the experimental preformation factors.

The present model also gives good results in the alpha decay arround the magic nucleus ^{208}Pb [7]. In the literature there are already different possible explanations of the absolute values in alpha decay which may explain this puzzle [11-17]. However, we consider that only a collective description is the most appropriate formulation for alpha decay. Alpha particles were described as solitons on the nuclear surface of the daughter nucleus. In this way we introduced a new large collective amplitude motion in nuclei which may contain the four-body correlations which are not included in the standard shell model. This assumes the existence of preformed alpha particles in the initial state. This is analogous with the microscopic treatment from ref. [17].

2. Hydrodynamical Nonlinear Model and One-Soliton Solution

The hydrodynamic model is based on the fact that on the surface of a layer of an ideal fluid, solitonic waves can occur if the conditions for the "shallow-water" model are fulfilled, i.e. the depth of the layer to be small compared to the half-width of the soliton. When we describe these perturbations of the nuclear surface we ignore the dissipative effects since they are slower than the effects of dispersion and nonlinearity. The linear hydrodynamic Euler equations give in this case the normal modes of oscillation described perfectly by the spherical harmonics. By introducing the nonlinear terms in the hydrodynamical equations and by considering the nonlinearities in the boundary conditions of the spherical shapes (inner and outer shapes of the layer) we get nonlinear waves as solutions. These could be one-soliton solutions [6-10] or many-soliton solutions. Here the dispersion is provided by the surface pressure of the fluid due to the curvature of the sphere. The theory of nonlinear waves in dispersive media is based mainly on the K-dV equation. According to this idea, we can regard the parameters of nonlinearity and dispersion as being of the same order of smallness when we expand the Euler and surface equations of the general form into series in powers of the deformation parameters of the surface. When we take into account these opposite effects on an equal basis, we can reveal essential features of the phenomena under consideration and obtain a foundation for explicit classification of them.

We present in the following an one-dimensional picture of such solitons (i.e. moving on a circle) stressing that the three-dimensional system could be treated on the same line via separation of spherical coordinates θ, ϕ into normal ($\phi_\perp$) and the parallel one ($\phi_\parallel$) [18-19]. Thus one can simplify the three-dimensional wave equation still retaining its fundamental properties.

Consequently let us consider a small perturbation propagating on the surface of a sphere of radius R. For an axial symmetry, with the symmetry axis in the direction of the perturbation, the problem reduces to a small perturbation propagating on a circle with the shape $r = R + \eta(\theta, t)$.

In the following we make three assumptions: first that the amplitude of the perturbation η is small compared with the radius R of a spherical nucleus so that we can introduce a small parameter $\xi = (r - R)/R \ll 1$, second that the spherical daughter nucleus is unperturbed up to the radius $r = R - h$ with $h \ll R$ and third that we have the case of an ideal, incompressible and irrotational fluid which leads to a field of velocities $\vec{V} = \nabla \Phi$ given by a scalar potential Φ satisfying the Laplace equation $\triangle \Phi = 0$. Writing the solution as a power series in ξ

$$\Phi(r, \theta, t) = \sum_{n \geq 0} \xi^n f_n(\theta, t), \tag{1}$$

we obtain an infinite system of recurrence relations for the unknown functions f_n given by:

$$f_{k+2} = -\frac{\sum_{n=1}^{k+1}(-1)^{k-n+1}nf_n + \sum_{n=0}^{k}(n-k-1)^{k-n}f_{n,\theta\theta}}{(k+1)(k+2)} \tag{2}$$

for $k = 0, 1, \ldots$ In the second order in ξ the radial $v = \Phi_r$ and the tangential $u = \Phi_\theta/r$ velocities depend only on $f_{0,\theta}(\theta, t)$ and $f_1(\theta, t)$ denoted in the following by $g(\theta, t)$ and $j(\theta, t)$, respectively

$$u = \frac{1}{R}[g + \xi(-g + j_\theta) + \frac{1}{2}\xi^2(2g - g_{\theta\theta} - 3j_\theta)], \tag{3}$$

$$v = \frac{1}{R}[j + \xi(-j - g_\theta) + \xi^2(j - \frac{1}{2}j_{\theta\theta} - \frac{3}{2}g_\theta)]. \tag{4}$$

In order to describe the collective dynamics we have to find the corresponding equations for shapes and velocities. The first equation is given by the total time derivative of the radial coordinate on the surface Σ

$$v|_\Sigma = \frac{dr}{dt}|_\Sigma = \eta_t + \frac{1}{r}u\eta_\theta. \tag{5}$$

By imposing the condition that $v = 0$ at $r = R - h$ we obtain in the first order in η, a relation between j and g_θ

$$j = -\frac{h}{R}g_\theta. \tag{6}$$

The second equation is given by the Euler equation

$$\rho_m(\frac{\partial \vec{V}}{\partial t} + (\vec{V}\nabla)\vec{V}) = -\nabla P + \nabla \Phi_e, \tag{7}$$

where ρ_m is the constant mass density, P is the surface pressure and Φ_e the electrostatic potential.

The pressure P at the surface Σ, in the second order in η, is given by

$$P|_\Sigma = \frac{\sigma}{R}[1 - \frac{1}{R}(\eta + \eta_{\theta\theta}) + \frac{1}{R^2}(\eta^2 + \frac{1}{2}\eta_\theta^2 + 2\eta\eta_{\theta\theta})]. \tag{8}$$

The electrostatic potential is given by the Poisson equation

$$\triangle \Phi_e = -\frac{\rho_{el}}{\epsilon_0}, \tag{9}$$

where ϵ_0 is the vacuum dielectric constant and ρ_{el} an uniform charge density which is ρ_d of the spherical daughter nucleus or an uniform charged perturbation ρ_α propagating over this sphere with the shape $r = \eta(\theta, t) + R$. We denote by $\lambda = \rho_\alpha/\rho_d$ the ratio of these charged densities. By using the same method as for Φ, the electrostatic potential Φ_e, up to the second order in η, is given by

$$\Phi_{el}|_\Sigma = \frac{\rho_d R^2}{3\epsilon_0}\left(1 - \frac{\eta}{R} + \frac{\eta^2}{R^2}\right) - \frac{\rho_\alpha \eta^2}{2\epsilon_0}. \tag{10}$$

From the continuity equation

$$\frac{\partial \rho_m}{\partial t} + \nabla(\rho_m \vec{V}) = \rho_m \nabla \vec{V} = 0, \tag{11}$$

and the condition of irrotationality $\nabla \times \vec{V} = 0$ we can put the Euler eq. (7) into a gradient form which leads to the following equation for Φ at the surface Σ

$$\Phi_t + \frac{1}{2}(\nabla \Phi)^2 + \frac{1}{\rho_m}P + \frac{\rho_{el}}{\rho_m}\Phi_e = N(t), \tag{12}$$

where $N(t)$ is a constant of integration depending only on time. Using the expressions for u, v, P and Φ_e and the relation (6) the first equation becomes

$$g_t + \frac{1}{R^2}gg_\theta - \frac{\sigma}{\rho_m R^2}(\eta_\theta + \eta_{\theta\theta\theta}) - \frac{\rho_\alpha \rho_d R \eta_\theta}{3\epsilon_0 \rho_m} + \frac{\rho_\alpha}{\rho_m \epsilon_0}\left(\rho_\alpha - \frac{\rho_d}{3}\right)\eta\eta_\theta = 0. \tag{13}$$

The second equation is obtained from rels. (5) and (6)

$$R\eta_t + \frac{h}{R}g_\theta + \frac{1}{R}(\eta g)_\theta + \frac{h}{R^2}\eta g_\theta = 0. \tag{14}$$

This system of partial nonlinear third order differential equations in the unknown functions $\eta(\theta, t)$ and $g(\theta, t)$ is solved by making succesively the transformations

$$g = \chi\eta + \psi(\theta, t) \tag{15}$$

and

$$\psi(\theta,t) = \frac{k\chi\eta - \frac{1}{R}\chi\eta^2 - R\int \eta_t d\theta + \psi^{(1)}}{k - \frac{1}{R}\eta}, \tag{16}$$

where χ is an arbitrary real parameter and $\psi^{(1)}(\theta,t)$ an arbitrary function. Choosing $\psi^{(1)}$ so that

$$\psi_t^{(1)} - \frac{\beta}{\chi}\psi_\theta^{(1)} + \frac{1}{R^2}\psi^{(1)}\psi_\theta^{(1)} + \left(\frac{\chi}{R^2} + \frac{\beta}{\chi Rk}\right)(\eta\psi^{(1)})_\theta = 0 \tag{17}$$

we obtain the Korteweg-de Vries equation for $\eta(\theta,t)$

$$A\eta_t + B\eta\eta_\theta + C\eta_{\theta\theta\theta} = 0 \tag{18}$$

with the coefficients

$$A = \chi + \frac{R^2\beta}{h\chi} \qquad B = \frac{\chi^2}{R^2} + \alpha + \frac{2\beta}{h} \qquad C = -\gamma. \tag{19}$$

One of the solutions of K-dV equation is

$$\eta(\theta,t) = \eta_0 \mathrm{sech}^2\left(\frac{\theta - Vt}{L}\right) \tag{20}$$

characterized by the half-width $L = \sqrt{12C/B\eta_0}$ and the angular velocity $V = B\eta_0/3A$. In terms of our parameters the amplitude of the soliton η_0, the layer depth h, parameters which are small compared with the radius R and χ which plays the role of the coupling constant between the soliton shape η and its dynamics characterized in principal by the velocity u, we have

$$L(\eta_0, h, \tilde{\chi}) = \eta_0^{-1/2}\left[N_1\frac{(\lambda^2 - \frac{\lambda}{3})Z^2}{A^{4/3}} + \left(\frac{1}{6} + N_2\frac{\lambda Z^2}{A}\right)\frac{1}{h} - \tilde{\chi}\right]^{-1/2} \tag{21}$$

$$V(\eta_0, h, \tilde{\chi}) = \eta_0 \frac{\tilde{\chi}V_0}{A^{2/3}}\left[\frac{\tilde{\chi}^2 - N_1\frac{(\lambda^2 - \frac{\lambda}{3})Z^2}{A^{4/3}} - \frac{1}{2h}\left(\frac{1}{6} + N_2\frac{\lambda Z^2}{A}\right)}{\tilde{\chi}^2 - \frac{1}{4h}\left(\frac{1}{6} + N_2\frac{\lambda Z^2}{A}\right)}\right], \tag{22}$$

where

$$N_1 = \frac{3e^2}{64\pi^2\sigma\epsilon_0 r_0^4}, \qquad N_2 = \frac{e^2}{32\pi^2\sigma\epsilon_0 r_0^3}, \qquad N_3 = \frac{\rho_m}{12\sigma} \tag{23}$$

$$\tilde{\chi} = \sqrt{N_3}\chi, \qquad \lambda = \frac{\rho_\alpha}{\rho_d}, \qquad R = r_0 A^{1/3}, \tag{24}$$

with $r_0 = 1.3$ fm and $V_0 = \frac{2}{r_0^2}\sqrt{\frac{\sigma}{3\rho_m}}$.

We should like to mention that such shapes could not be expanded in multipoles $\eta(\theta,t) = \sum_{l\geq 0} C_l(t) P_l(\cos(\theta))$. Even at the early stage of the soliton formation we have at least $l = 10$ and the highest weight (C_l) corresponds to $l = 4$. Evidently higher multipoles are appearing especially around larger values of the overlap with the corresponding sphere.

Due to the fact that the shape of the perturbation is represented by a travelling wave profile (the solitonic shape) we look only for solutions of the velocity with a special dependence on t and θ which allow to write the proportionality between the time and angle derivatives, i.e. $\frac{\partial}{\partial t} = -V\frac{\partial}{\partial \theta}$. This is equivalent to the assumption $\psi^{(1)}(\theta,t) = f(\theta - Vt)$. Since $\eta(\theta,t)$ fulfills this condition we ask also for a similar dependence of the function $\psi(\theta,t)$ and respectively for $\psi^{(1)}(\theta,t)$. Eq. (17) becomes after integration an algebraic equation for $\psi^{(1)}$ which gives

$$\psi^{(1)} = 2R^2\left[V + \frac{\beta}{\chi} - \left(\frac{\chi}{R^2} + \frac{\beta}{kR\chi}\right)\eta\right]. \tag{25}$$

Introducing succesively this solution in eqs. (15),(14) and (5) we get the tangential (eq. (2)) and radial (eq. (3)) velocities.

Knowing the shape and the velocities field we are able to evaluate the total energy of the system, arround the soliton shape as a sum of the surface E_S, Coulomb E_C and centrifugal E_{cf} energies. The shell effects are introduced phenomenologically by two overlaps between a variable shape connected with the parent nucleus and two fixed shapes corresponding to the daughter and cluster nuclei in their ground states.

The total potential energy must describe the transition from the ground state of the initial spherical nucleus of radius R_0 to a soliton moving on the surface of a layer situated above or below the ground state of the daughter nucleus of radius R, also considered as a sphere. Evidently the total potential energy must be separated into two parts with different descriptions, one arround the soliton and one arround the ground state. Due to the fact that a soliton could have a very small amplitude and a larger angular width close to π we may interpolate the potential energy of such a soliton in the range of the parameters where also the potential energy for the ground state is valid. In this way we can obtain an unique expression for the total potential energy. The corresponding dynamics is simply given by the mass of the emitted fragment if we assume that the daughter nucleus is a rigid core (ground state) unperturbed by the appearance of the soliton.

The surface energy is given by

$$E_S = 2\pi\sigma R^2 \int_0^\pi \left[2\frac{\eta}{R} + \frac{1}{R^2}\left(\eta^2 + \frac{\eta_\theta^2}{2}\right)\right]\sin\theta d\theta. \tag{26}$$

The Coulomb energy is

$$E_C = 0.665007(Z^2\tilde{A}^{-1/3} + \lambda^2 Z_{cl}^2 A_{sol}^{-1/3})$$

$$+3.32303Z^2\lambda\tilde{A}^{-1/3}\int_0^\pi\left(2\frac{\eta+h}{R} + \frac{(\eta+h)^2}{R^2}\right)\sin\theta d\theta - 0.665Z_0^2 A_0^{-1/3}. \tag{27}$$

The centrifugal energy is

$$E_{cf} = \frac{1}{3R}\int_0^\pi \frac{\eta}{(h+\eta)^4}\left[\tilde{\chi}\eta_\theta(h-\eta)^2 + (h^2\chi + hRV)\eta_\theta + \frac{1}{3}\chi h\eta\eta_\theta\right]^2\sin\theta d\theta. \tag{28}$$

We define the volume of the soliton as the volume of the matter situated between the surface of the daughter nucleus $r = R$ and the external envelope $r = R + \eta$

$$V_{sol} = \frac{2\pi}{3} R^3 \int_0^\pi \left(1 + \frac{\eta}{3}\right)^3 \sin\theta d\theta - \frac{4\pi}{3} R^3. \tag{29}$$

We choose the layer inside the daughter nucleus. We should like to mention that the perturbations connected with the soliton are existing only locally and that, due to the irrotationality of the motion and the fact that the field of velocities is symmetric relative to the soliton axis, the total angular momentum of the nucleus is zero. This is one of the conditions to be satisfied in alpha and cluster decays since the even-even nuclei decay mainly in the ground states. Evidently in a much more extended theory we have to include the vibrations and eventually the rotations of the daughter nuclei.

The shell effects E_{shell} for alpha decay around the double magic nucleus ^{208}Pb are introduced phenomenologically by introducing the shell effects for ^{208}Pb and $A_{cluster}$. We choose, like in the previous formulation of cluster decays as solitons on the surface of ^{208}Pb [6-10] the same analytical dependence of the shell corrections on the overlap integral V_{over} and two given volumes V_a and V_b

$$I(V_a, V_b, V_{over}) = \frac{V_{over}}{V_a + V_b + V_{over}}. \tag{30}$$

For ^{208}Pb shell correction we choose the overlap $V_{over}^{(208)}$ between the volume of the core $V_a = V_{core}$ and the volume of ^{208}Pb, $V_b = V_{208}$. The volume of the core V_{core} was defined by eliminating the volume of the travelling profile from the dynamical evolution of the parent nucleus from a sphere to a soliton of mass $A_{soliton}$ moving on the surface of the daughter nucleus taken also as a sphere. In the following we give an example of an analysis for the alpha particle as a soliton. At the beginning of the formation of the soliton the core is equal with the volume of the parent nucleus and at the final stage, when the soliton is a stable configuration, the core is equal with the volume of the daughter nucleus. This implies that the overlap $V_{over}^{(208)}$ between the volume of the core and ^{208}Pb for parent nuclei with $A \geq 212$ are equal with the volume of ^{208}Pb and for parent nuclei with $A < 212$ are equal with the volume of the corresponding daughter nuclei. Consequently we have a dependence of the parent nucleus with maximum effects for ^{212}Po with the daughter nucleus ^{208}Pb.

For alpha particle shell corrections we choose the overlap $V_{over}^{(4)}$ between the volume of the soliton $V_a = V_{sol}$ and the volume of the alpha particle $V_b = V_4$. Evidently the shell effects are increasing during the dynamical formation of the soliton being maximum when the maximum overlap with the free alpha particle is realized. This leads to a new minimum in the potential energy surface as function of our collective coordinates.

The total shell energy corrections could be written in the form

$$E_{shell} = \alpha I(V_{core}, V_{208}, V_{over}^{(208)}) + \beta I(V_{sol}, V_4, V_{over}^{(4)}), \tag{31}$$

where the two amplitudes represent the weights of the two terms in the total potential energy.

We determine the two amplitudes α and β by imposing two conditions. First that the potential energy for the parent nucleus with no centrifugal energy $E_{cf} = 0$ and no soliton shell effects ($\beta = 0$) is equal with the sum of the phenomenological terms E_S, E_C and $E_{shell} = -S_{jk}$ from the Seeger formula [20]. Second that the total energy including also the centrifugal energy (E_{cf}) and the soliton effects ($\beta \neq 0$) is degenerate in energy with the previous value. We introduce this condition due to the fact that alpha decay is a spontaneous decay.

The expression of the potential energy around the ground state minimum can be written in terms of normal modes

$$\eta(\theta, t) = \eta_0 \sum_{l \geq 2} \alpha_l(t) P_l(\cos \theta) \tag{32}$$

with

$$\alpha_l(t) = \int_{-\pi}^{\pi} P_l(\cos \theta) \eta(\theta, t) d\theta. \tag{33}$$

This leads to the following expression of the normalized potential energy in terms of normal modes

$$E_P = \frac{1}{2} \sum_{l \geq 2} C_l \alpha_l^2 \tag{34}$$

with

$$C_l = \frac{l-1}{4\pi} \left[E_S(0)(l+2) - 10 E_C(0) \frac{1}{2l+1} \right]. \tag{35}$$

As mentioned before, by interpolating the two expressions of the potential energy valid for soliton and for the ground state, we obtain an unique expression of E_P. The dynamics is simply given by the outside nucleons, i.e. the reduced mass of the emitted cluster.

We should like to mention that in the present model the dynamics of the nuclear matter is governed by a system of two nonlinear hydrodynamical equations (13) and (14). After some transformations the system reduces to a nonlinear equation for the nuclear shapes and an algebraic equation for the velocities field. The nonlinear equation reflects the balance between two opposite tendencies: the dispersion of the monochromatic wave components (normal modes of vibration) due to the contribution of the surface pressure (the term in $\eta_{\theta\theta\theta}$) and the nonlinear behaviour of the fluid (the term in $\eta \eta_\theta$). This equation leads to a solution with the travelling wave profile of solitonic type (eq. (20)). The algebraic equation, obtained in the hypothesis that the velocities field follows the above travelling profile, determines this field eq. (25). We stress that only the profile is travelling. No matter is transported at the surface of the nucleus.

3. Two-Soliton Solutions

We investigate in the following the possibility of extending the above calculations to higher order in our parameters. In such a way we can solve also the problem of the center of mass. The angular momentum of the soliton configuration could be calculated starting from the definition:

$$\vec{L} = \int d\vec{L} = \int_V \vec{r} \times d\vec{p} \tag{36}$$

which gives:
$$\vec{L} = \rho \oint \Phi(\vec{r} \times d\vec{S}), \tag{37}$$

where ρ is the mass density. Taking in eq. (37) the equations of the normal vector at the soliton shape in polar coordinates:

$$n_x = -\frac{(R'\cos\theta - R\sin\theta)\cos\phi}{\sqrt{R^2 + R'^2}}, \tag{38}$$

$$n_y = -\frac{(R'\cos\theta - R\sin\theta)\sin\phi}{\sqrt{R^2 + R'^2}}, \tag{39}$$

$$n_z = \frac{R'\sin\theta + R\cos\theta}{\sqrt{R^2 + R'^2}}, \tag{40}$$

where $R = R(\theta, \phi)$, we obtain $\vec{L} = 0$ due to the cancellation of the even-odd functions in θ occurring in the integral, with respect to any point on the symmetry axis.

The present calculations were performed with respect to the center of mass of the core + soliton system. Consequently in order to write the soliton solution with respect to the geometric center of the core is not rigorous in the first order in η_0/R. If we extend the calculations to the third order we obtain, in a natural way, that the center of mass still coincides with the geometrical center of the core if we put instead of one-soliton a two-soliton solution. That is, instead of using eqs. (13-14) we put:

$$j - R\eta_t - \frac{\eta}{R}(j + g_\theta) - \frac{\eta_\theta}{R}g - \frac{\eta\eta_\theta}{R^2}(j_\theta - 2g)$$
$$-\frac{\eta^2}{R^2}(j - \frac{5}{2}j_\theta + 3g - \frac{1}{2}j_{\theta\theta} + \frac{3}{2}g_\theta - \frac{1}{2}g_{\theta\theta}) = 0$$

and, respectively:

$$Rg_t + \frac{1}{R^2}(jj_\theta + gg_\theta) + \eta[g_t + j_{\theta t} - g_{\theta t} - \frac{1}{R^3}(j(j+g_\theta))_\theta + \frac{1}{R^3}(g(j_\theta - g))_\theta] + \eta_t j_\theta + \frac{2}{R}(j_\theta - g) +$$
$$+\frac{1}{R}\eta^2(j-g)_{\theta t} + \frac{1}{R^3}\eta_\theta[g(j_\theta - g) - j(j + g_\theta)] + \frac{\sigma}{R}(-\eta_\theta + \eta_{\theta\theta\theta} + 2\eta\eta_\theta + 3\eta\eta_{\theta\theta} + 2\eta\eta_{\theta\theta\theta}) -$$
$$-\frac{\rho_{sol}\rho_{sph}}{3\epsilon_0\rho_m}R\eta_\theta + \frac{\rho_{sol}}{\rho_m\epsilon_0}(\rho_{sol} - \frac{\rho_{sph}}{3})\eta\eta_\theta = 0, \tag{41}$$

where we use the same notation as in [6], i.e. $j = f_1$ and $g = f_{0,\theta}$. As we know from the analysis of previous calculations the function j is one order of magnitude smaller than g so that we can neglect the terms in order three which contain j or its derivatives. Consequently in eq. (41) we can eliminate j algebraically:

$$j = \frac{R\eta_t + \frac{1}{R}\eta_\theta j - \frac{2}{R^2}\eta\eta_\theta g + \frac{3}{R^2}\eta^2 g + \frac{3}{2R^2}\eta^2 g_\theta - \frac{1}{2R^2}\eta^2 g_{\theta\theta}}{1 - \frac{\eta}{R}}. \tag{42}$$

Introducing rel. (43) in eq. (42) we obtain a nonlinear partial differential equation in the functions g and η. This equation is bilinear with respect to η and consequently admits soliton solutions [18]. We can choose one-soliton solution for η but in this case we must add the contribution of the center of mass in the potential energy. The exact set of solutions for eqs. (41-43) can be obtained by different methods, e.g., by using the inverse scattering transform (IST), the Lie group theory, by constructing a certain completely integrable finite dimensional dynamic system whose solutions determine the exact solution of this modified K-dV equation etc. Due to the fact that the general case is investigated in literature [21] we limited ourselves to verify directly that the above equations admit as solution the two-soliton solution in the form:

$$R = R_0 + \eta(\theta, t) \tag{43}$$

with

$$\eta(\theta, t) = \eta_{01} sech^2 \frac{\theta - V_1 t}{L_1} + \eta_{02} sech^2 \frac{\theta - V_2 t}{L_2} \tag{44}$$

with two supplementary conditions:
1 - The coincidence of the center of mass with the geometrical center:

$$\frac{\int_0^\pi (1 + \eta_1(\theta, t))^4 \sin\theta \cos\theta d\theta}{2 + \int_0^\pi (1 + \eta_1(\theta, t))^3 \sin\theta d\theta} = \frac{\int_0^\pi (1 + \eta_2(\theta, t))^4 \sin\theta \cos\theta d\theta}{2 + \int_0^\pi (1 + \eta_2(\theta, t))^3 \sin\theta d\theta}, \tag{45}$$

where η_1 and η_2 represent the first and the second terms in the RHS of eq. (45).
2 - The equality of the angular velocity of the two solitons:

$$\eta_{01}^{\frac{3}{2}} L_1 = \eta_{02}^{\frac{3}{2}} L_2. \tag{46}$$

If this last condition is fulfilled we obtain a symmetric picture with two solitons placed in opposite diametrical positions (one with high amplitude and small width representing the cluster and one with very small height and large width moving with the same angular velocity) These two opposite solitons have the same volume and due to the eq. (46) they assure that the center of mass is in the center of the spherical core. Such solutions are allowed in principle but, due to the introduction of two new restrictions, eqs. (46) - (47) one needs to re-examine the problem of the degeneration of the second minimum with the ground state.

4. Numerical Results

First we remark that all the above mentioned quantities depend on three parameters (η_0 the amplitude of the soliton, $\tilde{\chi}$ the coupling constant between the soliton shape η and its dynamics, and k the ratio between the layer depth h and the radius of the initial nucleus R_0) and on the function $\eta(\theta, t)$. Because of the stable dependence of the propagating wave profile of the function η in the variables θ and t, we can calculate all these quantities at a given moment of time, i.e. $t = 0$. Consequently we can write them only in terms of the above three parameters. During the process of forming and growing the soliton, the parameter k is a function of η_0 and $\tilde{\chi}$.

First we make an analysis of a light cluster like an alpha particle.

The maximum and the minimum values for $\tilde{\chi}$, corresponding to $L = \pi$ and $V = \infty$ respectively, are given, as a function of k which defines the allowed values for this parameter. In reality we must restrict ourselves to a much narrower strip because we need compact solitons ($L < \pi$) and the angular velocity V much smaller then the velocity of light ($RV \ll c$). This picture is restricted to the solitons with the mass close to alpha particle mass. We can see a strong dependence of the parameter $\tilde{\chi}$ on the layer depth.

In order to keep the soliton volume constant during its formation we have to change continuously k in the plane $\tilde{\chi} - \eta_0$. For alpha decay we start with $\eta_0 \approx 5 \times 10^{-2}$, $\tilde{\chi} = 48$ and $k = 5 \times 10^{-4}$ up to $k = 2 \times 10^{-4}$. The final values of our parameters are fixed by the conditions mentioned before for determining ^{208}Pb and ^{4}He shell effects. This leads to $0.3 \leq \eta_0 \leq 0.4$; $0.18 \leq L \leq 0.35$; $40 \leq \tilde{\chi} \leq 180$ and $2 \times 10^{-4} \leq k \leq 10^{-3}$. For larger soliton mass we obtain other ranges for the limiting values like in the following table:

$A_{sol} = 14$	$A_{sol} = 32$
$0.2 < \eta_0 < 0.35$	$0.1 < \eta_0 < 0.22$
$0.18 < L < 0.6$	$1.03 < L < 1.6$
$35 < \chi < 50$	$5.1 < \chi < 18$
$10^{-2} < k < 10^{-3}$	$10^{-2} < k < 10^{-1}$

In the above range of the parameters we compute the corresponding total potential energy surface with a fixed volume of the soliton $V_\alpha = A_{sol}$ in the plane $\tilde{\chi}-\eta_0$, which implies variable k, i.e. variable depths of the layer, with the corresponding contour plots and the potential barrier between the minima corresponding to the ground and soliton states. Evidently an interpolation between the eq. (34) valid for the ground state and the potential energy E_P valid for the soliton state at half-width $L = 180°$ was made.

In this soliton description the preformation factors are given by the ratio of the two wave amplitudes in the corresponding wells, evaluated with the help of barrier penetrability between the two minima. The preformation factors are given by the penetrabilities of the above potential barriers

$$S = exp\left[-\frac{2}{\hbar}\int_0^{\eta_{0f}}(A_{red}(A_{cl})E_P(\eta_0))^{1/2}d\eta_0\right] \quad (47)$$

where η_{0f} is the coordinate of the final state.

The experimental values are given by the ratio between the experimental decay constant λ_{exp} and the Gamow decay constant λ_G calculated as a product between the barrier assault frequency ν and the penetrability P through a Coulomb and a nuclear barrier $\lambda_G = \nu P$.

Let us briefly discuss the influence of the parameters α and β from eq. (29), representing the weights of the shell corrections for ^{208}Pb respectively for ^{4}He, on the barrier shapes for the formation of the alpha particle as a soliton. If we consider only the shell effects related to alpha particle we have $\alpha = 0$ and β uniquely determined from the condition that the final state is degenerate in energy with the initial state. No empirical values of the total energy is needed. We should like to stress again that the second minimum was obtained by choosing the maximum overlap between the soliton and the alpha particle volumes. At this point we should like to mention that there are large ambiguities related to the values of the preformation factors due to the different choices of nuclear potentials. In order to describe corectly the preformation factors for Po-isotopes we have to introduce also shell effects for ^{208}Pb ($\beta \neq 0$) determined as we mentioned before.

5. Conclusions

In the present paper we formulated a large amplitude collective description of alpha and cluster decays in which the preformation factors are related to the existence of the solitons on the nuclear surface. This represents a new large collective amplitude motion in nuclei due to the nonlinearities in the hydrodynamical description of the nuclei. In this way a new coexistence model consisting of the usual shell model and of the cluster-like model with clusters moving as solitons on the nuclear surface was introduced. The ratio of the square of the corresponding cluster-like and shell model amplitudes gives the preformation factors.

On the other hand, many years ago, we described [22] alpha decay as a superasymmetric fission process. We assumed that the initial nucleus considered as a sphere is divided spontaneously into two other spheres, one large describing the daughter nucleus and one small describing the alpha particle. Later on [23], by introducing the neck degree of freedom we have shown that for such large asymmetries this parametrization is correct. We use the argument that maximum barrier penetrability gives the two spheres parametrization. Evidently this is another description of the alpha decay as a large collective motion in nuclei.

In the present paper we related the initial stage of the formation of clusters with the existence of the solitons on the nuclear surface. By the introduction of higher terms in the hydrodynamic equations we obtained also many-soliton solutions. Choosing a two-soliton solution with the opposite solitons moving with the same angular velocity we succeed to fix the mass center of the system in the center of the core. We hope that by this introduction of higher corrections we can describe the whole decay process including the separation of the solitons from the rest of the nucleus.

References

[1] A. Sandulescu, D. N. Poenaru and W. Greiner, *Soviet J. Part. Nucl.* **11**, 528 (1980)
[2] H. J. Rose and G. A. Jones, *Nature* **307**, 245 (1984)
[3] P. B. Price, *Ann. Rev. Nucl. Part. Sci* **39**, 19 (1989); A. Sandulescu, *J. Phys. G: Nucl. Part. Phys.* **15**, 529 (1989); E. Hourani, M. Hussonnois and D. N. Poenaru, *Ann. Phys. Fr.* **14**, 311 (1989)
[4] A. Sandulescu and W. Greiner, *Rep. Prog. Phys.* **55**, 1423 (1992)
[5] Y. S. Kivshar and B. A. Malomed, *Rev. Mod. Phys* **61**, 763 (1989)
[6] A. Ludu, A. Sandulescu and W. Greiner *Int. J. Modern Phys. E* **1**, 169 (1992)
[7] A. Ludu, A. Sandulescu and W. Greiner, *Int. J. Modern Phys. E* **3**, in press (1993)
[8] A. Sandulescu, A. Ludu and W. Greiner, *WE-Heraeus-Seminar on Nuclear Physics Concepts in Atomic Cluster Physics, Bad Honneff, Lect. Notes Phys.* **404**, 72 (1991)
[9] A. Sandulescu, A. Ludu and W. Greiner *Conf. on Nucl. and At. Clusters*, p. 262, Turku, Finland, edited by M. Brenner, T. Lönnroth and F. B. Malik, (Springer Verlag, Berlin, 1991)
[10] A. Ludu, A. Sandulescu and W. Greiner, *XXX Int. Winter Meeting on Nuclear Physics*, p. 493, Jan. 1992, Bormio, Univ. Milano, edited by I. Iori (1992)
[11] T. Fliessbach and H. J. Mang *Nucl. Phys. A* **263**, 75 (1976)
[12] A. Sandulescu, I. Silisteanu and W. Wunsch *Nucl. Phys. A* **305**, 205 (1978)
[13] I. Tonozuka and A. Arima *Nucl. Phys. A* **323**, 45 (1979)
[14] F. A. Janouch and R. J. Liotta, *Phys. Rev. C* **25**, 2123 (1982)
[15] A. Insolia, R. J. Liotta and E. Moglione *Europhys. Lett.* **7**, 209 (1988)
[16] S. Okabe *J. Phys. Soc. Jpn. Suppl.* **58**, 516 (1989)
[17] K. Varga, R. G. Lovas and R. J. Liotta *Phys. Rev. Lett.* **69**, 37 (1992)
[18] V. I. Petviashvili, *Plasma Physics*, p. 122, edited by B. Kadomtsev, MIR, Moscow (1981)
[19] V. I. Petviashvili, *Fizika Plazmy* **2**, 469 (1976)

[20] P. A. Seeger *Nucl. Phys.* **25**, 1 (1961)
[21] R. K. Bullogh and P. J. Caudrey, *Solitons*, (Berlin, Springer, 1980); R. K. Dodd, J. C. Eilbeck, J. D. Gibbon and H. C. Morris, *Solitons and Nonlinear Waves*, (New York Academic Press, 1982)
[22] D. N. Poenaru, M. Ivascu and A. Sandulescu *J. Phys. G: Nucl. Phys.* **5**, L169 (1979)
[23] K. Depta, J. A. Maruhn, Wang Hou-Ji, A. Sandulescu and W. Greiner, *Int. J. Modern Phys.* **5**, 3901 (1990)

ALPHA AND ^{12}C RADIOACTIVITIES OF ^{114}Ba

G. Popa[1], A. Sandulescu[1], I. Silisteanu[1] and W. Scheid[2]

[1] *Department of Theoretical Physics, Institute of Atomic Physics, Bucharest, POB MG-6, Romania*
[2] *Institut für Theoretische Physik der Justus-Liebig-Universität, Heinrich-Buff-Ring 16, D-35392 Giessen, Germany*

The spontaneous decay is a result of an interplay of nuclear structure effects dominated by two-body interactions and resonance effects occurring in the dynamical evolution of the system [1]. In this paper the two independent stages of reaction, first clustering formation and second tunneling, are described with two independent Hamiltonians. In order to calculate the formation probability of the cluster we make use of the superfluid model. First we solve numerically the Schrödinger-type equation:

$$[-\frac{\hbar^2}{2D}\frac{d^2}{d\xi^2} + V(\xi)]\Phi(\xi) = E\Phi(\xi), \tag{1}$$

where the initial mass parameter (superfluid) is [2]:

$$D = -\frac{\hbar^2}{2v(\Delta\xi)^2}. \tag{2}$$

It depends on the two-body effective pairing interaction $v = -2.9$ MeV connecting the neighboring shapes and on the mesh spacing $\Delta\xi$. The potential $V(\xi)$ is connected with the tridiagonal matrix Hamiltonian E_i by

$$E_i = V(\xi_i) + 2v. \tag{3}$$

The value of $V(\xi_i)$ is known at $\xi = 0$ when it becomes equal to the Q-value of the reaction energy and at $\xi = 1$ where it is given by the cluster-daughter scattering potential. After solving eq.(1) we calculate the cluster formation probability (CFP) at the touching configuration $(r = r_t)$.

$$W(r_t) = |\Phi(\xi=1)|^2 = \frac{\alpha}{\sqrt{\pi}} e^{-\alpha^2} \tag{4}$$

In the next step we calculate the penetration probability which in fact is exactly the Gamow one-body width Γ_0 [3]:

$$\Gamma_0 = 2\pi \left| \frac{<u_0(r) \mid u_0^r(r)>}{<u_n(r) \mid u_0^r(r)>} \right|^2, \tag{5}$$

where $u_0^r(r)$ is the resonance wave function of the system (cluster + daughter), $u_0(r)$ and $u_n(r)$ are solutions of the system of equations:

$$(T + V(r) - Q)u_n(r) = u_0^r(r) \tag{6}$$

$$(T + V(r) - Q)u_0(r) = 0. \tag{7}$$

The decay width is a product of formation and penetration probabilities:

$$\Gamma = W\Gamma_0. \tag{8}$$

Table 1. *Summary of the α and ^{12}C decay calculations with the parameters of potential taken from Ref.* [5].

Decay	Q	D	W	Γ_0	Log $T_{1/2}$	Log $T_{1/2}$ Ref. [4]
	MeV	$\hbar^2 \text{MeV}^{-1}$		MeV	s	s
$^{114}Ba(\alpha)$	3.601	2.75	1.490E-04	6.735E-20	1.63	1.72
$^{114}Ba(^{12}C)$	20.600	17.24	6.597E-07	3.050E-21	5.33	5.43

The present results are close to results of ref [4] obtained with the potential parameters extracted from optical analysis of reaction cross section data. One may conclude that the decay rates are not strongly dependent on potential parameters and especially on the inner nuclear part of potential. We get also a branching ratio relative to α-decay of $B = 10^{-3.7}$ in good agreement with the first experimental result $B_{exp} = 10^{-4}$ [6].

The work is supported by the Kernforschungszentrum Karlsruhe.

References

[1] A. Sandulescu, *Cluster decays*, J. Phys. G: Nucl. Phys. **A15**, 529 (1989); M. Ivascu, I. Silisteanu, *Heavy fragment radioactivity*, Sov. J. Part. Nucl. **21**(6), 599 (1990)
[2] F. Barranco, G. Bertsch, R. A. Broglia, E. Vigezzi, *Cluster radioactivity as a superfluid tunneling phenomenon*, Nucl. Phys. A **512**, 253 (1990)
[3] M. Ivascu, I. Silisteanu, *The microscopic approach to the rates of radioactive decay by emission of heavy clusters*, Nucl.Phys. A **485**, 93 (1988)
[4] W. Scheid, L. Silisteanu, A. Sandulescu, *Lifetimes of cluster radioactivity of neutron deficient trans-tin isotopes*, Rom. J. Phys. Rapid Communication **3**, 22 (1993)
[5] P. R. Cristensen and A. Winther, *Phys.Lett. B* **65**, 19 (1976)
[6] S. P. Tretyakova, V. L. Mikheev, Y. T. Oganessian, *Experimental evidence of ^{12}C decay of ^{114}Ba, Private Communication* (1993)

FINE STRUCTURE IN THE CLUSTER DECAYS

Ovidiu Dumitrescu

Department of Theoretical Physics, Institute of Atomic Physics, PO Box MG-6, RO-76900 Bucharest, Romania

Introduction

Recently Hourani and his co-workers [1] experimentally discovered the fine structure in the ^{14}C radioactivity [2], [3], [9]. The theoretical studies of alpha [7] (see also the review papers [4] [5], [6] and the references therein) and heavy cluster (e.g. ^{14}C) decay [8] (see also the recent review paper [9] and the references therein) have very much in common. The theoretical models of heavy cluster decay are based, essentially, on Gamow's theory [10] which was the first success of quantum mechanics when applied to the α-decay phenomenon. The differences in approaches are related to the way of calculating the potential barrier defined by the (nuclear plus Coulomb) interaction potential acting between the emitted cluster and the residual nucleus. All these theoretical treatments fit to a law for favored cluster transitions, analogous to the Geiger-Nuttall [11] law for favored α-decay, which emerges directly from the simplest JWKB expression of the penetrability determined by the square well plus Coulomb interaction potential.

The unfavored transitions do not follow the Geiger-Nuttall law, because of the large variations of the reduced widths [4] [5], [6], [7], which have a key role in the understanding of the decay process and require a precise knowledge of the structures of the initial and final quantum states. From such transitions we can learn much about the structure of atomic nuclei. In describing these transitions almost all the nowadays nuclear models fail, and it does not matter whether they are models for the structure of nuclear states or reaction mechanisms, or whether they are specific models, for the decay mechanism [4], [5], [13], [12], [9].

While the fine structure in α-decay has been more or less understood [4], few studies [14], [15] of the fine structure in heavy cluster decay are available.

It is the aim of this paper to calculate the hindrance factors for several α- and ^{14}C-cluster decays in the translead region. The calculations will be performed within the one level - one channel R - matrix approximation. The cluster residual nucleus scattering wave functions are generated by the Coulomb potential plus the realistic M3Y double folding potential [20]. The Pauli antisymmetrization kernel is used as proposed in Refs. [12], [7].

Several favored and weak - hindered α - transitions from the ground state of ^{243}Am to some excited states of ^{239}Np are calculated within the enlarged superfluid model (ESM) [18] and compared with those obtained by Mang, Poggenburg and Rasmussen [6] within the ordinary superfluid model and with experimental data [22].

The situation is not as simple as previous one in the case of α - and ^{14}C-decays of soft Ra and Th isotopes. In this transitional region between spherical nuclei near the N = 126 shell closure and the well deformed heavier isotopes unusual phenomena are observed, in particular the occurrence of a sequence of a very low - lying negative-parity states, strong E1 transitions and the existence of parity mixed doublets [27] in several odd-A nuclei, including the ground and the first excited state of ^{223}Ra. Such experimental evidences can be described by assuming either α-cluster models [24] or strong octupole correlations [25].

The experimentally observed [1] transitions from the ground state of ^{223}Ra to some excited states of ^{209}Pb are estimated and discussed in comparison with previous works [14], [15]. As a nuclear structure model for the ^{223}Ra we use a hybrid model discussed in detail latter.

Hindrance Factors

The experimental hindrance factor (HF_{exp}) of any cluster decay is defined as a ratio between the Geiger-Nuttall [11] (Γ_{GN}) width and the width of the radioactive transition we are interested in [1], [4]. The theoretical hindrance factor (HF_{theo}) is defined by an analogous ratio in which the widths are replaced by their theoretical expressions ($\Gamma_l^{(I_i^{\pi_i} K_i \to I_f^{\pi_f} K_f)} = \sum_l P_l(Q) \, (\gamma_l^{(I_i^{\pi_i} K_i \to I_f^{\pi_f} K_f)})^2$) : [16]):

$$HF_{exp} = \frac{\Gamma_{GN}(Q)}{\Gamma(Q)} \; ; \; lg\Gamma_{GN}(Q) = A + \frac{B}{\sqrt{Q}} \; ;$$

$$HF_{theo} = [\sum_l F_l d_l^2]^{-1} \; ; \; d_l = \frac{\gamma_l^{(I_i^{\pi_i} K_i \to I_f^{\pi_f} K_f)}}{\gamma_0^{(00+(g.s.e-e) \to 00+(g.s.e-e))}} \quad (1)$$

Here Q stands for the energy release of the studied decay. A and B are constant quantities for the isotopes of one element. $\gamma_l^{(I_i^{\pi_i} K_i \to I_f^{\pi_f} K_f)}$ and $\gamma_0^{(0+0(g.s.e-e) \to 0+0(g.s.e-e))}$ are the amplitudes of the reduced width corresponding to the studied transition and to the favored transition between the ground states of the neighboring double even nucleus, respectively. F_l is the ratio of the penetrabilities for $l \neq 0$ and $l = 0$. Its expression within the JWKB approximation is:

$$F_l = \frac{P_l(Q)}{P_0(Q)} = \exp \frac{2}{\hbar} \int_{R_c}^{r_0} (q_{l=0}(r) - q_l(r)) dr \; ; \; q_l(r) = \sqrt{2m_0 A_{red}(V_l^{coul+nucl} - Q)} \quad (2)$$

where "r_0" and "R_c" stand for the outer and inner turning points, respectively, $A_{red} = aA/(a+A)$, m_0 is the nucleon mass. The $V_l^{coul+nucl}$ is the sum of the Coulomb and nuclear one body potential acting between the cluster a and the daughter nucleus and it is calculated as a cluster-nucleus double-folding model potential obtained with the M3Y interaction [20], [12]. In the approximation, that the Coulomb part of this potential is replaced by a point-like Coulomb potential and the nuclear part by a Saxon-Woods one [4], [9], [5], F_l has a simple expression (e.g. in the case of α-decay: $F_l = \exp\left(-2.027 l(l+1) Z^{-\frac{1}{2}} A^{-\frac{1}{6}}\right)$ [21]).

Following the prescriptions of the Ref. [7] we may write down the relation between the amplitude of the reduced width and the internal wave function f_{int}, which is joined [7] to the irregular scattering wave function G_l:

$$\gamma_{LK}^{(I_i^{\pi_i}K_i \to I_f^{\pi_f}K_f)}(R_c) = \sqrt{\frac{\hbar^2}{2\mu R_c}} f_{int}^{(I_i^{\pi_i}K_i \to I_f^{\pi_f}K_f; L\,K)}(R_c)$$

$$= \sqrt{\frac{\hbar^2}{2\mu R_c}} \sum_N \theta_{NL}^{(I_i^{\pi_i}K_i \to I_f^{\pi_f}K_f)} R_c \mathcal{R}_{NL}(R_c, a_r) \qquad (3)$$

Here $\theta_{NL}^{(I_i^{\pi_i}K_i \to I_f^{\pi_f}K_f)}$ stands for the spectroscopic amplitude for the cluster a a generalization [8] of the α-spectroscopic amplitude defined, e.g., in the Ref. [7].

In the Refs. [7] and [8] an approach is given for the calculation of the spectroscopic amplitudes, $\theta_{NL}^{(I_i^{\pi_i}K_i \to I_f^{\pi_f}K_f)}$, by inserting an intermediate set of valence wave functions, which describe the nucleons that after decay belong to the outgoing cluster. Thus, these amplitudes are expanded in terms of two factors: - the so-called cluster overlap [19] (a generalization of the coefficient of fractional parentage), and - the intrinsic overlap integral (a generalization of the Mang's overlap integrals [5], see e.g. the eq. (8) of Ref. [7] and the eq. (17) of Ref. [8]). The channel radius dependence is extensively discussed in Ref. [7] and we shall follow the approach given in this paper. In the present work, however, we are interested in calculating the HF's, quantities which have a weaker dependence on the channel radius then the widths themselves.

Numerical Calculations

When observing small (close to unity) HF's, we are dealing with favored decays. The rough explanation of such decays is based on the picture according to, the cluster is built from the fermions just situated at the Fermi surface, for which strong collective correlations (e.g. pairing correlations [18]) occur. When the transition takes place between odd A nuclei, the unpaired nucleon remains in the single particle state that essentially characterizes the mother nucleus ground state. This state during the decay process may change its deformation, however, we may find it again among the excited states of the daughter nucleus.

By using ESM [18], we calculated the quasiparticle-phonon structure of the ^{243}Am and ^{239}Np ground and excited states (see Table 1) and the HF's for the favored and some unfavored α - decays of ^{243}Am to ground and some excited states in ^{239}Np nucleus. The expressions of the reduced widths within the superfluid model are given in Ref. [16]. The results have been compared with the calculations of Ref. [6] and the experimental data [22] (see Table 2). They are not far from our previous calculations [4]. A relatively good agreement with the experimental data is obtained.

In these calculations the used ESM parameters are: $G_p = 0.143$ MeV, $G_n = 0.103$ MeV, $G_4 = 0.268$ keV. The parameters of the average field are taken from Ref. [17]. The used deformation parameters are: $\beta_{20} = 0.24$ and $\beta_{40} = 0.06$. The used particle-hole quadrupole and octupole parameters (see Ref. [18]) are: $\kappa_{n\tau}^{\lambda\mu} = \kappa_{0\tau}^{2\mu} = 0.667$ keV fm^{-4}; $\kappa_{n\tau}^{\lambda\mu} = \kappa_{1\tau}^{2\mu} = 0.062$ keV fm^{-4}; $\kappa_{n\tau}^{\lambda\mu} = \kappa_{0\tau}^{3\mu} = 0.011$ keV fm^{-6} $\kappa_{n\tau}^{\lambda\mu} = \kappa_{1\tau}^{3\mu} = 0.001$ keV fm^{-6}. The used particle-particle quadrupole parameter (see Ref. [18]) are: $G_{n\tau}^{\lambda\mu} = G_\tau^{2\mu} = 15$ eV fm^{-4}. The rest of the terms in eq. (14) of Ref. [18] not mentioned above has been neglected.

From Tables 1 and 2 we conclude that the α - decay of ^{243}Am ground state to ^{239}Np $\frac{5}{2}^-$; $E_x = 74.65$ keV and $E_x = 666$ keV-states can be considered as favored and weak unfavored, respectively, α-transitions.

Table 1. The calculated, within ESM [18], structure of some ground and excited states entering the α transition: ^{243}Am (g.s) $\rightarrow \alpha + ^{239}Np$.

Nucleus	I^π	K	E_{exp} [MeV]	E_{theo} [MeV]	Structure
^{243}Am	$\frac{5}{2}^-$	$\frac{5}{2}$	0.	0.	**98.9** % [523] $\frac{5}{2}^-$ + 0.1 % [523] $\frac{5}{2}^-$ Q_{20}
^{239}Np	$\frac{5}{2}^+$	$\frac{5}{2}$	0.	0.	**82.7** % [642] $\frac{5}{2}^+$ + 2.1 % [642] $\frac{5}{2}^+$ Q_{20}
^{239}Np	$\frac{5}{2}^-$	$\frac{5}{2}$	0.075	0.06	**88.9** % [523] $\frac{5}{2}^-$ + 0.09 % [512] $\frac{5}{2}^-$ + 4.04 % [523] $\frac{5}{2}^-$ Q_{20} + 0.5 % [642] $\frac{5}{2}^+$ Q_{30}
^{239}Np	$\frac{5}{2}^-$	$\frac{5}{2}$	0.666	1.058	0.88 % [523] $\frac{5}{2}^-$ + 0.03 % [512] $\frac{5}{2}^-$ + **94.01** % [523] $\frac{5}{2}^-$ Q_{20} + 0.04 % [642] $\frac{5}{2}^+$ Q_{30}
^{239}Np	$\frac{1}{2}^-$	$\frac{5}{2}$	-	0.458	**91** % [530] $\frac{1}{2}^-$ + 4.09 % [530] $\frac{1}{2}^-$ Q_{20} + 9.04 % [523] $\frac{5}{2}^-$ Q_{22} + 1.05 % [642] $\frac{5}{2}^+$ Q_{32}

Table 2. The calculated, within ESM [18], hindrance factors for favored, weak unfavored and unfavored α-transitions from ^{243}Am (g.s.) to the members of the rotational bands of several intrinsic states of ^{239}Np. Theses results are compared with the calculated HF's by Mang, Poggenburg and Rasmussen [6] and experimental data [21], [22].

E_f [keV]	$I_f^{\pi f}$	HF_{exp}	HF_{MPR}	HF_{ESM}	E_f [keV]	$I_f^{\pi f}$	HF_{exp}	HF_{MPR}	HF_{ESM}
74.67	$\frac{5}{2}^-$	1.12	1.38	1.15	666	$\frac{5}{2}^-$	6.5	1.38	5.03
118	$\frac{7}{2}^-$	5.1	6.65	4.75		$\frac{7}{2}^-$		6.65	20
172	$\frac{9}{2}^-$	22	18.71	20		$\frac{9}{2}^-$		18	65
241	$\frac{11}{2}^-$	1600	943.62	1150		$\frac{11}{2}^-$		944	2167
320	$\frac{13}{2}^-$	1100	1406	1170		$\frac{13}{2}^-$		1400	5155
411	$\frac{15}{2}^-$	2300	2761	2916		$\frac{15}{2}^-$		2671	9171
0.0	$\frac{5}{2}^+$	1700	1432	1503		$\frac{1}{2}^-$		121500	128355
31.14	$\frac{7}{2}^+$	1500	1851	1905	267	$\frac{3}{2}^-$	1500	3233	3105
	$\frac{9}{2}^+$		3272	3341	359	$\frac{5}{2}^-$	1000	2824	3252
	$\frac{11}{2}^+$		7754	8428	327	$\frac{7}{2}^-$	1100	1236	1328
	$\frac{13}{2}^+$		531902	550401	427	$\frac{9}{2}^-$	2900	4039	4527
	$\frac{15}{2}^+$		27314	28890	438	$\frac{11}{2}^+$	5200	1328	1522

The ^{223}Ra nucleus belongs [26] to the well known region of soft nuclei with $Z \approx 88$ and $N \approx 134$, with strong octupole correlations in the ground and low lying excited states, where the $1j_{\frac{15}{2}}$ intruder orbital interacts strongly with the $2g_{\frac{9}{2}}$ natural parity orbital. The HF's for both the α- and ^{14}C-decays of the ground state of ^{223}Ra are very difficult to be calculated at the moment, due to the unknown accurate structure of the mother and daughter nuclei. Studying the experimental HF for α-decays to ^{219}Rn ground and low lying excited states [23] we learn that $\approx$ fifteen [23] transitions have small ($\leq$ 100) HF's and from these transitions five have HF's $\leq$ 10. The corresponding excited states have very different structure and this tells us that the structure of the ground state of ^{223}Ra is not as simple, as e. g. in the ^{243}Am case, and it may contain many more or less equal components of single quasi-particle or quasi-particle -phonon structure. Unfortunately, not all the spins and parities of the ^{219}Rn-excited states, populated by α - decay, are known. Thus, it is a real difficult problem to describe the quantum states involved in the α- and ^{14}C-decay of ^{223}Ra. In our opinion, it is not sufficient to describe these states within an independent particle model only [14], [15]. Residual interactions could play an important role [4].

To understand this situation we construct a very simple model, which proves to deserve attention by itself and to suggest the highly nontrivial behavior of any realistic model.

Assume, for a moment, that the structure of the ground state of the ^{223}Ra-nucleus consists of a spherical core described by an independent particle model. Above the core there exists a deformed single particle neutron orbital only. The wave function for this orbital can be expanded in terms of spherical orbitals. In this case the spectroscopic amplitude entering the expression of the HF can be factorized according to:

$$\theta_{Nl\, K_i - K_f}^{(K_i^{\pi_i} K_i \to K_f^{\pi_f} K_f)} = a_{N_i l_i j_i}^{\Omega_i = K_i} a_{N_f l_f j_f}^{\Omega_f = K_f} \sqrt{2I_f + 1} \begin{pmatrix} I_i & l & I_f \\ K_i & K & K_f \end{pmatrix} \theta_{spherical}^{(j_i \pi_i \to j_f \pi_f)} \qquad (4)$$

where a_{Nlj}^{Ω} are the Nilsson-like amplitudes, $\begin{pmatrix} I_i & l & I_f \\ K_i & K & K_f \end{pmatrix}$ stands for the 3-j symbol and $\theta_{spherical}^{(j_i \pi_i \to j_f \pi_f)}$ acts as a spectroscopic amplitude between many body spherical $|j_{i(f)} \pi_{i(f)} >$ states, including both the cluster overlaps [19], [8] and the intrinsic overlap integrals [8]. For instance, let us calculate the ratio between the HF's: $R(\frac{9}{11}) = \frac{HF(^{223}Ra(g.s) \to ^{209}Pb(g.s))}{HF(^{223}Ra(g.s) \to ^{209}Pb(779keV))}$ experimentally equal to 200 [1]. In the Refs. [14], [15] it has been considered the only different quantities in the expressions of the numerator and denominator - the following Nilsson-like amplitudes $a_{1i_{11/2}}^{\Omega_i = \frac{3}{2}}$ and $a_{2g_{9/2}}^{\Omega_i = \frac{3}{2}}$. Within this approximation the above ratio in Ref. [14] is $R(\frac{9}{11}) \approx 50$, while in Ref. [15] is $R(\frac{9}{11}) \approx 1$. Considering in addition to the Nilsson-like amplitudes of Ref. [14] the $\theta_{spherical}^{(j_i \pi_i \to j_f \pi_f)}$ quantities, we obtained for the above ratio a large value [8] ($R(\frac{9}{11}) \simeq 1000$), showing how important are all quantities entering the hindrance factors for cluster decay. Analogous estimation for the ratio between the HF's $R(\frac{15}{11}) = \frac{HF(^{223}Ra(g.s) \to ^{209}Pb(1423keV))}{HF(^{223}Ra(g.s) \to ^{209}Pb(779keV))}$, experimentally equal to 1 [1], within the above prescriptions is approximately 50, i.e. not very hindered.

A few more comments may be in order here. First of all our hybrid model with a spherical core and only one deformed orbital, when calculating the spectroscopic amplitudes, is not to be taken too seriously for very complex structures as in the case of ^{223}Ra. This should be not true even for structures close to single quasiparticle states, because the assumption of a spherical core is not realistic. On the other hand, when having realistic structures for both the initial and final states, calculations within shell models like OXBASH are practically impossible for nowadays computers. Therefore simple schematic models like above presented would be useful.

Conclusions

In this work we reported some calculations performed, within the enlarged superfluid model [18], for some selected (favored and weak hindered) α transitions in the - ^{243}Am (g.s) $\to \alpha + ^{237}$Np-process. A schematic model has been applied for ^{223}Ra (g.s) $\to$ ^{14}C + ^{209}Pb-process. In this case difficulties arise due to the unknown structure of ^{223}Ra ground state and due to the impossibility to calculate truly microscopically the spectroscopic amplitude. Nevertheless simple schematic models could help us in understanding the heavy cluster decay.

References

[1] E. Hourani et al., *Phys. Rev. C* **44**, 1424 (1991); L. Brillard et al., *C.R. Acad. Sci. Paris* **309**, 1105 (1989)

[2] H. J. Rose and G.A. Jones, *Nature* **307**, 245 (1984)

[3] P. B. Price, *Ann. Rev. Nucl. Part. Sci.* **39**, 19 (1989)

[4] O. Dumitrescu, *Fiz. Elem. Chastits At. Yadra* **10**, 377 (1979) (*Sov. J. Part. Nucl.* **10**, 147 (1979))

[5] H. J. Mang, *Ann. Rev. Nucl. Sci.* **14**, 1 (1964); *Z. Phys.* **148**, 572 (1957); *Phys.Rev.* **119**, 1069 (1960); *Proc. Second International Conf. Clust. Phen. Nucl.* (College Park, Maryland, 1975)

[6] H. J. Mang, M. K. Poggenburg and J. O. Rasmussen, *Phys. Rev.* **181**, 1697 (1969)

[7] M. Grigorescu, B. A. Brown and O. Dumitrescu, *Phys. Rev. C* **47**, 2666 (1993)

[8] O. Dumitrescu, *Fine Structure of Cluster Decays*, Preprint ICTP Trieste IC/93/164 (1993)

[9] A. Sandulescu and W. Greiner, *Rep. Progr. Phys.* **55**, 1423 (1992)

[10] G. Gamow, *Zeit. Phys.* **51**, 24 (1928)

[11] H. Geiger and J. M. Nuttall, *Phylos. Mag.* **22**, 613 (1911); *Phylos. Mag.* **23**, 613 (1911)

[12] A. Bulgac et al., *Nuovo Cimento* **70 A**, 142 (1982); F. Carstoiu et al., *Nucl. Phys. A* **441**, 221 (1985)

[13] S. G. Kadmensky and V. I. Furman, *Alpha decay and related nuclear reactions*, Energoatomizdat (Moscow, 1985)

[14] R.S. Sheline and I. Ragnarsson, *Phys. Rev. C* **43**, 1476 (1991)

[15] M. Hussonnois et al., *Phys. Rev. C*, **44**, 2884 (1991); G. Ardisson et al., *Proc. Int. School Seminar Heavy Ion Physics* (Dubna 1989), 336 (1990); G. Ardisson and M. Hussonnois, *C.R. Acad. Sci. Paris* **310**, 367 (1990); M. Hussonnois et al., *J. Phys. G* **16**, 177 (1990)

[16] O. Dumitrescu and A. Sandulescu, *Nucl. Phys. A*, **100**, 456 (1967)

[17] F. A. Gareev et al., *Nucl. Phys. A* **171**, 134 (1971); JINR (Dubna), **P4 - 5457** (1970)

[18] O. Dumitrescu, *Fiz. Elem. Chastits At. Yadra* **23**, 430 (1992) (*Sov. J. Part. Nucl.* **23**, 187 (1992)); *Nuovo Cimento A* **104**, 1057 (1991)

[19] B. A. Brown, *Nucl. Phys. A*, **522**, 221c (1991); B. A. Brown and B. H. Wildenthal, *Ann. Rev. Nucl. Part. Sci.* **38**, 29 (1988); B. A. Brown et al., *MSU-NSCL Report* **524** (1985)

[20] G. Bertsch, J. Borisowicz, H. McManus and W. G. Love, *Nucl. Phys. A* **284**, 399 (1977)

[21] J. O. Rasmussen, *Phys.Rev.* **113**, 1593 (1959); I. Perlman, J. O. Rasmussen in *Handbuch der Physik* **42**, 109 (1957)

[22] Y. A. Ellis and M. R. Schmorak, *Nuclear Data Sheets B* **8**, 345 (1972)

[23] C. Maples, *Nuclear Data Sheets* **22**, 223 (1977)

[24] H. J. Daley and F. Iachello, *Phys. Lett. B*, **131**, 281 (1983)

[25] G. A. Leander et al., Nucl. Phys. A **388**, 452 (1982); *Nucl. Phys. A* **413**, 375 (1984)

[26] S. Aberg, H. Flocard and W. Nazarewicz, *Ann. Rev. Nucl. Part. Sci.* **40**, 439 (1990)

[27] O. Dumitrescu and G. Clausnitzer, *Nucl. Phys. A* **552**, 306 (1992)

MICROSCOPIC CALCULATION FOR ALPHA AND ^{12}C EMISSIONS FROM PROTON-RICH Ba AND Ce ISOTOPES

A. Florescu[1] and A. Insolia[2]

[1] Institute of Atomic Physics, Bucharest, Romania
[2] I.N.F.N., Sezione di Catania, Catania, Italy

Recently the region of the proton-rich nuclei situated above the double-magic ^{100}Sn came into the focus of many experimental and theoretical studies, as new types of rare and exotic decays like heavier cluster emissions are expected to be observed here.

We attempted to evaluate the total decay widths and the reduced widths for alpha and ^{12}C emissions from some proton-rich Ba and Ce isotopes. The microscopic calculations were done in the frame of the R-matrix theory, with harmonic oscillator and alternatively Woods-Saxon shell model wave functions.

Considering the a-cluster emission through the decay $B \to A + a$ we wrote the cluster preformation amplitude as

$$F_L(R) = C_{antisym} \int d\xi_a d\xi_A d\Omega_R \left[\phi_a(\xi_a)\Psi_A(\xi_A)Y_L(\hat{R})\right]^*_{J_B M_B} \Psi_B(\xi_B)$$

where the factor $C_{antisym} = \left[\binom{N_B}{N_a}\binom{Z_B}{Z_a}\right]^{1/2}$ takes into account the antisymmetrization in the exit channel between the cluster nucleons and the initial nucleus. The overlap integral for emission of alpha and carbon clusters was calculated earlier[1] for spherical harmonic oscillator wave functions and resulted, for pure shell model configurations of pairs of nucleons coupled to J = 0, as a finite expansion in terms of the radial wave functions for the relative motion between the cluster and the final nucleus.

The shell model configuration mixing was taken into account through the BCS formalism[2]. In the spherical approximation for the nuclear mean field and with BCS-type wave functions for the initial and final nuclei we obtained the cluster preformation amplitude

$$F_0^{(BCS)}(R) = C_{antisym}(f_n)^q (f_p)^r \sum_{\{s\}} \left(\prod_{i=1}^q h(j_{ni})\right)\left(\prod_{k=1}^r h(j_{pk})\right)$$

$$\cdot \sum_N A_{(a)}^{(N)}(\{s_n\},\{s_p\}) R_{N_0}\left(A_a \alpha R^2\right)$$

where $f_n = \prod_{j_n} \left(U_{j_n}^i U_{j_n}^f + V_{j_n}^i V_{j_n}^f\right)^{\Omega_{j_n}}$ with $\Omega = j + 1/2$ and i, f superscripts standing for the initial and final nuclei. Also $\{s\} = \{j_{n1}...j_{nq}, j_{p1}...j_{pr}\}$ and $q = r = 1$ for alpha and $q = r = 3$ for the ^{12}C case. The factors f_p have quite similar expressions and

$$h(j) = \frac{\Omega_j}{\hat{l}} \cdot \frac{V_j^i U_j^f}{U_j^i U_j^f + V_j^i V_j^f}$$

Above the ^{100}Sn core we utilized newly measured experimental single-particle energies[3]. The cluster internal wave functions were built in the frame of the cluster model for nuclear structure. The oscillator size parameters employed were 0.58 fm^{-2} for alpha and 0.352 fm^{-2} for ^{12}C in order to reproduce the cluster nuclear mean square radii, while for the parent nuclei they were from 0.197 fm^{-2} for A=114 to 0.195 fm^{-2} for A=118.

The penetrability factor P for the barrier between the final fragments was evaluated within the WKB approximation, with modified Coulomb functions as solutions of the one-body radial Schrödinger equation corresponding to an optical (e.g. Christensen-Winther) potential. The Q-values were obtained from the Janecke-Masson mass tables[4]. The results for the decay widths Γ for the alpha and ^{12}C emission from ^{114,116,118}Ba and ^{116,118}Ce are given in the table below.

The present calculations can be easily extended to deformed nuclei, and also to more realistic internal wave functions for the emitted clusters. Another problem which needs further study is the influence of the antisymmetrization effects on the cluster preformation factors, as we considered here these effects only through a crude ad-hoc factor.

	Q (MeV)	R_c (fm)	Γ (MeV)	$B = \Gamma_c/\Gamma_\alpha$
$^{114}Ba \to ^{12}C$	20.62	7.1	$1 \cdot 10^{-28}$	
$\to \alpha$	3.44	7.3	$7 \cdot 10^{-23}$	$1 \cdot 10^{-6}$
$^{116}Ba \to ^{12}C$	17.00	7.2	$9 \cdot 10^{-38}$	
$\to \alpha$	2.96	7.3	$1 \cdot 10^{-26}$	$9 \cdot 10^{-12}$
$^{118}Ba \to ^{12}C$	15.10	7.2	$8 \cdot 10^{-43}$	
$\to \alpha$	2.41	7.4	$2 \cdot 10^{-32}$	$4 \cdot 10^{-11}$
$^{116}Ce \to ^{12}C$	18.87	7.2	$2 \cdot 10^{-32}$	
$\to \alpha$	3.59	7.4	$1 \cdot 10^{-23}$	$2 \cdot 10^{-9}$
$^{118}Ce \to ^{12}C$	17.77	7.2	$2 \cdot 10^{-34}$	
$\to \alpha$	3.19	7.4	$4 \cdot 10^{-26}$	$5 \cdot 10^{-9}$

References

[1] A. Florescu, S. Holan and A. Sandulescu, *Rev. Roumaine de Physique* **33**, 243 (1988) ; **34**, 595 (1989)
[2] D. Delion, A. Insolia and R. J. Liotta, *Phys. Rev. C* **46**, 1346 (1992)
[3] H. Grawe et al., *Prog. Part. Nucl. Phys.* **28**, 281 (1992)
[4] J. Janecke and P. J. Masson, *At. Data and Nucl. Data Tables* **39**, 266 (1988)

CLUSTERING AND THE CONTINUUM IN NUCLEI

R. J. Liotta

KTH - Physics at Frescati, S-10405 Stockholm, Sweden

The process of clustering of nucleons and subsequent decay of the already formed cluster has a long and rich history in nuclear physics. To go through this development in detail is well beyond the scope of the two one-hour lectures that I will give here. Instead I plan to present my own experience in the field as I went through different etages in the study first of alpha decay and then the continuum in nuclei.

This story starts in 1977 when experimentalists at the Research Institute of Physics in Stockholm asked me to calculate the alpha decay of ^{212}At(gs). That Institute in itself has its own rich history: it was up to 1964 the Nobel Institute of Physics, after 1988 the Manne Siegbahn Institute and since July 1993 it is divided in three sections. The nuclear physics section belongs to the Royal Institute of Technology or in short KTH.

I perform that calculation using the nuclear field theory, which was at the time a popular theory. For the alpha decay process I used the Gamow theory[1] as developed by Wigner[2] and Thomas[3], which assumes that the decay width is the residues of the S-matrix. Moreover, it is also assumed that the decay process is stationary and, as a result, both the energies, i. e. the poles of the S-matrix, and the decay width are in general complex quantities. In the case of alpha-decay the fact that a quantity proportional to a probability (the decay width) is complex is not a practical problem because usually the width is very small and the residue of the S-matrix is practically real. In fact, in the one-level case (when one can neglect any overlap among resonances) one can show that the residue of the S-matrix is strictly real. But in the neutron decay of giant resonances this is a serious drawback which usually is ignored, as I will show later.

In the Wigner-Thomas theory the alpha decay width is given by

$$\Gamma_L(R) = 2P_L(R)\frac{\hbar^2 R}{2M}F_L^2(R),$$

where M is the reduced mass, $P_L(R)$ is the Coulomb penetration factor with angular momentum L and $F_L(R)$ is the formation amplitude of the α-particle at the point R.

In my calculations I used for R the sum of the radii of the α-particle and the daughter nucleus while to describe the formation amplitude I used only one shell-model configuration, as it was standard at the time, although mostly to evaluate relative decay width[4]. I was dismayed to find that my calculated absolute decay width was wrong by several orders of magnitude. I first thought that it was the nuclear field theory which was deficient. Although indeed the NFT was eventually shown to be plagued by divergencies, in the case of alpha-decay the theory was actually equivalent to the shell-model[5-7].

The weakest point of this calculation was not its total disagreement with experiment but that the penetration P was so strongly dependent on R that one could find a rather acceptable value of the distance for which that agreement improved drastically. It was then obvious that something fundamental was missing since one should expect that Γ_L is independent of R for distances outside the daughter nucleus, where the alpha particle is already formed. The calculated formation amplitude was vanishing small just outside the nuclear surface, indicating that the only configuration included in my calculation was not enough to describe the alpha decay process in the important region where the alpha cluster penetrates the Coulomb barrier. That is, high lying configurations, which are relevant at large distances, should be included in the calculations as had been suggested many years before[7]. This was rather easy to do within the NFT as well as with the multi-step shell-model, as later calculations showed[8]. We found that indeed the effect of high lying configurations was both to make the alpha decay width independent of R in a region close to the nuclear surface and, at the same time, to increase the value of Γ by many orders of magnitude. We also found that the physical reason of this increase is that, through the high lying configurations, the nucleons that eventually become the alpha particle are clustered. Yet, the calculated alpha decay width was more than one order of magnitude too small. We then thought that the neutron-proton interaction, which had been excluded in our and other previous calculations[8,9], was important to cluster nucleons of different isospin. This we introduced through a "giant pairing resonance", that is a high lying collective pairing state scalar in isospin[8,10]. But within reasonable limits for the mixing of this giant resonance in the mother nucleus wave function the alpha decay width could not be increased more than a few per cent. A proper treatment of the neutron-proton interaction would have required a full shell model calculation, as was later performed within a cluster shell-model configuration mixing model[11]. I will come back to this point at the end of my lectures. But at the time when our calculations where deficient we thought that the continuum was still not properly taken into account by the large number of configurations used in our calculations.

To treat the continuum exactly would have required computation capabilities beyond what is possible even within the powerful computers of today. An approximate approach to this problem was to replace the proper continuum by the set of outgoing solutions of the Schrödinger equation, which are either bound states or resonances in the continuum. But my concern was at that time mainly related with alpha decay and therefore more than a formalism to study the continuum in general I was interested in solving the problem of alpha clustering. Eventually I got involved in both problems, collaborating with many scientists. I will present briefly our work below. For details an abundant list of references is also given.

To present the method that we developed to treat the continuum it is convenient to start with the response function corresponding to an external field f, i.e.

$$R(E) = \int d\vec{r}\, d\vec{r}'\, f(\vec{r})^* G(\vec{r}\vec{r}'; E) f(\vec{r}')$$

where G is the Green function which describes the propagation of the nuclear system.

The response function $R(E)$ is real below threshold. It has poles on the real axis corresponding to the bound states and shows a resonant behaviour above threshold whenever the Green function has a complex pole close enough to the real energy axis. Therefore the influence of the continuum is controlled by the Green function.

For a spherical symmetric potential the exact form of the single-particle Green function is $g(r, r', k) = -u(r_<)v(r_>)/W$ where u(r) and v(r) are the regular and irregular solutions, respectively, of the single particle Hamiltonian H_0 and W denotes their Wronskian. The smaller (larger) between r and r' is denoted by $r_<(r_>)$. The solutions u and v satisfy the boundary conditions $u(r=0) = 0$ and $\lim_{r\to\infty} v(r) = e^{ikr}$.

Newton[12] found that the single-particle Green function can also be written in a spectral expansion form, i. e. in terms of the eigenvalues of H_0 as

$$g(r,r';k) = \sum_n \frac{w_n(r,k_n)w_n(r',k_n)}{k^2 - k_n^2} + \frac{2}{\pi}\int_0^\infty dq \frac{u^{(+)}(r,q)u^{(+)}(r',q)}{k^2 + i\epsilon - q^2},$$

where w_n are the wave functions of the bound single-particle states and $u^{(+)}(r,q)$ are scattering states, i.e. they represent the partial wave components of a wave consisting of an incoming plane wave plus an outgoing spherical wave as $r \to \infty$.

One can generalize the definition of "eigenvectors" of the single particle Schrödinger equation by requiring the boundary conditions [13,14] $\lim_{r\to 0} w_n(r,k_n) = 0$ and $\lim_{r\to\infty} w_n(r,k_n) = N_n e^{ik_n r}$ where k_n is the asymptotic momentum of the state with energy eigenvalue $\mathcal{E}_n$, i. e. $\mathcal{E}_n = \frac{\hbar^2}{2\mu}k_n^2$.

The eigenvalues $\mathcal{E}_n$ can now be complex. Writing $k_n = \kappa_n - i\gamma_n$ the eigenvectors belonging to those eigenvalues can be classified in four classes, namely: (a) bound states, for which $\kappa_n = 0$ and $\gamma_n < 0$; (b) antibound states with $\kappa_n = 0$, $\gamma_n > 0$; (c) decay resonant states (Gamow resonances) with $\kappa_n > 0$, $\gamma_n > 0$ and (d) capture resonant states with $\kappa_n < 0$, $\gamma_n > 0$. From the asymptotic behaviour of the solutions one sees that only the bound wave functions do not diverge.

With the standard definition of scalar product only the bound states can be normalized in an infinite interval. Therefore this definition has to be generalized in order to be able to use the generalized "eigenvectors". This can only be done if one uses a bi-orthogonal basis and apply some regularization method for calculating the resulting integrals. Bi-orthogonality means that in bra position one should use the mirror state $\tilde{w}_n$ instead of w_n, i.e. the solution which corresponds to $\tilde{k}_n = -k_n^*$. As regularization method we use the complex rotation suggested in ref. [15]. Within this method one rotates the radial distance r with a suitable angle only beyond the distance where the nuclear interactions die out.

The Newton representation of the single-particle Green function can be generalized by changing the path of integration on the continuum states. The resulting expression, due to Berggren[16], is

$$g(rr';k) = \sum_n \frac{\tilde{w}_n^*(r,\tilde{k}_n)w_n(r',k_n)}{k^2 - k_n^2} + \frac{2}{\pi}\int_{L^+} dq \frac{\tilde{u}^*(r,q)u_n(r',q)}{k^2 + i\epsilon - q^2}$$

where the sum runs over bound states plus the decaying resonant states which lie between the right hand half of the path L (denoted by L^+) and the real axis. Although there is not any real physical system without continuum states, the approximation of neglecting the integral contribution to the Green function is justified only in the analysis of bound and, in some cases, even quasi-bound states. But when properties closely linked with the continuum are to be analysed (e. g. escape widths of giant resonances) the bound representations show their limitations. The advantage of introducing the L contour is that though we neglect the integral in the Berggren representation of the Green function and use only the finite sum of the first term, we still include the most important (from a physical point of view) process occurring in the continuum, i. e. the resonant part. The numerical evaluation of that integral would imply the discretization of the wave number k along the contour L^+ and the solution of the radial equation for each of these complex k-values. Though we avoid this formidable task we are able to estimate its effect by taking the difference between the exact results and those obtained within the truncated spectral expansion.

In the Berggren expansion of the Green function we then consider only bound states and Gamow resonances as a single-particle representation to describe excitations lying in the continuum. It has been shown that Gamow resonances have large overlaps with certain wave packets centered at the resonance energy[17]. Therefore, the use of Gamow resonances would correspond to the use of wave packets, which is a proper procedure to describe processes in the continuum.

Another representation of the single-particle Green function is given by the Mittag-Leffler expansion[13,18-20], i. e.

$$g(r,r^{'};k) = \sum_n \frac{w_n(r,k_n)\tilde{w}_n^*(r^{'},\tilde{k}_n)}{2k_n(k-k_n)}$$

where the sum runs now over all classes of poles. It only contains a countably infinite set of discrete states but for practical purposes the series must be truncated and in this way the method becomes an approximate one. If there were some long range potential in the problem the truncation to a finite number of states would imply the neglect of the contribution of an integral along a complex contour[13,18].

The advantage of the Berggren expansion is that it is a natural extension of the set of states corresponding to the shell-model representation. The bound states are the same in both sets while the unbound resonances of the Berggren representation is what one expects to find in the continuum part of the odd nucleus that defines the single-particle states. This similarity temps one to consider the Berggren set of states as a basis, generalization of the shell-model basis, to describe quantum processes in the continuum.

Also the Mittag-Leffler set of states can be considered as a representation to describe the continuum, but in this case one has to be careful with the use of that set as a basis because it is not a "complete" basis in a standard sense. The corresponding completeness relation is thus called "reduced", to point out this limitation[20,21]. The advantage of the Mittag-Leffler representation is that, being of a mathematical nature, has all the proper symmetries built in. As a result, it describes well physical quantities within relatively small dimensions[20].

Within these representations one can study complex excitations in the continuum. Starting with the continuum RPA formalism[22] and using the Berggren representation, the particle-hole response function leads to a set of equations which are similar to the ones that one obtains within bound representations. The main difference is that the particles moves now in resonances states. The corresponding formalism was thus called "resonant RPA"[23,24] (RRPA). Using the RRPA properties of giant resonances could be described well. It was found that the position of the poles of the Green function corresponding to narrow resonances are well determined within a basis consisting of narrow elements. this explains why calculations done within bound (e. g. harmonic oscillator) representations have been so successful in predicting the position of giant resonances[25] since it is only narrow resonances which can be detected experimentally (note that "narrow" here means small *escape* width). The limitations of bound representations become apparent when properties closely related to the continuum are studied. For instance, partial decay widths cannot be directly calculated without including the proper continuum in some way[26,27]. In fact, not only the calculation but also the experimental measurement of partial decay widths is a difficult undertaking[28,29]. These quantities can rather easily be calculated within the Berggren and Mittag-Leffler representations and the calculated quantities agree well with exact results[26] and other calculations[27]. One important outcome of this study is that we found that in neutron decay the sum of the partial decay widths *is not* the same as the total escape width (the imaginary part of the complex energy). Only in the case of a very narrow resonance that equality is approximately fulfilled. This is discussed in detail by T. Vertse in his lecture at this School.

An important point in relation to the different expansions discussed here is whether the set of states in the expansions can be considered as representations in general. This is a subject that we are investigating at the moment. Preliminary results[20] indicate that the Mittag-Leffler expansion indeed provides a good description of regular functions while the description of such a function using the Berggren expansion is approximate within about 10%.

But the original reason to introduce the approximate treatment of the continuum was to study the alpha-particle formation amplitudes. Although most of the work done so far with the Berggren and Mittag-Leffler expansions is related to giant resonances, the Berggren "representation" was also used to describe the wave function of ^{212}Po and the corresponding alpha formation amplitude[30]. It was found that the $^{212}Po \rightarrow \alpha + ^{208}Pb$ decay width is independent of the distance R up to distances well beyond the nuclear surface. This result was just what we expected to obtain by introducing the influence of the continuum in the formalism.

Even complex excitations of a thermal nature[31] and spreading widths of giant resonances[32] have been studied with the Berggren representation. One may go even further and calculate also high lying two-phonon states built upon giant resonances within these representations.

The other attempt to solve the problem of the decay of the alpha particle was to describe clustering microscopically but within a basis which is a combination of shell- and cluster-model set of elements[11]. As mentioned before, within this basis the neutron-proton excitation could be taken into account properly and a good account of the experimental data could be given[11]. This is described in detail by R. Lovas in his lecture at this School.

Finally, also the description of alpha decay from deformed nuclei shows the importance of the continuum part of the spectrum to induce clustering[33-35]. Surprisingly enough (and in contrast to the spherical case) good agreement with experimental data is obtained by including a large shell-model representation. The importance of alpha decay from deformed nuclei is that one in this way expects to obtain information about nuclear shapes[35].

Here I presented the work of many scientists, as seen in the list of references. I would like to express my gratitude to all of them, but particularly to T. Vertse with whom I performed most of the work on the continuum.

References

[1] G. Gamow, *Z. Phys.* **51**, 204 (1928)
[2] T. Teichmann and E. P. Wigner, *Phys. Rev.* **87**, 123 (1952)
[3] R. G. Thomas, *Prog. Theor. Phys.* **12**, 253 (1954)
[4] M. A. Radii, A. A. Shibab-Eldin and J. O. Rasmussen, *Phys. Rev. C* **15**, 1917 (1977)
[5] H. J. Mang, *Ann. Rev. Nucl. Sci.* **14**, 1 (1964)
[6] J. O. Rasmussen, *Alpha, beta and gamma spectroscopy*, p. 701, edited by K. Siegbahn (North-Holland, Amsterdam, 1965)
[7] K. Harada, *Prog. Thor. Phys.* **26**, 667 (1961)
[8] G. Dodig-Crnkovic, F. A. Janouch and R. J. Liotta, *Nucl. Phys. A* **501**, 533 (1989), and references therein

[9] I.Tonozuka and A.Arima, *Nucl. Phys. A* **323**, 45 (1979)
[10] M. Herzog, O. Civitarese, L. Ferreira, R. J. Liotta, T. Vertse and L. J. Sibanda, *Nucl. Phys. A* **448**, 441 (1986)
[11] K. Varga, R. G. Lovas and R. J. Liotta, *Phys. Rev. Lett.* **69**, 37 (1992); *Nucl. Phys. A* **550**, 421 (1992)
[12] R. G. Newton, *Scattering theory of waves and particles*, p. 368, (McGraw-Hill, New York, 1966)
[13] W. J. Romo, *Nucl. Phys. A* **419**, 333 (1984)
[14] Y. B. Zel'dovich, *JETP (Sov. Phys.)* **12**, 542 (1961)
[15] B. Gyarmati and T. Vertse, *Nucl. Phys. A* **160**, 573 (1971)
[16] T. Berggren, *Nucl. Phys. A* **109**, 265 (1968)
[17] W. J. Romo, *Nucl. Phys. A* **398**, 525 (1983)
[18] J. Bang, F. A. Gareev, M. H. Gizzatkulov and S. A. Gonchanov, *Nucl. Phys. A* **309**, 381 (1978); N. Elander, C. Carlsund, P. Krylstedt and P. Winkler, in: *Resonances, Lectures Notes in Physics*, Vol. 325, p. 383, edited by E. Brändas and N. Elander (Springer Verlag, New York, 1989)
[19] T. Vertse, P. Curutchet and R. J. Liotta, *Phys. Rev. C* **42**, 2605 (1990)
[20] P. Lind, E. Maglione, R. J. Liotta and T. Vertse, *Z Phys. A*, in press
[21] P. Lind and T. Berggren, *Phys. Rev. C*, in press
[22] S. Shlomo and G. Bertsch, *Nucl. Phys. A* **243**, 507 (1975)
[23] T. Vertse, P. Curutchet, O. Civitarese, L. S. Ferreira and R. J. Liotta, *Phys. Rev. C* **37**, 876 (1988)
[24] P. Curutchet, T. Vertse and R. J. Liotta, *Phys. Rev. C* **39**, 1020 (1989)
[25] K. Goeke and J. Speth, *Annu. Rev. Nucl. Sci.* **32**, 65 (1982)
[26] T. Vertse, P. Cururchet, R. J. Liotta, J. Bang and N. Van Giai, *Phys. Lett. B* **264**, 1 (1991)
[27] E. Maglione, R. J. Liotta and T. Vertse, *Phys. Lett. B* **298**, 1 (1993)
[28] S. Brandenburg, W. T. A. Borghols, A. G. Drentje, L. P. Ekström, M. N. Harakeh, A. van der Woude, A. Hakanson, L. Nilsson, N. Olsson, M. Pignanelli and R. De Leo, *Nucl. Phys. A* **466**, 29 (1987)
[29] A. Bracco, J. R. Beene, N. Van Giai, P. F. Bortignon, F. Zardi and R. A. Broglia, *Phys. Rev. Lett.* **60**, 2603 (1988)
[30] S. Lenzi, O. Dragún, E. E. Maqueda, R. J. Liotta and T. Vertse, *Phys. Rev. C*, in press
[31] G. G. Dussel, H. Sofia, R. J. Liotta and T. Vertse, *Phys. Rev. C* **46**, 558 (1992)
[32] G. G. Dussel, H. Sofia, R. J. Liotta and T. Vertse, in progress
[33] A. Insolia, P. Curutchet, R. J. Liotta and D. S. Delion, *Phys. Rev. C* **44**, 545 (1991)
[34] D.S.Delion, A.Insolia and R.J.Liotta, *Phys. Rev. C* **46**, 884 (1992)
[35] D. S. Delion, A. Insolia and R. J. Liotta, *Phys. Rev. C* **46**, 1346 (1992)

CLUSTER FORMATION AND DECAY IN A MICROSCOPIC MODEL: THE ALPHA DECAY OF ^{212}Po

R. G. Lovas[1], K. Varga[1] and R. J. Liotta[2]

[1] *Institute of Nuclear Research of the Hungarian Academy of Sciences, Debrecen, P. O. Box 51, H-4001, Hungary*
[2] *Royal Institute of Technology, Frescativägen 24, S-104 05 Stockholm, Sweden*

Introduction

The purpose of this lecture is to outline a fundamental and apparently satisfactory description of cluster decay. The main title is meant to express this general aim. But the approach has as yet been applied[1] only to the α-decay transition leading to the ground state (g.s.) of ^{208}Pb. Therefore, for simplicity, I shall deal with an α-decay leading to a doubly-closed-shell core, to be denoted by c. To adapt this approach to more complicated cases is straightforward but needs a lot of labour. Nevertheless, since our formulae will not be very detailed, it would only require trivial changes if α were substituted by another cluster. If, however, the daughter were different, then the formulae should be modified to express that the residual state is not a 'core state' at the same time, i.e., it contains valence nucleons.

An exhaustive microscopic theory of decay should contain the description of the parent state, of the residual nuclear states as well as of the decay process itself. In our view the parent state and the process are indivisible, and, therefore, the decay is to be described as a single quasistationary state. This implies that the formation of a cluster is not an event, which has a history or a time evolution; it thus cannot be attributed to a transition operator or to a mechanism. Cluster formation is essentially a static property of the parent state: the word expresses an innate propensity of the nucleus to behave like a cluster–core system. A state always has such a property if it contains a component in the cluster–core subspace, i.e., in the subspace, in which the wave function can be written as an antisymmetrized product of the intrinsic states of the fragments and an arbitrary function of the relative coordinate, $\phi(\boldsymbol{r}_{c\alpha})$: $\mathcal{A}\{\Phi_\alpha(\xi_\alpha)\Phi_c(\xi_c)\phi(\boldsymbol{r}_{c\alpha})\}$.

This subspace is spanned by a basis $\{\Psi^{(\boldsymbol{r})}\}$ whose elements describe the two clusters pinned down at a relative displacement $\boldsymbol{r}$:

$$\Psi^{(\boldsymbol{r})} = \mathcal{A}\{\Phi_\alpha(\xi_\alpha)\Phi_c(\xi_c)\delta(\boldsymbol{r} - \boldsymbol{r}_{c\alpha})\}. \tag{1}$$

This basis is not orthogonal,

$$A(\boldsymbol{r}, \boldsymbol{r}') \equiv \langle \Psi^{(\boldsymbol{r})} | \Psi^{(\boldsymbol{r}')} \rangle \neq \delta(\boldsymbol{r} - \boldsymbol{r}'), \tag{2}$$

because the antisymmetrization mixes up the coordinates between the two fragments. The overlap of the basis states, $A(\boldsymbol{r}, \boldsymbol{r}')$, plays the role of a 'metric matrix', but since its 'labels' are continuous, it is in fact a *function* of two variables. A multiplication by such a 'matrix' amounts to applying the integral operator $\hat{A}$ whose kernel is $A(\boldsymbol{r}, \boldsymbol{r}')$. This operator is called the norm operator[2].

Then the cluster formation can be characterized by the amplitudes

$$g(\boldsymbol{r}) = \langle \Psi^{(\boldsymbol{r})} | \Psi \rangle \quad \text{or} \quad G(\boldsymbol{r}) = \hat{A}^{-1/2} g(\boldsymbol{r}), \tag{3}$$

or by their norm squares,

$$S = \int d\boldsymbol{r} |g(\boldsymbol{r})|^2, \quad \mathcal{S} = \int d\boldsymbol{r} |G(\boldsymbol{r})|^2. \tag{4}$$

The former, S, is the (conventional) spectroscopic factor, which enters into direct reaction theories[3], while the latter, $\mathcal{S}$, is the amount or probability of clustering[4], which is the weight of the cluster–core component in Ψ. The amplitude $g(\boldsymbol{r})$ is the usual spectroscopic amplitude, which is more often called the (pre)formation amplitude in decay studies. The clustering amplitude $G(\boldsymbol{r})$ plays an important role in Fliessbach's prescription for the α-decay matrix element[5], which we will not use here; this function will only be used in interpreting the results. It should be emphasized that probability meaning may only be attributed to $\mathcal{S}$ and to $|G(\boldsymbol{r})|^2$.

The microscopic model for cluster-decaying nuclei is essentially the shell model, and, indeed, it has been extensively used to describe α-decay. The problem with the application of the shell model to decaying states is that its wave function can only be correct in the nuclear interior, whereas the decay rate can only be extracted from the tail of the wave function. Nevertheless, there have been many attempts to deduce *absolute* α-decay rates from the shell model in the past decades, but the heroic efforts failed to bring full success.

Let me illustrate this with the transition ^{212}Po(g.s.) $\to$ ^{208}Pb(g.s.)+α. Mang's early shell-model calculations[6] undershot the experimental decay width Γ by four orders of magnitude. To be able to extract a decay rate, he constructed an outer solution and continued it inwards to the spatial domain in which the shell-model wave function is reliable. In the mid-seventies Fliessbach[5] pointed out that the Pauli principle was not treated correctly in such conventional calculations. To remedy this, he proposed an alternative treatment and showed that improvements of several orders of magnitude can be expected from it. Nevertheless, Tonozuka and Arima's most extensive shell-model calculations[7] still fell short of the experimental value by a factor of 23, in which the gain from Fliessbach's recipe was merely a factor of 3. They extended the basis up to convergence in Γ at the expense of omitting configurations with two like nucleons on different orbits and neglecting the proton–neutron (p–n) interaction. The latter approximation implies a considerable simplification: the four-valence-particle problem boils down to two independent two-valence-particle problems. Although the neglect of the p–n interaction seemed justifiable on the grounds of former results[8], it has been very difficult to understand why the p–n interaction is not more important. The matter is that all shell-model calculations produce some surface α-clustering in α-decaying states, and, considering that the free α-particle would not be bound without the p–n interaction, it is hard to believe that the p–n force is idle just in forming these surface clusters. All the same, even the recent calculations[9] which are based on an extensive multistep shell model[10] and have produced the best shell-model result hitherto ($\Gamma_{\text{theor}}/\Gamma_{\text{exp}} \approx 1/13$) failed to bring about the improvement expected from the p–n interaction.

From the foregoing survey it should be clear that a thorough increase of the shell-model state space should be combined with the inclusion of the p–n interaction. However, when the p–n interaction is included, the protons and the neutrons have to be treated *in the same shell-model calculation*, which results in an uncontrollable dimensional explosion. Therefore, we propose to complement a shell-model basis of reasonable size with unconventional basis states that allow the wave function to be correct far from the centre just in the decay channel. Such basis states are cluster-model basis states. Attempts to combine the shell model with a cluster model to describe α-decay were made by Steinmayer, Sünkel and Wildermuth[11] as well as by Okabe[12]. The results are not fully convincing, however, essentially because in these models the technical difficulties are avoided by introducing *ad hoc* elements[1].

In the following we outline the decay model, the combined shell and cluster model of the parent state, its application to the structure of ^{212}Po and of some of its subsystems and to the decay properties of ^{212}Po. We shall conclude with an analysis of the working mechanism of the model.

Decay Model

The decay rate and half-life are quantities defined for a purely exponential decay. If in the Schrödinger equation $i\hbar \partial \Upsilon/\partial t = \mathcal{H}\Upsilon$ we substitute the ansatz of exponential decay, $\Upsilon = \Psi \exp[(-iE - \frac{1}{2}\Gamma)t/\hbar]$, we end up with a time-independent Schrödinger equation for a complex energy:

$$\mathcal{H}\Psi = (E - i\tfrac{1}{2}\Gamma)\Psi. \tag{5}$$

This clearly shows that the decay cannot be purely exponential, but the fact that the experimental data are consistent with exponential decay confirms that (5) must approximate the physical situation very well. The function Ψ, the so-called Gamow wave function, should describe an outgoing wave in the decay channel:

$$\Psi = \mathcal{A}\{\Phi_\alpha(\xi_\alpha)\Phi_C(\xi_C)r_{C\alpha}^{-1}\mathcal{O}_L(k, r_{C\alpha})Y_{LM}(\hat{r}_{C\alpha})\} \quad \text{(for large } r_{C\alpha}\text{)}, \tag{6}$$

where $\mathcal{O}_L$ is an outgoing Coulomb wave. By solving the boundary-condition problem (5,6), one obtains the complex eigenvalue, $E - i\tfrac{1}{2}\Gamma$, and with this, one has the width Γ. Unfortunately, however, to determine the complex energy $E - i\tfrac{1}{2}\Gamma$ so that its tiny imaginary part ($\Gamma/E \sim 10^{-10} - 10^{-30}$) may be accurate enough requires a tremendous accuracy, which is hopeless to achieve. What can be done instead is to extract Γ from Ψ, whose asymptotic behaviour in the decay channel does carry the same information and whose calculation is not too demanding numerically.

A typical Gamow solution of a cluster-decay-like potential problem is shown schematically in Fig. 1. Outside the ranges of both the nuclear forces (R_N) and of Pauli exchanges (R_P) it has a long neck up to the outer turning point (R_{T_2}). This neck is proportional to the irregular Coulomb function $G_L(r)$ and thus the phase of the function can be chosen so as to make it almost exactly real within R_{T_2}. Within that limit the Gamow state can thus be approximated by a bound state, which is amenable to a variational approximation. For a many-particle state it is the amplitudes $g(\boldsymbol{r})$ or $G(\boldsymbol{r})$ that behave like the one in Fig. 1; in fact, beyond R_P, which is actually the range of the norm operator, $g(\boldsymbol{r}) = G(\boldsymbol{r})$. It turns out[13] that, for a narrow state of our concern, the solution depends on the choice of the trial function very slightly provided

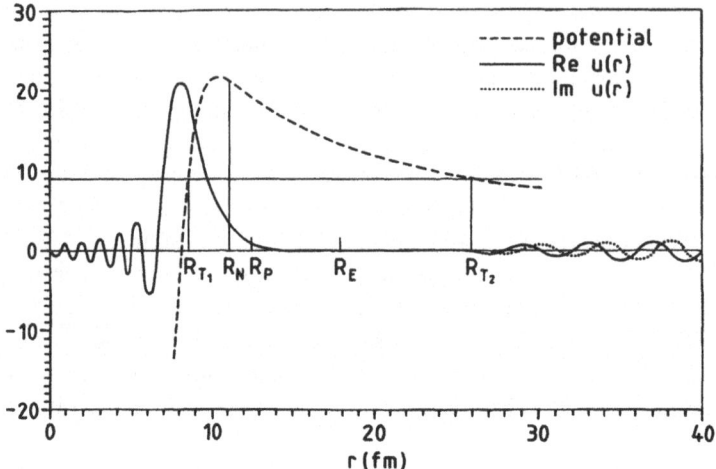

Figure 1. *Regions in the cluster–core channel represented by a local potential. The boundaries are the classical turning points (R_{T_1}, R_{T_2}) and the ranges of the nuclear interaction (R_N), of the Pauli exchanges (R_P), and of the trial function (R_E). Plotted is also a typical Gamow wave function, $u(r)$.*

it is chosen so as to cover a radial range $(0, R_E)$, with $R_E \in [\max(R_N, R_P), R_{T_2}]$. Then Ψ should be identified with an R-matrix eigenfunction; this can always be done by suitably choosing a channel radius a_L and a boundary condition at that point. The amplitude $g(\mathbf{r})$ will thus be proportional to the corresponding R-matrix partial-width amplitude around a_L, and the width can be calculated by the single-level R-matrix formula for the partial width[14],

$$\Gamma = 2 P_L(a_L) \frac{\hbar^2}{2 M_\alpha a_L} g_L^2(a_L), \tag{7}$$

where M_α is the reduced mass, P_L is the Coulomb penetration factor and

$$g_L(r) = r \int d\hat{\mathbf{r}}\, Y_{LM}(\hat{\mathbf{r}}) g(\mathbf{r}). \tag{8}$$

This is valid and hence independent of a_L, provided $a_L \in [\max(R_N, R_P), R_E]$.

Though the philosophy I have propounded here is unconventional, the final formula for the decay width is fairly standard. The main departure from the conventional method will be in its application: we are intent on producing a wave function Ψ, with which Eq. (7) is *really* applicable. That is why we can avoid matching $g_L(r)$ with an external solution, discussing recipes for the choice of a_L, and pondering on whether $g(r)$ or $G(r)$ is the correct amplitude. [Since $g(a_L) = G(a_L)$, both are correct.]

Nuclear Structure Model

How to construct such a parent wave function? As we mentioned in the Introduction, such a state in the conventional shell model would require astronomical dimensions. Therefore, we choose the variational trial function to be a superposition of ordinary shell-model and cluster+core-type cluster-model terms:

$$\Psi = \Psi^{\text{sh.}} + \Psi^{\text{cl.}} = \mathcal{A}\{\Phi_{\text{v}}(\mathbf{r}_{\text{c}1},...,\mathbf{r}_{\text{c}4})\Phi_{\text{c}}(\xi_{\text{c}})\}, \tag{9}$$

where the vectors $\mathbf{r}_{\text{c}i}$ point from the core centre of mass (c.m.) to the valence nucleons i. The valence wave function is also a sum of shell-model and cluster-model terms, $\Phi_{\text{v}} = \Phi^{\text{sh.}} + \Phi^{\text{cl.}}$, with the cluster-model term being a product of a relative-motion function and of a cluster intrinsic wave function:

$$\Phi^{\text{sh.}} = \sum_i C_i \Phi_i^{\text{sh.}}(\mathbf{r}_{\text{c}1},...,\mathbf{r}_{\text{c}4}), \quad \Phi^{\text{cl.}} = \phi_{LM}(\mathbf{r}_{\text{c}\alpha})\Phi_\alpha(\xi_\alpha). \tag{10}$$

The cluster intrinsic wave function may be chosen to be a Slater determinant $\tilde{\Phi}_\alpha$ of harmomic-oscillator (h.o.) single-particle (s.p.) states [for an α-particle just 0s states $\varphi^{(\alpha)}(\mathbf{x}) = (\alpha/\pi)^{3/4}\exp(-\frac{1}{2}\alpha x^2)$] divided by the c.m. factor[15] $\varphi^{(4\alpha)}(\mathbf{r}_{\text{c}\alpha})$:

$$\tilde{\Phi}_\alpha(\mathbf{r}_{\text{c}1},...,\mathbf{r}_{\text{c}4}) = \varphi^{(4\alpha)}(\mathbf{r}_{\text{c}\alpha})\Phi_\alpha(\xi_\alpha). \tag{11}$$

The antisymmetrization in Eq. (9) makes $\Psi^{\text{cl.}}$ extremely involved. To make such a wave function still tractable, we should express it in terms of determinant wave functions. To this end, we can use the old trick of the generator-coordinate version of the cluster model[2], which consists in expanding $\phi_{LM}(\mathbf{r}_{\text{c}\alpha})$ in terms of angular-momentum projected shifted Gaussians:

$$\phi_{LM}(\mathbf{r}_{\text{c}\alpha}) = \sum_k f_k \int d\hat{\mathbf{s}}_k Y_{LM}(\hat{\mathbf{s}}_k)\varphi^{(b)}(\mathbf{r}_{\text{c}\alpha} - \mathbf{s}_k). \tag{12}$$

By choosing the centres of the overlapping Gaussians $\varphi^{(b)}$ dense enough, one can, in this way, approximate any well-behaved function with arbitrary accuracy, no matter what the width parameter b is. Then let $b = 4\alpha$, so that each relative-motion Gaussian multiplied by the intrinsic function Φ_α be equal to a determinant of a shifted centre [cf. Eq. (11)]:

$$\varphi^{(4\alpha)}(\mathbf{r}_{\text{c}\alpha} - \mathbf{s})\Phi_\alpha(\xi_\alpha) = \tilde{\Phi}_\alpha(\mathbf{r}_{\text{c}1} - \mathbf{s},...,\mathbf{r}_{\text{c}4} - \mathbf{s}). \tag{13}$$

In the heavy-core approximation ($M_\alpha/M_{\text{C}} \approx 0$), both the core c.m. and the total c.m. can be pinned down at the origin, so that the total c.m. wave functions can be chosen $\phi_{\text{c.m.}} \approx \delta(\mathbf{R}_{\text{c.m.}})$, and the displacements with respect to the core c.m., $\mathbf{r}_{\text{c}i}$, will reduce to s.p. coordinates, $\mathbf{x}_i$, and hence the determinants will become Slater determinants:

$$\mathcal{A}\{\tilde{\Phi}_\alpha(\mathbf{r}_{\text{c}1} - \mathbf{s},...,\mathbf{r}_{\text{c}4} - \mathbf{s})\Phi_{\text{c}}(\xi_{\text{c}})\phi_{\text{c.m.}}(\mathbf{R}_{\text{c.m.}})\}$$
$$\approx \mathcal{A}\{\tilde{\Phi}_\alpha(\mathbf{x}_1 - \mathbf{s},...,\mathbf{x}_4 - \mathbf{s})\tilde{\Phi}_{\text{c}}(\mathbf{x}_5,...,\mathbf{x}_A)\}. \tag{14}$$

The heavy-core approximation is invariably used in the shell model of heavy nuclei, and we adopt it for the cluster-model term as well. Such an approximation amounts to violating translation invariance, but since in the decay process no high momentum is transferred, it is a very good approximation[15]. A cluster-model Slater determinant can be rewritten by replacing the shifted Gaussians with functions from which the s.p. states occupied in the core have been projected out:

$$\begin{align}
\mathcal{A}\{\tilde{\Phi}_\alpha \tilde{\Phi}_c\} &= (A!)^{-1/2}\det\{\varphi^{(\alpha)}(\boldsymbol{x}_1 - \boldsymbol{s}_k)\chi_1(1)...\varphi^{(\alpha)}(\boldsymbol{x}_4 - \boldsymbol{s}_k)\chi_4(4)\psi^{(5)}(5)...\psi^{(A)}(A)\} \\
&= (A!)^{-1/2}\det\{\psi^{(1)}(\boldsymbol{x}_1 - \boldsymbol{s}_k)...\psi^{(4)}(\boldsymbol{x}_4 - \boldsymbol{s}_k)\psi^{(5)}(5)...\psi^{(A)}(A)\}, \tag{15}
\end{align}$$

where χ_i are spin–isospin eigenvectors and

$$\psi^{(i)}(\boldsymbol{x}_i - \boldsymbol{s}_k) = P_i \varphi^{(\alpha)}(\boldsymbol{x}_i - \boldsymbol{s}_k)\chi_i(i), \tag{16}$$

with

$$P_i = 1 - \sum_{j=5}^{A} |\psi^{(j)}(i)\rangle\langle\psi^{(j)}(i)|. \tag{17}$$

The two determinants are equal because only the first four columns are changed by adding to them linear combinations of the rest.

Since in the second line of Eq. (15) all valence orbits are orthogonal to all core orbits, all matrix elements between functions $\Psi_k = \mathcal{A}\{\tilde{\Phi}_{\alpha_k}\tilde{\Phi}_c\}$ of this type with a fixed core state $\tilde{\Phi}_c$ can be formally reduced to those of 4×4 Slater determinants. To show this for the Hamiltonian $\mathcal{H}$, one should write it as a sum of a core term, a valence term, and a core–valence interaction term:

$$\mathcal{H} = \sum_{i=1}^{A} T(i) + \sum_{1 \le i < j \le A} V(i,j) \equiv H_C + H_V + V_{VC}. \tag{18}$$

With standard techniques[2] one can show that a matrix element of $\mathcal{H}$ between Ψ_k and $\Psi_{k'}$ is then

$$\langle\Psi_k|\Psi_{k'}\rangle^{-1}\langle\Psi_k|\mathcal{H}|\Psi_{k'}\rangle = \langle\tilde{\Phi}_C|H_C|\tilde{\Phi}_C\rangle + \langle\tilde{\Phi}_{\alpha_k}|\tilde{\Phi}_{\alpha_{k'}}\rangle^{-1}\langle\tilde{\Phi}_{\alpha_k}|\left[H_V + \sum_{i=1}^{4} U'(i)\right]|\tilde{\Phi}_{\alpha_{k'}}\rangle, \tag{19}$$

where

$$U'(i) = \langle\tilde{\Phi}_C|\sum_{j=5}^{A} V(i,j)(1 - P_{ij})|\tilde{\Phi}_C\rangle \quad (i = 1, ..., 4), \tag{20}$$

with P_{ij} being an operator exchanging particle labels. The core term is a constant, and the s.p. potential is non-local, like a Hartree–Fock potential, owing to its exchange terms. In the exchange terms the core and valence orbits are still mixed up. The reduction to a four-particle problem will be complete if we approximate the s.p. potential U' with a local potential U. The four-particle Hamiltonian will then look like

$$H = H_V + \sum_{i=1}^{4} U(i) = \sum_{i=1}^{4}[T(i) + U(i)] + \sum_{1 \le i < j \le 4} V(i,j). \tag{21}$$

This problem can be solved variationally. All variational parameters, C_i and f_k, are linear, so the problem boils down to an ordinary diagonalization over a non-orthogonal basis. The hybrid model obtained in this way can be called the 'cluster-configuration shell model'. By making all these simplifications, the microscopic character is not lost: all core orbits are still present in the s.p. functions $\psi^{(i)}(\boldsymbol{x}_i - \boldsymbol{s}_k)$ through the Pauli projectors. This also implies that much of the technical difficulties survive. These are overcome by calculating all matrix elements fully analytically[16]. The calculations are made analytically by choosing $\psi^{(5)},...,\psi^{(A)}$ to be h.o. s.p. states, by expressing the Woods–Saxon+spin–orbit+Coulomb-shaped potential U as a combination of Gaussians, and by choosing the residual interaction $V(i,j)$ also to have Gaussian form factors. The significance of introducing the phenomenological U instead of the microscopically derived potential U' is that thereby we can make the problem more realistic. By choosing $U(r)$ to be a realistic s.p. potential, we ensure that the barrier that the α-particle has to penetrate, and which emerges out of the s.p. barriers, will also be realistic.

Application to ^{212}Po(g.s.) $\to$ ^{208}Pb(g.s.)+α

Nuclear Structure Aspects

To set the parameters of the model, the description of α-decay has to be accompanied by the description of the structure of the subsystems involved. The details of the parameter choice are given in our longer paper[1]. We included all s.p. states belonging to the first major shell above the core, both for protons and neutrons. The s.p. potentials were chosen to be consistent with the known charge and matter distributions of ^{208}Pb. Two potential sets, labelled A and B, have been chosen, the latter having 1.5% smaller radii. The depths were set so as to fit the lowest-lying s.p. energies best. The resulting s.p. energies are given in Table 1. The levels close to the particle emission threshold are somewhat less accurate since in our model the s.p. energies are defined as expectation values of $T + U$ in h.o. states.

Table 1. *Single-particle energies in MeV*

Proton orbits	nlj	$0h_{9/2}$	$1f_{7/2}$	$0i_{13/2}$	$1f_{5/2}$	$2p_{3/2}$	$2p_{1/2}$	
	Potential A	-3.75	-2.96	-2.05	-0.04	-0.25	0.06	
	Potential B	-3.81	-3.00	-2.01	-0.12	-0.70	0.09	
	Experiment	-3.77	-2.89	-2.17	-0.96	-0.67	0.53	
Neutron orbits	nlj	$1g_{9/2}$	$0i_{11/2}$	$0j_{15/2}$	$2d_{5/2}$	$3s_{1/2}$	$1g_{7/2}$	$2d_{3/2}$
	Potential A	-3.93	-3.23	-2.38	-2.51	-1.60	-0.90	-1.03
	Potential B	-4.01	-3.04	-2.40	-2.57	-1.33	-1.45	-1.02
	Experiment	-3.94	-3.15	-2.53	-2.36	-1.91	-1.45	-1.42

The residual interaction has to be suitable not only for the valence nucleons but for the free α-particle as well, and that requires effective interactions used to describe bulk properties, like the forces used in cluster-model calculations. Surprisingly, both the Volkov 1 (V1)[17] and the Brink–Boeker B1[18] forces have proved to be very good residual interactions. When potentials A and B were used together with the V1 and the B1 force, respectively, the yrast states of the core+two-nucleon systems were reproduced excellently (Table 2). Thus we used potential A with V1 and potential B with B1 throughout. The agreement with experiment is slightly worse for ^{210}Po because we neglected the Coulomb force between the valence protons.

The four-particle shell-model basis was taken the tensorial product of the two-proton and two-neutron bases, with including two-particle angular momenta up to 8 and the dimension

Table 2. Energies in MeV of the yrast states of the two-valence-particle subsystems

J	^{210}Po A	^{210}Po B	^{210}Po Exp.	^{210}Pb A	^{210}Pb B	^{210}Pb Exp.	^{210}Bi A	^{210}Bi B	^{210}Bi Exp.
0	−8.91	−8.73	−8.78	−9.06	−8.96	−9.12	−8.33	−8.25	−8.36
1							−8.18	−8.20	−8.40
2	−8.05	−7.90	−7.60	−8.32	−8.21	−8.32	−8.02	−7.93	−8.08
3							−7.95	−7.90	−8.06
4	−7.73	−7.67	−7.36	−8.06	−8.00	−8.02	−7.87	−7.82	−7.90
5							−7.89	−7.86	−7.96
6	−7.64	−7.61	−7.31	−7.99	−7.95	−7.93	−7.81	−7.77	−7.85
7							−7.92	−7.89	−7.97
8	−7.60	−7.59	−7.22	−7.96	−7.94	−7.84	−7.78	−7.76	−7.82
9							−8.17	−8.17	−8.13

Table 3. Energies of ^{212}Po and the α particle in MeV

	E_{Po} [a] Shell model	E_{Po} [a] Cluster model	E_{Po} [a] Shell+cluster model	E_α	$\varepsilon_\alpha = E_{\text{Po}} - E_\alpha$
A	−18.61	−16.47	−18.88	−27.79	8.91 [b]
B	−18.96	−16.41	−19.18	−28.18	9.00 [b]
Exp.			−19.35	−28.30	8.95

[a] Energy counted from the threshold of the removal of four nucleons.
[b] Calculated with E_{Po} of the shell+cluster model.

totalling 538. The cluster-model basis was chosen so as to cover the interval (0, 20 fm) in the relative motion with 40 elements. We see in Table 3 that the g.s. energy in the pure shell model is some 2–2.5 MeV deeper than in the pure cluster model, and the inclusion of the cluster model deepens the energy by a mere 20 keV with respect to the pure shell model. The energy in the mixed model is consistent with the energy of the free α-particle in the sense that the same amount of binding is missing from both, so that the α-decay energy is nearly correct just in the mixed model.

Alpha Decay

Figure 2 compares the radial parts of the formation amplitudes of the pure models and of the mixed model calculated with potentials A. Note that the shell-model amplitude is multiplied by 10. One sees the enormous importance of the cluster-model admixture. It shifts the outermost maximum outwards by 1 fm while enhancing it by an order of magnitude. Since the non-Coulomb effects become negligible only beyond 10 fm, the pure shell model does not contain enough information on the decay width; the width could only be extracted from it by continuing a phenomenological external solution inwards, the ambiguity[5] of which we would like to avoid.

Figure 3 shows the radial clustering amplitudes corresponding to the formation amplitudes displayed in Fig. 2. The cluster-model $G(r)$ is, of course, normalized to unity. The shell-model amplitude also shows a non-negligible clustering, and it peaks inside the nucleus. The clustering amplitude of the mixed model counters common prejudices in two respects, both of which are due to the contribution of the ordinary shell model. First, it exhibits non-negligible clustering inside the nucleus, and, second, it does not have the shape of a radial wave function. The latter finding is relevant to direct-reaction studies; together with Fig. 2 it suggests that the conventional spectroscopic amplitude $g(r)$ is easier to approximate by a scaled eigenfunction of a local potential.

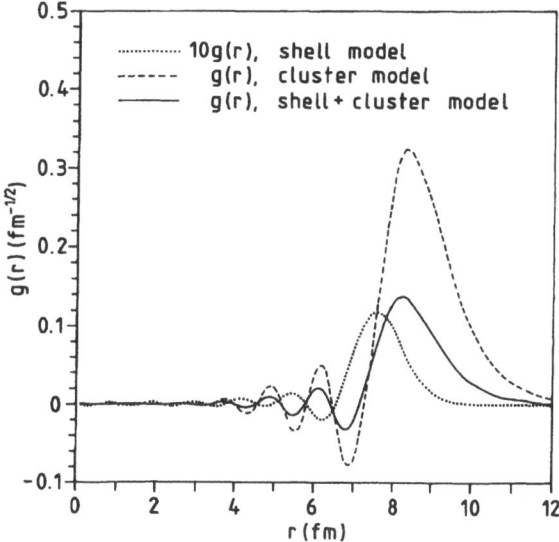

Figure 2. Radial formation amplitudes produced by potential A in the three models. Note the magnification of the shell-model curve.

With potentials A and B, respectively, the model predicts $\Gamma = 1.5 \times 10^{-15}$ and 2.1×10^{-15} MeV, while the experimental value is 1.52×10^{-15} MeV. By correcting for the deviation from the experimental decay energy, we would get 1.2×10^{-15} and 1.6×10^{-15} MeV, respectively. The error due to the 'graininess' of the shifted-Gaussian basis is about 10–20%. To assess the parameter dependence, the parameters should be changed such that all known subsytem properties as well as the decay energy should be unchanged. This restricts freedom to the extent of the uncertainties in the known properties. From among these properties, the nucleonic distributions are only uncertain, and that allows one to change $U(r)$ to some extent. Even the uncertainty of $U(r)$ turns out to be small, however, provided that the radii of U and of the h.o. involved in the basis are changed in accord. All in all, we think that the prediction of the model is reliable within a factor of 2.

The values of the spectroscopic factors and amounts of clustering are collected in Table 4. All values obtained in the mixed model are substantially higher than any previous estimates. In view of the correctness of the predicted Γ, these values must be considered realistic.

Table 4. ^{208}Pb+α spectroscopic factors S and amounts of clustering $\mathcal{S}$ in ^{212}Po

Potential		Shell model	Cluster model	Shell+cluster model
A	S	1.5×10^{-4}	0.15	0.025
	$\mathcal{S}$	3.7×10^{-2}	1.	0.30
B	S	1.7×10^{-4}	0.19	0.023
	$\mathcal{S}$	4.2×10^{-2}	1.	0.23

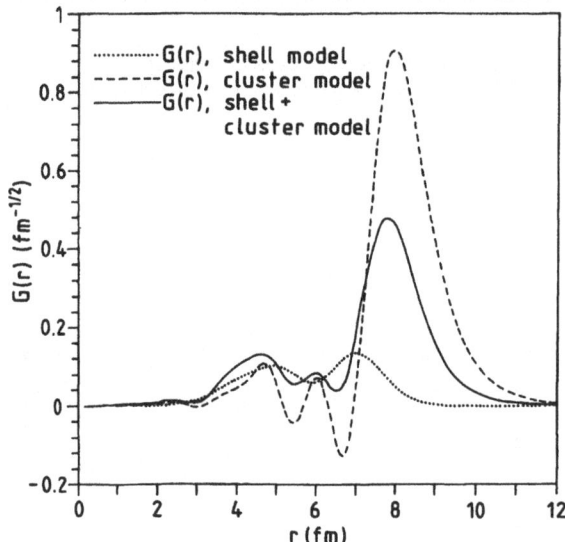

Figure 3. Radial clustering amplitudes produced by potential A in the three models.

Thus one is forced to accept that the g.s. of ^{212}Po contains a ^{208}Pb+α configuration with a probability of 0.23–0.30. This explains the success of the phenomenological extreme cluster model[19] in reproducing α-widths. Indeed, since this model reproduces the systematics of a whole range of favoured α-decay data, the large amount of α-clustering is likely to be typical in α-decaying states. But the extreme cluster model does seem to work even for heavy-cluster decay[20], which, in turn, indicates that the weights of the core+heavy-cluster configurations must be substantial in such radioactive nuclei. Note also the ratio $\mathcal{S}/S \sim 10 - 12$, which is by far smaller than the values (300–500) prevailing in the literature. These large values were obtained from shell-model calculations, and are now borne out by our pure shell model. The difference between the shell-model and the realistic mixed-model results is caused by the fact that the mixed-model amplitude peaks farther out of the nucleus, where the Pauli effects embodied in the operator $\hat{A}$ are substantially smaller.

Discussion

We made various tests to understand the working mechanism of the model. The results hardly depend on the actual basis size, thus the state space may be considered saturated. For this saturation it is, however, necessary to have shell-model basis elements with energies near the g.s. energy of the cluster model. Thus, for example, the deepest-lying single shell-model configuration is not enough to get the cluster configurations admixed sufficiently.

Table 5. *Shell-model results for potential B, with and without p–n interaction*

Quantity	With p–n interaction	Without p–n interaction
E_{Po} (MeV)	-18.96	-17.92 [a]
$\mathcal{S}$	1.7×10^{-4}	9.2×10^{-5} [a]
S	4.2×10^{-2}	2.7×10^{-2} [a]

[a] The same as in the hybrid model without p–n interaction.

In Table 5 the role of the p–n interaction is tested. One sees that upon omission of the p–n interaction from the shell model the binding is decreased by 1 MeV, but the clustering properties are not changed dramatically. This is in full accord with the findings about the p–n interaction in former shell-model studies of α-decay, but still goes against common sense. To investigate further, we included the cluster-model basis while the p–n interaction was still switched off. The result was dramatic: no change whatsoever! This finding was elucidated by a pure cluster-model calculation with no p–n interaction. In this calculation the energy was enhanced by more than 30 MeV with respect to the case of p–n included, and that is why the cluster-model state could not mix with the shell model. Thus the p–n interaction does play a crucial role in making it favourable to form a correlated structure in the nuclear surface, but the basis of an ordinary shell-model of tractable size is not flexible enough at large distances to accommodate such a correlation. The part of the p–n correlation not allowed for by our pure shell model hardly adds to the binding, yet provides a substantial fraction of the wave function. This wave-function term peaks at large core–α separations, where nuclear saturation no longer hinders the strong attraction between the four nucleons. The gain in the binding energy is still so meagre because the strong intracluster attraction is balanced by the strong Coulomb repulsion since this region overlaps with the Coulomb barrier.

It is important to see clearly that the substantial clustering found in our calculations cannot be an artifact. Our calculations are governed by a variational principle, which is equivalent to the Schrödinger equation, and the potentials used are realistic. Of course, since our cluster-model basis is virtually complete in the cluster–core subspace, while the basis orthogonal to this subspace is not, the amount of clustering obtained is likely to be an upper bound. But the almost complete insensitivity of the mixed-model results to enlargements of the shell-model basis shows that no qualitative change can be expected from more complete calculations. Even if it should turn out that, say, 20% of the wave function is to be redistributed to higher-lying shell-model configurations that are orthogonal to the cluster-model subspace, this would probably decrease the amount of clustering just by 20%.

Conclusion

In summary, the approach proposed in this lecture is based on a quasistationary-state decay model. Its essential point is to complement the shell-model basis with cluster-model terms. With this basis we get a realistic microscopic model, for the parent state, which is correct not only in the nuclear interior but in the decay channel as well. We managed to cast the cluster-model basis elements into a form similar to the shell-model configurations, except that the cluster configurations involve excentric orbits. This enabled us to eliminate the core degrees of freedom like in an ordinary shell model. The core effects are included in the form of a s.p. potential and of Pauli projectors, and what remains to be solved is to treat Pauli-projected excentric orbits. So far the analytical formulae have only been developed for 0s excentric orbits, and the approach has only been applied to the decay of the g.s. of ^{212}Po.

The results are very encouraging. We have successfully reproduced, for the first time, the α-decay width of a heavy nucleus by a microscopic model of no free parameter. We found that the uncertainties in the prediction of the model are moderate, but this hinges mostly on the fact that the nuclei involved in the example are known very well. The essential ingredients responsible for the success are the inclusion of the p–n interaction, without which there is no four-particle correlation, and the inclusion of the sector of the state space that allows the four nucleons to correlate, viz. cluster-model states around the nuclear surface. We found that there is an appreciable α-clustering, mainly but not exclusively, in the surface region. We conclude that clustering is likely to be appreciable in other cluster-decaying nuclear states as well.

This work was supported by OTKA, Hungary (Grant No. 3010)

References

[1] K. Varga, R. G. Lovas and R. J. Liotta, *Phys. Rev. Lett.* **69**, 37 (1992); *Nucl. Phys. A* **550**, 421 (1992)
[2] H. Horiuchi, *Prog. Theor. Phys. Suppl.* **62**, 90 (1977)
[3] R. G. Lovas, *Z. Phys. A* **322**, 589 (1985)
[4] R. Beck, F. Dickmann and R. G. Lovas, *Ann. Phys. (N. Y.)* **173**, 1 (1987)
[5] T. Fliessbach and H. J. Mang, *Nucl. Phys. A* **263**, 75 (1976)
[6] H. J. Mang, *Phys. Rev.* **119**, 1069 (1960)
[7] I. Tonozuka and A. Arima, *Nucl. Phys. A* **323**, 45 (1979)
[8] N. K. Glendenning and K. Harada, *Nucl. Phys.* **72**, 481 (1965)
[9] G. Dodig-Crnkovic, F. A. Janouch and R. J. Liotta, *Nucl. Phys. A* **501**, 533 (1989)
[10] R. J. Liotta and C. Pomar, *Nucl. Phys. A* **382**, 1 (1982)
[11] T. Steinmayer, W. Sünkel and K. Wildermuth, *Phys. Lett.* **125B**, 437 (1983)
[12] S. Okabe, *Suppl. J. Phys. Soc. Japan* **58**, 516 (1989)
[13] R. G. Lovas and M. A. Nagarajan, *J. Phys. A* **15**, 2383 (1982)
[14] R. G. Thomas, *Prog. Theor. Phys.* **12**, 253 (1954)
[15] K. W. Schmid, contribution to this volume
[16] K. Varga, in preparation
[17] A. B. Volkov, *Nucl. Phys.* **74**, 33 (1965)
[18] D. M. Brink and E. Boeker, *Nucl. Phys. A* **91**, 1 (1967)
[19] B. Buck, A. C. Merchant and S. M. Perez, *Phys. Rev. Lett.* **65**, 2975 (1990)
[20] B. Buck, A. C. Merchant and S. M. Perez, *J. Phys. G* **17**, L91 (1991)

THE ALPHA-PARTICLE MEAN FIELD AND CONSISTENT PRE-EQUILIBRIUM AND STATISTICAL EMISSION

M. Avrigeanu[1], V. Avrigeanu[1] and P.E. Hodgson[2]

[1] *Institute of Atomic Physics, P.O. Box MG-6, Bucharest, Romania*
[2] *Nuclear Physics Laboratory, Department of Physics, Oxford, U.K.*

A set of global optical model parameters has been derived starting from alpha-particle elastic scattering with energies higher than 80 MeV where the continuous ambiguities in the optical model analyses are eliminated. The geometry and mass number dependence derived by Nolte et al.[1] and the energy dependence found by Put and Paans[2] were thus adopted. Moreover, variation of the imaginary well depth at lower energies has been adopted by taking into account results derived from the study of the elastic scattering on ^{90}Zr[2] and ^{24}Mg[3].

Although generally the optical model parametrizations of elastic scattering data at high energies do not work well at low energies for either complete fusion or elastic scattering[4], this global parameter set describes alpha-particle emission near the Coulomb barrier in the mass range $A \simeq 50$ (Figure 1). It can be concluded that the transmission coefficients for alpha-particle evaporation are rather strongly related to the fusion cross sections and elastic scattering at energies above 80 MeV, but not to elastic scattering at lower energy where specific features (e.g. the transparency effect) are comprised within complex optical potential parametrizations. The lack of agreement for the residual nuclei ^{45,47}Ca could be due to anomalously lower diffuseness parameters for the calcium isotopes[5,6] relative to the normal increase with $A^{1/3}$.

The alpha-particle pre-equilibrium emission has been analyzed in the frame of the Geometry-Dependent Hybrid (GDH) model modified to take into account the energy dependence of the single-particle state density and different values of this density at the Fermi level for holes and excited particles, respectively. The particle-hole state density formalism also includes corrections for finite nuclear potential well, Pauli principle and pairing effects as well as shell effects. Both the energy-dependent single-particle state densities and the nuclear surface effects are thus properly matched within the GDH model. Under these circumstances the GDH intranuclear transition rates have been related to the averaged imaginary optical model potentials, while the same parameter sets have been involved within the pre-equilibrium and statistical models. The comparison with the available experimental (n,α) reaction cross sections and energy spectra in the mass range $A \simeq 50$ supports therefore the alpha-particle global optical potential from both sides.

Furthermore, when both the statistical emission and the GDH intranuclear transition rates were thus established, the confident investigation of the single particle state density for α-particles has become possible. A comparative analysis of pre-equilibrium and statistical alpha-particle emission proves the value[7] $g_\alpha = (A/10.36)$ MeV^{-1} as the most appropriate one.

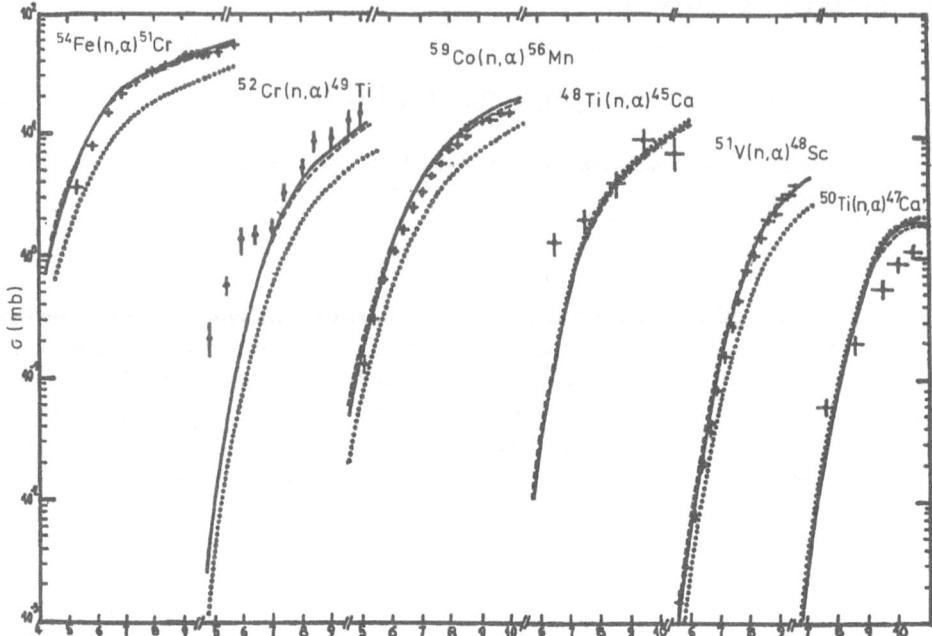

Figure 1. Comparison of experimental and calculated (n,α) reaction cross sections by using the alpha-particle optical potentials from this work (full curves), of Huizenga and Igo[8] (dash-dotted curves), and the McFadden and Satchler parameter set Ca9 for residual nuclei ^{45,47}Ca and the average four-parameter set[9] for the rest of nuclei (dotted curves).

References

[1] M. Nolte, H. Machner and J. Bojowald, *Phys. Rev. C* **36**, 1312 (1987)
[2] L. W. Put and A. M. J. Paans, *Nucl. Phys. A* **291**, 93 (1977)
[3] P. P. Singh and P. Schwandt, *Nukleonika* **21**, 451 (1976)
[4] M. A. McMahan and J. M. Alexander, *Phys. Rev. C* **21**, 1261 (1980)
[5] B. A. Brown, S. E. Massen and P. E. Hodgson, *J. Phys. G* **5**, 1655 (1979)
[6] A. Andl, K. Bekk, S. Goring, A. Hauser, G. Nowicki, H. Rebel, G. Schatz and R. C. Thompson, *Phys. Rev. C* **26**, 2194 (1982)
[7] E. Gadioli and E. Gadioli Erba, *Z. Phys. A* **299**, 1 (1981)
[8] J. R. Huizenga and G. Igo, *Nucl. Phys.* **29**, 462 (1962)
[9] L. McFadden and G. R. Satchler, *Nucl. Phys.* **84**, 177 (1966)

NEUTRON MULTIPLICITIES IN SPONTANEOUS FISSION AND NUCLEAR STRUCTURE STUDIES

J. H. Hamilton[1], J. Kormicki[1], Q. Lu[1], D. Shi[1], K. Butler-Moore[1], A. V. Ramayya[1], W.-C. Ma[1], B. R. S. Babu[1], G. M. Ter-Akopian[2], Yu. Ts. Oganessian[2], G. S. Popeko[2], A. V. Daniel[2], S. Zhu[3], M. G. Wang[3], J. Kliman[4], V. Polhorsky[4], M. Morhac[4], J. D. Cole[5], R. Aryaeinejad[5], R. C. Greenwood[5], N. R. Johnson[6], I. Y. Lee[6] and F. K. McGowan[6]

[1] *Department of Physics and Astronomy Vanderbilt University, Nashville, TN, 37235, USA*
[2] *Joint Institute for Nuclear Research, Dubna, Russia*
[3] *Physics Department, Tsinghua University, Beijing, P.R. China*
[4] *Institute of Physics SASc, Bratislava, Slovak Republic*
[5] *Idaho National Engineering Lab., Idaho Falls, ID, 83415, USA*
[6] *Oak Ridge National Laboratory, Oak Ridge, TN 37831, USA*

Introduction

New insights into the fission process can be gained by better quantitative knowledge of how the energy released in fission is distributed between the kinetic energy of the two fragments, the excitation energy of the two fragments and the number of neutrons emitted. Studies of prompt gamma-rays emitted in spontaneous fission (SF) with large arrays of Compton suppressed Ge detector arrays are providing new quantitative answers to long-standing questions concerning fission as well as new insights into the structure of neutron-rich nuclei. For the first time the triple gamma coincidence technique was employed in spontaneous fission studies. Studies of SF of ^{252}Cf and ^{242}Pu have been carried out. These γ-γ-γ data provide powerful ways to identify uniquely gamma rays from a particular nucleus in the very complex gamma-ray spectra given off by the over 100 different nuclei produced. The emphasis of this paper is on the first quantitative measurements of the multiplicities of the neutrons emitted in SF and the energy levels populated in the fragments. Indeed, in the break up into Mo-Ba pairs, we have identified for the first time fragments associated with from zero up to ten neutrons emitted and observed the excited energy states populated in these nuclei. The zero neutron emission pairs like ^{104}Mo - ^{148}Ba, ^{106}Mo - ^{146}Ba and ^{104}Zr - ^{148}Ce observed in this work are particularly interesting because they represent a type of cold fission or a new mode of cluster radioactivity as proposed by Greiner, Sandulescu and co-workers[1,2]. These data provide new insights into the processes of cluster radioactivity and cold fission.

These SF studies also continue to be a rich source of information on neutron rich nuclei. The triple gamma coincidence technique was used to uniquely identify levels in ^{136}Te and new higher spin states in its $N = 84$ isotones ^{138}Xe and ^{140}Ba.[3] Earlier evidence for octupole deformation was observed in the even-even ^{144,146}Ba[4]. Recently we have discovered similar evidence for octupole deformation in odd-A ^{143}Ba and observed a new band to high spin in ^{145}Ba.[5]. These nuclei can give more direct information on the orbitals involved in the octupole deformation.

Experimental Methods

A spontaneous fission source of ^{252}Cf (0.1 μg) with a strength of about $6 \cdot 10^4$ fissions/sec and with a 250 μm Be window was used for the Oak Ridge study. This source was placed in the center of the 20 Compton-suppressed Ge-detector Compact Ball at the Holifield Heavy Ion Research Facility at Oak Ridge National Laboratory. Approximately $2 \cdot 10^9$ γ-γ coincidences were collected during a five day run. Both double- and triple-coincidence events were recorded and sorted in data analysis. A comparable strength source and the same experimental arrangements were used to study the SF of ^{242}Pu. The coincidences obtained for the γ-rays of fission fragment pairs can be used as triggers for the fission events of a given nucleus against the background of β-decay and fission of other nuclides. The triple coincidence technique is particularly powerful in eliminating or reducing all types of background, especially those produced by the process of two or more gamma-rays from different nuclei falling in an energy gate set on a single γ-ray in a γ-γ coincidence experiment.

Neutron Multiplicities and Cold Fissions (Cluster Radioactivity)

The distribution (relative intensities) of the different-mass partners formed through the emission of different numbers of neutrons are of particular interest in understanding the fission process. Of special interest is the relative probability of zero neutron emission. This case may correspond to a new type of cluster radioactivity. The phenomena of cluster radioactivity was initiated by Sandulescu, Poenaru and Greiner[6] who proposed this new class of radioactivity in which a nucleus decays into two fragments with both masses heavier than the alpha particle but one lighter than the usual fission fragments without the emission of other light particles like neutrons. Greiner, Ivascu, Poenaru and Sandulescu[1] have recently reviewed this field. Greiner and Sandulescu[2] have emphasized that another prediction of their two-center shell is cold fission in addition to cluster radioactivity. In this case a nucleus splits into two "unexcited" nuclei. Cold fission means both fragments are left in low excited states (small excitation energy), so neutron emission is not allowed. The energy goes mainly into kinetic energy of the fragments. In induced fission Gönnenwein and his colleagues (see elsewhere in these proceedings) have observed events in which the kinetic energy of the fragments essentially equals the Q value. In our SF fission studies we carry this evidence one step further.

Table 1. *Normalized relative yields of the correlated fragment pairs for $Z_L/Z_H = 40/58$.*

Ce partner	Zr^{98}	Zr^{99}	Zr^{100}	Zr^{101}	Zr^{102}	Zr^{103}	Zr^{104}	<A>
^{148}Ce	0.05	0.15	1.00	0.52	0.64	0.14	0.06	100.9
^{146}Ce		0.20	0.60	0.55	1.00	0.26	0.20	101.4

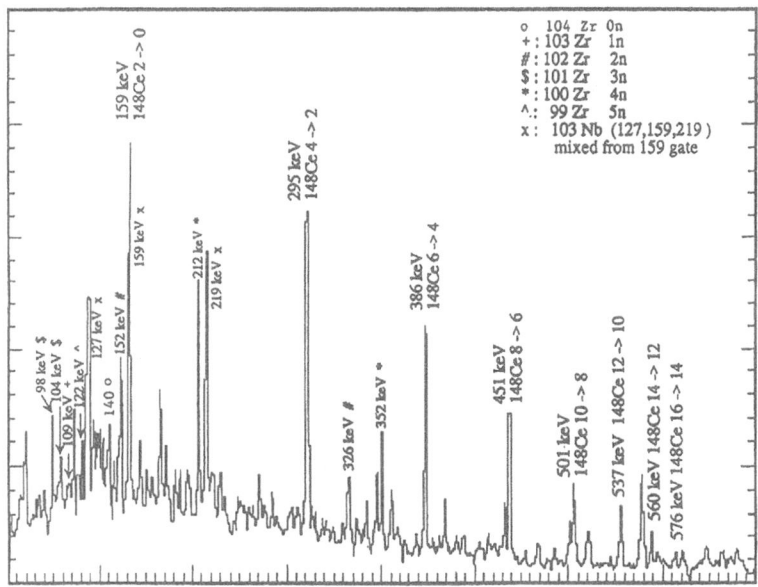

Figure 1. *Gamma rays in coincidence with sum of yrast cascades.*

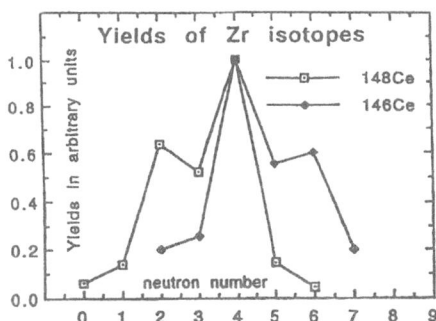

Figure 2. *Yields of Zr isotopes (= Table 1).*

The ^{252}Cf SF data have been analyzed to determine the different neutron emission channels and the relative intensities of these channels. The procedure is to gate on a gamma transition in for example ^{148}Ce and look for the gamma rays associated with the different mass zirconium isotopes. The difference between the masses of the two partners and $A = 252$ gives the number of neutrons emitted. In Fig. 1 is shown the low energy region of the gamma ray spectrum in coincidence with the first six transitions in the yrast cascade in ^{148}Ce. Gamma rays corresponding to transitions in ^{104}Zr down to ^{98}Zr are observed in coincidence with ^{148}Ce. Then gates were set on each zirconium isotope and the transitions in the different cerium isotopes were observed.

To establish that a particular partner pair was being observed, the right transition had to be seen in the other partner when pulling gates from each partner. Table 1 and Fig. 2. show the mass distributions of Zr isotopes obtained in coincidence with ^{146}Ce and ^{148}Ce. These distributions are deduced from the detected intensities of transitions between the

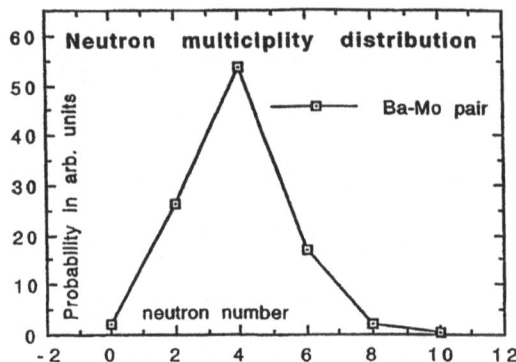

Figure 3. *Relative neutron distribution for Ba-Mo-pairs.*

lowest excited and ground states of the Zr fragments corrected for the known γ-ray energy dependence of the detection efficiency ε (E_γ) of the Close-packed Ge detector Ball. Including the small variation of the γ-ray multiplicity with the mass number of the Zr fragments, we deduced from the data of Table 1 that the most probable Zr mass numbers are shifted by two mass units (from ^{100}Zr to ^{102}Zr) in going from ^{148}Ce to ^{146}Ce, whereas the mean value of the Zr fission fragment mass varies rather slowly, 0.5 mass units between ^{148}Ce and ^{146}Ce. This comes about because of the difference in the intensities of the higher and lower neutron emission channels in the two nuclei as seen in Fig. 2.

Table 2. *Relative yields of the correlated fragment pair masses $Y(A_L, A_H)$ for $Z_L/Z_H = 42/56$. The matrix is normalized to 100.*

	^{138}Ba	^{140}Ba	^{142}Ba	^{144}Ba	^{146}Ba	^{148}Ba	Σ
^{100}Mo			<0.4	<0.4	<0.4	<0.6	
^{102}Mo		<0.4	0.6	2.9	4.7	1.7	9.9
^{104}Mo	0.2(1)	0.9	7.9	26.2	8.2	1.8	45.2
^{106}Mo	0.5	5.3	16.8	12.3	0.1		35.1
^{108}Mo	0.2	4.7	3.0	1.8			9.6
Σ	0.9	10.9	28.3	41.5	13.0	3.5	

Using the same gating procedures as outlined above, a complete set of relative yields of different pairs of fission fragments is presented in Table 2 for the charge division of ^{252}Cf into $Z_L/Z_H = 42/56$ (Mo/Ba). This is the first time such detailed distributions as those given in Tables 1 and 2 have been obtained in a fission study. The last row and column of Table 2 give the sums of the correlated relative independent yields of Mo and Ba isotopes which are the independent yields according to the known evaluated data tables.[7] The distributions presented in Tables 1 and 2 give us a deeper insight into the multiplicity distributions of prompt fission neutrons. Note the ^{106}Mo-^{146}Ba and ^{104}Mo-^{148}Ba pairs have zero neutron emission, and the ^{104}Mo-^{138}Ba pair has 10 neutron emission. Figure 2 shows the neutron multiplicity distributions obtained in coincidence with ^{148}Ce and ^{146}Ce fission fragments, and Fig. 3 shows the neutron multiplicity distribution for the Mo/Ba charge division of the californium nucleus. These are the real distributions deduced directly from our experimental data corrected only for the γ-ray detection efficiency. This is a specific positive feature of our results. Neutron multiplicities obtained previously were deduced from the experimental data via a sophisticated reduction procedure. For the first time we obtained clear data about the long sought fractions of californium spontaneous fission events with zero on to ten prompt neutrons. For the first time the charge asymmetry dependence of the mean number of prompt neutrons can be deduced directly from an experiment. For the 42/56 charge asymmetry we obtained the average neutron multiplicity $< \nu > = 3.85$. Similar data for the various fragments pairs are currently being extracted.

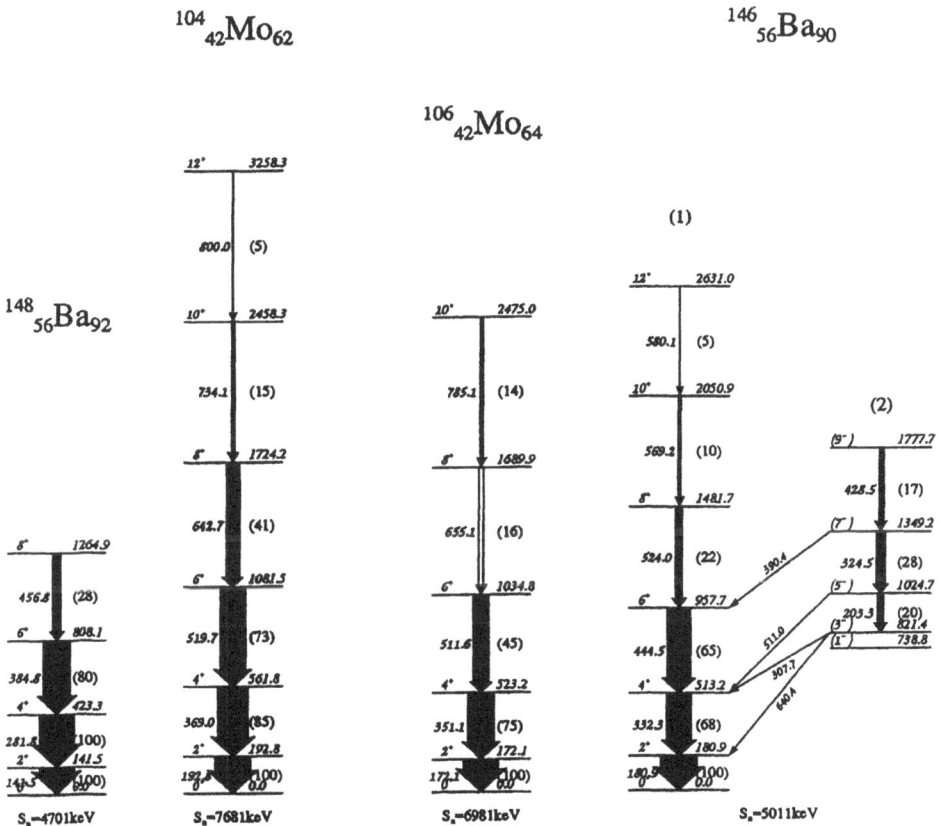

Figure 4. *Levels observed in zero neutron emission (cold fission) correlated pairs* ^{108}*Mo-*^{148}*Ba and* ^{106}*Mo-*^{146}*Ba.*

Three definite zero neutron emission channels have been observed, ^{104}Zr - ^{148}Ce, ^{104}Mo - ^{148}Ba and ^{106}Mo - ^{146}Ba. The energy levels of these nuclei observed in this work are shown in Fig. 4 and 5. The single neutron binding energies are given below each nucleus. From momentum conservation, the kinetic energies are roughly equivalent for each fragment.

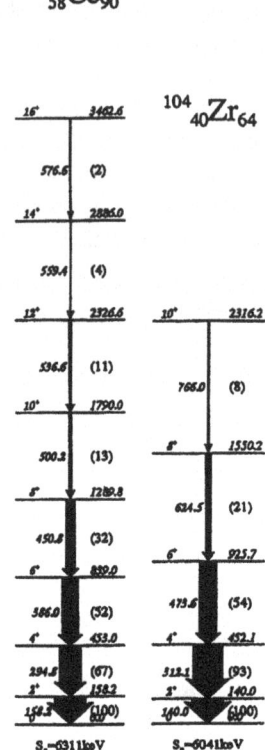

Figure 5. *Levels observed in zero neutron emsission (cold fission) correlated pair* ^{104}Ce-^{106}Ce.

Since in these channels zero neutrons are emitted, the nuclei must be left after fission with less than the neutron binding energies which vary from 4.7 to 7.7 MeV. The large fission energy observed in the 4n up to 10n channels thus must essentially all go into kinetic energy of the fragments. Note in Figs. 4, 5, the maximum excited state energies observed in these nuclei go from 1265 up to 3558 keV, so a small fraction of the fission energy goes into excitation of the low lying excited states in these nuclei with zero neutron emission. Thus these three cases correspond to cold fission or a new type of cluster radioactivity as earlier predicted by the Frankfurt-Bucharest group.[1,2] Earlier in cluster radioactivity, decays to excited states of the heavy fragment were observed. Here, however, we see cluster radioactivity or cold fission where both fragments are left in relatively low energy excited states.

Mean Angular Momentum Distribution

The intensities of the transitions within the yrast cascade, normalized to 100 for the $2^+ \rightarrow 0^+$ transitions, are shown in Table 3 for some complementary fission fragments in the SF of ^{252}Cf, namely Zr-Ce (Zr 40,58) and Mo-Ba (Zr 42,56). From these data on the population of the states as a function of angular momentum, one can extract the average values of the angular momentum $<j>$ for the levels populated in each nucleus. For the first two rows ^{146,148}Ce, the data represent average over all neutron channels to different Zr isotopes

Table 3. Relative γ-ray intensities are given for the transitions in the fragment in column one in coincidence with the $2^+ - 0^+$ transition in the fragment or in a particular partner and the mean angular momenta $<j>$ of the levels populated in the fission for the fragment. One σ errors are in parentheses.

Fission fragment	4^+-2^+	6^+-4^+	8^+-6^+	10^+-8^+	12^+-10^+	14^+-12^+	$<j>$
^{146}Ce[1]	100(5)	41(3)	11(2)	4(2)			5.1(1.2)
^{148}Ce[1]	100(5)	85(5)	58(4)	36(4)	21(2)	7(3)	8.1(1.3)
^{148}Ce[2]	100(5)	78(4)	46(3)	36(3)	20(2)	9(4)	7.8(1.2)
^{148}Ce[3]	100(5)	75(4)	46(3)	31(3)	19(2)	7(3)	7.7(1.2)
^{100}Zr[4]	72(4)	61(3)	21(3)				5.1(0.9)
^{100}Zr[5]	77(4)	70(4)	26(3)				5.5(0.9)
^{102}Zr[4]	78(4)	50(4)	21(3)				5.0(0.9)
^{102}Zr[5]	92(5)	61(4)	24(3)	5(2)			5.6(1.0)
^{144}Ba[6]	92(6)	85(7)	52(6)	15(5)			6.7(1.4)
^{104}Mo[7]	76(6)	48(5)	15(5)	7(3)			4.9(1.5)

[1] Average over Zr isotopes (in this case, the gate is the 2^+-0^+ transition in ^{146,148}Ce, respectively), [2] with ^{100}Zr, [3] with ^{102}Zr, [4] with ^{148}Ce, [6] with ^{104}Mo, [7] with ^{144}Ba, [2-7] these results are derived from two-gates in the triple coincidence spectra (see text).

for the population intensities in these two nuclei. The next two rows give data for ^{148}Ce for specific four and two neutron-out channels to ^{100}Zr and ^{102}Zr, respectively. These data are obtained from the γ-γ-γ data by double gating on the 2-0 transition in ^{100}Zr or ^{102}Zr and the 2-0 transition in ^{148}Ce. The next six rows were then from spectra double gated on the 2-0 transition in each partner. Table 3 gives the angular moments of different specific fission fragments in coincidence with the individual partner nuclei associated with different numbers of emitted neutrons. Our data give a better detailed picture of the correlation of the angular momentum with the mass and Z of the fission fragments compared to other results.[8-11] These data are directly related to the initial fragment excitation or, in other words, to the elongation of the fissioning nucleus and neck of the two fragments at the scission point.

Levels in ^{136}Te and High Spin States in Neutron-rich N=84 Isotones

Nuclei near the doubly magic $^{132}_{52}$Sn$_{82}$ are especially interesting with regard to the coupling of the single particle states to the collective deformation. The light $N = 84$ isotones with two neutrons outside the $N = 82$ closed shell are produced in ^{252}Cf SF. Recently we reported the levels in ^{136}Te together with a new high-spin states in the $N = 84$ isotones, ^{138}Xe and ^{140}Ba.[3] The levels in ^{136}Te were simultaneously reported first by Cizewski et al.[12] At the same time we studied levels in the $^{112-116}$Pd partners.[13]

When ^{252}Cf scissions into a heavy fragment and a light fragment, approximately four neutrons and approximately ten γ rays are emitted on the average per scission. All the transitions of a particular light fragment and those of its complementary heavy partners (and vice-versa) are in coincidence with each other. So, coincidence spectra from gating a transition in a particular nuclide exhibits peaks not only for the transitions in that nuclide, but also for those in its several heavy or light fission partners. In addition there is a high probability that a gate on a particular energy transition will overlay not only the transition of interest, but nearly the same energy transitions (one or more) in different isotopes. These complexities make it difficult to identify unknown transitions in new nuclides in double-coincidence spectra.

To eliminate most of the γ-rays not belonging to a particular isotope of interest, we employed for the first time in SF, the triple-coincidence technique. This allows us to see only the transitions in the two fission partners or emphasize the transitions in one partner. The triple γ-ray coincidence technique has been employed in SF studies to identify uniquely states in ^{136}Te, and also to resolve doublets in ^{138}Xe and ^{140}Ba. Let us illustrate the power of this technique.

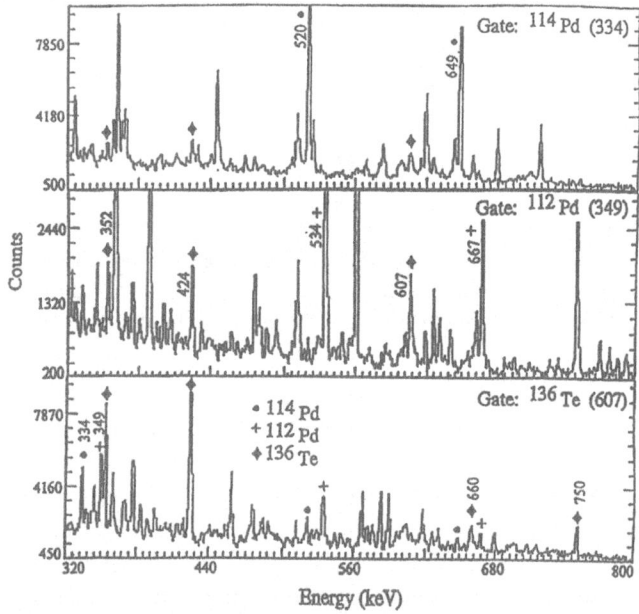

Figure 6. *Spectra in coincidence with a single gamma ray in one nucleus as shown.*

Figure 6 shows three double coincidence spectra. The top spectrum is gated on the $2^+ \to 0^+$ transition in ^{114}Pd, and 2n complement to ^{136}Te. The middle spectrum is gated on the $2^+ \to 0^+$ transition in ^{112}Pd, the 4n complement of ^{136}Te. In both the Pd gates we see other Pd yrast transitions and several peaks that are common in both these gates, these are transitions in Te isotopes. The bottom spectrum is gated on the most intense peak in common with the top and middle spectra, the 607 keV transition. In this gate we see $gamma$-rays of ^{112,114}Pd and candidate γ-rays for transitions in ^{136}Te. However, in all three spectra there are a number of unidentified peaks. Figure 7 includes three triple coincidence spectra. The top two are gated with the 607 keV transition assigned to ^{136}Te and the $2^+ \to 0^+$ transitions of its ^{112,114}Pd complements. The bottom spectrum consists of gates on two of the transitions newly assigned to ^{136}Te. The other transitions in ^{136}Te are now seen as prominent peaks.

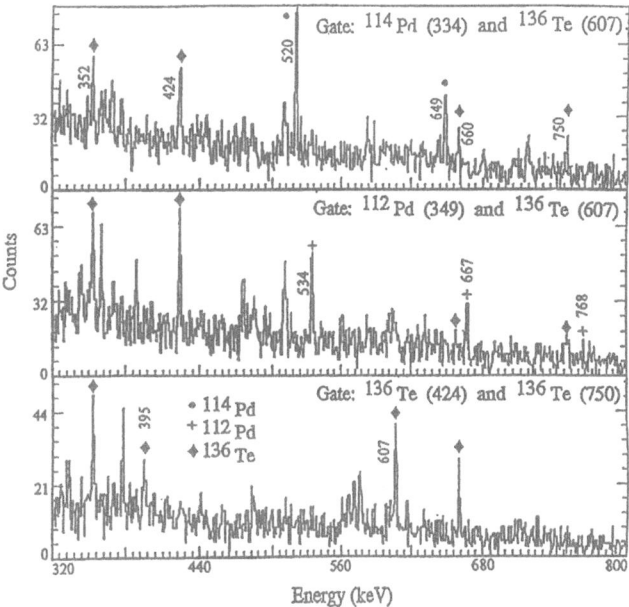

Figure 7. *Spectra in coincidence with double gated gamma rays as shown.*

The same new transitions in ^{136}Te were simultaneously reported first by Cizewski et al.[12] However, in that work, the doublet nature of the 423 keV in the double-coincidence spectra prevented them from obtaining a precise energy for the 4^+ and higher levels in ^{136}Te.

The level schemes of ^{138}Xe and ^{140}Ba, known previously only up to 4^+, were extended to 12^+ and 8^+, respectively. These are shown in Fig. 8. Note that in both ^{138}Xe and ^{140}Ba the $6^+ \to 4^+$ and $4^+ \to 2^+$ transitions are very close doublets in energy. These doublets are established clearly in the respective isotopes with the triple coincidence data. Another interesting feature in the high spin data of ^{136}Te and ^{138}Xe is the energies of the $8^+ \to 6^+$ transitions. These transitions have energies significantly greater than the energies of either the $10^+ \to 8^+$ or the $6^+ \to 4^+$ transitions in these two nuclei. This indicates a definite change in structure between the 6^+ to 8^+ levels in these nuclei.

Band Crossing in Neutron-rich Pd Isotopes

Another example of the detailed physics one can obtain is our observation of new excited states and band crossings in ^{112}Pd, ^{114}Pd and ^{116}Pd[13]. In the previous experiments the levels up to spin $J^\pi = 6^+$ have been reported in these isotopes. Now we have extended the excited states of ^{112}Pd, ^{114}Pd and ^{116}Pd to 10^+, 12^+ and 12^+, respectively.

Shown in Fig. 6 single gamma gates contain peaks not only in coincidence with yrast transitions, but also in coincidence with several γ-rays in complementary Te fragments, as well as with transitions in other isotopes with the overlapping gates. To eliminate most of the γ-rays not associated with the isotopes of interest, we employed the triple-coincidence technique as discussed above. The level schemes of ^{112}Pd, ^{114}Pd, ^{116}Pd, deduced from our data, are shown in Fig. 9 with identified excited states extended to 10^+, 12^+ and 12^+, respectively. We carried out cranked shell model (CMS) calculations to determine whether

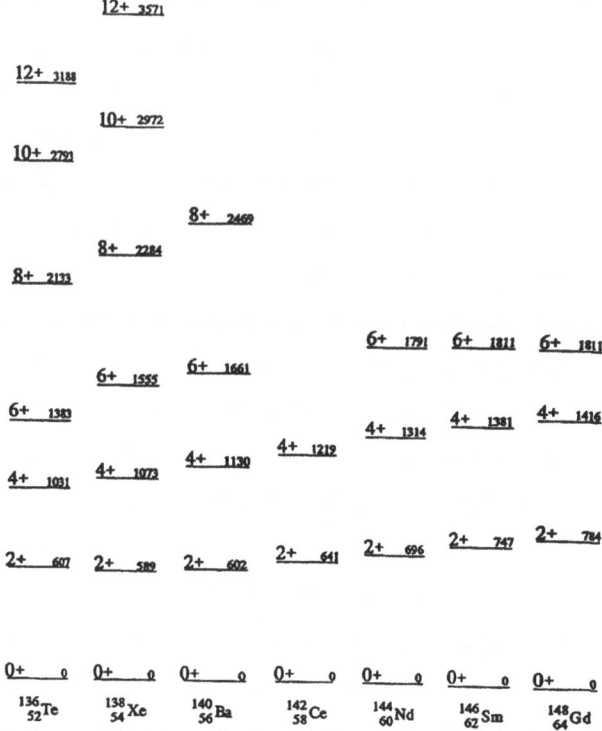

Figure 8. Levels in $n = 84$ isotone. The levels in ^{136}Te and above 4^+ in ^{138}Xe and ^{140}Ba are new.

the proton orbital ($\pi g_{9/2}$) or neutron orbital ($\nu h_{11/2}$) is responsible for band crossings at the observed rotational frequencies. The calculations predict a band crossing related to the alignment of two $g_{9/2}$ protons at $\hbar\omega = 0.35$ MeV which is very close to the experimental values of 0.32 - 0.34 in ^{112,114}Pd. The crossing related to the alignment of two $h_{11/2}$ neutrons is predicted to occur at $\hbar\omega = 0.42$ MeV, which is much higher than the observed values. Hence, we propose that the $g_{9/2}$ proton orbital is responsible for backbending in this region.

Octupole Deformation in Ba Nuclei

From studies of spontaneous fission of ^{252}Cf, Phillips et al[4] found evidence for two intertwining positive and negative parity levels connected by fast E1 transitions in ^{144,146}Ba as shown in Fig. 4 for ^{146}Ba. Their level patterns and E1 transition rates indicated evidence for the onset of octupole deformation with increasing spin.

We have studied the level structure of odd A ^{143}Ba from the spontaneous fission of ^{252}Cf. New level structures are found built on the known $9/2^-$ level at 117 keV as shown in Fig. 10. One cannot help but be struck by the strong similarity of these new intertwined levels in ^{143}Ba and the intertwined positive and negative parity levels in ^{144}Ba. Note ^{142}Ba does not exhibit the simple two intertwined band structure of ^{144}Ba. In ^{143}Ba already by the 1409.3 keV level, tentatively assigned $(19/2^+)$ one sees level spacings characteristics of a band associated with the rotation of an octupole deformed shape. This suggests that the odd neutron is important in helping to drive the octupole deformation in ^{143}Ba.

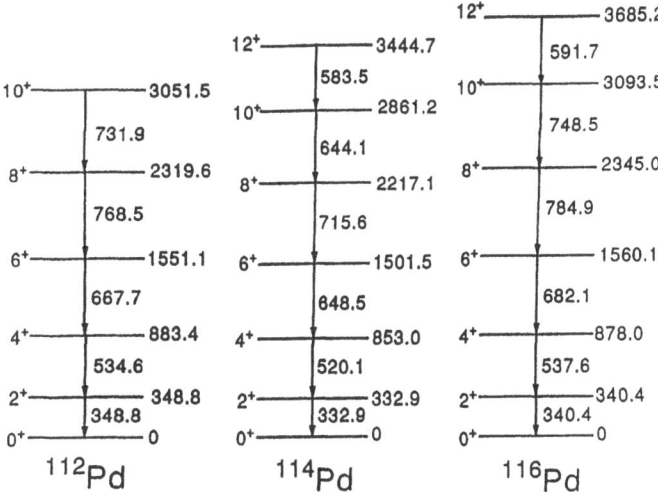

Figure 9. Levels in $^{112-116}$Pb. The levels above 6^+ are new.

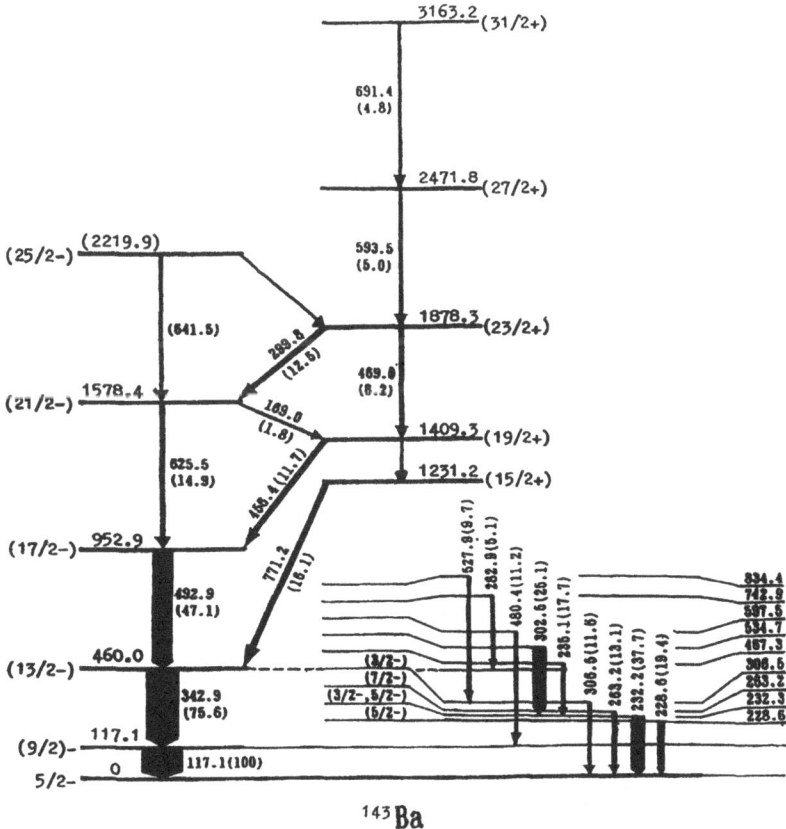

Figure 10. New levels reported above 9/2 - 117.1 keV level that indicate octupole deformations in ^{143}Ba.

Acknowledgement

I. Y. Lee is now at Lawrence Berkeley Laboratory. Work is supported at Vanderbilt University by U.S. Department of Energy under grant No. DE-FG-05-88ER40407, at Idaho by U.S. Department of Energy under contract No. DE-AC07-76ID01570 and at Oak Ridge National Laboratory by Martin Marietta Energy Systems under contract No. DE- AC05840R21400 with the U.S. Department of Energy.

References

[1] W. Greiner, M. Ivascu, D. N. Poenaru, A. Sandulescu, *Treaties on Heavy Ion Science*, Vol. 8, *Nuclei Far From Stability*, edited by D. A. Bromley, (Plenum Press, 1989)

[2] W. Greiner and A. Sandulescu, *Scientific American* **262**, 58 (1990)

[3] K. Butler-Moore, J. H. Hamilton, A. V. Ramayya, S. Shu, X. Zhao, W. C. Ma, J. Kormicki, J. K. Deng, W. B. Gao, J. D. Cole, R. Aryaheinejad, I. Y. Lee, N. R. Johnson, F. K. McGowan, G. Ter-Akopian and Y. Oganessian, *J. Phys. G: Nucl. Part. Phys.* **19**, L121-L126 (1993)

[4] W. R. Phillips et al., *Phys. Rev. Lett.* **57**, 3257 (1986)

[5] S. Zhu, D. B. Wang, Q. H. Lu, D. Shi, J. H. Hamilton, W. C. Ma, A. V. Ramayya, R. S. Babu, G. M. Ter-Akopian, Yu. Ts. Oganesian, J. D. Cole, R. Aryaheinejad, R. C. Greenwood, N. R. Johnson, I. Y. Lee and F. K. McGowan, to be published

[6] A. Sandulescu, D. N. Poenaru and W. Greiner, *Sov. J. Part. Nucl.* **11**(6), 528 (1980)

[7] A. C. Wahl, *Atomic Data and Nuclear Data Tables* **39**, 1 (1988)

[8] J. B. Wilhelmy et al., *Phys. Rev. C* **5**, 204 (1972)

[9] J. R. Leigh et al., *Phys. Lett. B* **159**, 9 (1985)

[10] Y. Abdelrahman et al., *Phys. Lett. B* **199**, 504 (1987)

[11] M. Ogihara et al., *Z. Phys. A* **335**, 203 (1990)

[12] J. Cizewski, M. A. C. Hotchkis, J. L. Durell, J. Copnell, A. S. Mowbray, J. Fitzgerald, W. R. Phillips, L. Ahmad, M. P. Carpenter, R. V. F. Janssens, T. L. Khoo, E. F. Moore, L. R. Morss, Ph. Benet and D. Ye, *Phys. Rev. C* **47**, 1294 (1993)

[13] R. Aryaeinejad, J. D. Cole, R. C. Greenwood, S. S. Harrill, N. P. Lohsteter, K. Butler-Moore, J. H. Hamilton, A. V. Ramayya, X. Zhao, W. C. Ma, J.Kormicki, J. K. Deng, W. B. Gao, I. Y. Lee, N. R. Johnson, F. K. McGowan, G. Ter-Akopian and Yu. Oganessian, *Phys. Rev. C* **48**, 566 (1993)

COLD FISSION

Friedrich Gönnenwein

Physikalisches Institut, Universität Tübingen, D-72076 Tübingen, Germany

Introduction

The energy liberated in a nuclear fission process, the Q-value of the reaction, is shared between the total kinetic energy, TKE, and the total excitation energy, TXE, of the fission fragments. At the very instant of scission, both, TKE and TXE, may be broken down into several contributing terms. At scission, besides a contribution to TKE due to the prescission kinetic energy, the lion's share of the eventual TKE of the fission fragments is still tied up as Coulomb repulsion energy between the fragments. Likewise, only part of the eventual TXE of the fragments is already present at scission as intrinsic excitation, E_X^{Sci}, while another part is still bound as potential energy of deformation. While the fragments are flying apart, the deformation energy V_{Def}, relaxes into intrinsic excitation. Experimentally the total excitation energy TXE is readily found either as the difference between the Q-value and the kinetic energy TKE measured, or from the number and energies of neutrons and gammas emitted from the fragments. In contrast, it is rather difficult to assess precisely the individual contributions to both, TKE or TXE. This is unfortunate since, for example, the intrinsic excitation energy of the fragments at scission should carry information on the damping of the fission mode between the saddle and the scission point due to the viscosity of nuclear matter.

In recent years special attention has been given to those limiting cases of the fission process where the excitation energies of the fragments are close to nil. The most stringent condition in this context is to search for events where the total fragment excitation energy vanishes, or nearly so: $TXE \approx 0$. Under this constraint the fragments have to be born both in a cold and undeformed state ($E_X^{Sci} = 0$ and $V_{Def} = 0$). The phenomenon has indeed been observed and named "cold fission".[1] In view of the compact scission configurations corresponding to this process, a more descriptive denomination is "cold compact fission". However, the above condition for cold fission may be relaxed by imposing merely $E_X^{Sci} \approx 0$, i.e., by requesting the excitation energy of the fragments to vanish at scission only. This more general condition encompasses cold compact fission. A generic name for processes with the constraint $E_X^{Sci} \approx 0$ is "cold scission". These processes should be of physical interest since they are probing the limits of phase space being accessible at scission. It is noteworthy to point out that in cold scission the fragments will in general get excited at infinity in case they are deformed at scission.

Besides cold compact fission, the present review surveys cold scission phenomena having been brought into evidence in recent experimental studies. More specifically, cold deformed, cold mass asymmetric, cold shape asymmetric and cold ternary fission will be discussed by turns.

Frontier Topics in Nuclear Physics, Edited by W. Scheid and A. Sandulescu, Plenum Press, New York, 1994

Cold compact fission

Recalling that, by definition, in cold compact fission the total excitation energy of the fragments is vanishingly small ($TXE \approx 0$), energy conservation tells that the fragments' kinetic energy TKE should come close to the Q-value ($TKE \approx Q$). There is, therefore, a clear-cut experimental signature for this process to occur. Since the Q-value of the fission reaction depends on the mass and charge fragmentation of the parent nucleus, it has to be checked fragment mass by mass, and fragment charge by charge, whether TKE exhausts Q. Several fragment detection schemes fitting this purpose have been developed these last years. A simplifying feature in cold compact fission studies is due to the fact that no neutrons are emitted. Hence, in a binary fission event, it is sufficient to identify one single fragment and to measure its kinetic energy or velocity; the fission reaction is then fully specified by mass-, charge- and momentum conservation. On the other hand, the low count rates in cold fission represent a major difficulty.

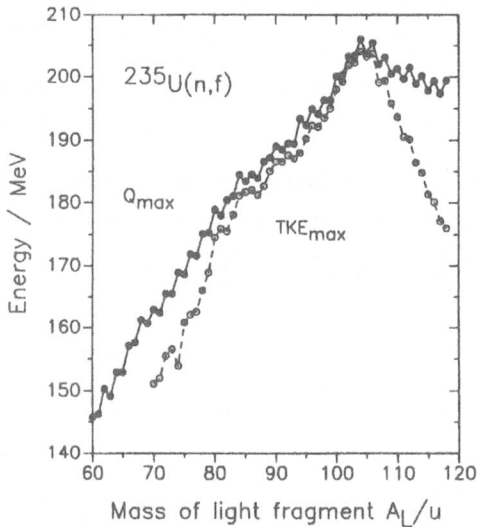

Figure 1. *Maximum total kinetic energy TKE_{max} (open points) and maximum Q-value Q_{max} (full points) as a function of light fragment mass A_L for $^{235}U(n_{th},f)$.*

Without going any further into technical details, a typical experimental result for cold compact fission is shown in Figure 1. For the standard reaction $^{235}U(n,f)$ induced by thermal neutrons, the maximum total kinetic energy TKE_{max} to be observed is plotted as a function of the light fragment mass A_L (open points).[2] The mass range extends from very asymmetric mass splits with $A_L = 70$, corresponding to heavy masses $A_H = 166$, to mass symmetric splits with $A_L = A_H = 118$. For comparison, the maximum Q-values having been calculated from mass tables are given as a function of A_L (full points). For each mass ratio A_L/A_H, the charge ratio Z_L/Z_H maximising the Q-value has been selected. It is evident from the Figure that for a quite broad mass range from $A_L \approx 80$ up to $A_L \approx 108$ the maximum kinetic energy TKE_{max} comes close to the respective Q_{max}-value within a few MeV. In other words, for the above mass range cold compact fission is obtained. Evoking the famous mass asymmetry in low energy fission of the lighter actinides, we note in passing that precisely for light fragment masses from $A = 80$ up to $A = 107$ the mass yields $Y(A_L)$ exceed 0.1%. In contrast to the above findings, in regions of lower mass yields, i.e., for very asymmetric fission and, even more conspicuously, for symmetric fission the kinetic energy release TKE_{max} falls behind the available energy Q_{max}. This means that for fragment masses being not favored in yield also cold compact fission does not show up.

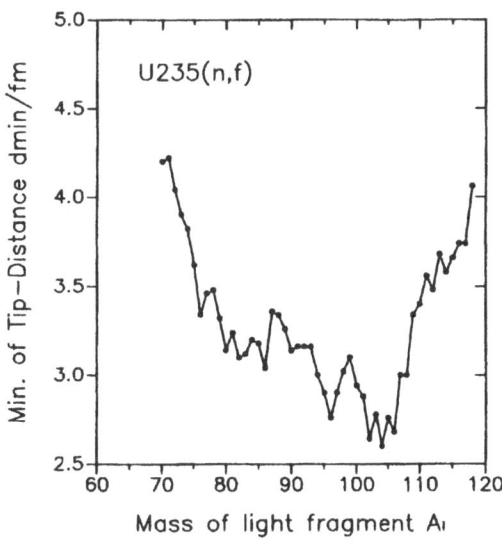

Figure 2. *Minimum tip distances d_{min} as a function of light fragment mass A_L for $^{235}U(n_{th},f)$.*

Several models have been proposed for interpreting the above results for ^{235}U(n,f) and similar results for other fission reactions.[3] Based on the liquid drop model and in close analogy to the theory of cluster radioactivity, some of the models start from the notion that cold fission should be characterized by most compact scission configurations with the fragments being born in their ground states. It has thereby proved crucial to take the ground state deformations of the nascent fragments into account. In a formalism developed by Shi and Swiatecki[4], the potential energy of Coulomb and nuclear interaction between fragments may readily be calculated as a function of the separation distance. The chance for cold fission to occur is then either found from the tunnel probability through the potential barrier as the fragments separate[3] or from the minimum distance between the tips of the fragments facing each other as they emerge from the barrier. These minimum tip distances d_{min} for the reaction ^{235}U(n,f), induced by thermal neutrons, are plotted in Figure 2 as a function of light fragment mass A_L.[5] From the comparison between Figures 1 and 2 it is to be noted that cold fission corresponds to tip distances $d_{min} \approx 3$ fm ($A_L \approx 80$ through $A_L \approx 108$). In view of the leptodermous surface of nuclei, minimum tip distances of about 3 fm appear to be a reasonable guess for a scission configuration. Therefore, the correlations between the actual observation of cold fission and tip distances in the above range are making sense. It should also be noticed from Figure 2 that optimum conditions for cold fission are predicted for masses around $A_L = 104$, as indeed observed in experiment (s. Figure 1). On the other hand, minimum tip distances in excess of about 3 fm can no longer be considered as a valid picture for a scission configuration. In these cases the fragments have necessarily to be deformed compared to their ground state shapes in order to bring down the Coulomb repulsion between fragments, thereby allowing for acceptable tip distances without violating energy conservation. However, being deformed the fragment will pick up deformation energy V_{Def} and, hence, cold fission should be outruled. In fact, again comparing Figures 1 and 2, for too large minimum tip distances in both, very asymmetric and symmetric mass fragmentations, cold fission is not reached. Due to this limitation in phase space being accessible at scission, the total yield for both, very asymmetric and symmetric fission, is furthermore anticipated to be small. This is in full agreement with experiment.

It is intriguing to observe in cold fission studies of an (e,e)-compound nucleus like ^{236}U* in Figure 1 that even and odd mass splits are about equally probable. Before starting the experiments it had been conjectured that the superfluidity of nuclear matter should become observable under optimum conditions in cold fission. In fact, in case fragments are born in their ground states, the chances for nuclear pairing to be preserved in the proton and neutron system should be highest. The conjecture has been disproven most convincingly by Simon et al.[6] Identifying both, masses and charges of fragments in cold fission of ^{236}U* (and some other reactions), it was found that not only odd masses compete with even ones, but that, starting from an (e,e)-compound nucleus, fragmentations into two (o,o)-nuclei come closest in total kinetic energy to the respective Q-value leaving, hence, the nascent fragments with the least total excitation energy $TXE = Q - TKE$.

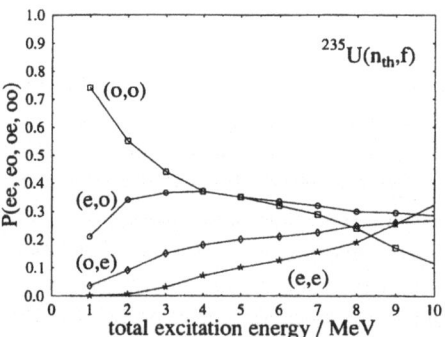

Figure 3. *Relative yields of (o,o)-, (e,o)-, (o,e)- and (e,e)-fragments as a function of total excitation energy for $^{235}U(n_{th},f)$.*

This is demonstrated in Figure 3 where the probability for (o,o)- (e,o)-, (o,e)- and (e,e)-fragments from ^{235}U(n,f) is plotted as a function of the total excitation energy TXE. Surprisingly, at the lowest TXE, even mass fragments with odd proton and odd neutron numbers show up with highest probability. The conclusion to be drawn is that, even in cold fission, nucleon pairing is not fully preserved in the course of the fission process. The relative probabilities in Figure 3 evoke level densities of fragment nuclei to come into play.[7] The argument is often advanced and based on the fact that, most pronounced at low excitation energies, the level density of (o,o)-nuclei is largest, while for odd-mass and especially for (e,e)-nuclei the densities are much smaller. It is, however, not obvious how to put the reasoning with densities of final states into more quantitative terms. One difficulty is that at scission some angular momentum is in general imparted to the fragments and it is not known whether this is still true in cold fission. In any case, provision should be made to correct level density formulas for given angular momentum. A more serious concern is that so far no formalism has been developed to show how in a fission reaction intrinsic nuclear structure interacts with the collective variables describing the fissioning process. The question of nuclear dissipation is closely connected to the above problem.

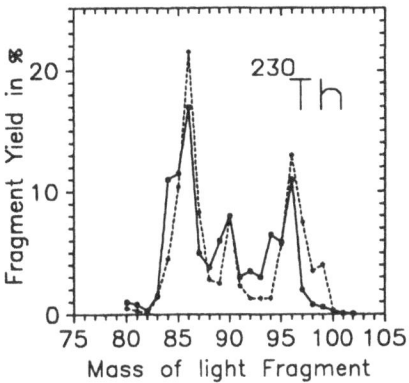

Figure 4. *Fragment yield as a function of light fragment mass for $^{229}T(n_{th},f)$ at a given kinetic energy of light fragment $E_L = 108$ MeV.*

An interesting attempt to reproduce mass distributions in the cold fission regime has recently been suggested by Florescu et al.[8] It is assumed that any (small) excitation energy TXE of the fragments exclusively leads to β-vibrations. The rms-amplitude of the vibration will then directly be linked to the TXE, but also depends on the deformabilities of the fragments. The deformability may be expressed in terms of vibration frequencies and mass parameters. Experimental data for these latter quantities have been collected and fitted to simple expressions taking the ground state deformations of nuclei explicitly into account. In addition, special allowance has been made for odd-mass fragments to be more stretched than even-mass fragments. The idea behind is that nuclei with high level densities should be more easily deformable. It thus appears that arguments similar to those given in the preceding paragraph are invoked here but translated into the language of a collective property. The definite advantage is that now both, excitation energies and nuclear structure effects, are subsumed under a deformation. A standard calculation of tunnel probabilities through the potential barrier having been discussed in the foregoing finally yields mass (and charge) distributions as a function of excitation energy TXE. An example is provided in Figure 4 for the reaction ^{229}Th(n,f) induced by thermal neutrons[9]. Mass distributions having been measured at fixed kinetic energies of the light fragments (corresponding mass-by-mass to varying TXEs) are compared to model calculations. The agreement between theory and experiment is quite good.

The phenomenological models having been discussed are useful because the calculational effort is not very large and, therefore, many details of cold compact fission phenomena may be scrutinized. On the other hand, a very ambitious microscopic model of fission has been put forward by Berger et al.[10] Here, only fission of the compound nucleus ^{240}Pu* could be analyzed. But especially the study of cold fission in the framework of this theory has been rewarding. It was shown that for cold fission to occur a scission barrier in deformation space has to be overcome. The scission barrier separates the so-called fission valley, in which a single deforming nucleus is moving, from the so-called fusion valley for two separated fragments. Cold fission is accounted for in this theory by coupling the elongation mode of the nucleus to a transverse necking-in mode. Vibrational excitations of the necking-in mode are calculated to have sufficient energy to pass over the scission barrier without any tunneling having to intervene. The probability for cold fission is rather well predicted by this mechanism. It is further argued that, at least in cold compact fission, pairbreaking as a prerequisite for (o,o)-fragments to show up in ^{240}Pu fission can only take place at the very instant of scission. Obviously, in several respects the microscopic and phenomenological approaches to cold fission are at variance.

Cold deformed fission

Besides cold compact fission with $TXE \approx 0$, scission point models predict that also cold

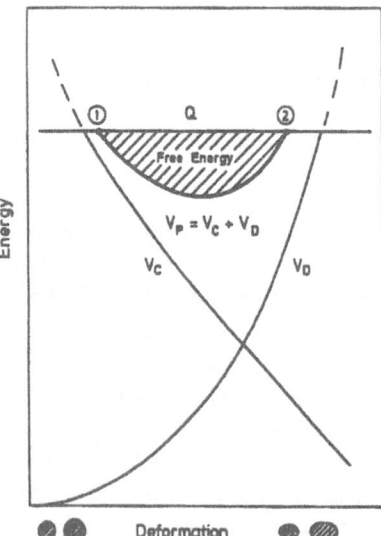

Figure 5. *Scheme of energies coming into play at scission as a function of deformation in a scission point model. For explanation of symbols see text.*

deformed fission should exist.[11] As already outlined in the introduction, a more appropriate name for this process should be "cold deformed scission", since now it is only required that the excitation at scission vanishes: $E_X^{Sci} \approx 0$. The basic idea of scission point models is that all of the characteristic features of fragment distributions may be understood by assuming the collective degrees of freedom to be equilibrated and only weakly coupled to the intrinsic degrees of freedom. The probability for a given fragmentation is then found from a Boltzmann factor, with the collective potential and kinetic energies playing a key role. The potential energies of Coulomb interaction and fragment deformation at scission are calculated in a two-spheroid approximation with the tip distance between the two fragments being kept fixed. These models have proved to be quite successful.

Figure 5 shows schematically the Coulomb energy V_C, the deformation energy V_D and their sum V_P as a function of fragment deformation. For a fixed mass and charge split of the fissioning nucleus, the Q-value is setting constraints on the phase space. Two limiting cases show up, labelled in the Figure by 1 and 2. In both cases all of the available energy is tied up as potential energy with no "free energy" being left to excite the fragments, i.e. $E_X^{Sci} = 0$. Case 1 corresponds to compact scission configurations with large Coulomb repulsion V_C and, hence, large eventual fragment kinetic energies TKE. This is precisely the signature for cold compact fission already discussed in the foregoing. On the other hand, for case 2 the Coulomb interaction V_C and consequently the kinetic energy TKE are small, but the deformation energy V_D of the fragments is large. Fission events from case 2 are, therefore, characterized as being cold at scission ($E_X^{Sci} = 0$) and having the smallest possible kinetic energies TKE. This case is called "cold deformed scission". It should be stressed that ultimately the fragments are heated up when the deformation energy relaxes into intrinsic excitation through damped collective vibrations.

The challenge then is to identify cold deformed scission in experiment, the task being to assess $E_X^{Sci} = 0$ at low kinetic energies of the fragments. Extensive studies of fragment charge distributions have revealed that the even-odd effect in the charge yields may serve as a thermometer for the excitation energy at scission.[12] Quantitatively this even-odd effect is defined as $\delta = (Y_e - Y_o)/(Y_e + Y_o)$, with Y_e and Y_o being the yields of fragments with even and odd charge numbers Z, respectively. For compound nuclei with even Z it is observed that even-Z fragments prevail compared to odd-Z fragments, i.e. δ is generally positive. It tells that, at least to some extent, proton pairing in the fissioning nucleus survives fission and this means that, in the low energy fission processes studied here, dissipation is not very strong. The measured dependence of δ on the excitation energy and the fissility Z^2/A of the compound nucleus on one hand, and on the kinetic energy of the fragments on the other hand, all fit into this interpretation.

The latter dependence is given in Figure 6 for fission of the target nuclei ^{232}U, ^{235}U and ^{239}Pu induced by thermal neutrons.[13] The charge even-odd effect δ (given in percent) is plotted as a function of the kinetic energy of the light fragment. For fragment energies larger than about 90 MeV the δ-effect is seen to increase with kinetic energy. Since for increasing kinetic energy the energy available for excitation is diminishing, pairing is anticipated to have an enhanced chance of survival and, therefore, the observed trend of δ with kinetic energy is readily understood. It should be noted that, upon approaching the regime of cold compact fission at still larger energies than shown in Figure 6, the presentation chosen gets biased since more and more fragmentations are running out of energy. Eventually, the mass-charge

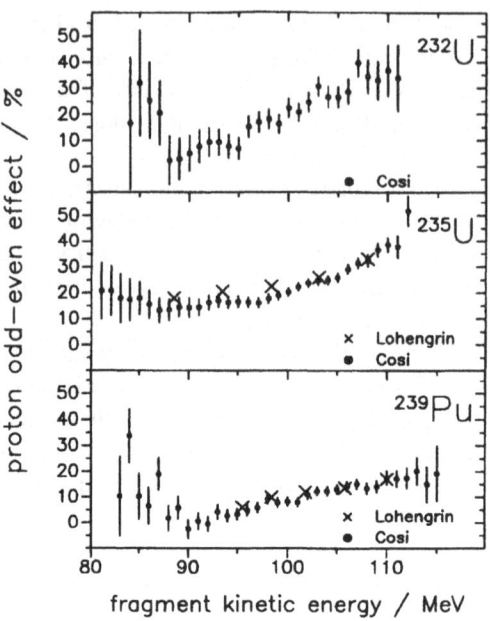

Figure 6. Proton even-odd effect δ_p for $^{232}U(n_{th},f)$, $^{235}U(n_{th},f)$ and $^{239}Pu(n_{th},f)$ as a function of light fragment kinetic energy.

split with the largest Q-value stands out isolated and it makes no longer sense to figure out an even-odd effect. An example of a better suited presentation of data for this limiting case has been given in Figure 4. There the relative yields of even and odd mass-charge divisions are portrayed as a function of total excitation energy, the energy being evaluated for each mass-charge split individually and, hence, treating all fragmentations on the same footing. The preponderance of (o,o)-fragmentations for the very coldest compact fission events in Figure 4 should, however, not be construed as a proof against the notion of superfluidity in fission since, according to Berger et al.[10], even a perfect pairing up to the scission point could be masked by pair-breaking at the instant of scission and, in any case, the marked differences in density of final states at the lowest excitation energies will strongly favor odd compared to even fragmentations.

Coming back to the issue on cold deformed scission, Figure 6 exhibits a striking increase of the even-odd effect for decreasing fragment energies below about 90 MeV. This reversal of the trend of δ vs. kinetic energy is taken as a strong indication that for very low TKE events the excitation energy E_X^{Sci} at scission is getting smaller, which is just the fingerprint for cold deformed scission. Even-odd effects as a function of fragment kinetic energy have been analyzed in the framework of a scission point model by Märten.[14]. At least qualitatively the pattern of Figure 6 is reproduced by these calculations.

Cold mass-asymmetric fission

Discussing cold compact fission it has been argued that the mass ranges of fragments, where cold fission comes into reach, roughly coincide with mass ranges of sizeable yield. In the spirit of scission point models, a decrease in yield has to be traced to a narrowing of the

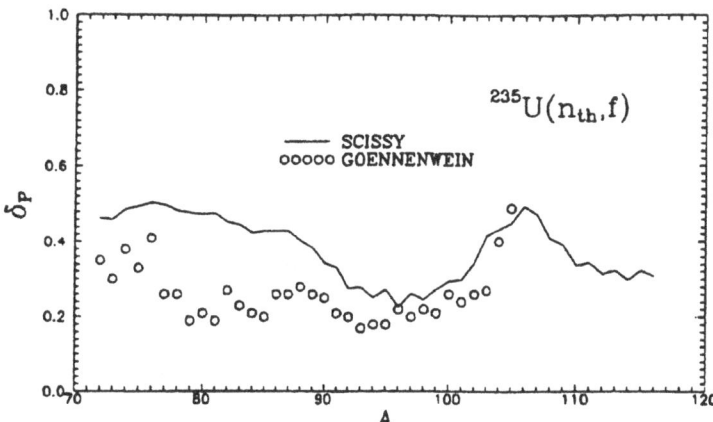

Figure 7. Proton even-odd effect δ_p as a function of light fragment mass A for $^{235}U(n_{th},f)$.

available phase space at scission. The reasoning may be visualized in Figure 5 by imagining a decrease in the Q-value of the reaction, thereby squeezing the hatched area labelled "free energy" to a smaller size. For very asymmetric mass divisions with, say, light masses $A_L \lesssim 80$ and heavy masses $A_H \gtrsim 156$ in $^{235}U(n,f)$ the Q-values indeed drop dramatically (s. Figure 1). The hatched zone in Figure 5 should correspondingly become rather shallow and fission should be cold at scission. Again making use of the charge even-odd effect for probing the temperature of nuclear matter at scission, the δ-effect is expected to rise for very mass-asymmetric fission.

The so-called local even-odd effect in the charge yields is plotted in Figure 7 as a function of light fragment mass, again for the reaction $^{235}U(n,f)$ with thermal neutrons. The open points stem from an evaluation[15] of several studies of charge distributions in $^{236}U^*$-fission.[16,17] Notably in the work of Sida et al.[17] charge measurements have been extended into mass regions with very low yield. The expected increase of the δ-effect for very asymmetric fission is seen in Figure 7 to be indeed observed in experiment. Meanwhile this phenomenon of "cold mass asymmetric fission" (or more precisely scission) has been corroborated for other fissioning systems.[18]

A similar increase in δ as the one just discussed shows up in Figure 7 for light fragment masses around $A_L = 105$. Although, when moving to larger masses towards symmetric mass divisions, also here the mass yield drops significantly, arguments as those just outlined should be put forward more cautiously. For $A_L = 105$ the complementary heavy fragment has a high probability to carry the exceptionally stable charge number $Z_H = 50$ and this shell effect could contribute to the rise of the even-odd effect. It is further apparent from Figure 7 that charge data for symmetric fission with $A_L \approx A_H \approx 118$ are conspicuously absent. This is due to the poor charge resolution of experimental techniques in the mass range in question. Computational results by Märten[14] for the local even-odd effect in a scission point model are given in Figure 7 as a solid line. Though the agreement between experiment and theory is not everywhere perfect, the close resemblance in the overall shape of the plot δ vs. A_L has to be considered as a success. The lack of data for symmetric fission is regrettable, particularly because theory predicts quite low δ-effects in that mass region. In case small even-odd effects were to be confirmed by actual measurements, this should give credit to the view that symmetric fission belongs to a mode being distinct from asymmetric fission. Symmetric fission is conjectured to pass over a separate saddlepoint and to exhibit exceptionally long descent paths from saddle to scission. The intrinsic excitation accumulated in the descent is the reason for the small even-odd effect calculated in theory.

Cold shape asymmetric fission

In the schematic illustration of scission point models in Figure 5, potential energies being relevant at scission are given as a function of deformation of the fragments. The deformation should be understood here as an average over the sum of the deformations of the two complementary fragments, the deformations being expressed in any suitable parametrization of nuclear shapes. Actually, the deformation energy has to be determined for given pairs of deformation for the individual fragments. In a standard fission reaction (i.e. away from both, cold compact and deformed scission) the total deformation may be shared in different amounts between the two fragments. In the limiting case of an unbalanced share, one fragment stays virtually undeformed while the other fragment takes on all of the deformation, up to a maximum set by the available energy Q for the sum of Coulomb and deformation energy. In other words, again a limiting scission configuration is attained where the intrinsic excitation energy of fragments at scission E_X^{Sci} has to vanish. Since under these conditions the shapes of the two fragments will be disparate, the name "cold shape asymmetric fission" (or better scission) has been given to this type of fission.

The question then is how to pin down cold shape asymmetric scission in experiment. Since the asymmetry in shape entails a corresponding asymmetry in deformation energy, and since eventually the deformation energy is converted into intrinsic excitation energy of the fragments, the asymmetry shows up in the neutron emission numbers from the two fragments. In experiment, in addition to the fragment characteristics, the numbers of evaporated neutrons have therefore to be measured. Two studies of this kind have been performed.[19,20] Though charges were not measured whence to deduce an even-odd effect signalling cold scission, fine structures in the fragment mass distributions were observed for asymmetric partitions of neutron numbers. The structures in mass yields are about 5 mass units apart and traced to an underlying even-odd effect in charge yield (note that in heavy nuclei the mass/charge ratio is $A/Z \approx 5/2$). This is a strong indication for cold shape asymmetric scission to exist as a limiting case of fission. Very recently Märten[21] has calculated the expected size of the charge even-odd effect in a scission point model. The effect is present and supports from theory the interpretation given to the experimental findings.

Cold ternary fission

In ternary fission, besides two main fission fragments, a third and usually much lighter charged particle is emitted. The ratio T/B for ternary/binary yield is small and typically $T/B \approx 1/400$. In the majority of cases (about 90%) the third particle is an α-particle. All hydrogen and helium isotopes taken together account for 99% of the ternary yield. In the angular correlation between the three outgoing particles, there is a pronounced structure with most of the light particles being focused into an equatorial plane perpendicular to the flight directions of the main fragments. For these ternary particles the distributions of kinetic energy are Gaussian in shape and, most noteworthy, the distributions appear to extend down to zero energy. The characteristics of both, the angular correlations and the kinetic energy distributions, point to an emission mechanism where the ternary particles are born in between the two main fragments. The joint Coulomb repulsion by the two fission fragments will eject the third light particle perpendicularly to the fragments' flight paths while, in extreme cases, the Coulomb repulsion by the fragments can balance each other and leave the ternary particle with acceleration zero. Intuitively, ternary particles may then be visualized as clusters being formed out of nucleons from the neck in a stretched scission configuration. This picture is corroborated in several approaches to the theory of ternary fission.

Compared to binary fission, the kinetic energy release of the two main fragments is on average some 12 to 14 MeV smaller in ternary fission.[22] However, the decrease is more than compensated if the energy of the ternary particle is included in the total kinetic energy TKE. For α-particles representing the bulk of ternary clusters the mean kinetic energy is 16 MeV, rather independent from the fission reaction considered. On the other hand, the total available energy is calculated from mass tables to be smaller for ternary than for binary fission. In the case of α-emission the difference in Q-value is $(Q_{bin} - Q_{ter}) \approx 5$ MeV. The total excitation energy TXE is thus seen from the difference $(Q - TKE)$ to fall behind by almost 10 MeV when contrasting ternary with binary fission. The experimental observation that, on average, ternary fission is colder than binary fission may be understood by reckoning that in a ternary fission event part of the deformation and intrinsic excitation energy close to scission is consumed by the energy it costs to create a third particle and to place it in the Coulomb field of the two remaining fission fragments. Ternary fission therefore appears to be a propitious candidate for a further instance of cold fission. The limiting case of vanishing total excitation energy TXE may be conjectured to be approached here by searching for unusually heavy ternary clusters with large energy costs draining all of the TXE.

A first experiment where this question could be studied has been performed on the mass separator Lohengrin of the Institut Laue-Langevin in Grenoble.[22] The reaction ^{242}Am(n,f) induced by thermal neutrons was investigated. For the masses of the reaction products it was required that they should be heavier than ^{4}He and lighter than the lightest fission fragments ($A \approx 60$). This should be the mass window of interest for ternary clusters. The separator allows to identify unambiguously the masses and charges of decay products from the fission reaction and to measure their energy distributions. It is, however, not possible to take coincidences between decay products. It had, therefore, first to be verified that the kinetic energy distribution of clusters are Gaussians, with energies ranging to zero as witnessed in correlation studies for ternary particles. For the heaviest clusters with poor statistics of detection it could only be checked that the data are compatible with a Gaussian energy distribution. All cluster events having been observed so far comply with this condition and are attributed to ternary fission.

Of course, the kinetic energy distributions were also scanned for binary decays of ^{243}Am* with one of the decay products coming in the mass range set for the clusters. The decay would correspond to an extremely asymmetric fission event closely akin to cluster radioactivity.[3] In the spontaneous decay mode the two particles are known to end up in their respective ground states, with only few exceptions to the rule. Cluster radioactivity is thus a perfect example of cold fission. If similar conditions were to hold in induced cluster decay studied here, monoenergetic clusters should come into sight at much larger energies than being typical for ternary particles. No such events were seen up to the limit of detection corresponding to a cross section of about 10 nb.

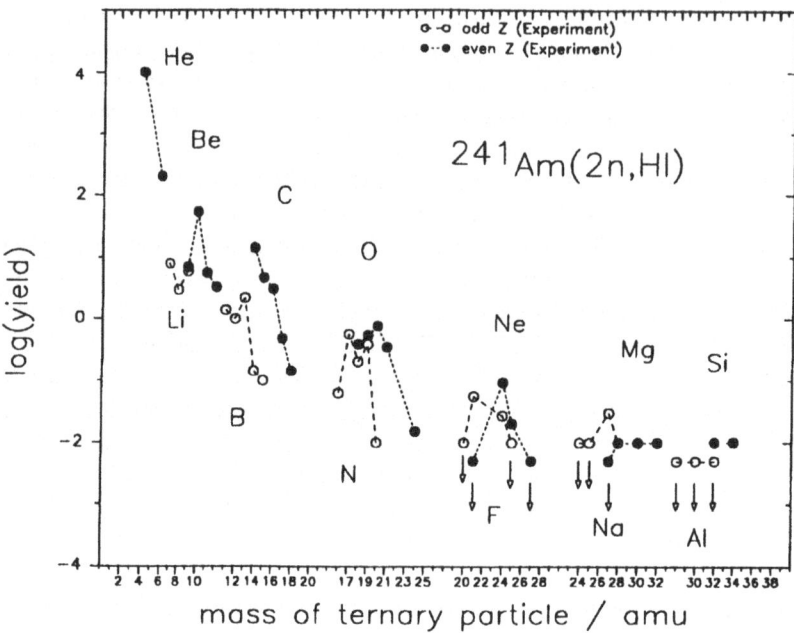

Figure 8. Yields of ternary particles as a function of mass for ^{242}Am(n_{th}, f). Full (open) points: even (odd) Z elements. Data points labelled by an arrow are upper bounds.

Cluster yields integrated over energy are on display in Figure 8 for the reaction ^{242}Am(n,f). Yields for the isotopes of carbon and heavier elements are from the present study, while the yields for the lighter clusters have been taken from the literature.[21,23] All yields have been normalized to the ^{4}He-yield $Y(^4\text{He}) = 10^4$. Ternary emission of clusters with surprisingly large masses and charges has been disclosed, the heaviest isotope having been seen so far being ^{34}Si. The yield decreases quite rapidly with increasing charge and mass of the clusters. Relative to ^{4}He the yield for ^{34}Si is down by 6 orders of magnitude. Roughly, this corresponds to one ^{34}Si-cluster per 3×10^8 fission events. There is in addition a pronounced even-odd effect in the yields, clusters with even charges being favored as compared to odd charges. A similar observation holds concerning yields for even and odd neutron numbers. For any element only few isotopes if not a single isotope, carry the main contribution to the yield. In the list of these isotopes viz. ^{14}C, ^{17}N, ^{20}O, ^{21}F, ^{24}Ne, ^{27}Na, ^{28,30,32}Mg and ^{32,34}Si, it is amazing to recognize precisely those nuclei also seen in cluster radioactivity. In cluster radioactivity it is a good approximation to guess that, in the decay of a parent nucleus heavier than ^{208}Pb, the cluster isotopes will be those leaving the stable ^{208}Pb as the complementary daughter nucleus. Evidently, this reasoning does not apply to the present case of ternary cluster emission. Nevertheless, also here it is apparent from the systematics of Q-values for ternary fission that the isotopes of an element being created with high yields are just those characterized by large Q-values. By the way, the data presented in Figure 8 are far from being complete. For example, the isotope ^{23}F is still missing though it is expected to have a measurable yield; it has not yet been searched for.

As a last step it remains to be shown that especially for heavy cluster emission the ternary fission process is indeed cold. To this purpose ideas having been put forward many years

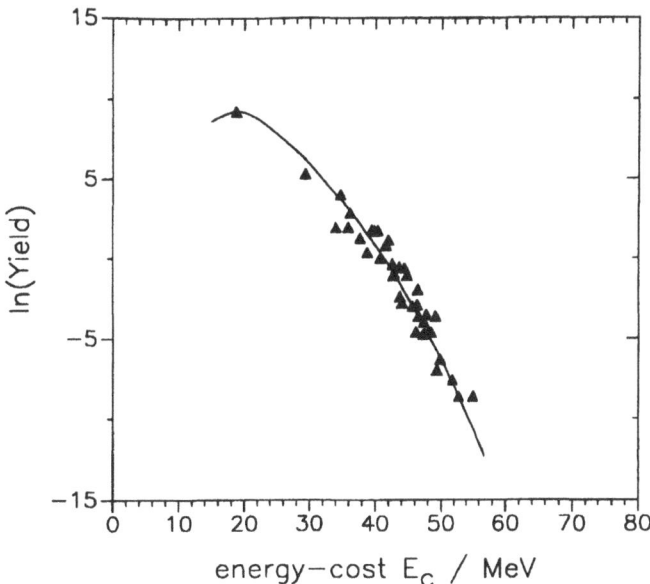

Figure 9. *Logarithm of ternary yields from experiment as a function energy costs from a model calculation.*

ago by Halpern[24] have been reconsidered. Halpern established a correlation between the yield of ternary particles and the energy necessary to generate these particles at scission. More specifically it is assumed that ternary fission is a sequential process in the sense that, first, a binary fragmentation develops and, second, a ternary cluster is set free from one of the fragments while all three particles are still close to each other. A first energy term contributing is then the binding energy of the cluster to one of the primary fragments. The energy is readily found from mass tables for any given fragmentation. In keeping with the above view of ternary fission, the binding energy was averaged over the binary mass and charge distribution. A second contribution to the energy costs for the cluster is the increase in Coulomb energy when the third particle is placed in between the two main fragments. As already pointed out, the choice for this scission configuration is stipulated by the angular correlations observed in experiment. A free parameter of the model is the overall elongation of the configuration. The parameter was varied in a range which was judged to be meaningful in physical terms.

Without going into computational details, a typical result from the model outlined is given in Figure 9. Plotting the natural logarithm of the measured cluster yields for $Z \geq 2$ from Figure 8 as a function of the calculated energy costs, a quite convincing correlation comes into view. The lowest energy costs correspond to ^{4}He with the largest yield, while the highest energy costs tally with clusters where only an upper bound for the yield could be determined. Within reasonable bounds of the above elongation parameter of the scission configuration the maximum energy costs for the heaviest detected clusters always turn out to be around 50 MeV. Usually the correlation in Figure 9 is approximated by a straight line.[24] However, a slightly better fit to the logarithmic plot is obtained with an inverted parabola. These characteristics of the energy costs have to be contrasted with those for the total excitation energy TXE. Excitation energies for the ^{242}Am(n,f) reaction with thermal neutrons are not precisely known, but the distribution is a Gaussian and a reasonable estimate seems to be $\overline{TXE} \approx 25$ MeV for the average total excitation energy and $\sigma_{TXE} \approx 8$ MeV for the standard deviation. The comparison tells that the generation of the heaviest clusters with energy costs of more than 50 MeV require an energy amount in excess of $(\overline{TXE} + 3\sigma)$ to be drained from the total excitation energy. This is certainly a limiting case where all three particles have to be born close to cold and undeformed states. Therefore, ternary fission with heavy third particles appears to be a further example of cold fission.

Conclusion

The present survey discussing the experimental evidence for cold fission and cold scission phenomena has brought into focus a variety of situations where the fission process explores the limits of phase space. The common feature of all these limiting cases is that the intrinsic excitation of nascent fragments at scission is vanishingly small. It should, indeed, be anticipated that, as a rule, fission invades all phase space. Starting from this conjecture, the occurrence of cold scission does not come as a surprise. The surprise is that, in spite of the necessarily small probabilities, the existence of cold scission could be disclosed in experiment. Yet, there is more behind the investigation of cold scission than the satisfaction of having overcome technical difficulties. The study of these processes has been triggered by the discovery of cold compact fission as a special case of cold scission. There, a quite unexpected feature has been revealed, viz. the inversion of the common even-odd effect. The phenomenon has so far not found a generally accepted interpretation from theory. The investigation of the other instances of cold scission has not yet been pushed to the same level of accuracy and it remains a challenge to search also here for possibly new physics in future experiments. Anyhow, already at the present stage, the study of cold scission events has been rewarding by imposing experimentally proven conditions at the limits of observation for any theory of fission. So far only scission point models have met these conditions. A particular example is provided by the measurement of heavy cluster yields in ternary fission. Here it will be interesting to see whether still heavier clusters meeting the energy requirements of phase space limit are to be observed, or whether other physical constraints are cutting off the yields.

The present work has been supported by the Federal Ministry for Science and Technology, BMFT Bonn, under contract no. 06 TÜ 656.

References

[1] C. Signarbieux, M. Montoya, M. Ribrag, C. Mazur, C. Guet, P. Perrin and M. Maurel, *J. Physique Lett.* **42**, L437 (1981)
[2] C. Signarbieux, private communication
[3] A. Sandulescu and W. Greiner, *Rep. Progr. Phys.* **55**, 1423 (1992)
[4] Y.-J. Shi and W. J. Swiatecki, *Nucl. Phys.* A **464**, 205 (1987)
[5] F. Gönnenwein and B. Börsig, *Nucl. Phys.* A **530**, 27 (1991)
[6] G. Simon, J. Trochon and C. Signarbieux, Proc. Conf. "50 Years with Nuclear Fission", Gaithersburg, USA, 1989, *Amer. Nucl. Soc.* Vol. I, p. 313
[7] F. Gönnenwein, Proc. "Seminar on Fission", Pont d'Oye, Belgium 1986, edited by C. Wagemans, *SCK/CEN Mol BLG 586*, p. 106
[8] A. Florescu, A. Sandulescu, C. Cioaca and W. Greiner, *J. Phys.* G **19**, 669 (1993)

[9] N. Boucheneb, P. Geltenbort, M. Asghar, G. Barreau, T. P. Doan, F. Gönnenwein, B. Leroux, A. Oed and A. Sicre, *Nucl. Phys. A* **502**, 261c (1989)
[10] J. F. Berger, M. Girod and D. Gogny, *Nucl. Phys. A* **502**, 85c (1989)
[11] R. W. Hasse, *GSI Report* 87-24, Darmstadt, Germany, (1987)
[12] F. Gönnenwein in: "The Nuclear Fission Process", edited by C. Wagemans, (CRC Press, USA, 1991)
[13] F. Gönnenwein, J. Kaufmann, W. Mollenkopf, P. Geltenbort and A. Oed, *Ricerca Scient. ed Educacione Perm., Suppl.* **84**, 338 (1991)
[14] H. Märten, Proc. "Seminar on Fission", Pont d'Oye II, Belgium 1991, edited by C. Wagemans, IRMM Geel Belgium, 1991, p. 15
[15] F. Gönnenwein, *Nucl. Instr. Meth. A* **316**, 405 (1992)
[16] W. Lang, H.-G. Clerc, H. Wohlfarth, H. Schrader and K.-H. Schmidt, *Nucl. Phys. A* **345**, 34 (1980)
[17] J. L. Sida, P. Armbruster, M. Bernas, J. P. Bocquet, R. Brissot and H.R. Faust, *Nucl. Phys. A* **502**, 233c (1989)
[18] R. Hentzschel, H. R. Faust, H. O. Denschlag, B. D. Wilkins and J. Gindler, *Nucl. Phys. A*, to be published
[19] I. D. Alkhazov, A. V. Kuznetsov and V. I. Shpakov, Proc. Conf. "Fiftieth Anniversary of Nuclear Fission", Leningrad 1989, (Nova Science, Commack, N.Y., 1994)
[20] I. Düring, M. Adler, H. Märten, A. Ruben, B. Cramer and U. Jahnke, Proc. Conf. "Dynamical Aspects of Nuclear Fission", Smolenice, Slovakia 1993, Dubna 1994, to be published
[21] H. Märten, Proc. Conf. "Dynamical Aspects of Nuclear Fission", Smolenice, Slovakia 1993, Dubna 1994, to be published
[22] F. Gönnenwein, B. Börsig, U. Nast-Linke, S. Neumaier, M. Mutterer, J.P. Theobald, H. Faust and P. Geltenbort, *Inst. Phys. Conf. Ser. No.* **132**, 453 (1993)
[23] A. A. Vorobev, V. T. Grachev, I. A. Kondurov, Yu. A. Miroshnichenko, A. M. Nikitin, D. M. Seliverstov and N. N. Smirnov, *Sov. J. Nucl. Phys.* **20**, 461 (1974)
[24] I. Halpern, *Ann. Rev. Nucl. Science* **21**, 245 (1971)

THEORY OF SUPER-ASYMMETRIC COLD FISSION AND CLUSTER-DECAY

Raj K. Gupta[1]

Physics Department, Panjab University, Chandigarh-160014, India (Permanent address)
and
Institut für Theoretische Physik, Justus-Liebig-Universität, Gießen, Germany

1. Introduction

As early as in 1975, Săndulescu, Gupta, Scheid and Greiner[1-5] at Frankfurt showed that the fission or fusion valleys arise due to shell stabilization effects. This result was the key basis for the prediction of cluster radioactivity theoretically by Săndulescu, Poenaru and Greiner[6] in 1980, which was later confirmed experimentally in 1984 by Rose and Jones[7]. In other words, cluster radioactivity is not an isolated phenomenon but is very closely related to cold fission and cold fusion, which are also established experimentally[8,9].

Another related question, that was indicated already in the early calculations[1-5], but is recently studied in detail by Gupta et. al[10,11], is the role of deformed closed shells. The deformed shell stabilization effects are predicted to exist for nuclei with Z=N=38 and in the neighbourhood of Z=50 and 82 spherical- and N=108 deformed -shells. Specifically, $^{186}_{80}Hg_{106}$ is predicted to decay via ^8Be, with a half-life time $T_{\frac{1}{2}} \sim 10^{28}s$, leaving behind a deformed closed- or nearly closed-shell daughter ^{178}Os. However, no experimental signatures exist at present for a deformed daughter cluster radioactivity.

In the following section, some observations are made from the experimental data. A brief look into the interpretation of Gamow theory for α-decay (the Geiger-Nuttall law) shows a kind of analogy between the two processes of α-decay and exotic cluster-decay. In particular, it demands an introduction of the concept of cluster preformation or the like. For more details, we refer the reader to a recent review by Gupta and Greiner[12].

2. Experimental Data – Some Observations

^{34}Si is the heaviest cluster observed so far, which is far from the lightest spontaneous fission fragment ($A_{SF} \sim 80$) or the cold fission fragment ($A_{CF} \sim 60$) measured to-date. Thus, we still do not know: where does the cluster decay stops or the fission process begins? It is possible that the two processes overlap for some range of cluster masses. We shall see in the following that cold fission (CF) studies of Kumar, Gupta and Scheid[13] predict an overlap of two processes (cluster-decay and cold fission) for cluster masses $A_2 > 42$. Cold fission means the fission products with maximum kinetic energy.

Spontaneous fission (SF) measurements, made along with heavy cluster (c) decay, give the decay constant $\lambda_{SF} \sim 10^2 \lambda_c$. Here, SF means either the most probable symmetric or

---

[1] UGC National Fellow

nearly-symmetric fission fragments or the fission fragments averaged over all masses and kinetic energies. We present here the (cold) fission calculations[13] that are made for the first time in the super-asymmetric mass region of cluster radioactivity. No experimental data are available, except for the one presented at this School for the first time by Gönnenwein.

2.1 The Geiger-Nuttall Law

Analogous to the fine-structure of α-decay, the fine-structure of ^{14}C radioactivity from Ra nuclei is also observed. In view of this analogy, the Gamow theory of α-decay is applied to the observed data on the exotic cluster-decay. However, the resulting Geiger-Nuttall (G.N.) plots ($log_{10}T_{\frac{1}{2}}(s)$ vs. $ln P$, where $T_{\frac{1}{2}}$ are the measured decay half-life times and P, the calculated Coulomb barrier penetrabilities) present some striking differences with the ones for α-decay. The G.N. plots for α-decay are for each element and have exactly the same slope. On the other hand, the same for exotic cluster-decays are for each cluster, having somewhat different slopes and, of-course, different intercepts. This means that the equation of straight line (the G.N. law) for each cluster is different and we associate this difference to their having different preformation factors P_0. Thus, for heavy cluster-decays, the Gamow α-decay constant ($\lambda_\alpha^G = \nu_0 P$) gets modified as

$$\lambda_c = \nu_0 P P_0 \qquad (\lambda = \frac{ln 2}{T_{1/2}}), \tag{1}$$

with $P_0 = 1$ for α-decay. Here, ν_0 is the barrier assault frequency which is practically constant for all cluster-decays, including the α-decay. Thus, if ν_0 is considered[14] as the frequency of existence of α-particles at the barrier, then $\nu_0 P_0$ is the frequency of occurrence for heavy-clusters. Thus, different P_0-values identify different clusters in Gamow theory.

3. Cluster-Decay Theories

The available theories can be classified as the unified "fission" models (UFM) and the preformed-cluster models (PCM). The UFM's [15-20] use the more realistic nuclear potentials, instead of the simple square-well, but still define the decay constant, as in Gamow theory,

$$\lambda_c^{UFM} = \nu_0 P. \tag{2}$$

Though some of these models are the fission models[15,16], for heavy cluster decays they all do not use the fission process explicitly, but are used to explain the spontaneous fission as well as the α- and heavy cluster-decays by simply varying the parameters of the nuclear potential used. Thus, UFM do not distinguish between cluster-decays and SF.

The other class of models, called PCM, uses not only the more realistic nuclear potentials but also associates with it the concept of cluster preformation probability P_0. Thus, the decay constant in PCM is defined, as in eq. (1),

$$\lambda_c^{PCM}(= \lambda_c) = \nu_0 P P_0. \tag{3}$$

Since the concept of pre-formation of clusters in parent nuclei demands the inclusion of nuclear structure information, two approaches have been used. These are the works of Blendowske, Fliessbach and Walliser[21] (and later of Silisteanu-Ivascu[22], and Kadmensky et.

al[23]) using the shell model basis, and that of Malik and Gupta[24] (and later of Rubchenya et. al[25]) using the collective model basis. Thus, the PCM treat the α- and heavy-cluster decays alike, but distinguish them from SF.

Then, the saddle- or scission-point fission models[26,27] consider the parent nuclei to deform continuously and arrive at the saddle or scission configuration. These are the true fission models, whose use can, of-course, be extended to the super-asymmetric mass region of cluster radioactivity. This means treating the cluster radioactivity as a fission process.

In the following, we confine ourselves to the PCM of Malik and Gupta[24], using the collective model basis, and the saddle-point fission (SPF) model of Gupta et. al[26].

4. Preformed Cluster Model using Collective Model Basis

Malik and Gupta[24] considered a coupled motion in dynamical collective coordinates of mass-asymmetry $\eta(=\frac{A_1-A_2}{A})$ and relative separation R and solved the stationary Schrödinger equation (the charge-asymmetry is fixed by minimizing the potential in this coordinate)

$$H(\eta, R)\psi(\eta, R) = E\psi(\eta, R), \tag{4}$$

with the Hamiltonian constructed as

$$H(\eta, R) = V(\eta) + V(R) + V(\eta, R) + \frac{1}{2}B_{\eta\eta}\dot{\eta}^2 + \frac{1}{2}B_{RR}\dot{R}^2 + \vec{B}_{R\eta}\dot{R}\dot{\eta}. \tag{5}$$

For the collective potentials calculated in the Strutinsky method ($V = V_{LDM} + \delta U$; δU are the shell corrections obtained from the two-centre shell model (TCSM) energies renormalized to an appropriate liquid-drop model (LDM)) and B_{ij} as the cranking masses, the effects of both the coupling- potential and -mass are small. This reduces the problem to one in a decoupled approximation. Then, for λ_c given by eq. (3), $P_0 \propto |\psi(\eta)|^2$ and $P \propto |\psi(R)|^2$.

For η-motion, the stationary Schrödinger equation (for fixed R) is

$$[-\frac{\hbar^2}{2\sqrt{B_{\eta\eta}}}\frac{\partial}{\partial\eta}\frac{1}{\sqrt{B_{\eta\eta}}}\frac{\partial}{\partial\eta} + V(\eta)]\psi^\nu(\eta) = E_\eta^\nu\psi^\nu(\eta), \tag{6}$$

whose numerical solution gives the ground-state ($\nu = 0$) preformation probability

$$P_0(A_2) = |\psi^0(\eta)|^2\sqrt{B_{\eta\eta}(\eta)}\frac{2}{A}. \tag{7}$$

Choosing $R = R_1 + R_2(= R_t)$ or $= C_1 + C_2(= C_t)$, C_i being the Süssmann central radii, the potential in two spheres approximation is defined as the sum of experimental binding energies $B_i(A_i, Z_i)$, the Coulomb and nuclear proximity potentials. For mass parameters, the classical hydrodynamical model of Kröger and Scheid[28] is used.

For R-motion, instead of solving the corresponding radial Schrödinger equation, the WKB method is used. Then, for the penetration path shown in Fig. 1,

$$P = P_i W_i P_b, \tag{8}$$

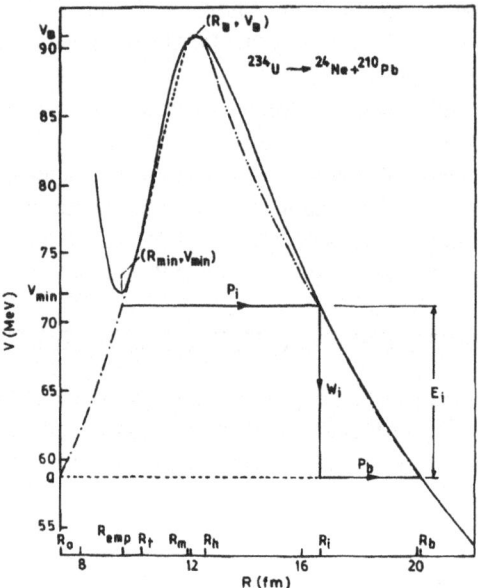

Figure 1. *The scattering potential V(R) for $^{234}U \rightarrow\ ^{24}Ne + ^{210}Pb$, calculated as a sum of Coulomb and nuclear proximity potentials (solid line). The dot-dashed, dashed, dotted and double-dot dashed curve is the analytical fit. The tunneling path is shown with the first turning point radius $R_a = R_{emp} (\approx R_{min})$.*

where, for simplicity, the internal de-excitation probability W_i is taken as unity, and the WKB penetrabilities are

$$P_i = exp[-\frac{2}{\hbar} \int_{R_t}^{R_i} \{2\mu [V(R) - V(R_i)]\}^{\frac{1}{2}} dR], \qquad (9)$$

and

$$P_b = exp[-\frac{2}{\hbar} \int_{R_i}^{R_b} \{2\mu [V(R) - Q]\}^{\frac{1}{2}} dR]. \qquad (10)$$

Both these integrals are solved analytically by parametrizing the potential V(R) suitably (see dotted line in Fig. 1).

For the clusters formed in the ground-state, the assault frequency is defined simply as

$$\nu_0 = \frac{v}{R_0} = \frac{(2E_2/\mu)^{\frac{1}{2}}}{R_0}, \qquad (11)$$

with $E_2 = \frac{A_1}{A} Q$ and $E_1 = Q - E_2$.

The model is applied to the observed heavy-cluster decays, up to ^{30}Mg. Fig. 2 shows the results of this calculation[29] for $R = C_t$ and R_t. Apparently, the experimental data lie in between the two calculations, with lighter clusters preferring $R = R_t$ and the heavier clusters preferring $R = C_t$. Recently, Kumar and Gupta[30] have shown that for a best empirical fit to most of these data, $R_{emp} = R_t - 0.7 fm$ and that this value corresponds, within $\pm 0.15 fm$,

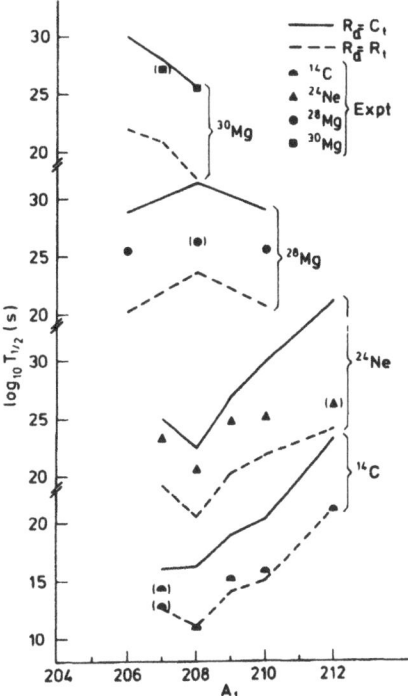

Figure 2. *The calculated and experimental logarithms of the cluster-decay half-life times as a function of mass number A_1 of daughter nuclei plotted for different clusters. The calculations are made on the PCM of Malik and Gupta and are for two different values of initial turning point R_a.*

to R_{min}, i. e. $R_{emp} = R_{min} \pm 0.15 fm$, where R_{min} is the position of the pocket in V(R), if proximity potential for $R \leq R_t$ were calculated. Here, all nuclei are spheres.

Fig. 3 shows the calculated cluster preformation probabilities P_0, relative to α-particle preformation factors, compared with the empirical estimates of Blendowske and Walliser[31] (for $A_2 \leq 28$) that fit the other PCM calculations of Blendowske, Fliessbach and Walliser[21] based on shell model approach. Interesting enough, the average behaviour predicted by the model of Malik and Gupta (marked MG) match the Blendowske-Walliser (marked BW) estimates. For $A_2 > 28$, the collective model approach of Malik and Gupta predict an increase in P_0, which ultimately becomes constant for $A_2 > 42$. This may be a real effect, since for heavier clusters SF becomes important.

Following the earlier work of Gupta et. al[11] for ^{120}Ba, and the recently planned experiments by Milano group[32], Kumar and Gupta[33] have extended these calculations to many Ba-isotopes. Fig. 4 (and Table 1 published[33] in the book of abstracts of this School) shows that many decays up to ^{20}Ne lie below the limit of present experimental methods and that, other than α-decay, ^{12}C decay of ^{112}Ba is most probable. This result is a manifestation of the doubly closed shell Z=N=50 ^{100}Sn daughter nucleus.

Recently, the model of Malik and Gupta is extended[30] by including the deformation efffects of both the cluster and daughter nuclei. The proximity potential is derived for a necked configuration, in terms of the parameters of an asymmetric TCSM nuclear shape. The result of this calculation is illustrated in Fig. 5. We notice that the barrier is lowered

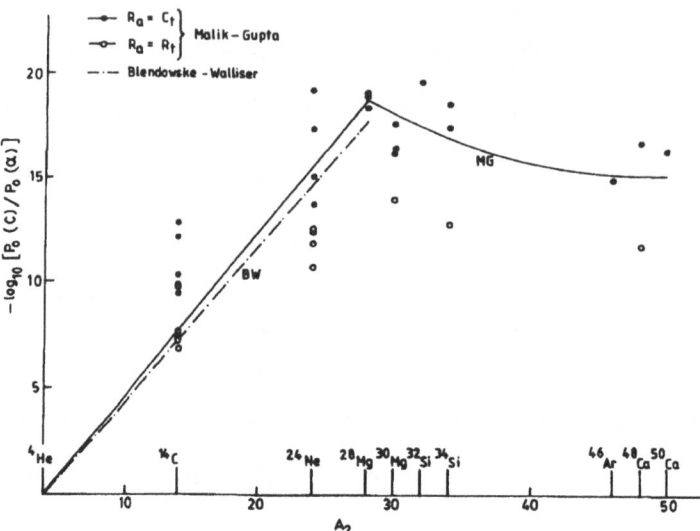

Figure 3. *The calculated logarithms of the cluster preformation probabilities relative to the α-particle as a function of mass number A_2 of the emitted cluster for the PCM of Malik and Gupta (solid line, marked MG), compared with the empirical estimates of Blendowske and Walliser (dashed line, marked BW). The solid line gives the average behaviour of the calculated solid and open circles, for two different values of the initial turning point $R_a = C_t$ and R_t.*

significantly and the proximity potential minimum shifts down exactly at the Q-value but at an $R(=R_{min}) > R_0$. This means that the deformation effects of both the cluster and daughter nuclei assimilate the empirical choice of R-value (in the neighbourhood of R_t) in Figs. 2 and 3 for the spherical nuclei.

Fig. 6 shows the empirical preformation factor $P_0 (= \frac{\lambda_{expt}}{\lambda_{cal}})$, plotted as a function of the cluster mass, for the decay constant λ_{cal} calculated for the potential with deformation effects included (solid curve in Fig. 5). Notice that the two turning points are now defined by $V(R_a) = Q = V(R_b)$, instead of the ones defined in Fig. 1. The empirical results of Blendowske and Walliser[31] and the preformation factors of another UFM model based on M3Y potential (with deformation of cluster included)[20] are nicely reproduced by the calculations of Kumar and Gupta[30]. For $A_2 > 32$, these calculations also predict an increase of P_0, which seem to become constant for $A_2 > 48$. Further experimental data are required.

5. Saddle-Point Fission Model

As already stated, this is a fission model where the nucleus is considered to deform continuously, to penetrate the barrier as a whole with probability P and to reach $R = R_{saddle}$, where both the mass and charge distributions are decided. In other words, the fragments are born (or formed) with probability P_0 at R_{saddle}. Then, just as in PCM, the decay constant for the fission process is defined as[26]

$$\lambda_f = 2\nu_0 P P_0. \qquad (12)$$

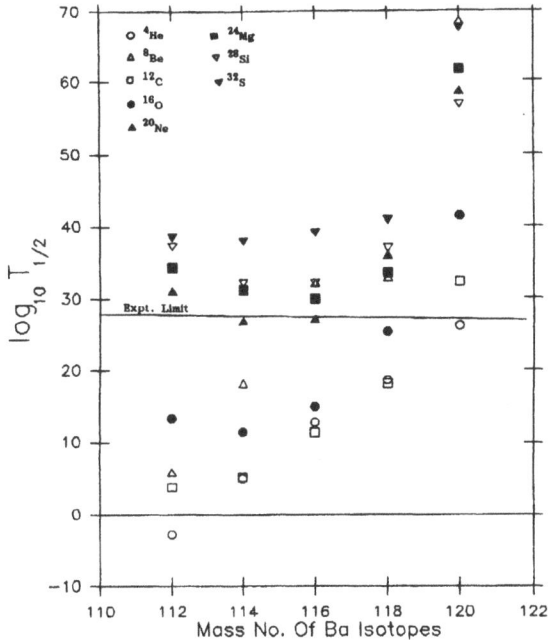

Figure 4. *The logarithms of the calculated half-life times for various cluster-decays plotted as a function of mass of Ba-isotopes, for the PCM of Malik and Gupta. The calculations are for $R_a = C_t$ and hence represent upper limiting values. The present limit of experiments is also shown.*

The factor of 2 comes in here because in most of the fission experiments, both light and heavy fragments are measured simultaneously. The saddle shape is parametrized as TCSM nuclear shape with the scattering potential (illustrated in Fig. 7) calculated in Stutinsky method. This differs from the cluster-decay model potential (Fig. 1) in that here the scission configuration lies outside the barrier. Thus, use of fission model here means simply carrying out the fission calculations in the super-asymmetric mass region of cluster radioactivity.

Apparently, P_0 is given by eq. (7), calculated from eq. (6) at $R = R_{saddle}$, with the TCSM potential $V(\eta) = V_{LDM} + \delta U$ and the cranking masses $B_{\eta\eta}$. It is not straight-forward to calculate P in this model, though one could reasonably well fix $\nu_0 = 10^{21} s^{-1}$. However, since both ν_0 and P are same for all fission fragments, one can define the branching ratios

$$B_f = \frac{\lambda_f^{max}}{\lambda_f^c} = \frac{P_0^{max}}{P_0^c}, \tag{13}$$

where, λ_f^{max} refers to the most probable symmetric or nearly-symmetric fission fragment. This is to be compared with the corresponding cluster-decay branching ratio

$$B_c = \frac{\lambda_\alpha}{\lambda_c}, \tag{14}$$

since α-particle is the most probable product in cluster-decay.

The model is applied[13] to fission of ^{234}U, ^{238}Pu, ^{241}Am and ^{252}Cf where the cluster-decay data are also availabe for, at least, the first two nuclei. Kumar et. al[13] have first calcu-

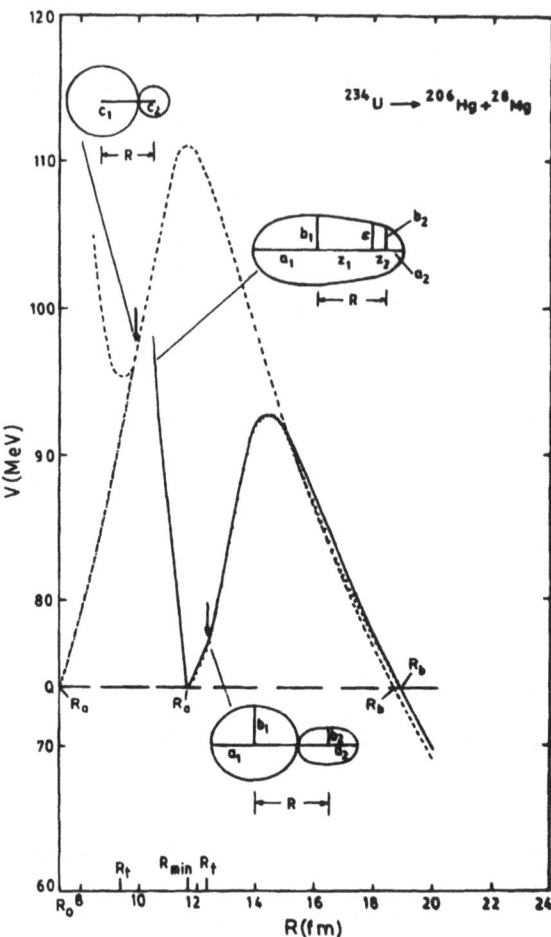

Figure 5. *The scattering potential V(R), calculated as a sum of the Coulomb plus nuclear proximity potentials, for $^{234}U \to {}^{206}Hg + {}^{28}Mg$, with both the outgoing fragments taken spherical or deformed (dashed and solid lines, respectively). In the case of deformed fragments, the potential for the overlap region ($R \leq R_{min}$) is calculated by using the proximity potential for a necked TCSM nuclear shape. The decay path (R_a and R_b values) as well as the analytical fits are indicated (dots) for the deformed case.*

lated the mass yield distributions in the high yield region of the symmetric or near-symmetric mass region and compared them with the available experimental data. The comparisons are good, which means a support for the idea of fragments being formed at R_{saddle}, already tested earlier for mass- and charge distributions[34–36], and through an explicit solution of time-dependent Schrödinger equation[37]. The present calculations are then extended for the first time to the super-asymmetric mass region. Fig. 8 illustrates this calculation for fission of ^{234}U at two different energies. We notice that strong peaks occur at ^{24}Ne, ^{26}Ne and ^{28}Mg, which are exactly the clusters observed in cluster-decay studies. This is also true for other cases. Thus, cold fission of nuclei prefer the same light fragments as are observed in cluster radioactivity. Also, some new clusters are predicted, which are not yet seen in

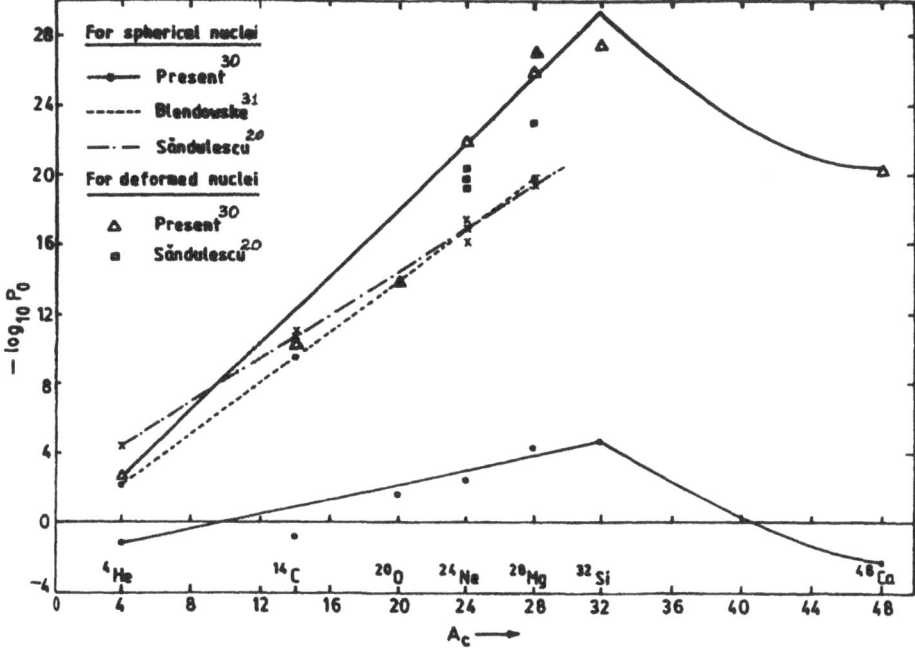

Figure 6. *The logarithms of the empirical cluster preformation probabilities P_0 plotted as a function of the cluster mass, for the cases of both spherical and deformed nuclei.*

cluster radioactivity. These are $^{24-26}Ne$ decays of ^{238}Pu and ^{36}Si, ^{37}P and ^{38}S decays of ^{241}Am. The branching ratio calculations show that cold fission is, in general, a more probable process, particularly, for the newly predicted decays. This is best illustrated in Fig. 9 where the cluster-decay experimental data are also plotted. We also notice in Fig. 9 that the two processes of cold fission and cluster-decay seem to overlap for $A_2 > 42$. However, there are no experimental data on binary fission for the super-asymmetric mass region of cluster radioactivity. The thermal neutron induced fission measurements of Gönnenwein, presented at this School, give negative results but our calculations are for SF or at more than 20MeV of excitation energies.

6. Conclusions

Cluster radioactivity, cold fission and cold fusion are related phenomena[38] in the sense that they all arise due to (spherical) shell closure effects. This is also supported for Z=50 daughter in the decay of Ba-isotopes. Deformed closed shell effects are also predicted for which experiments could be planned for decay of Hg-isotopes.

The concept of cluster preformation in nuclei is shown to be essential even for the simple application of Gamow theory to heavy cluster decays. The preformed cluster model of Malik and Gupta (MG)[24] and, in particular, its extension by Kumar and Gupta[30] to include deformations of both the cluster and daughter nuclei, give the preformation probabilities P_0 that match the shell model based empirical estimates of Blendowske and Walliser (BW)[31]. Thus, the shell model and collective model approaches predict similar orders of P_0. For cluster

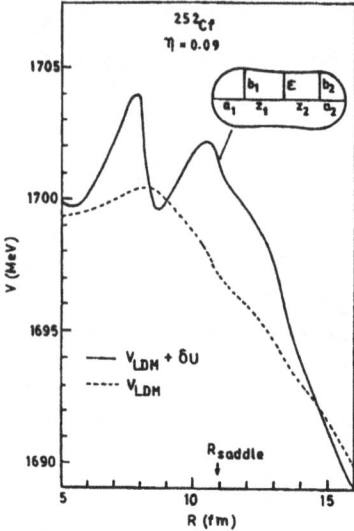

Figure 7. *The scattering potential V(R) for fixed mass asymmetry $\eta = 0.09$ of ^{252}Cf, calculated for both the liquid drop (V_{LDM}) and the liquid drop plus shell correction ($V_{LDM} + \delta U$)) potentials. The nuclear shape at the saddle point is also shown along with the associated collective coordinates.*

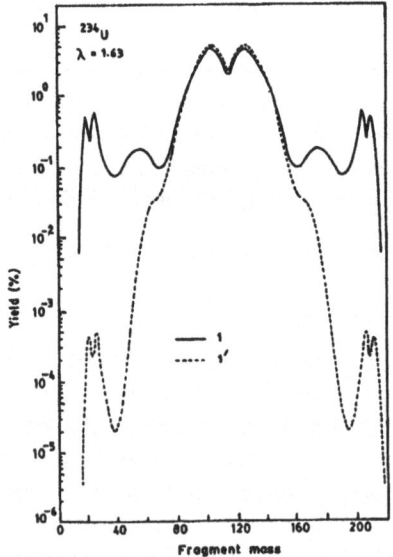

Figure 8. *The calculated fission yields for ^{234}U, including the super-asymmetric mass region of cluster radioactivity. The calculations are for SF (dashed line) and for an excitation energy of 21.64MeV (solid line).*

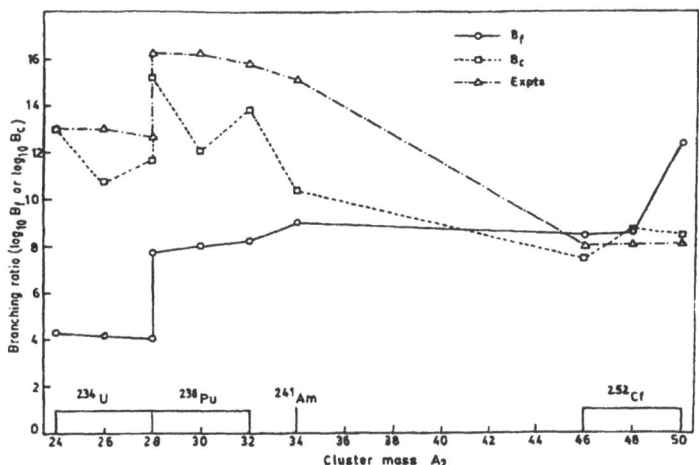

Figure 9. *The calculated branching ratios $log_{10}B_f$ for cold fission and $log_{10}B_c$ for PCM, as well as the experimental data on exotic cluster decays, as a function of the cluster mass A_2.*

masses $A_2 > 32$, P_0 is predicted to increase and finally becomes constant for $A_2 > 48$. This may be due to the fission process becoming important beyond $A_2 > 48$.

Fission calculations of radioactive nuclei in the super-asymmetric mass region[13] are reported for the first time. It is shown that cold fission also prefers the same products as are already observed in cluster radioactivity and is a more probable process experimentally, as compared to the cluster-decay. Also, cold fission and cluster decay are predicted to become indistinguishable for $A_2 > 42$. This means predicting a possible overlap of the two processes (cold fission and cluster-decay) for $A_2 > 42$. However, to-date there are no experimental data to compare with these calculations.

Acknowledgements

The author is thankful to Professor Werner Scheid for the kind hospitality at his Institute and to Alexander von Humboldt-Stiftung for the support of his visit on a revisit invitation.

References

[1] A. Săndulescu, R. K. Gupta, W. Scheid and W. Greiner, *Phys. Lett.* **60B**, 225 (1976)
[2] R. K. Gupta, A. Săndulescu and W. Greiner, *Phys. Lett.* **67B**, 257 (1977); *Z. Naturforsch.* **32a**, 704 (1977)
[3] R. K. Gupta, C. Párvulescu, A. Săndulescu and W. Greiner, *Z. Phys. A* **283**, 217 (1977)
[4] R. K. Gupta, *Sovt. J. Part. Nucl.* **8**, 289 (1977); *Nucl. Phys. and Solid St. Phys. (India) A* **21**, 171 (1978); R. K. Gupta, D. R. Saroha and N. Malhotra, *J. Physique (Paris) Coll.* **45**, C6-477 (1984)
[5] J. A. Maruhn, W. Greiner and W. Scheid, *Heavy Ion Collisions*, Vol. II, Chap. 6, edited by R. Bock (North Holland, Amsterdam, 1980)
[6] A. Săndulescu, D. N. Poenaru and W. Greiner, *Sovt. J. Part. Nucl* **11**, 528 (1980)
[7] H. J. Rose and G. A. Jones, *Nature* **307**, 245 (1984)
[8] H.-G. Clerc, W. Lang, M. Mutterer, C. Schmitt, J. P. Theobald, U. Quade, K. Rudolph, P. Armbruster, F. Gönnenwein, H. Schrader and D. Engelhardt, *Nucl. Phys. A* **452**, 277 (1986)
[9] G. Münzenberg, *Rep. Prog. Phys.* **51**, 57 (1988)

[10] R. K. Gupta, W. Scheid and W. Greiner, *J. Phys. G: Nucl. Part. Phys.* **17**, 1731 (1991)
[11] R. K. Gupta, S. Singh, R. K. Puri and W. Scheid, *Phys. Rev. C* **47**, 561 (1993)
[12] R. K. Gupta and W. Greiner, *Int. J. Mod. Phys. E*, to be published (1993)
[13] S. Kumar, R. K. Gupta and W. Scheid, *Int. J. Mod. Phys. E*, to be published (1993)
[14] Y. Hutsukawa, H. Nakahara and D. C. Hofman, *Phys. Rev. C* **42**, 674 (1990)
[15] D. N. Poenaru, M. Ivascu, A. Săndulescu and W. Greiner, *Phys. Rev. C* **32**, 572 (1985); *J.Phys. G: Nucl. Phys.* **10**, L183 (1984); D. N. Poenaru, D. Schnabel, W. Greiner, D. Mazilu and R. Gherghescu, *At. Data and Nucl. Data Tables* **48**, 231 (1991)
[16] G. A. Pik-Pichak, *Sovt. J. Nucl. Phys.* **44**, 923 (1986)
[17] Y.-J. Shi and W. J. Swiatecki, *Phys. Rev. Lett.* **54**, 300 (1985); *Nucl. Phys. A* **438**, 450 (1985).
[18] G. Shanmugam and B. Kamalaharan, *Phys. Rev. C* **38**, 1377 (1988); *Phys. Rev. C* **41**, 1184 (1990); *Phys. Rev. C* **41**, 1742 (1990)
[19] B. Buck and A. C. Merchant, *J. Phys. G: Nucl. Part. Phys.* **15**, 615 (1989); *Phys. Rev. C* **39**, 2097 (1989); *Europhys. Lett.* **8**, 409 (1989); *Phys. Rev. C* **45**, 2247 (1992).
[20] A. Săndulescu, R. K. Gupta, F. Cărstoiu, M. Horoi and W. Greiner, *Int. J. Mod. Phys. E* **1**, 379 (1992); R. K. Gupta, M. Horoi, A. Săndulescu, M. Greiner and W. Scheid, *J.Phys. G: Nucl. Part. Phys.* **19**, 2063 (1993)
[21] R. Blendowske, T. Fliessbach and H. Walliser, *Nucl. Phys. A* **464**, 75 (1987); *Z. Phys. A* **339**, 121 (1991); *Handbook of Nuclear Decay Modes*, edited by D. Poenaru (1992)
[22] M. Ivascu and I. Silisteanu, *Nucl. Phys. A* **485**, 93 (1988)
[23] S. G. Kadmensky, W. I. Furman and Yu. M. Chuvilśkii, *Sovt. J. Izv. Akad. Nauk. SSSR, ser. Fiz.* **50**, 1786 (1986)
[24] S. S. Malik and R. K. Gupta, *Phys. Rev. C* **39**, 1992 (1989); R. K. Gupta, Invited Review Talk at Seventh National Symposium on Radiation Physics, Mangalore, India, Nov. 16-20, 1987; Proc. Vth Int. Conf. on Nucl. Reaction Mechanisms, Varenna, Italy, 1988, p.416; *IANCAS Bull. (India)* **6**, 2 (1990)
[25] V. A. Rubchenya, V. P. Eysmont and S. G. Yavshits, *Izv. Akad. Nauk. SSSR, ser. Fiz.* **50**, 1017 (1986)
[26] S. Kumar, S. S. Malik and R. K. Gupta, *Nucl. Phys. Symp. (India)* **31B**, 61 (1988); R. K. Gupta, S. Kumar, H. Kumar, W. Scheid and W. Greiner, 7th Adriatic Int. Conf. on Nucl. Phys.: *Heavy Ion Physics-Today and Tomorrow*, Brioni, Yugoslavia, May 27-June 1, 1991, p.5; S. Singh, *Ph. D. Thesis*, Panjab University, Chandigarh, India (unpublished, 1992).
[27] F. Barranco, R. A. Broglia and G. F. Bertsch, *Phys. Rev. Lett.* **60**, 507 (1988); *Phys. Rev. C* **39**, 2101 (1989); *Nucl. Phys. A* **512**, 253 (1990)
[28] H. Kröger and W. Scheid, *J. Phys. G* **6**, L85 (1980)
[29] R. K. Gupta, S. Singh, R. K. Puri, A. Săndulescu, W. Greiner and W. Scheid, *J. Phys. G: Nucl. Part. Phys.* **18**, 1533 (1992)
[30] S. Kumar and R. K. Gupta, Contri. Int. Conf. on Nuclear Structure and Nuclear Reactions, Dubna, Russia, Sept. 15-19, 1992; Panjab University, Chandigarh Preprint, to be published (1993); S. Kumar, *Ph.D. Thesis*, Panjab University, Chandigarh, India (unpublished, 1992).
[31] R. Blendowske and H. Walliser, *Phys. Rev. Lett.* **61**, 1930 (1988)
[32] A. Guglielmetti and R. Bonetti, *Private communication*, April 1993.
[33] S. Kumar and R. K. Gupta, Contri. this Institute, published in the book of Abstracts
[34] J. Maruhn and W. Greiner, *Phys. Rev. Lett.* **32**, 548 (1974)
[35] R. K. Gupta, W. Scheid and W. Greiner, *Phys. Rev. Lett.* **35**, 353 (1975)
[36] D. R. Saroha and R. K. Gupta, *Phys. Rev. C* **29**, 1101 (1984)
[37] D. R. Saroha, R. Aroumougame and R. K. Gupta, *Phys. Rev. C* **27**, 2720 (1983)
[38] R. K. Gupta, S. Singh, W. Scheid and W. Greiner, 6th Int. Conf. on Nucl. Reaction Mechanisms, Varenna, Italy, June 10-15, 1991; S. Singh, R. K. Gupta, W. Scheid and W. Greiner, *J. Phys. G: Nucl. Part. Phys.* **18**, 1243 (1992)

SUPERASYMMETRIC FISSION TRAJECTORIES IN A TRIDIMENSIONAL CONFIGURATION SPACE

M. Mirea

Tandem Laboratory, Institute of Atomic Physics, PO Box MG-6, RO-76900 Bucharest, Romania

Introduction: All the phenomenological models[1] which were used to obtain cluster emission predictions were characterized by one degree of freedom, namely the elongation. Some attempts were tried to develop the numerical superasymmetric fission model (NSAFM) by determining the behaviour of the necking-in shape dependence[2] during the process. In the present paper, an investigation of the optimum fission trajectories in a tridimensional configuration space with respect to elongation, necking-in and mass-asymmetry is realised.

Action Integral Minimization: The nuclear parametrization[3] was obtained by smoothly joining two intersected spheres of radius R_1 and R_2 with a neck surface generated by the rotation of a circle of radius R_3 around the axis of symmetry. The distance between the center of this circle and the axis of symmetry is given by ρ_3. By imposing the condition of volume conservation the surface is perfectly determined by the values of the parameters R (distance between the centers of spheres), $C = S/R_3$ ($S = +1$ when $\rho_3 - R_3 > 0$ and $S = -1$ when $\rho_3 - R_3 < 0$) and R_2 which characterize the elongation, necking and mass-asymmetry, repectively. Diamond-like shapes are obtained for $S = -1$ and necked-in shapes for $S = +1$. Assuming a system of two touching spheres at the scission point, approximation available for cold fission processes, the penetrability through the barrier can be expressed as a product of two probabilities

$$P = P_{ov}P_s = \exp\left[-(K_{ov} + K_s)\right]. \qquad (1)$$

Here K_s represents the action integral for separated spherical fragments and K_{ov} corresponds to the overlaping region in the WKB-approximation:

$$K_{ov} = \frac{2}{\hbar}\int_{R_i}^{R_s} \sqrt{2B(R,C,R_2,\frac{\partial C}{\partial R},\frac{\partial R_2}{\partial R})E(R,C,R_2)}dR = \frac{2}{\hbar}\int_{R_i}^{R_s} FdR \qquad (2)$$

where $R_i = 1.16(A_0^{1/3} - A_2^{1/3})$ is a turning point, $R_s = 1.16(A_1^{1/3} + A_2^{1/3})$ is the scission point, A_0, A_1, A_2 being the mass-numbers of the parent, the daughter and emitted nuclei, respectively. The effective mass $B(R,C,R_2,\partial C/\partial R,\partial R_2/\partial R)$ along the trajectory is computed[4] using the Werner-Wheeler approximation of an irrotational, nonviscous, hydrodynamical motion. The deformation energy $E(R,C,R_2)$ contains terms like nuclear energy, Coulomb energy and symmetry energy in the framework of the Yukawa-plus-exponential model extended for binary systems with different charge densities[5]. A phenomenological correction which allows to obtain the experimental Q-value is added to the energy. The optimum trajectory which minimizes the K_{ov} integral among all possible trajectories which connect the two end-points in the configuration space can be obtained by solving the two associated Euler-Lagrange equations

$$\frac{\partial F}{\partial C} = \frac{d}{dR}(\frac{\partial F}{\partial \dot{C}}) \; ; \; \frac{\partial F}{\partial R_2} = \frac{d}{dR}(\frac{\partial F}{\partial \dot{R_2}}). \qquad (3)$$

The shooting theorem allows to transform the two end-point conditions in one condition imposed at the scission point where R_2 and C are known. Different trajectories are obtained by changing the derivatives $\partial R_2/\partial R$ and $\partial C/\partial R$. Only the best results are retained.

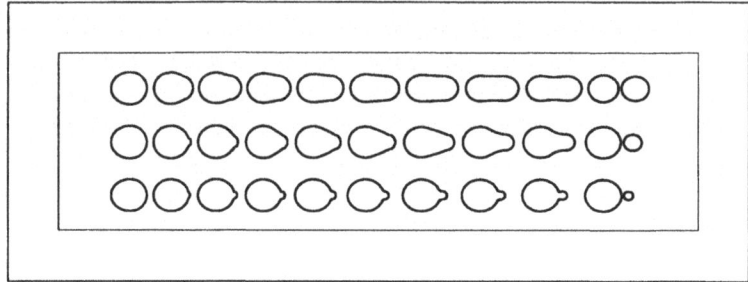

Figure 1. *Nuclear shapes along optimum fission paths for cold fission (top), cluster emission (middle) and alpha decay (bottom).*

Results: Calculations for alpha-decay, ^{28}Mg-radioactivity and cold fission with the light fragment ^{100}Zr of ^{234}U (the first nucleus for which these three groups of decay modes have been experimentally detected) are performed. The optimum fission trajectory for alpha-decay shows a very small neck radius. The mass asymmetry is kept nearly constant during the disintegration process. A different situation appears to be for Mg-radioactivity and cold fission where in the first stage of the desintegration we have diamond-like shapes. For cold fission the neck radius decreases in the vicinity of the scission point. A general increase of R_2 before reaching the final value characterizes the cluster emission while in cold fission a small decrease is observed. The sequences of shapes during these three decay modes are plotted in figure 1.

References

[1] D. N. Poenaru, M. Ivascu, W. Greiner in: *Particle Emission from Nuclei*, Vol. III, p. 203, edited by D. N. Poenaru and M. Ivascu (CRC, Boca Raton, Florida, 1989)
[2] M. Mirea, D. N. Poenaru, W. Greiner, *Nuovo Cimento* **105A**, 4, 571 (1992)
[3] K. Depta, R. Hermann, J. A. Maruhn, W. Greiner in: *Dynamics of Collective Phenomena*, p. 29, edited by P. David (World Scientific, Singapore, 1987)
[4] D. N. Poenaru, M. Ivascu, I. Ivascu, M. Mirea, W. Greiner, K. Depta, W. Renner in: *50 Years with Nuclear Fission*, p. 617 (American Nuclear Society, Lagrange Park, 1989)
[5] D. N. Poenaru, M. Ivascu, D. Mazilu, *Comp. Phys. Comm.* **19**, 205 (1980)

QUANTUM TUNNELING SPECTRUM AND APPLICATION TO COLD FISSION

E. Stefanescu[1] and A. Sandulescu [2]

[1] *Research Institute for Electronic Components, 72996 Bucharest, Romania*
[2] *Institute of Atomic Physics, 76900 Bucharest, Romania*

Tunneling is essentially a quantum process standing at the basis of important applications in electronics, chemistry, biology and nuclear physics. In principle, we can conceive it as a transition from a "localized" state Ψ_0 of the compound nucleus to a state Ψ_t of the reaction channel. If the physical system is closed, the energy is conserved within the limits of Heisenberg's uncertainty principle. However, in nuclear physics, the experimental spectra of some fission modes with cold fragments, display important line shifts and broadenings[1]. This means that the proper fragmentation process is assisted by dissipative processes, leading to energy loss and diffusion. Consequently, an open physical system must be considered.

In a previous paper[2] we have studied quantum tunneling in open systems based on Lindblad's master equation:

$$\frac{d\rho}{dt} = -\frac{i}{\hbar}[H,\rho] + L(\rho) \qquad (1)$$

where $L(\rho)$ is a function of the system operators q, p and of the friction and diffusion coefficients $\lambda, D_{qq}, D_{pp}, D_{pq}$. Evidently, from the first term, Gamow's tunneling rate with energy conservation is obtained. From the openess operator $L(\rho)$, additional terms of the tunneling rate with energy transfer to the dissipative environment are obtained.

In this paper, we report the calculation of the tunneling spectrum as a function of the barrier characteristics: the height U_M, the coordinate q_M of U_M, the zero-point vibration energy E_0, the Q-value and the distance R_0 between the two partners. At thermal equilibrium[3], the spectral density of the tunneling rate, depending only on λ and the temperature T, becomes:

$$\Gamma_\omega(t) = \Omega_\omega^2 \frac{\sin^2(\omega t/2)}{t(\omega/2)^2} +$$
$$+ \lambda C_\omega \Omega_\omega \frac{1}{\omega}[1 - \cos\omega t + \coth\frac{E_0}{kT}(\cos\omega t - \frac{\sin\omega t}{\omega t})] +$$
$$+ \lambda[-2q_\omega s_\omega + \coth\frac{E_0}{kT}(\alpha q_\omega^2 + \frac{s_\omega^2}{\alpha})], \qquad (2)$$

where ω is the transition frequency, $\alpha = m\omega_0/\hbar$, $\Omega_\omega, q_\omega, s_\omega$ are the transition matrix elements of the tunneling operator, the coordinate and respectively the momentum, and C_ω is the overlap integral[2].

The first term is the very narrow Gamow's peak, the second term is a "tail" of a width proportional to E_0 and a height proportional to t, and the third term is a large spectrum, like a "pedestal". For small enough values of ω_0/λ the pedestal becomes dominant, i.e. a very large width is obtained. We found that this large spectrum appears if some dissipation processes occur in the Coulomb region of the nuclear barrier. It can be understood in a very simple way: the dependence of the matrix elements on the energy E_i is given by the channel wave functions $\Psi_i \sim \exp(-\frac{2}{\hbar}\int_q^{q_i}\sqrt{2m[U(q)-E_i]}dq)$, which for the very small interval of integration close to the outer turning point q_i, has a weak dependence on E_i. As an example, we give the tunneling spectrum for the cold fission of $^{236}U \rightarrow ^{138}Xe + ^{98}Sr^4$ in the figure below.

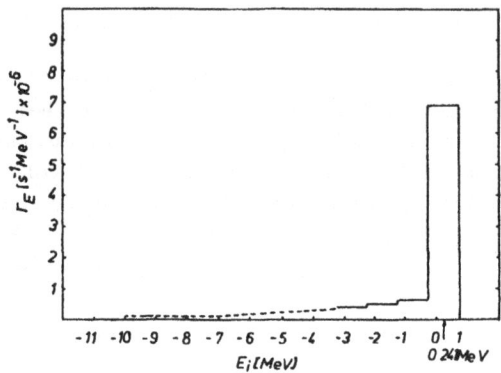

References

[1] A. Sandulescu, *A new radioactivity*, J. Phys. G **15**, 529 (1989)

[2] E. Stefanescu, A. Sandulescu and W. Greiner, *Quantum tunneling in open systems*, Int. J. Mod. Phys. E **2**, 233 (1993)

[3] A. Sandulescu and H. Scutaru, *Open quantum systems and the damping of collective modes in deep inelastic collisions*, Ann. Phys. **173**, 277 (1987)

[4] A. Sandulescu, A. Florescu and W. Greiner, *Cold fission as emission of fragments*, J. Phys. G: Nucl. Phys. **15**, 1815 (1989)

FUSION AND QUASI-ELASTIC REACTIONS AT NEAR-BARRIER ENERGIES

L. Corradi[1], D. Ackermann[1], S. Beghini[2], G. Montagnoli[2], L. Mueller[2], D. R. Napoli[1], C. Petrache[3], G. Pollarolo[4], N. Rowley[5], F. Scarlassara[2], G. F. Segato[2], C. Signorini[2], P. Spolaore[1], F. Soramel[2], A. M. Stefanini[1]

[1] INFN-Laboratori Nazionali di Legnaro, Legnaro (Padova,Italy)
[2] Dipartimento di Fisica dell'Universita' and INFN, Padova (Italy)
[3] Institute of Atomic Physics and Engineering, Bucharest (Romania)
[4] Dipartimento di Fisica dell'Universita' and INFN, Torino (Italy)
[5] SERC, Daresbury Laboratory, Warrington WA44AD (UK)

The recent availability of high precision data of heavy-ion fusion and quasi-elastic reactions at energies close to the Coulomb barrier allows detailed comparisons with microscopic nuclear theories. Two experiments have been recently performed in Legnaro : A) the detailed measurement of the fusion excitation function for the system ^{58}Ni+^{60}Ni and B) the measurement of the elastic scattering and transfer cross sections for ^{32}S+^{208}Pb. Both experiments have the common aim of enlightening the role of nucleon correlations, in the fusion process (A) and in the multinucleon transfer (B). In the following a very short preliminary report is given.

A) In recent years the role of inelastic and transfer channels as doorway states to fusion is being investigated by examining the second derivative of the fusion excitation function[1]. We chose to study the system ^{58}Ni+^{60}Ni in order to identify the characteristic barrier distribution it should show if a channel with Q=0 is dominant, corresponding in this case to the elastic transfer of 2 neutrons. The experiment has been made with an electrostatic separator and the fusion excitation function has been measured from $\simeq$ 8 MeV below to $\simeq$ 13 MeV above the Coulomb barrier. The extracted second derivative of the fusion cross section is shown in Fig.1 together with a coupled channel (CC) calculation made by coupling only one channel with Q=0 and a suitable strength. The reasonable agreement with the data obtained for such a channel and the completely different structure calculated by coupling Q$\neq$0 channels, support the idea that the 2 neutron channel is important for the fusion process. Other calculations, including the inelastic channels as well, are in progress.

B) The reasons for measuring transfer reaction cross sections for the system ^{32}S+^{208}Pb are many, but mainly we wanted a combination of spherical nuclei to test specific microscopic theories[2,3]. While the DWBA or CC theory has been used so far for reproducing the transfer of one nucleon, no microscopic theory has been applied yet in the case of multinucleon transfer. For this purpose differential cross sections have been measured at barrier energies for a variety of transfer channels and pure elastic scattering has been extracted as well. Fig.2 shows the experimental results at 173.4 MeV for the elastic scattering and the transfer of one particle. The lines are the theoretical calculations where CC theory has been used for the elastic scattering and CWKB theory[2] has been applied for the transfer. As can be seen the theory agrees with the data quite well. On this basis a schematic model has then been

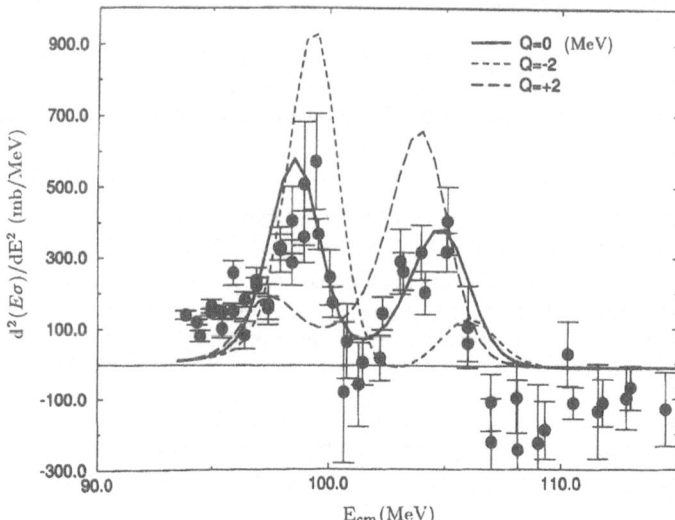

Figure 1. *Second derivative of $E\sigma$ for the fusion excitation function of $^{58}Ni+^{60}Ni$. The lines are CC calculations (see text).*

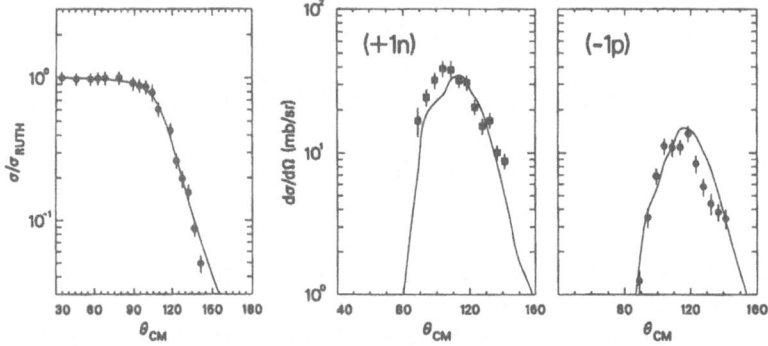

Figure 2. *Elastic and one particle transfer angular distributions for the reaction $^{32}S+^{208}Pb$ at 173.4 MeV (+1n is the pick up of 1 neutron, -1p the stripping of 1 proton). The lines are calculations as described in the text.*

applied for the multinucleon transfer channels[3], and preliminary results show that the main features of the angular and Q-value distributions are reproduced. Further refinements of the calculations are in progress.

References

[1] N. Rowley, I. J. Thompson and M. Nagarajan, *Phys. Lett.* B **282**, 25 (1992)
[2] E. Vigezzi and A. Winther, *Ann. of Phys.* **192**, 432 (1989)
[3] A. Winther, *Fisika* **22**, 41 (1990); NBI preprint nbi-93-26

A POCKET IN COLD FUSION POTENTIAL BARRIER

Radu Gherghescu

Institute of Atomic Physics, PO Box MG-6, RO-76900 Bucharest, Romania

A pocket in the potential barrier of large systems represents a necessary condition to produce giant nuclear molecules wich could be used to test the electrodynamics of strong fields.[1] The purpose of the present paper is to study the symmetrical cold fusion and fission barrier shapes for nuclei with Z protons and N neutrons in the range $80 \leq Z \leq 120$ and $Z \leq N \leq 200$. The calculations are made within the macroscopic Yukawa plus Exponential Model (Y+EM). The two fragments are supposed to be spherical.

Due to the finite range of nuclear forces, attraction between two separated nuclei begins to manifest before the touching point configuration is reached. The nuclear energy E_Y, replacing the usual liquid drop surface energy, is active as long as the distance between the fragment surfaces is smaller than the range of the nucleon-nucleon interaction. This quantity is the double folded Y+EM deformation energy for nuclear shapes starting from two identical separated spherical fragments, going through intersected spheres and ending to one spherical nucleus for cold fusion and viceversa for cold fission. One assumes the fragment volume V_f=constant.

The outer zone (postscission for fission), where $R > 2R_f$ (R_f is the fragment radius) corresponds to the separated fragments. For the assumed spherical shapes analytical relations are available:

$$E_C = \frac{Z_f^2 e^2}{R} \; ; \; E_Y = -4\left(\frac{a}{r_0}\right)^2 \sqrt{a_{21}a_{22}} \left[g_1 g_2 \left(4 + \frac{R}{a}\right) - g_2 f_1 - g_1 f_2\right] \frac{e^{-R/a}}{R/a} \quad (1)$$

where E_C is the Coulomb energy, E_Y is the (Y+EM) energy and f_k, g_k are functions of R_k/a.

In the inner zone (prescission for fission), where the two spheres intersect, the involved energies[2-4] are given by:

$$E_C = E_C^0 B_C \; ; \; B_C = \left(\frac{\rho_{1e}}{\rho_{0e}}\right)^2 B_{C1} + \left(\frac{\rho_{2e}}{\rho_{0e}}\right)^2 B_{C2} + \frac{\rho_{1e}\rho_{2e}}{\rho_{0e}^2} B_{C12} \quad (2)$$

where E_C^0 is the Coulomb energy for a spherical nucleus. The charge densities ratios $\frac{\rho_{ie}}{\rho_{0e}}$ (i=1,2) become equal for identical fragments. The Y+EM energy is expressed as:

$$E_Y = E_Y^0 B_Y \; ; \; B_Y = \frac{c_{s1}}{c_s} B_{Y1} + \frac{c_{s2}}{c_s} B_{Y2} + \frac{\sqrt{c_{s1}c_{s2}}}{c_s} B_{Y12} \quad (3)$$

where E_Y^0 is the appropiate term for a sphere, $c_s = a_s(1 - \kappa_s I^2)$ and $I = (N - Z)/A$. We have used the following values of parameters: $a_s = 21.13$ MeV, $\kappa_s = 2.3$, $a = 0.68$ fm. If the fragments are identical, we set $c_{s1} = c_{s2} = c_s$. The deformation dependence of the nuclear shape in the fusion process is contained in B_C and B_Y. The involved multiple integrals are computed numerically by Gauss-Legendre quadratures. As a result of the

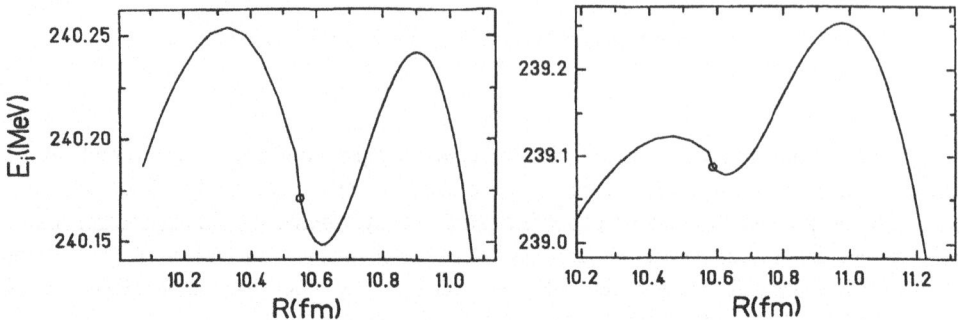

Figure 1. *Cold fusion barriers of ^{94}Rh (left hand side) and of ^{95}Rh (right hand side).*

computation, several regions characterized by different shapes of the fusion barrier have been obtained (see Fig.1 of Ref.5). The studied (N, Z) map is divided in four regions. Region (1), (3) and (4) present one maximum of the barrier shape. Region (1) has the maximum in the outer zone. Regions (3) and (4) have the maximum in the inner zone; the difference between them consists in the crossing point with the released energy Q axis. The region (2) is the most interesting one. In this region a second minimum appears in the barrier shape. If one takes a closer look at the barrier shape from this region one can observe (see for example Fig.1) that, for the same compound nucleus (Th here) the first maximum in the inner zone, decreases with the increase of the neutron number. Also the second maximum, in the outer zone, decreases and the first one increases when the neutron number decreases. Usually the second minimum only appears as a result of adding of the shell corrections to the macroscopic part of the potential. This apparently strange behaviour within a macroscopic calculation, certainly due to the finite range of the nuclear forces, will be further explored.

References

[1] W. Greiner, B. Müller and J. Rafelski, *Quantum Electrodynamics of Strong Fields*, (Springer, Berlin, 1985)
[2] H. J. Krappe, J. R. Nix, A. J. Sierk, *Phys. Rev. Lett.* **42**, 215 (1979)
[3] D. N. Poenaru, M. Ivaşcu, D. Mazilu, *Comp. Phys. Com.* **19**, 205 (1979)
[4] D.N. Poenaru in: *Handbook of Nuclear Decay Modes*, edited by D.N. Poenaru and W. Greiner, (CRC Press, Boca Raton, to be published)
[5] R. Gherghescu, D. N. Poenaru, *Roum. J. Phys.* **37**, 1005 (1992)

III.

Heavy Elements

PROPERTIES OF HEAVIEST NUCLEI

R. Smolańczuk, J. Skalski and A. Sobiczewski

Sołtan Institute for Nuclear Studies, Hoża 69, PL-00-681 Warszawa, Poland
and
GSI, D-64220 Darmstadt, Germany

Introduction

The objective of the present paper is to present some of recent theoretical results on the ground-state properties of the heaviest nuclei. Main attention is paid to the alpha-decay and spontaneous-fission half-lives and in particular to the effects of the neutron deformed shell, predicted at the neutron number $N=162$, in these half-lives. Long lifetimes expected theoretically for nuclei around ^{270}Hs (i.e. the isotope with $N=162$ of the element hassium with atomic number $Z=108$) give a chance of observation of these very heavy deformed nuclei.

A more extensive discussion of the properties of the heaviest nuclei may be found e.g. in the review papers[1-4].

Importance of Shell Effects

Shell effects are important for all nuclei. Their significance for heaviest nuclei is, however, essential, as they are crucial for the half-lives of these nuclei. Some of the heaviest nuclei

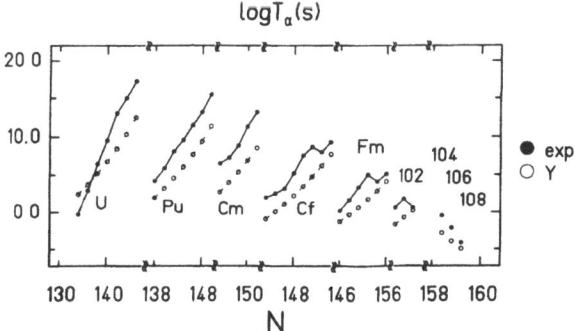

Figure 1. *Logarithm of experimental (exp) and calculated macroscopically (Y) alpha-decay half-lives T_α (Ref.[5]).*

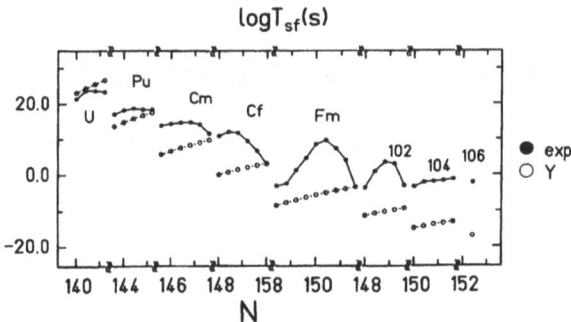

Figure 2. *Same as in Figure 1, but for the spontaneous-fission half-lives T_{sf} (Ref.[5]).*

could not exist without these effects. To illustrate these effects, Figure 1 gives[5] logarithm of the alpha-decay half-life T_α: experimental and calculated within a macroscopic model[6] without any shell effects. Thus, the difference between the two values is the shell effect in T_α. However, instead of showing this difference directly, we intentionally show the half-life, itself, to see its values. In particular, to see the strong dependence of T_α on the neutron N and proton Z numbers. One can see in Figure 1 that for all analyzed nuclei, except two lightest isotopes of uranium, the shell effect delays alpha decay. The delay is up to about 5 orders of magnitude. This fact is usually not realized and stresses the importance of shell effects in T_α also for well deformed nuclei, located far from spherical magic (or near magic) nuclei for which the effects are commonly discussed.

Figure 2 gives similar illustration for the spontaneous-fission half-life T_{sf}. The macroscopic calculation of T_{sf}, done here, consists in exploiting the widely used macroscopic Yukawa-plus-exponential model[6], for the calculation of the fission barrier, and a smooth, phenomenological model[7-9] of the inertia parameter, which describes the inertia of a nucleus with respect to changes of its deformation.

One can see in Figure 2 that the shell effect delays the fission process in all considered nuclei, except only few lightest ones (isotopes of uranium). The delay increases from few orders (Pu isotopes) to about 15 orders of magnitude for the heaviest even-even nucleus with measured T_{sf} (^{260}106). For such a heavy nucleus like ^{260}106, with T_{sf} of the order of few milliseconds, this elongation of T_{sf} makes up practically the whole half-life of these nuclei. In other words, they would not exist without shell effects.

Theoretical Methods

In the present paper, we are interested in description of the main properties of the heaviest nuclei, i.e. mass, and alpha-decay and spontaneous-fission half-lives.

Mass is calculated by the macroscopic-microscopic method, with the Yukawa-plus-exponential model[6] taken for the macroscopic part of the mass. The Strutinski shell correction, used for the microscopic part, is based on the Woods-Saxon single-particle potential[10].

The alpha-decay half-life T_α is calculated by the phenomenological formula of Viola and Seaborg[11] with the four adjustable parameters refitted[12] to account for new data.

Finally, the spontaneous-fission half-life T_{sf} is calculated in the dynamical way (e.g. Refs.[13-15]). It consists in the search for a one-dimensional fission trajectory in a multi-dimensional deformation space, which minimizes the action integral corresponding to the penetration of the fission barrier. The inertia tensor appearing in the integral and describing the inertia of the nucleus with respect to its deformation is calculated in the cranking approach (e.g. Ref.[16]).

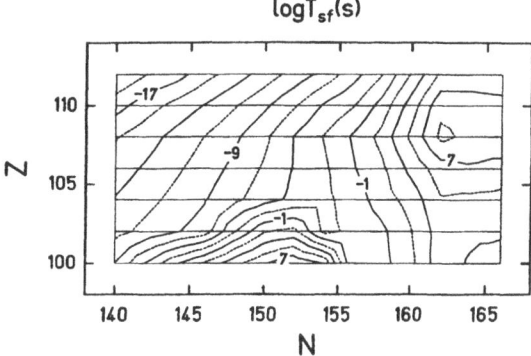

Figure 3. *Contour map of logarithm of spontaneous-fission half-life T_{sf} (given in seconds), calculated as a function of the proton Z and neutron N numbers. The difference in value of $\log T_{sf}$ between neighbouring solid lines is 4. Dashed lines divide this difference by two*[15].

Some of Theoretical Results

Figure 3 gives a contour map[15] of logarithm of the spontaneous-fission half-life T_{sf}. The half-life is calculated dynamically for nuclei with $Z=100-112$ and $N=140-166$.

A rather complex structure of the map is seen. Two maxima of T_{sf} are obtained: one for the known nucleus ^{252}Fm and the other for the not yet observed nucleus ^{270}Hs. The maxima are connected with the strong deformed shells at $N=152$ (and a weaker at $Z=100$) and at $N=162$ and $Z=108$, appearing for the ground-state configuration of these nuclei. The shells result in relatively high fission barriers for these nuclei. The shells at $Z=108$ and $N=162$, obtained in the calculations, are especially strong, qualifying the nucleus ^{270}Hs to be a good candidate for a double-magic deformed nucleus[17,18]. An interplay between the effects of two neutron deformed shells, the known shell at $N=152$ and the predicted one at $N=162$, seen in Figure 3, makes the dependence of the half-lives on the neutron number N especially interesting and complex. We will illustrate this dependence for both the alpha-decay, T_α, and spontaneous-fission, T_{sf}, half-lives for the elements 104 and 106. They have been calculated using the 7-dimensional deformation space $\{\beta_\lambda\}$, $\lambda = 2, 3, 4, ..., 8$, where β_λ are the usual deformation parameters. The half-lives shown in Figure 3 have been obtained in a smaller space $\{\beta_\lambda\}$, $\lambda = 2, 3, 4, 5$.

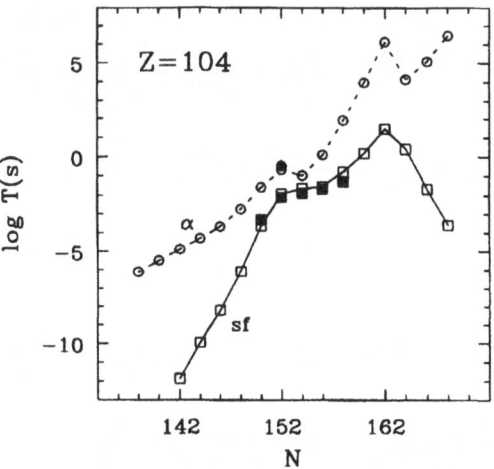

Figure 4. *Dependence of logarithm of the alpha-decay and spontaneous-fission half-lives (given in seconds) on the neutron number N, for the element $Z=104$ (Ref.[19]).*

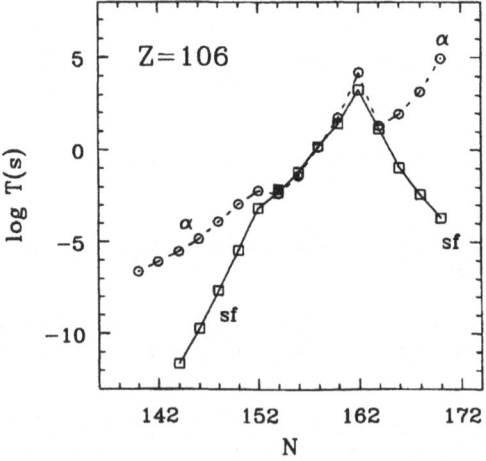

Figure 5. *Same as in Figure 4, but for the element $Z=106$.*

Figure 4 shows[19] the half-lives for $Z=104$. One can see the effects of both shells, at $N=152$ and 162, in the both half-lives. The effect of the shell at $N=152$ is smaller. This shell manifests itself more strongly in lighter elements (around Fm). The effect of the shell at $N=162$ is larger. One can also see that for all isotopes of the element 104, the calculated T_{sf} is smaller than T_α. It is smaller by only less than one order of magnitude for the isotope with $N=154$, but by as much as about 7 orders for the lightest ($N=142$) and about 10 orders of magnitude for the heaviest ($N=168$) isotope, for which both half-lives have been calculated. Thus, only spontaneous fission is practically expected to be observed for these light and heavy isotopes.

Figure 5 gives the results for the element 106. Here, the effect of the $N=162$ shell is even stronger than that for $Z=104$, probably because we are closer to the expected double-magic deformed nucleus ^{270}Hs, and the effect of the proton closed shell at $Z=108$ also contributes. The calculated fission half-life T_{sf} for the nucleus ^{268}106 is larger than the half-life T_{sf} of any isotope of the element 104. In particular, it is much larger than T_{sf} of the nucleus ^{266}104. Thus, due to large shell effects in these deformed nuclei, strong deviations from the rule that the fission half-life T_{sf} decreases with increasing atomic number Z are expected.

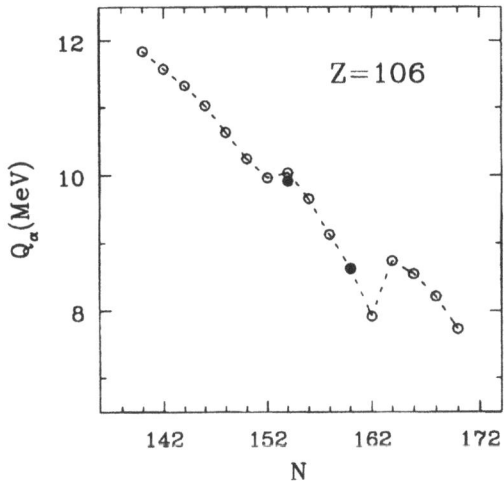

Figure 6. *Same as in Figure 5, but for the alpha-decay energy Q_α.*

Finally, Figure 6 shows the alpha-decay energy Q_α for the element 106. Effects of both shells, at $N=152$ and 162, are visible. The recent experimental value, obtained for ^{266}106 is also shown[20] and appears very close to the theoretical value.

In conclusion, one can say that the theoretical analysis of the properties of heaviest nuclei, performed in a multidimensional deformation space, reveals the essential role of shell effects in these properties. The effects make the half-life systematics, expected for these nuclei, much complex and unusual.

The authors would like to thank P. Armbruster, S. Hofmann, G. Münzenberg and Z. Patyk for helpful discussions. A support by the Polish State Committee for Scientific Research, grant no. 209 549 101 is gratefully acknowledged.

References

[1] S. Hofmann, *New results on heavy element research and plans for the future* in: *Proc. 24th Zakopane School on Physics, Vol. 1: Selected Topics in Nuclear Structure*, edited by J. Styczeń, and Z. Stachura, (World Scientific, Singapore, 1990)

[2] A. Sobiczewski, *Nuclear structure analyzed in a multidimensional deformation space* in: *Predeal International Summer School: New Trends in Theoretical and Experimental Nuclear Physics*, edited by A. A. Raduta, D. S. Delion and I. I. Ursu, (World Scientific, Singapore, 1992)

[3] P. Möller and J.R. Nix, *Stability of heavy and superheavy elements*, J. Phys. G, in press (1993)

[4] A. Sobiczewski, *Multidimensional analysis of the properties of heavy and superheavy nuclei*, Particles and Nuclei, in press (1993)

[5] Z. Patyk, A. Sobiczewski, P. Armbruster and K.-H. Schmidt, *Shell effects in the properties of the heaviest nuclei*, Nucl. Phys. A **491**, 267 (1989)

[6] H. J. Krappe, J. R. Nix and A.J. Sierk, *Unified nuclear potential for heavy-ion elastic scattering, fusion, fission, and ground-state masses and deformations*, Phys. Rev. C **20**, 992 (1979)

[7] J. Randrup, C. F. Tsang, P. Möller, S. G. Nilsson and S. E. Larsson, *Theoretical predictions of fission half-lives of elements with Z between 92 and 106*, Nucl. Phys. A **217**, 221 (1973)

[8] J. Randrup, S. E. Larsson, P. Möller, S. G. Nilsson, K. Pomorski and A. Sobiczewski, *Spontaneous-fission half-lives for even nuclei with $Z \geq 92$*, Phys. Rev. C **13**, 229 (1976)

[9] A. Sobiczewski, *Mass parameters in nuclear fission*, Particles and Nuclei **10**, 1170 (1979)

[10] S. Ćwiok, J. Dudek, W. Nazarewicz, J. Skalski and T. Werner, *Single-particle energies, wave functions, quadrupole moments and g-factors in an axially deformed Woods-Saxon potential with applications to the two-centre-type nuclear problems*, Comput. Phys. Commun. **46**, 379 (1987)

[11] V. E. Viola, Jr. and G. T. Seaborg, *Nuclear systematics of the heavy elements-II*, J. Inorg. Nucl. Chem. **28**, 741 (1966)

[12] A. Sobiczewski, Z. Patyk and S. Ćwiok, *Deformed superheavy nuclei*, Phys. Lett. B **224**, 1 (1989)

[13] H. C. Pauli, *On the shell model and its application to the deformation energy of heavy nuclei*, Phys. Reports C **7**, 35 (1973); *Four lectures on fission: Fragments of a dynamic theory of collective motion in nuclei*, Nukleonika **20**, 601 (1975)

[14] A. Baran, K. Pomorski, A. Lukasiak and A. Sobiczewski, *A dynamic analysis of spontaneous-fission half-lives*, Nucl. Phys. A **361**, 83 (1981)

[15] Z. Patyk, J. Skalski, A. Sobiczewski and S. Ćwiok, *Potential energy and spontaneous-fission half-lives for heavy and superheavy nuclei*, Nucl. Phys. A **502**, 591c (1989)

[16] K. Pomorski, T. Kaniowska, A. Sobiczewski and S. G. Rohoziński, *Study of the inertial functions for rare-earth nuclei*, Nucl. Phys. A **283**, 394 (1977)

[17] Z. Patyk and A. Sobiczewski, *Main deformed shells of heavy nuclei studied in a multidimensional deformation space*, Phys. Lett. B **256**, 307 (1991)

[18] Z. Patyk and A. Sobiczewski, *Ground-state properties of the heaviest nuclei analyzed in a multidimensional deformation space*, Nucl. Phys. A **533**, 132 (1991)

[19] R. Smolańczuk, J. Skalski and A. Sobiczewski, *The isotopic spin dependence of half-lives of heaviest nuclei* in: *Proc. Int. School-Seminar on Heavy Ion Physics*, Dubna, in press (1993)

[20] Yu. A. Lazarev et al., *Search for enhanced nuclear stability near the new deformed shells N=162 and Z=108* in: *Proc. Int. School-Seminar on Heavy Ion Physics*, Dubna, in press (1993)

THE PROSPECTS OF HEAVY ELEMENT RESEARCH

Gottfried Münzenberg

Gesellschaft für Schwerionenforschung mbH, Planckstrasse 1, D-64291 Darmstadt, Germany

The investigation of the transactinide elements at the very limits of macroscopic stability led to the discovery of a region of shell stabilized nuclei. Their unambitious identification by single-atom decays was possible only with the technique of in-flight separation, a method now well established for the investigation of nuclei far-off stability and heavy-element research. The prospects for the synthesis of heavy elements considering the new possibilities with secondary beams and heavy ion storage rings will be discussed. A brief comment on the naming of the heaviest elements Nielsbohrium, Hassium, and Meitnerium will be given.

1. Introduction

The fundamental problem behind heavy element research is the question for the upper end of the periodic table of elements. Already Otto Hahn and his group wanted to create transuranium nuclei by neutron irradiation of uranium. Neutron capture followed by beta decay, a method invented by Fermi, was known to lead to the next heavier element above the target. As is well known Hahn was not successful to create transuranium nuclei but discovered the nuclear fission. Lise Meitner and Otto Frisch immediately gave the physics interpretation to Hahns result and concluded on the basis of the nuclear liquid drop model that nuclear fission would terminate the nuclear table near element 110. With the improvement of nuclear models shell corrections were included. The method of Strutinsky allowed to calculate deformed nuclei, an inevitable necessity to obtain fission barriers. On the basis of macroscopic-microscopic calculations it was predicted that the stabilization of the next doubly closed shell above the doubly magic lead, located at 114 protons and 184 neutrons, should be strong enough to create an island of purely shell stabilized superheavy nuclei in a region of macroscopic instability (Myers and Swiatecki 1966, Mosel and Greiner 1969). Halflive calculations (Fizet and Nix 1972) predicted half-lives of up to 10^{15}s for the most stable isotopes in this region. A large scale search for primordial superheavy nuclei in nature, on the earth as well as in exraterrestric samples of meteorites started. Simultaneously various experiments on the synthesis of superheavy nuclei were carried out, both without positive result (Herrmann 1990).

Hahn and collaborators also elaborated the principles of the experimental method for the identification of exotic nuclei where only small numbers of atoms are available: the chemical separation followed by identification based on nuclear decay measurement.

2. Experimental Methods

Various techniques were applied for the investigation of heavy elements up to the transactinide region (Seaborg and Loveland 1990). The classical chemical methods were useful only up to element 104. Only recently the chemical separation of a few atoms of element 105 was successful (Kratz 1992). Methods not involving any separation were the helium-jet system using a gasflow to transport the heavy nuclei from the target to silicon surface barrier detectors, which has been applied up to element 106 to identify a new element by its alpha decay using the parent-daughter correlation technique, and the rotating drum to detect fission

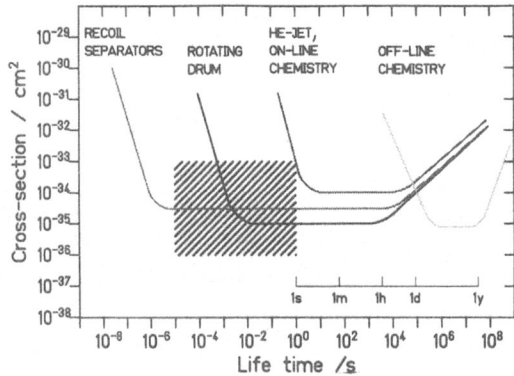

Figure 1. *Sensitivity in cross-section and half-life of in-flight separation compared to the rotating drum, heliumjet gas transport systems, and chemical separation methods. The hatched region corresponds to the sensitivity needed for transactinide nuclei.*

with nuclear track detectors (Armbruster 1985; Münzenberg 1988). The direct observation of an element beyond element 106 was only possible with the recoil separation in-flight. Fig. 1 compares time ranges and sensitivities of the various methods showing clearly that in-flight separation is the only method to observe and identify an element above element 106 directly. The rotating drum technique even being fast and sensitive enough has the severe drawback not to be suitable for the safe identification of a specific isotope as nuclear fission is not a characterisic decay signature. We know today that the known isotopes beyond element 106 do not decay by fission.

All transactinide elements have been synthesized by complete fusion of heavy ions. For this reaction most efficient is the kinematic separation: complete fusion products, created by the complete amalgamation of target and projectile, suffer the full mometum transfer

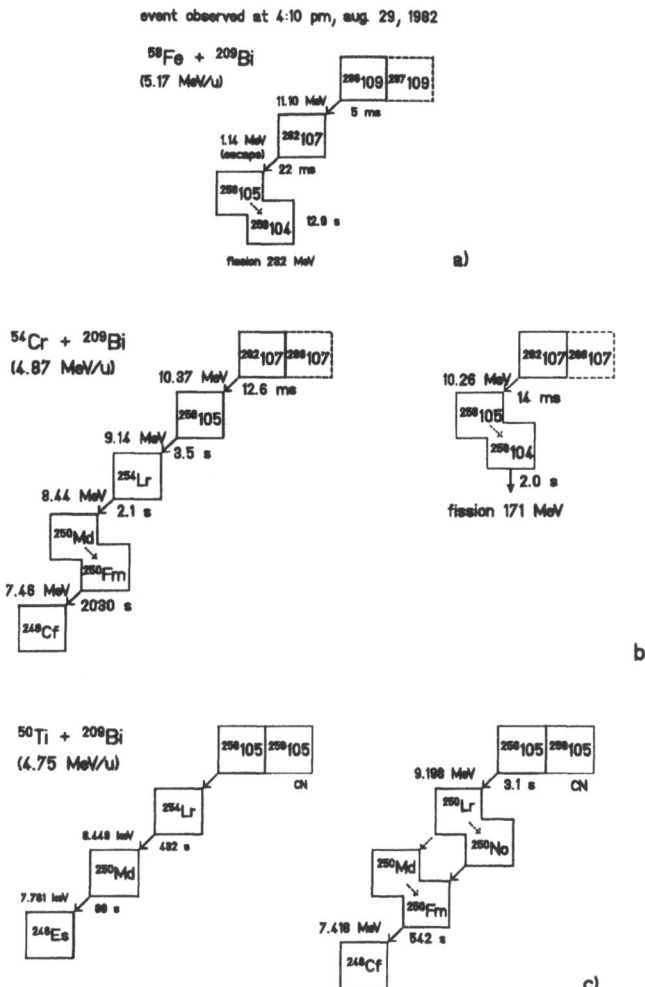

Figure 2. *Building up the decay chain for element 109. Examples of granddaughter, daughter single-atom decay chains, respectively, and one chain of element 109 are displayed.*

from the projectile and are the only group of reaction products moving with center of mass velocity. Therefore velocity selection with a velocity filter collecting all ionic charge states is the most sensitive separation. The energetic nuclei are then implanted into silicon surface barrier detectors where they decay.

The setup used for the discovery of the elements 107, 108, and 109 is the two stage velocity filter SHIP (Münzenberg 1979). After implantation in an array of position sensitive surface barrier detectors (Hofmann 1984) the new elements were identified by building up the alpha decay chains from known isotopes. Fig. 2 shows as an example how the element-109 decay chain (Münzenberg 1989) was built from the granddaughter isotope $^{258}105$ followed by the daughter $^{262}107$ (Münzenberg 1984). A number of decay chains from the elements 105 and 107, respectively, was collected and alpha decay energies as well as the statistical (Poisson) distributions of the measured time intervals between corresponding decays were used for our assignment.

2.1 Present Status and Naming

Recently the heaviest elements 107, 108, and 109 have been named, about ten years after their discovery. The "Transfermium Working Group" headed by Sir Denys Wilkinson and installed by the IUPAC and IUPAP, reviewed all publications on the transfermium elements to decide on the priorities of the discovery. As for the heaviest elements 107, 108, and 109 according to their report (Barber 1992) the priority of the discovery was given to the GSI group with the recommendation to honour the contribution of the Dubna group to the physics of transactinide research. The GSI group: P. Armbruster, F.P. Hessberger, S. Hofmannn, K. H. Schmidt, W. Reisdorf, M. E. Leino, a guest from the University of Yuvaskyla, and me proposed the names for these elements to JUPAC and IUPAP (Armbruster et al. 1993):

Nielsbohrium (Ns) for element 107, in honouring Niels Bohr. Originally Hahnium and Nielsbohrium were proposed for element 105 by the Berkeley and Dubna groups, respectively. To save both names and to give credit to the Russian contribution to heavy element reseach the name Nielsbohrium was taken for element 107 in agreement with the Dubna group.

Hassium (Hs) was given to element 108. Hassia is the old latin form of Hessen, the country of the discovery of these elements: This was to give credit to the strong support predominantly of the universities of Hessen to GSI and especially to heavy element research.

Meitnerium (Mt) was proposed for element 109 in honour of the physicist in Hahns Group, Lise Meitner, who immediately after the discovery of nuclear fission explained the physics of this new decay mode. She recognized the tremendous amount of energy created in this process and predicted the perodic table of elements to be limited by fission.

2.2 Groundstate Properties

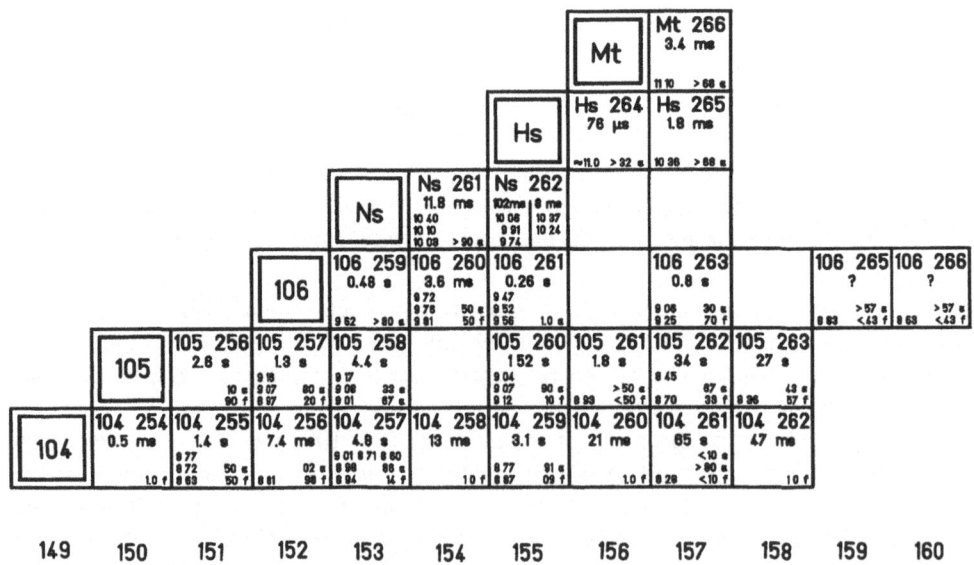

Figure 3. *The upper end of the periodic table.*

The upper part of the nuclear table is displayed in fig. 3 representing the status of experimental results. While for most of the known isotopes of the elements 104 and 105 spontaneous fission is observed – for the even-even isotopes of element 104 fission is the up to now only observed decay mode – the transactinide elements undergo primarily alpha decay, including the heaviest known even-even isotopes ^{260}106 and ^{264}108 (Münzenberg 1987). This is contrary to what one would expect in the frame of the macroscopic model and points towards a microscopic stabilization.

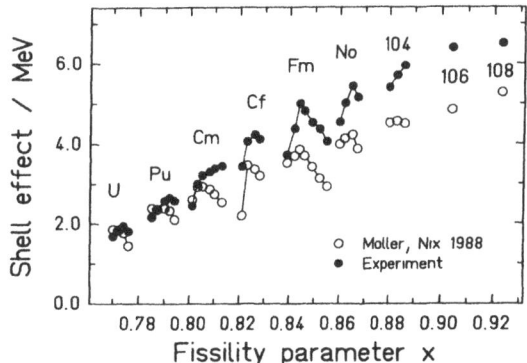

Figure 4. *Experimental shell corrections (dots) compared to the calculations of Möller (circles, Möller 1988). The lines connect isotopic chains.*

In the macroscopic-microscopic description the experimental shell effects can be obtained from the ground state masses – or Q_α values, respectively – with an appropriate macroscopic model. These experimental shell corrections are displayed in fig. 4 and compared to the calculations of Möller et al. They increase from uranium to elements 106 and 108 to values of about 6 MeV which is half the value of the shell correction for ^{208}Pb. This enhanced stability is due to an increasing hexadecapole deformation with a maximum near 162 neutrons (Möller and Nix 1992, Sobiczewski 1993). A nice proof gave the recently discovered most neutron-rich isotopes of element 106 with 159 and 160 neutrons, respectively. Despite their lower half-lives, predicted to 10-50 s by systematics from the experimental α-decay energy of 8.8 and 8.6 MeV, respectively, no fission was observed, even for the even-even isotope ^{260}106 (Lazarev 1993).

The fission barrier stabilizing nuclei against prompt disintegration can to a first approximation again be described as the superposition of a macroscopic barrier and the ground state shell correction. The contribution of the macroscopic barrier in the transactinide region drops below 1 MeV. The elements near 106 and above are completely shell stabilized. The dominance of fission in the vicinity of element 104 can be explained now: the macroscopic stablilization disappears while the microscopic stabilization is still weak. Another nice proof for the shell character of the transactinides can be taken from the barrier curvatures displayed in fig. 5. The actinides have thick, liquid drop dominated barriers with a small curvature parameter. Towards the transactinides the barrier curvature increases to 1 MeV, compatible with the barrier curvature measured for the inner barrier of the nuclei in the uranium to californium region (Münzenberg 1988). This strongly supports the picture that the outer barrier sitting on top of the liquid drop barrier at large deformation drops below the grondstate when the macroscopic barrier disappears. As a consequence only the thin, shell determined inner barrier remains.

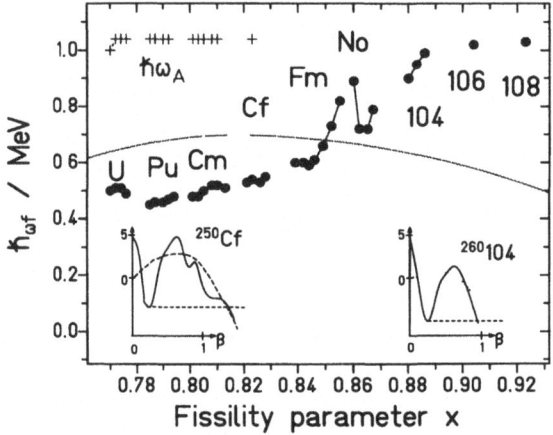

Figure 5. *Experimental barrier curvature parameters (dots) and curvatures of the inner barrier (crosses), the solid line indicates the curvature of the liquid-drop barrier. Inserts: qualitative shape of the actinide and transactinide barriers, respectively. Dashed line: contribution of the liquid-drop barrier.*

Details of a more profound theoretical description of the experimental results and their consequences for the stability of the heaviest elements are discussed in the contribution of A. Sobiczewski to this school (Sobiczewski 1993). Here we will only show the consequences for the half-lives predicted for the heaviest elements (Fig. 6): The hexadecapole stabilized nuclei near $N=164$ create a new region of deformed-shell stabilization connecting the upper end of the periodic table and the doubly magic shell closure near $Z=114$ and $N=184$. The half-lives

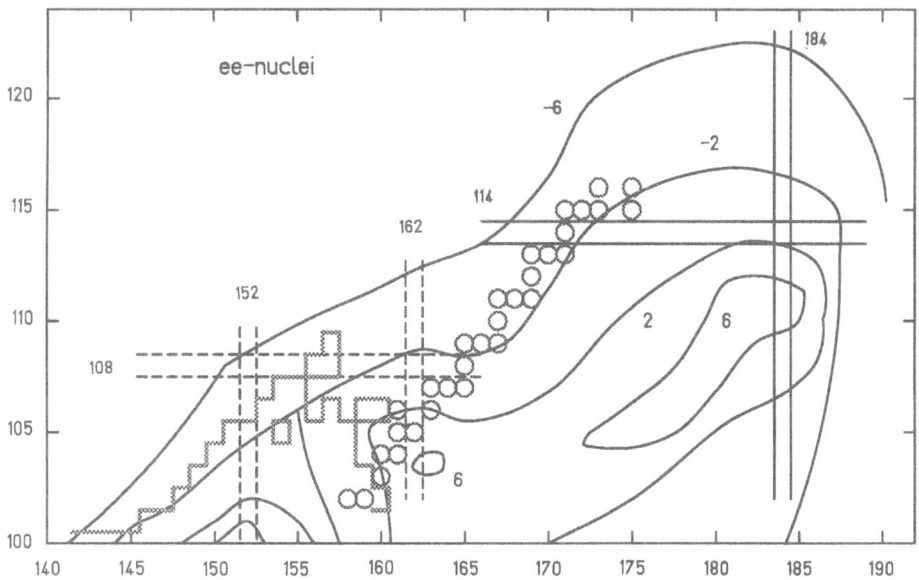

Figure 6. *Theoretical half-life contour plot for the heavy and superheavy elements (after Sobiczewski 1989). The half-lives are seconds given in exponentials of ten. The circles indicate possible landing points for a ^{254}Es target and stable beams.*

of the most stable nuclei in both regions are almost equal and predicted to 10^6 s (Sobiczewski 1989). Between the upper end of the periodic table and the spherical superheavies half-lives never drop below one second and are accessible to chemical separation. As a consequence experiments on the chemistry of the transactinides have been started (Kratz 1992).

3. Production

The only method sucessful up to now for the production of new elements in the transactinide region is the complete fusion of heavy ions (Armbruster 1985, Münzenberg 1988). As the projectile energy to overcome the Coulomb resp. fusion barrier, E_B, generally exceeds the Q-value necessary for its formation, the compound nucleus is excited and has to dissipate energy by particle emission to come to the ground state. There are two competing processes: the probability of the projectile to penetrate the fusion barrier which can be described in terms of the fusion cross section σ_{fus} and the survival probability of the excited system w_i which counts for each deexcitation step i. Both depend on the projectile energy E_P and the angular momentum l:

$$\sigma_{VR} = \sigma_{fus}\,(E_B - E_p\,,\,l)\,\sum w_i(E_p + Q\,,\,l)\,. \tag{1}$$

Two types of target-projectile combinations have been sucessfully used: actinide targets from plutonium to californium with correspondig beams around nitrogen or targets near lead and beams from chromium to iron. More symmetric target-projectile combinations have been tried but were unsuccessful in heavy element synthesis. Excitation energies at the Coulomb barrier for actinide reactions are near 40 MeV whereas with lead targets typical excitation energies are 15 MeV. Correspondingly the compound nuclei need four or one evaporation steps, respectively, to dissipate the excitaion energy. For the highly fissile heavy elements the chance to fission is about two orders of magnitude higher than to evaporate a neutron. This led to the idea, to try cold formation of the highly fissile elements at the top of the nuclear table, firstly published by Oganessian (Oganessian 1974) and theoretically proven by a number of calculations in the frame of the fragmentation theory (Gupta et al. 1977). We produced the elements 104 to Mt by irradiations of ^{208}Pb and ^{209}Bi with beams of ^{50}Ti, ^{54}Cr, and ^{58}Fe, respectively. Fig. 7 shows the production cross sections for both types of reactions leading into the sub-nanobarn region above element 105. They are of the same order which is surprising as one would expect the fusion-evaporation cross sections for the cold fusion to be much enhanced.

The explanation was found in the symmetric reactions leading to compound nuclei heavier than mercury (Schmidt and Morawek 1991): Massive systems do not fuse at the Coulomb barrier, their barrier is shifted to higher energies due to dissipative effects in the process of amalgamation, they need an "extra push" to undergo complete fusion (Bjornholm and Swiatecki 1982). The barrier shift is equivalent to a fusion hindrance. The fusion of massive systems is governed by the so called entrance fissility, an average of the fissilities (Blocki 1986) of the *fused* system scaling with the ratio of the Coulomb- to the nuclear energy for the compound nucleus and the *entrance* fissility depending on the ratio of the Coulomb force to the nuclear force of the two tangent spheres projectile and target. It could also been shown that nuclear structure effects such as shell effects in target or projectile reduce the dynamical heating (Berdichewski et al. 1989).

3.1 Towards Element 110

As for the cold fusion our extrapolation gives a cross section of one picobarn for element 110 (fig.7), which corresponds to the production of about one atom per week and is clearly the limit of our present possibilities (Hofmann et al. 1993).

The actinide based reactions certainly will not suffer so much from fusion hindrance (fig. 7), have however the drawback of the hot compound nucleus. Up to now is not clear how the balance between both processes will influence cross sections beyond element 106 as there are no data for fusion-evaporation reactions in this region. The questions are
- to which extent pre-compound processes will play a role and
- the time scale of the prompt fission of the hot compound nucleus.

Here we should keep in mind that for the very heavy systems in the region of vanishing liquid drop barriers we deal with a flat potential-energy surface with almost no pockets. For near-barrier energies the interactions between target and projectile will be rather slow and give the chance for non-equilibrium processes. There is some evidence that fission is much slower than accepted up to now (Hilscher 1992). As a consequence there are expectations for cross sections of the order of 50 picobarns for actinide based reactions leading to the elements 108 and 110, respectively (Oganessian 1992), an other idea is to rely on pre-compound particle emission, such as α-particles emission to cool down the hot compound system.

4. The New Generation of Experiments

In the meantime recoil separation has been established for heavy element research. Various methods for separation in-flight are used (Fujioka 1993, Münzenberg 1994). The energy filter VASSILISSA installed at Dubna uses electric deflection condensers for recoil separation.

Recoil-product mass separators, further developed Mattauch-Herzog type separators, such as the Argonne Fragment Mass Analyser FMA, are ready for heavy element research. Gasfilled magnetic deflection systems, up to now preferably used for the separation of fission fragments, are successfully used for heavy element research. The SASSY separator at LBL Berkeley was the first one to separate a transfermium element (Ghiorso 88). With the GNS installed at Dubna two new isotopes of element 106 were found recently (Lazarev 1993), and the GARIS separator in RIKEN (Japan) started running a heavy element program (Morita 1993). The recently commissioned RITU separator of the University of Jyväskylä will start heavy element research (Leino 1993).

As for the near future the search for element 110 is going on at SHIP, which has been improved and equipped with a new detector system to detect alpha decays with almost 4π geometry to avoid the escape-alphas. All in all the sensitivity of the GSI setup has been improved by a factor of ten (Hofmann 1993). The experiment on the synthesis of element 110 will rely on the cold heavy ion fusion reaction ^{208}Pb(^{62}Ni,n)^{269}110. The Dubna experiments will try actinide based reactions such as ^{249}Cf(^{26}Mg,4n)^{271}110.

Of fundamental importance for the understanding of the physics of the heavy elements would be a more detailed spectroscopic investigation of the transactinides to proceed towards the N=164 and possibly to the N=184 neutron shells with the aims to measure fission branches and to investigate α-decay to establish the shell stabilization in the region of the hexadecapole deformed nuclei and to establish ground state deformations from the 0^+ to 2^+ measured transitions obtained in α-decay. These experiments would help to further improve the nuclear models and their predictive power in the heavy element region.

The possibilities of new accelerator installations such as tandem cyclotrons and accelerator coupled heavy ion storage-rings have not yet been discussed, for instance to use them for direct mass measurement.

As for the production of the heaviest element clearly the
- fusion hindrance for actinide systems
- the survival of hot compound systems
- the time scale of compound fission and
- pre-equilibrium particle emission

would be the crucial questions.

Here it is noteworthy to point out that not only the investigation of the heavy evaporation residues but also to follow the dinuclear system on its way to complete fusion would be of interest, such as the investigation of nuclear transfer and the various modes of prompt fission and accompanied pre-compound particle emission in kinematically complete experiments.

The investigation of the chemistry of the transactinide elements being under way now would reveal the influence of relativistic effects of the inner electrons on the chemical behaviour.

5. Heavy Element Production with Energetic Beams of Exotic Nuclei

Fig. 7 shows the possibilities to reach the center of the deformed region near N=162 neutron-rich transactinides with the heaviest practicable target ^{254}Es and corresponding beams of Be and heavier. With the new possibilities to produce energetic secondary beams of instable nuclei by projectile fragmentation the question immediately arises whether such beams could be used for the production of heavy and superheavy elements. With complete fusion reactions

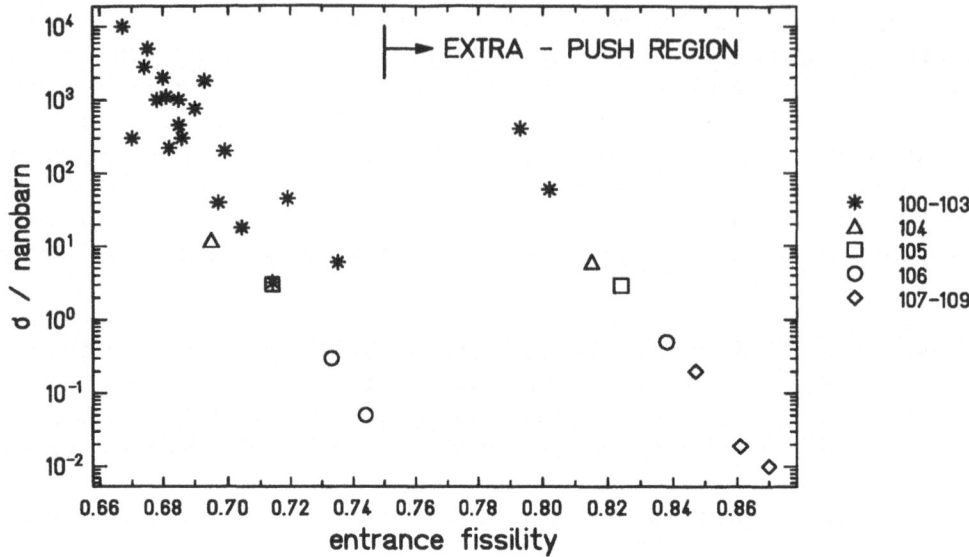

Figure 7. *Production cross sections for lead-based and actinide-based fusion-evaporation reactions. The experimental cross sections are plotted versus the entrance fissility.*

the doubly magic nuclei near $^{298}114$ are out of reach even for exotic secondary beams. About 10-20 neutrons are missing as compared to the stable species. This region can only be reached using a new reaction mechanism such as transfer of extremely neutron-rich clusters. Nevertheless it is worthwhile to start a first generation of experiments to investigate heavy ion fusion with secondary nuclear beams as they offer the unique possibility to use long isotopic chains to study the influence of nuclear structure on the complete heavy ion fusion and extra push, such as closed shells, deformations, or neutron skins. The temperature dependence of shell effects in the heavy compound system can be studied with the disintegration of relativistic secondary beams created by the fragmentation of heavy nuclei such as uranium beams. Excitation mechanisms are Coulomb excitation of the giant resonance (Schmidt 1993) or inelastic collisions e.g. with protons. In the latter case kinematically exclusive experiments would allow an exact determination of the excitation energy.

Finally we will discuss the question to which extent secondary beam intensities of light projectiles are sufficient for the production of heavy elements. Table 1 shows a compilation of beam intensities obtainable with the two most important schemes for the production of energetic secondary beams: The fragmentation and separation in-flight of relativistic heavy ions, as available at GANIL (France), Michigan State University (USA), and RIKEN (Japan) for intermediate energies, and at GSI for high energies; or the two accelerator scheme, the fragmentation of a heavy target by energetic protons or heavy ions and post acceleration, working now for light ions and low energies at Louvain-La-Neuve (Belgium) and planned for GANIL (Kienle 1993) and possibly CERN ISOLDE.

For projectile fragments not too far from stability production cross sections are of the order of millibarns, so we give the luminosity in $mb^{-2}s^{-1}$. Secondary beam intensities of 10^6/s to 10^7/s are within reach already now, 10^8/s to 10^9/s in the future with the two accelerator scheme. To calculate the luminosities for heavy ion fusion we have to take into account the small width of the excitation function for the formation of a specific isotope covering only 10

Table 1. *Heavy-element production with secondary beams.*

SECONDARY BEAMS		LUMINOSITIES
AVAILABLE (relativistic beam)	GANIL, RIKEN GSI	10^7 mb^{-1} s^{-1} 10^6 mb^{-1} s^{-1}
PROJECTS (post acceleration)	CERN-ISOLDE GANIL-PLUS	10^9 mb^{-1} s^{-1} 10^8 mb^{-1} s^{-1}
EVAPORATION RESIDUES (target 10^{18}/cm^2)		10^{-1}–10^2 μb^{-1} s^{-1}

MeV of projectile energy. The corresponding energy-loss of a projectile in the target covers about 10^{18} atoms/cm^2. The resulting luminosities with secondary beams are 0.1 μb d^{-1} to 100 μb d^{-1} values still at the limit to investigate transfermium isotopes, as the cross section compilation in fig 7. shows. Secondary beams in any case are strong enough to investigate fusion-fission.

The new capabilities of heavy ion storage rings have also been discussed in this context (Münzenberg 1992). The use of cooled and stored beams and internal targets would enhance luminosities tremendously as
- the storage will allow to use the recirculating beam coasting with 10^5 to 10^6 revolutions per second
- accumulation will enhance the available beam intensity

The inherent problem is the ionic charge exchange in the internal target which will create wide ionic charge distributions difficult to store. One way to come out of this problem would be an ion trap target with bare ions, a technical problem not solved yet.

6. Conclusion

The result of heavy element research is the discovery of the deformed region of stability centered at 108 protons and 164 neutrons forming a bridge between the upper end of the periodic table and the double shell closure at 114 protons and 184 neutrons. Improved theories predict half-lives of less than one year for the doubly magic superheavies, so to our present knowledge uranium will be the heaviest primordial element.

In-flight separation is now the established method for heavy element research, together with the new developed detection techniques, which permit to identify new elements and to extract gross nuclear properties from the decay of single atoms.

I gratefully acknowledge fruitful discussions with P. Armbruster, S. Hofmann and F. P. Hessberger. I thank E. Pfeng for arranging the manuscript.

References

[1] P. Armbruster, *Ann. Rev. Nucl. Part. Sci.* **35**, 135 (1985)

[2] P. Armbruster et al., *Prog. Part. Nucl. Phys.* **31**, 241 (1993)
[3] R. C. Barber et al., *Prog. Part. Nucl. Phys.* **29**, 253 (1992)
[4] D. Berdichewski et al., *Nucl. Phys. A* **499**, 609 (1989)
[5] S. Bjornholm and Swiatecki W. J., *Nucl. Phys. A* **391**, 471 (1982)
[6] J. P. Blocki et al., *Nucl. Phys. A* **459**, 145 (1986)
[7] M. Fujioka et al. eds., *Proc. 12th. int. Conf. on Electromagnetic Isotope Separators and Techniques Related to their Application EMIS 12, Nucl. Instrum. Meth. B* **70** (1992)
[8] R. K. Gupta et al., *Z. Phys. A* **283**, 217 (1977)
[9] E. O. Fiset and Nix, J. R., *Nucl Phys. A* **193**, 647 (1973)
[10] A. Ghiorso, *Nucl. Instrum. Meth. A* **269**, 194 (1988)
[11] G. Herrmann, *Proc. "Fifty Years with Transuranium Elements", The Robert A. Welch Foundation Conference on Chemical Research*, Houston (USA), 1990, 343 (1990)
[12] D. Hilscher and H. Rossner, *Ann. Phys. Fr.* **17**, 471 (1992)
[13] S. Hofmann et al., *Nucl. Instrum. Meth. A* **282**, 28 (1984)
[14] S. Hofmann, *Proc. "Actintides '93 Int. Conf"*, Santa Fe (USA), 1993, to be publ. in *Journ. of Alloys and Coumpounds* (1993)
[15] P. Kienle (chairman), *NUPECC (Nuclear Physics European Collaboration Committee) Report "European Radioactive Beam Facilities"* (1993)
[16] J. V. Kratz et al., *Radiochimica Acta* **48**, 121 (1992)
[17] Yu. Lazarev, *private communication* (1993)
[18] M. E. Leino, *private communication* (1993)
[19] T. Morita, *private communication* (1993)
[20] U. Mosel and Greiner, W., *Z. Phys.* **222**, 261 (1969)
[21] P. Möller and Nix J. R., *At. Data Nucl. Data Tables* **39**, 213 (1988)
[22] P. Möller and Nix J. R., *Nucl. Phys. A* **549**, 84 (1992)
[23] G. Münzenberg et al., *Nucl. Instrum Meth* **16**, 65 (1979)
[24] G. Münzenberg et al., *Z. Phys. A* **315**, 145 (1984)
[25] G. Münzenberg et al., *Z. Phys. A* **328**, 49 (1987)
[26] G. Münzenberg et al., *Z. Phys. A* **333**, 163 (1989)
[27] G. Münzenberg, *Reports on Progress in Physics* **51**, 57 (1988)
[28] G. Münzenberg, *Workshop on "Physics and Techniques of Secondary Nuclear Beams"*, Dourdan (France), Editions Frontiers, Gif-Sur-Yvette (France), 253 (1992)
[29] G. Münzenberg, *In-Flight Separation of Heavy Ions* in *Handbook of Nuclear Decay Modes*, edited by D. N. Poenaru, (CRC Press Inc., New York, in press) (1994)
[30] Yu. Ts Oganessian, *Lect. Notes in Physics* **33**, 221 (1974)
[31] Yu. Ts Oganessian, *private communication* (1992)
[32] G. T. Seaborg and Loveland, W. D., *The Elements Beyond Uranium*, (John Wiley and Sons Inc., New York, 1990)
[33] K. H. Schmidt and Morawek, W., *Rev. Prog. Phys.* **54**, 949 (1991)
[34] K. H. Schmidt, *Int. Worksh. on Gross Properties of Nuclei and Nuclear Excitations XXI*, Hirschegg, 1993, edited by H. Feldmeier; Int. Rep. Gesellschaft für Schwerionenforschung, 1993
[35] A. Sobiczweski et al., *Phys. Lett. B* **186**, 6 (1989)
[36] A. Sobiczewski, contribution to this school (1993)

MULTINUCLEON TRANSFER REACTIONS – AN ALTERNATIVE PATH TO HEAVY ELEMENT SYNTHESIS

M. T. Magda

State University of New York at Stony Brook, Department of Chemistry, Stony Brook, NY 11794-3400, USA

1. Introduction

The identification of transuranium, and particularly, transactinide ($Z > 104$) elements brings information on the properties of these nuclei and provides a strong test of the nuclear structure models adopted in the theoretical calculations. Over the years, considerable effort on both theoretical and experimental side has been dedicated to the discovery of heavy and superheavy elements, an excellent presentation of which can be found in ref. [1]. Since the first man-made heavy element (Np) has been prepared in 1940 by McMilan and Abelson[1], 17 transuranium elements have been synthesized, extending thus the limits of the periodic table to Z=109 [1-4] (and references therein). Two methods have been used to produce heavy elements: cold fusion reactions and multinucleon transfer reactions (hereafter abbreviated as MNTR). The object of this study are MNTR leading to heavy elements, with particular reference to those induced by exotic beams. After a review of the present status of the search for transuranium elements (Chapter 2) general features and production mechanism of MNTR are discussed in Chapter 3. Chapter 4 explores the possibilities of using radioactive nuclear beams (hereafter abbreviated as RNB) for the synthesis of heavy elements by MNTR.

2. Present Status of the Synthesis of Heavy Elements

A chart of the isotopes of elements with atomic number $Z = 101 - 109$ discovered so far is given in Fig. 1 (data used in this figure have been provided by S. Hofmann[5]).

Isotopes of elements up to $Z = 105$ have been synthesized in the laboratory by both heavy ion cold fusion reactions and MNTR. The heaviest elements $107 - 109$ have been produced by cold fusion evaporation reactions between ^{54}Cr and ^{58}Fe with targets of ^{208}Pb or ^{209}Bi.[2-4] The fusion products were separated by the velocity filter SHIP and identified by investigating the genetic relationship of the detected elements within the predominant decay mode, namely α-decay. The production cross sections were of $167pb$ for Ns, $19pb$ for Hs and $10pb$ for Mt,[2-4] this last nuclide being identified by three decaying nuclei. The analysis of α-decay energies led to a semiempirical evaluation of the microscopic term (shell correction) in the nuclear masses and thus to a realistic determination of binding energies and fission barriers.[4] These semiempirical results confirmed the theoretical predictions [6,7] of the existence of the shell stabilized deformed nuclei with Z around 108, the "peninsula" of heavy stable elements. Besides the discovery of the new elements at the upper end of

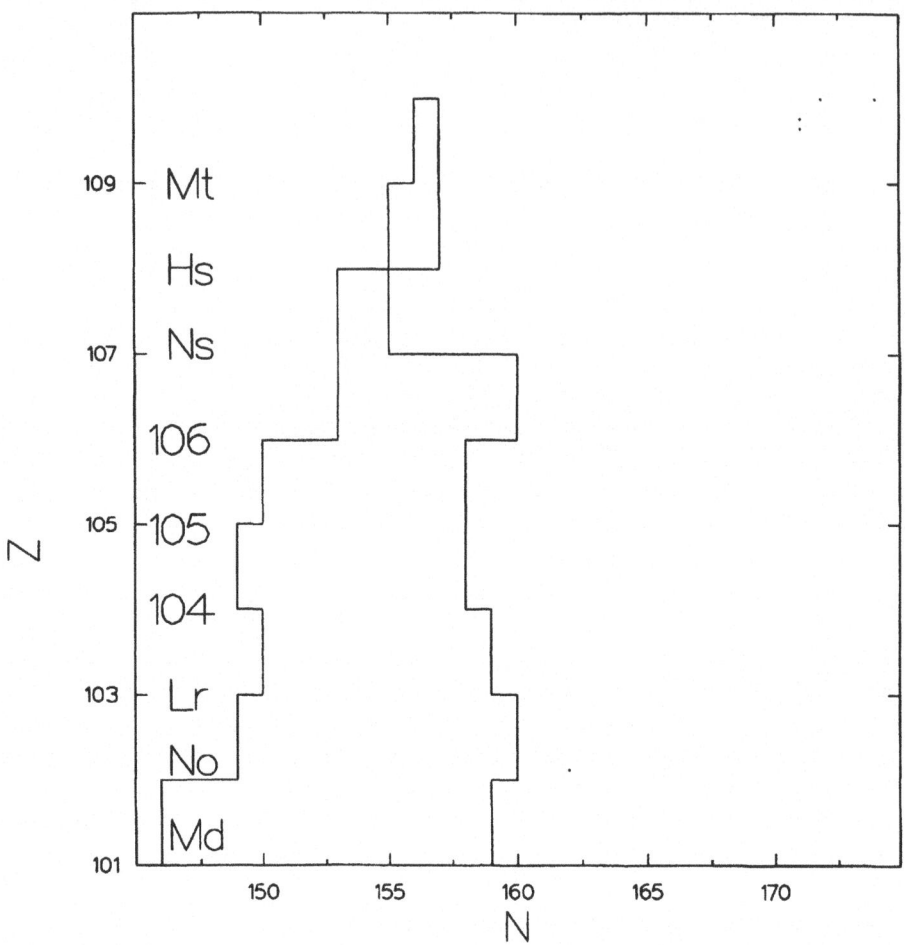

Figure 1. *Chart of discovered isotopes (full line) of elements $Z = 101 - 109$. The dotted line represents the possible extension to neutron-rich isotopes by using MNTR with RNB.*

the periodic table, the α-emission as the main decay mode of the nuclei with proton number $Z = 104 - 109$ along with the unexpected stability against spontaneous fission represents a remarkable finding of the recent search for heavy elements.[3,4]

As a result of the failure to produce superheavy elements the strategy of going up in Z step by step has been adopted and thus the element 110 has become the target of the latest investigations. Previous work on heavy element synthesis has shown that this is a complex process, in which the stability and therefore the survival probability of the synthesized products is only one of its many facets.[1-3,8,9] Another important aspect is related to the limitations of the production process: the dynamical hindrance in the entrance channel (the "extra-push" limitation) and the thermal hindrance in the exit channel (by fission losses in the evaporation cascade).[2] The choice of the "best" combination projectile-target and optimum bombarding energy must take into account the delicate balance between these factors. The experimental systematics of production cross sections of residues resulting in fusion reactions with evaporation of 1, 2 and 4 neutrons obtained in Berkeley, Darmstadt and Dubna [4,8] indicates a steep decrease of the cross sections versus the element number. The extrapolation of these data to $Z = 110$ predicts the highest cross sections in the case of reactions with ^{208}Pb and ^{209}Bi of the order of $1pb$. Therefore a detection sensitivity of one picobarn is a prerequisite of such experiments as recently shown by S. Hofmann.[8] Accordingly, the experiments planned at GSI-Darmstadt on identification of element 110 will use the reaction $^{62}Ni + ^{208}Pb \rightarrow ^{269}110 + 1n$. The predicted decay of $^{269}110$ points out α-emission as the predominant decay mode, with reasonable half-life between $(10 - 500)\mu s$. More measurements of excitation functions for reactions leading to $104 + 1n$ and $Hs + 1n$ will be performed in order to choose the optimum bombarding energy, as well as several improvements of the experimental method will be made aiming to increase the detection efficiency and to supress the background. The use of a ^{209}Bi target would lead to the compound nucleus $^{271}111$.[8]

A different system, namely $^{209}Bi(^{59}Co, n)^{267}110$ has been chosen for the experiments performed recently (August – September 1991) at the LBL SuperHILAC by a large experimental collaboration led by A. Ghiorso.[10] The predicted maximum cross section for the formation of $^{267}110$ is $3.9pb$, while the expected lifetime of the main decay mode (α-emission) is $17\mu s$. An improved version of the gas-filled magnetic spectrometer (SASSY2) has been used during the 38 days of beam but no clear evidence for the formation of element 110 has been found.

The use of RNB for heavy element synthesis seems promising and it has been emphasized as one of the important research to be performed at the radioactive beam facilities.[11-13] One expects that unstable neutron-rich projectiles will increase the probability of fusion in low energy collisions. In fact, due to the presence of large halos of loosely bound neutrons the height of the Coulomb barrier is diminished and effects that play an important role in heavy ion fusion at low energies, like formation of a neck[14] and neutron flow [15] between the colliding nucleus set in at larger distances than in the case of reactions with stable nuclei. Another advantage of the unstable neutron-rich projectiles is the increase of the neutron emission probability with respect to fission due to the decrease of neutron binding energies. There is not yet clear if the dynamical hindrance in the entrance channel is less effective in the case of unstable neutron-rich projectiles. Earlier estimates indicated smaller values of the extra-extra -push energy for increasing number of neutrons.[13] A recent study based on Swiatecki's model of nuclear collisions gives an opposite result, namely an increase of the extra-extra-push energy with the neutron number.[16] At this point experimental data too are needed to clarify the situation.

Estimates of the best combinations RNB-target undergoing cold fusion indicate neutron-rich doubly magic ($N = 50, Z = 28$) projectile and stable doubly magic heavy nucleus as a target, as for example $^{78}Ni + ^{208}Pb \rightarrow ^{286}110$.[13] One expects the compound nucleus to have a low excitation energy of $\approx 20 MeV$ and consequently non-negligible cross sections of the reactions $^{208}Pb(^{78}Ni, n)^{285}110$ and $^{208}Pb(^{78}Ni, \gamma)^{286}110$. The predicted cross sections are rather large and of course it would be interesting to examine in more detail various factors expected to change in the case of radioactive nuclear projectiles.

The possibilities of using RNB to produce heavy elements have been explored recently in detail by W. Loveland.[17] A semiempirical formalism has been used which includes the dynamical hindrance of fusion and treats the deexcitation of the composite system, assuming an energy independent ratio of the neutron emission and fission probabilities. The model is tested on existing data and used for the case of elements with $Z \geq 100$. The results indicate that production cross sections of new neutron-rich isotopes of elements 104 and element 105 are important but decrease strongly in the case of element 110 and above. Symmetric radiative capture reactions are suggested as the best synthesis reactions to be considered, as for example $^{138}Ba(^{144}Ba,\gamma)^{282}112$. The comparison of the production rates predicted for RNB to those resulting when using stable beams shows the advantage of RNB for the synthesis of the neutron-rich lighter transactinides. On the other hand the stable projectiles appear to be more suitable for the production of superheavy elements.

3. MNTR Leading to Heavy Elements

The MNTR constitute the second major class of heavy ion reactions that have been used to synthesize heavy elements.[1,18] The high production cross section (maximum cross sections of the order of $1 - 10mb$) is one important advantage of these reactions with respect to those of fusion. However, the strong decrease of the production rates as the transferred charge increases prevented the use of MNTR for the synthesis of heaviest elements, which have been observed in fusion-evaporation reactions (ch. 2). Isotopes of elements up to $Z = 103$ have been prepared by MNTR. Even using the highest-Z target available (^{254}Es) one only could go up to element 105.[19,20] In this context the question arises if the MNTR are of interest in the study of heavy elements? The answer is "yes" for the following reasons. The search for heavy elements has been mainly extensive, aiming to go as high as possible in atomic number Z. During that stage, complete fusion reactions offered the best approach of the problem. At present, the character of the investigation of heavy elements has changed, becoming also intensive. Now, that the upper end of the periodic table has been extended so far, it is of interest to produce many, particularly more *neutron-rich* isotopes of the new discovered elements and to study their properties. One can see in Fig. 1 that the isotopes discovered so far have small neutron numbers. Recalling that neutron-rich heavy elements are of a special interest because recent theoretical studies have indicated the existence of a region of deformed nuclei, stable against fission, around neutron number $N = 162$,[7] the need for the identification of such nuclides is obvious. MNTR are good producers of neutron-rich isotopes and offer thus an excellent tool to achieve this goal, at least up to a certain atomic number Z. The recent improvements of the detection methods have pushed the sensitivity of measurements to lower values (of the order of one picobarn) of the production cross sections, required also in heavy element production by fusion reactions.[8] We will discuss in the following the estimated production cross sections of elements up to $Z = 110$ and show which is the region of nuclei to be produced by MNTR with cross sections that can be measured.

3.1 Production Mechanism of MNTR

A comparative presentation of the existing models for MNTR leading to actinides can be found in the literature.[18,21] Here we will discuss one of them, namely the massive transfer model[18,22-25] that will be used later to explore the possibilities of MNTR in the synthesis of heavy elements.

The main hypotheses made in the massive transfer model for MNTR leading to heavy elements rely on the features of these reactions, resulting from the experimental data. The large cross sections observed for the transfer of many nucleons compared to the transfer of a single nucleon, suggest that a massive cluster (or aggregate) has been transferred *as a whole* and not *one-by-one*, which comes with higher transfer probabilities for the transfer of many nucleons. The massive transfer model assumes that the cluster is produced in the first step of the reaction by the fragmentation of the projectile nucleus in the Coulomb and nuclear field of the target nucleus. The capture of the cluster by the target nucleus leads to the primary distribution of isotopes which is subsequently changed by neutron emission and fission. The main features of MNTR, namely the scaling of the isotopic cross sections with the transferred charge ΔZ as well as the behaviour of the production cross sections in the reactions induced by various projectiles on a given target and / or by a given projectile on many targets, suggest such a picture of the production mechanism. We emphasize the fact that the projectile dependence of the individual isotopic distributions observed in reactions induced by various projectiles on the same target reflects the separation energies of the clusters in the projectile. The signature of the separation energies in the primary distributions is changed due to neutron emission and fission, but it is still transparent in the measured cross sections. The probability for the breakup of the projectile at bombarding energies down to energies slightly above the Coulomb barrier is smaller than at higher energies, but *not negligible* so that such processes have been observed in the last years[18] (and references therein). One expects this kind of reactions to be boosted by the Coulomb disruptive forces, which become important in the case of the high-Z targets used in the production of heavy elements.

The scheme considered in the massive transfer model is:

$$P + T \to F_1 + F_i + T \to F_1 + (F_i + T)^*.$$

Here P and T denote the projectile and target nuclei and F_1, F_i the two clusters in the projectile. The $*$ shows that the product nucleus has been formed in an excited state, the decay of which has to be considered further. The transfer cross section of a cluster $F_i(Z_i, N_i)$ can be written as the product of the probabilities for two processes to occur, namely the separation of the massive cluster F_i from the projectile and its capture by the target nucleus, $\gamma_P^{F_i}$ and $\sigma_C(F_i, T)$ respectively:

$$\sigma(F_i) = \gamma_P^{F_i} \sigma_C(F_i, T). \tag{1}$$

A wave function of the truncated Yukawa form has been adopted to describe the two clusters in the projectile together with the parametrization given by the Friedman model.[26] Then the factor $\gamma_P^{F_i}$ can be expressed as

$$\gamma_P^{F_i} = \frac{1}{\gamma_P} S_{F_i} \frac{e^{-2\mu_i x_{0F_i} b}}{x_{0F_i}^3 (1-b)} S'_{F_i} \tag{2}$$

where $\mu_i = \sqrt{2m_r E_s^{F_i}}/\hbar$, with m_r the reduced mass, $E_s^{F_i}$ the separation energy of the cluster and b a constant equal to $0.4 fm$. x_{0F_i} denotes the cutoff radius given by $r_0 A_{F_i}^{1/3}$, ($r_0 = 1.2 fm$). The spectroscopic factor[26] expresses the probability of finding together the right combination of Z_F protons and N_F neutrons which must be removed from the projectile to produce the cluster of interest.

We mention that in the case of particle unstable clusters ($^{2,5-7}He$, $^{4,5,8}Li$, $^{5,6,8}Be$, etc) a second spectroscopic factor S'_{F_i} has been introduced in eq. (2) in order to account for the depletion of the isotopic populations. The normalization factor:

$$\gamma_P = \sum \gamma_P^{F_i} \tag{3}$$

was introduced for calculating the cross sections in absolute units. The summation is performed over all the possible two-fragment splittings of the projectile.

A semiclassical approximation was used for the capture probability. The primary isotopic distribution (eq. (1)) is modulated by the decay of the excited products by particle emission and fission. Only neutron emission has been considered, because due to the large atomic numbers of the emitters in these particular MNTR, the emission of charged particles is strongly suppressed.Therefore the depletion of the primary isotopic distributions has been treated considering neutron evaporation and fission.The prescription of Sikkeland et al.[27] has been used to estimate the ratio of the partial widths for neutron emission and fission.

The final production cross section of a nuclide is given by the sum of two contributions, the one due to the cold transfer of a cluster and the sum of all "feedings" coming from neutron emission of higher mass excited nuclei formed by the transfer of heavier clusters.

The predictive power of the model has been tested by comparing calculated cross sections with the experimental ones comprising a wide range of systems.The model gives a good description of the shapes and centroids of the measured isotopic distributions. Absolute values of the production cross sections are obtained that agree within an order of magnitude with experiment, with a few exceptions. The massive transfer model does not reproduce the excitation functions of MNTR, particularly the decrease of the cross sections at the high energy end, and this is a shortcoming of the model. However, the model gives a consistent description of a whole set of experimental data and captures the physics of this class of reactions. It provides an effective tool for exploring a large number of systems in order to find out the "favorable" combinations to be used in the synthesis of heavy elements.

4. Heavy Element Production in MNTR induced by RNB

Recent advances in accelerator technology made possible the use of RNB which opened new axes of research on nuclear structure as well as on nuclear reactions.

Two methods[28] have been used to produce RNB. One is based on the fragmentation of the projectile at intermediate and high energies and delivers RNB of high energy too (GANIL,GSI,RIKEN,BERKELEY). The other alternative, the so called ISOL (Isotope Separator On Line) method was developed at CERN, then after adopted at Louvain-La-Neuve, Isospin Laboratory at Oak Ridge National Laboratory, University of Notre Dame, Grenoble (PIAFE) and consists in producing exotic nuclei as low energy products of spallation, fission or other types of reactions. These products are mass separated and accelerated in a second accelerator to the desired energy.

MNTR are potential producers of heavy neutron-rich nuclei and one expects that more neutron-rich nuclei will be produced in reactions induced by a neutron-rich beam. We have explored the possibilities of these reactions using the massive transfer model to estimate the production cross sections of heavy elements with atomic number $Z = 100 - 110$. We try to answer two questions: 1) Which is the highest Z one can reach in these reactions with cross sections large enough to be measured? 2) How much will the upper end of neutron numbers of the produced nuclides be expanded?

The most favorable target for the synthesis of heavy elements by MNTR, ^{254}Es has been chosen for this study. ^{254}Es is the highest-Z available target and MNTR induced on it by stable beams of ^{13}C, ^{18}O, ^{22}Ne have been already studied.[19,20] A large variety of exotic beams has been used in the calculations: $^{8}He, ^{9,11}Li, ^{11,14}Be, ^{15}B, ^{16}C, ^{20,22}O$, ^{26}Ne and ^{50}Ca. The energy of the beam has been chosen at values slighty higher than the interaction barrier, as it was done in previous MNTR experiments with stable beams. No amplification factors of the capture cross sections have been introduced throughout the calculations. Possible effects leading to an amplification of the cross sections as expected in the case of neutron-rich projectiles, might only act in the right direction and determine an increase of the estimated cross sections.

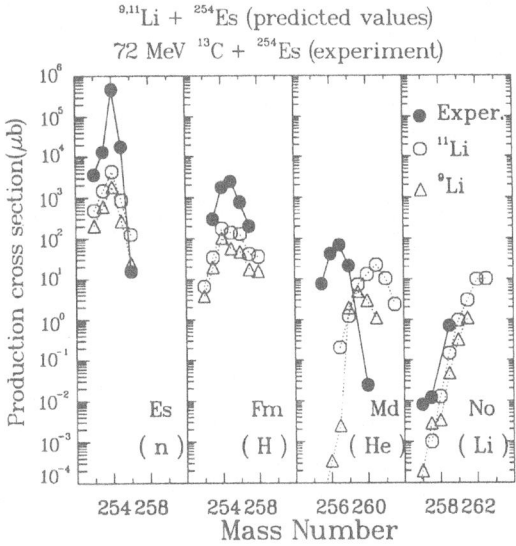

Figure 2. *Isotopic distributions of Es, Fm, Md and No isotopes from $50 MeV$ ^9Li, $61 MeV$ $^{11}Li +^{254}Es$ reactions compared to experimental data from $72 MeV$ $^{13}C +^{254}Es$.[19]*.

Results obtained in two extreme cases, namely for the systems $^{9,11}Li +^{254}Es$ and $^{50}Ca +^{254}Es$ are shown in detail. ^{11}Li nucleus has a special structure, with ^9Li as a core and two weakly bound neutrons that constitute a "halo".[29,30] The neutron halo extends out to distances of about $8.3 fm$, much larger than the radius of $2.5 fm$ of the ^9Li core. In the case of MNTR induced by RNB an important change is expected to come from the separation energies of various clusters in the neutron-rich projectiles due to the exponential dependence of the isotopic yields on these quantities(eq. (2)). The estimated production cross sections from $50 MeV^9Li +^{254}Es$ and $61 MeV^{11}Li +^{254}Es$ are shown on Fig. 2. As a reference experimental values obtained in the case of a close system $^{13}C +^{254}Es$ [19] are also given in

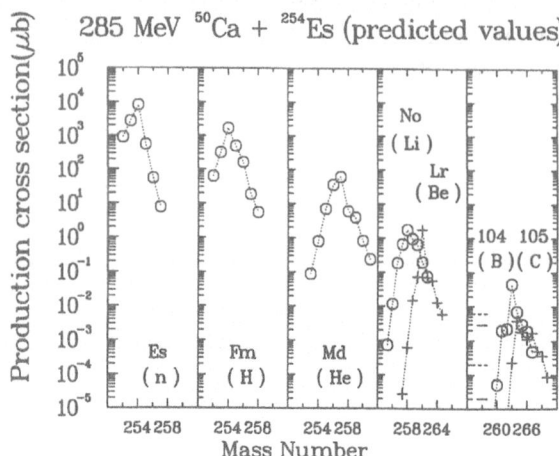

Figure 3. *Estimated cross sections of $Es - 105$ from the $285 MeV$ $^{50}Ca +^{254} Es$.*

Fig. 2. The symbols in the parentheses under each element indicate the type of transferred cluster which produced the element, for example Fm-isotopes have been produced by the transfer of ^{1-4}H, Md-isotopes by transfer of ^{2-7}He, etc. The transfer of neutrons is treated on equal footing with the one of charged clusters. Notice that particle - unstable nuclei such as ^2n, ^2He, ^5He, $^{4,5}Li$, etc., have been considered throughout the calculations. The possible existence of a dineutron and even of more complex states near the surface of the nucleus has been pointed out by Migdal.[31] These additional bound states of the particle pair appear as a special case of the three-body problem when one of the three particles has a mass much larger than the others, thus creating an external field for the two light particles. If the two particles are slightly unbound, the pairing effect may become important and sufficient to bind them to the nucleus. This idea has been recently used in the discussion of the neutron halo discovered in the extremely neutron-rich nuclei as ^{11}Li, ^{14}Be.[32] The necessity to take into account particle-unstable resonances has also been indicated in the case of particle emission from highly excited nuclear systems.[33] Existing mass tables have been used in the calculations.[34,35]

It is worth to notice that the maxima of the predicted production cross sections do not correspond to the most favored cluster, as also observed in the measured isotopic distributions of Es, Fm, Md in MNTR with stable beams, that peak at isotopes which do not correspond to the transfer of a n, ^1H, and ^4He respectively. This emphasizes the effect of neutron and fission decay of the primary products, which blurs the signature of the separation energies of various clusters in the projectile nucleus.

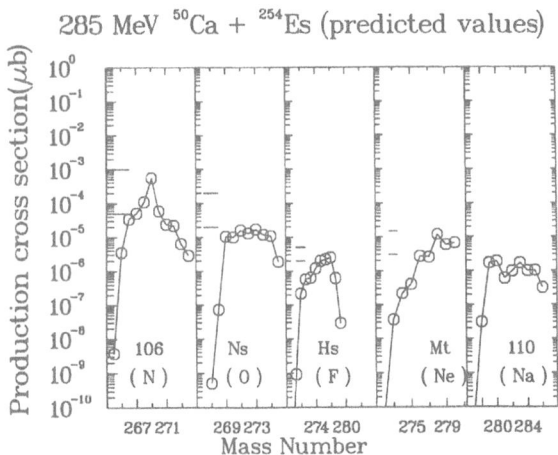

Figure 4. *Predicted cross sections of* 106 − 110 *from* $285 MeV$ $^{50}Ca +^{254}Es$.

The smaller binding energies of the neutrons in ^{11}Li are reflected in larger yields of the Es isotopes (Fig. 2). The expected shift of the isotopic distributions towards higher neutron numbers is obvious in the case of the more neutron-rich projectile ^{11}Li. Production cross sections of Md- and No-isotopes are different compared to those from the ^9Li induced reaction. An interesting result is the fact that the cross section values estimated in the reactions with RNB of $^{9,11}Li$ and those measured with the stable projectile ^{13}C are comparable, the only difference being that the isotopic distributions are located in the region of the more *neutron-rich* nuclei in the first case. Production cross sections of comparable values are obtained for ^8He, $^{11,14}Be$, ^{15}B, ^{16}C, $^{20,22}O$ and $^{26}Ne +^{254}Es$ systems, while the isotopic distributions are shifted continuously to more neutron-rich nuclei.

When increasing the mass number of the projectile heavier elements can be reached because more massive clusters are produced. The isotopic distributions of higher Z nuclei become wider, still extending to the region of neutron-rich nuclei. Figures 3 and 4 show such isotopic distributions predicted in the case of the heavier projectile ^{50}Ca with ^{254}Es. The horizontal lines placed at certain isotopes indicate the values of the cross sections measured in cold fusion reactions.[2,4] The formation of massive clusters as Be through Na in the $^{50}Ca +^{254}Es$ reaction makes possible the production of heavy elements up to $Z = 110$. As expected, the cross sections drop fast to around $1pb$ or less. However they are comparable to the cross sections measured in cold fusion reactions, as one sees in Figures 3 and 4. The necessity of a detection sensivity at the level of pb has been documented by S. Hofmann for the synthesis of 110.[8] Our calculations show that the same requirement appears if MNTR induced by RNB are used to synthesize elements $104 - 109$. It is interesting to compare production cross sections of heavy elements (up to Hs) estimated here for MNTR induced by RNB with those predicted for fusion reactions. Iljinov and Mebel[13] have estimated production cross sections of elements $Z = 102 - 108$ in reactions of radioactive nuclei $^{14,15}C$, ^{22}O, ^{23}F, $^{24,26}Ne$, $^{28,29}Mg$, ^{44}Ar, $^{63,64}Mn$, ^{66}Fe with ^{208}Pb, ^{232}Th, ^{244}Pu, ^{244}Pu, ^{249}Bk, and ^{254}Es targets. Their results give values close to the cross sections estimated by the massive transfer model for MNTR with RNB leading to elements $Z = 101 - 109$. However, the isotopes produced by MNTR with cross sections which can be measured are located in the region of high neutron numbers, covering $N = 162$ and extending as high as up to $N = 170$.

5. Conclusions

The possibilities of using MNTR to produce heavy actinides have been explored with particular emphasis on radioactive neutron-rich beams. Systematic estimates of the production cross sections of elements $Z = 100-110$ resulting in transfer reactions induced by radioactive nuclei ^8He, $^{9,11}Li$, $^{11,14}Be$, ^{15}B, ^{16}C, $^{20,22}O$, ^{26}Ne, ^{50}Ca on ^{254}Es show that the expected production cross sections are not larger than in the case of stable projectiles. The important achievement is the extension of the isotopic distributions produced in these reactions to isotopes with large neutron numbers: $N \geq 162$ (Fig. 1, dotted line). From this point of view MNTR in general, including those induced by stable beams, should be considered for the production of neutron-rich lighter heavy elements. The production cross sections decrease strongly when going up in Z, but for the elements up to $Z = 108$ the use of MNTR seems promising.

ACKNOWLEDGEMENTS

We would like to thank Drs. S. Hofmann and W. Loveland for sending their articles before publication and also Dr. J. M. Alexander for useful discussions.

References

[1] G. T. Seaborg and W. D. Loveland, *The Elements Beyond Uranium*, (John Wiley and Sons Inc., New York, 1990)

[2] P. Armbruster, *On the production of heavy elements by cold fusion: the elements 106 to 109*, Ann. Rev. Nucl. Part. Sci. **35**, 135 (1985)

[3] G. Münzenberg, *Recent advances in the discovery of transuranium elements*, Rep. Progr. Phys. **51**, 57 (1988)

[4] G. Münzenberg, *Heavy element production and limits to fusion*, GSI-preprint **89-47** (1989)

[5] S. Hofmann, *private communication* (1993)

[6] A. Sobyczewski, Z. Patyk, S. Cwiok, *Do the superheavy nuclei really form an island?*, Phys. Lett. B **186**, 6 (1987)

[7] S. Cwiok, A. Sobyczewski, *Potential energy and fission barriers of superheavy nuclei calculated in multidimensional deformation space*, Z. Phys. A **342**, 203 (1992)

[8] S. Hofmann, *Identification of rare heavy nuclei at cross-sections of one picobarn*, Contribution to the *International School-Seminar on Heavy Ion Physics*, Dubna, Russia, May 10–15, 1993, GSI-Preprint **93-37** (1993)

[9] W. Reisdorf and M. Schädel, *How well do we understand the synthesis of heavy elements by heavy-ion induced fusion?*, Z. Phys. A **343**, 47 (1992)

[10] W. Loveland, *private communication* (1993)

[11] T. Nomura, *Exotic nuclei arena in Japanese hadron project*, Proc. of First Int. Conf. on Ra-

dioactive Nuclear Beams, 16-18 Oct. 1989, p. 13, edited by W. D. Myers, J. M. Nitschke, E. B. Norman, (World Scientific, Singapore, 1990)

[12] A. S. Iljinov, V. M. Lobashev, Yu. Ts. Oganessian, *Prospects for studies with radioactive nuclear beams from the Moscow meson factory, Proc. of First Int. Conf. on Radioactive Nuclear Beams*, 16-18 Oct. 1989, p. 23, edited by W. D. Myers, J. M. Nitschke, E. B. Norman, (World Scientific, Singapore, 1990)

[13] A. S. Iljinov, M. V. Mebel, E. A. Cherepanov, *The possibilities of synthesizing heavy elements with radioactive nuclear beams, Proc. of First Int. Conf. on Radioactive Nuclear Beams*, 16-18 Oct. 1989, p. 289, edited by W. D. Myers, J. M. Nitschke, E. B. Norman, (World Scientific, Singapore, 1990)

[14] C. E. Aguiar, V. C. Barbosa, L. F. Canto, R. Donangelo, *Liquid drop effects in subbarier fusion reactions*, Phys. Lett. B **201**, 22 (1988)

[15] P. H. Stelson, *Neutron flow between nuclei as the principal mechanism in heavy ion subbarrier fusion*, Phys. Lett. B **205**, 190 (1988)

[16] C. E. Aguiar, V. C. Barbosa, *Dynamical hindrance of fusion from reactions involving neutron-rich nuclei*, Phys. Lett. B **289**, 12 (1992)

[17] W. Loveland, *Production of transuranium nuclides with radioactive nuclear beams*, to be published (1993)

[18] M. T. Magda and J. D. Leyba, *Production of heavy elements by transfer of massive clusters*, Int. J. of Mod. Phys. E **1**, 221 (1992)

[19] K. J. Moody, R. W. Lougheed, R. J. Dougan, E. K. Hulet, J. F. Wild, K. Sümmerer, R. L. Hahn, J. van Aarle, G. R. Bethune, *Actinide cross sections from the reaction of ^{13}C with ^{254}Es*, Phys. Rev. C **41**, 152 (1990)

[20] M. Schädel, W. Brüchle, H. Gäggeler, K. J. Moody, D. Schardt, K. Sümmerer, E. K. Hulet, A. D. Dougan, R. J. Landrum, R. W. Lougheed, J. F. Wild, G. D. O'Kelley, *Transfer cross sections from reactions with ^{254}Es as a target*, Phys. Rev. C **33**, 1547 (1986)

[21] J. D. Leyba, *A systematics study of actinide production from the interactions of heavy ions with ^{248}Cm*, Ph. D. Thesis, LBL – 29540 (1990)

[22] M. T. Magda, A. Pop, A. Sandulescu, *Large cluster transfer processes in reactions leading to heavy actinides*, J. I. N. R. Rapid Commun. Dubna, FSU, No. 17 - 86 (1986)

[23] M. T. Magda, A. Pop, A. Sandulescu, *Cross section estimates for multinucleon transfer reactions leading to heavy actinides*, J. Phys. G **13**, L127 (1987)

[24] M. T. Magda, *A model for large cluster transfer reactions leading to heavy actinides*, Bull. Am. Phys. Soc. **35**, 1660, BE1 (1990)

[25] M. T. Magda, *A Model for Transfer Reactions Leading to Heavy Actinides*, Symposium on "Nucleus-Nucleus Collision Mechanism", August 25 – 30, 1991, New York, (unpublished)

[26] W. A. Friedman, *Heavy ion projectile fragmentation: a reexamination*, Phys. Rev. C **27**, 569 (1983)

[27] T. Sikkeland, A. Ghiorso, M. J. Nurmia, *Analysis of excitation functions in $Cm(C, xn)No$ reactions*, Phys. Rev. **172**, 1232 (1968)

[28] See for ex. *Proc. of First Int. Conf. on Radioactive Nuclear Beams*, 16 – 18 Oct. 1989, edited by W. D. Myers, J. M. Nitschke, E. B. Norman, (World Scientific, Singapore, 1990)

[29] T. Kobayashi, O. Yamakawa, K. Omata, K. Sugimoto, T. Shimoda, N. Takahashi, I. Tanihata, *Projectile fragmentation of the extremely neutron-rich nucleus ^{11}Li at $0.79 GeV/nucleon$*, Phys. Rev. Lett. **60**, 2599 (1988)

[30] N. A. Orr, N. Anantaraman, S. M. Austin, C. A. Bertulani, K. Hanold, J. H. Kelley, D. J. Morrisey, B. M. Sherril, G. A. Souliotis, M. Thoennessen, J. S. Winifield, J. A. Winger, *Momentum distribution of ^9Li fragments following the breakup of ^{11}Li*, Phys. Rev. Lett. **69**, 2050 (1992)

[31] A. B. Migdal, *Two interacting particles in a potential well*, Sov. J. Nucl. Phys. **16**, 238 (1973)

[32] P. G. Hansen, B. Jonson, *The neutron halo of extremely neutron-rich nuclei*, Europhys. Lett. **4**, 409 (1987)

[33] M. A. Bernstein, W. A. Friedman, W. G. Lynch, *Emission of particle unstable resonances from compound nuclei*, Phys. Rev. **29**, 132 (1984)
[34] A. H. Wapstra, G. Audi, *The 1983 atomic mass evaluation*, Nucl. Phys. A **432**, 1 (1985)
[35] S. Liran, N. Zeldes, *A semiempirical shell-model formula*, Atomic Data Nucl. Data Tables **7**, 431 (1985)

MICROSCOPIC AND SEMI-MICROSCOPIC APPROACH TO THE PROPERTIES OF TRANSACTINIDE NUCLEI

L. Bitaud, J.F. Berger, J. Déchargé and M. Girod

Service PTN, CEA Bruyères–le–Châtel, BP 12, 91680 Bruyères–le–Châtel

The constrained Hartree-Fock-Bogolyubov (HFB) method employed with Gogny's finite range effective force [1] (D1S) has appeared as a very reliable tool for describing the properties of actinide nuclei and for understanding fission barriers and dynamics [2]. In view of the rapid experimental progress in the synthesis of very heavy elements and in the search for superheavy species, we have begun to apply this kind of approach to transactinide nuclei. Our goal is to derive the trend of the valley of stability beyond Z=104 and to predict the fission and alpha-decay stability of these very heavy nuclei.

As a starting point, constrained HFB calculations with axial and left-right symmetries have been done for even-even nuclei between Z=98 and Z=128, and for N=150 up to 214. In this microscopic approach of nuclear structure, the force D1S is used to generate not only the nuclear mean field but also the nuclear pairing field. In this way, all the parameters employed are independent of the nucleus studied. We use a variational procedure, where we look for an independent quasi-particle state $|\Phi\rangle$ such that

$$\delta[\langle\Phi|H - \lambda_n \widehat{N_n} - \lambda_p \widehat{N_p} - \sum_{lm} \lambda_{lm} \widehat{Q_{lm}}|\Phi\rangle] = 0 \qquad (1)$$

where $\widehat{N_n}, \widehat{N_p}$ are respectively the neutron and proton number operators and $\widehat{Q_{lm}}$ are external field operators. In this equation λ_n, λ_p are calculated to ensure the conservation of the mean number of particles, an additional condition necessary in the HFB theory. The mean value of $\widehat{Q_{lm}}$ is imposed to generate various types of deformation. In the present study, the mass quadrupole deformation $\widehat{Q_{20}}$ is the only constraint we prescribe. Futhermore, all the calculations employ 15 shell deformed harmonic oscillator basis, and include the exchange Coulomb field energy and the 2 body center of mass correction on kinetic energy.

The evolution of the binding energy as a function of $\widehat{Q_{20}}$ is given in figure 1, for some nuclei between A=256 and A=336. One can see on these plots the great sensitivity to the neutron number N. For small values of N (N $\lesssim$ 164), the ground state (GS) is strongly deformed and the fission barrier is composed of only one hump (B1). As N increases, from 156 to 184, the GS tends to a spherical shape. On the other hand, around N=164, we can see the emergence of an isomeric state (IS) involving the appearance of a second hump (B2) that rises while B1 decreases. Futhermore, the top of B1 regularly moves towards the region of strong deformation (with a "jump" in quadrupole moment for N $\approx$ 184, invisible on this plot) such as the IS deformation becomes more and more uncertain. From N around 198, the behaviour of the fission barrier against $\widehat{Q_{20}}$ is rather different; B1 disappears and therefore the GS deformation cannot be defined. Then one observes the reappearance of a GS well,

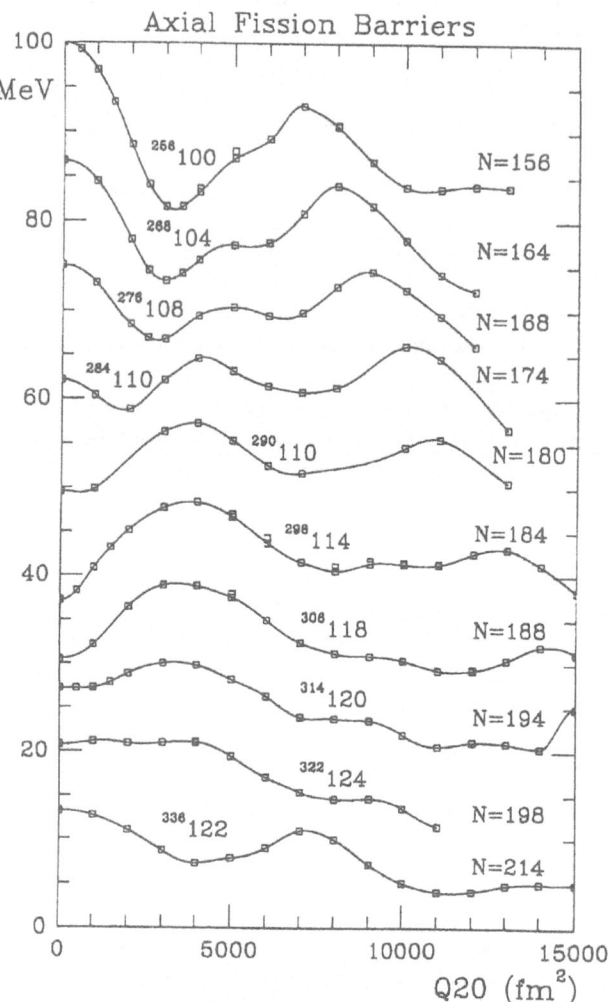

Figure 1. *Evolution of the binding energy as a function of the mass quadrupole deformation Q_{20} for nuclei between A=256 and A=336*

with a barrier height of about 4 MeV for $^{336}122$, that may be foreshadow some next island of deformed shell stability.

Figure 2 displays a map of all the even-even nuclei studied and the valley of maximum β- stability. The analysis of the GS binding energies in the Z=104-108 region shows that the higher stability appears at neutron numbers slightly bigger than those allowed by presently available experimental projectile-target combinations.

Shell effects of significant magnitude are found in single-particle levels at neutron numbers N=184 for spherical shapes and N=162 for deformed nuclei (cf. figure 3). As a consequence, nuclei having N=182 to 188 are all found spherical, while those having N around 160 are strongly deformed ($\beta \sim 0.25$). Much smaller shell effects are observed in single-particle spectra at proton numbers Z=114 and Z=120 (cf. figure 4).

Figure 2. *Valley of maximum β - stability for even-even (Z,N) nuclei with Z=98 to 128 and N=150 to 214 , as predicted from HFB calculations with axial and left-right symmetries. The most stable nuclei are shown as full boxes, while the others are represented as open circles in boxes. Futhermore, a full circle in a box means that experimental data exist, and the question mark in a box that ground state deformation cannot be defined.*

Alpha-decay properties of these nuclei have been also considered. A semi- phenomenological model reproducing lifetimes in rare earths and actinides shows that the nuclei between N=170 and N=184 have alpha-decay lifetimes larger than one minute and, for a few of them, beyond one hour. The α-decay probability is given by

$$\lambda = p \times \frac{v}{2R} \times e^{-\gamma} \qquad (2)$$

where p is the probability of a preformed α, $\frac{v}{2R}$ is the barrier hitting rate and $e^{-\gamma}$ is the barrier penetration factor. Here R is the radius of the decaying nucleus taken from HFB

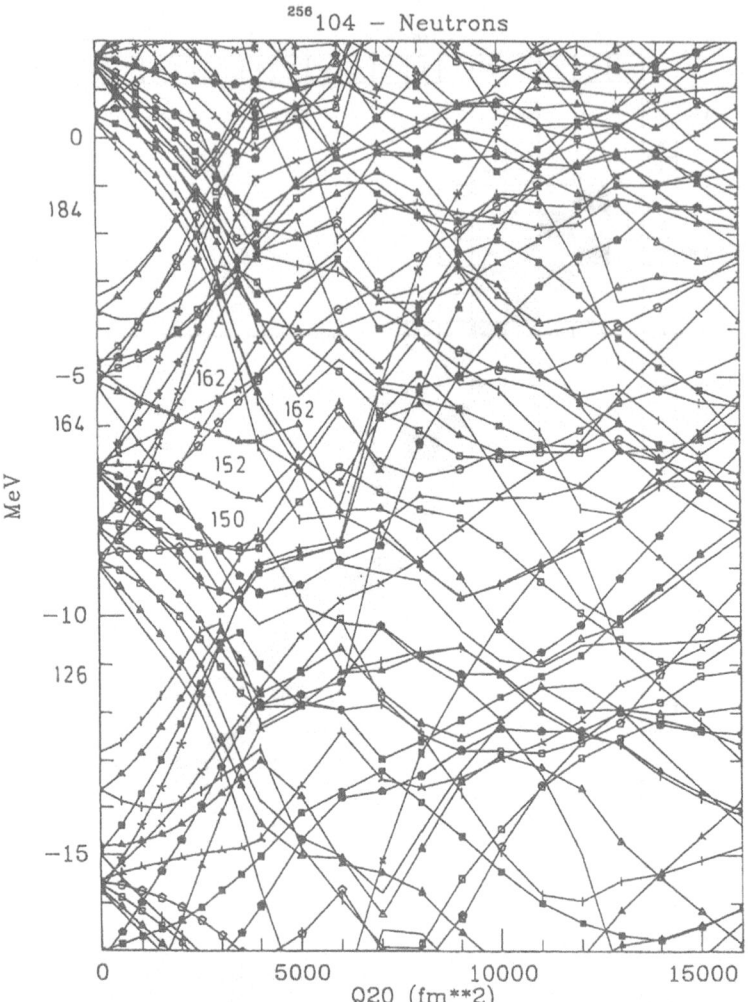

Figure 3. *Neutron single particle levels for $^{256}104_{152}$*

results and γ is computed from standard WKB, the Q value is coming from the HFB mass differences.

Putting together these alpha-decay and preliminary fission results, it seems that the best candidates for super-heavy synthesis would be $^{302}118$ ($B_f \approx 13.5$ MeV, $T_\alpha \approx 1$ mn), $^{298}114$ ($B_f \approx 11$ MeV, $T_\alpha \approx$ a few mn) and $^{294}112$ ($B_f \approx 10$ MeV, $T_\alpha \approx 8$-10 years), but also nuclei with slightly less neutrons as $^{288}110$ ($B_f \approx 7$ MeV, $T_\alpha \approx$ a few mn).

In order to improve this analysis, one has to check that the fission barriers of these nuclei are not lowered when left-right asymmetric and/or non-axial shapes are allowed. Since full HFB calculations become extremely time-consuming in this case, another approach has been developed (First results from this original technique will soon be published). It consists in

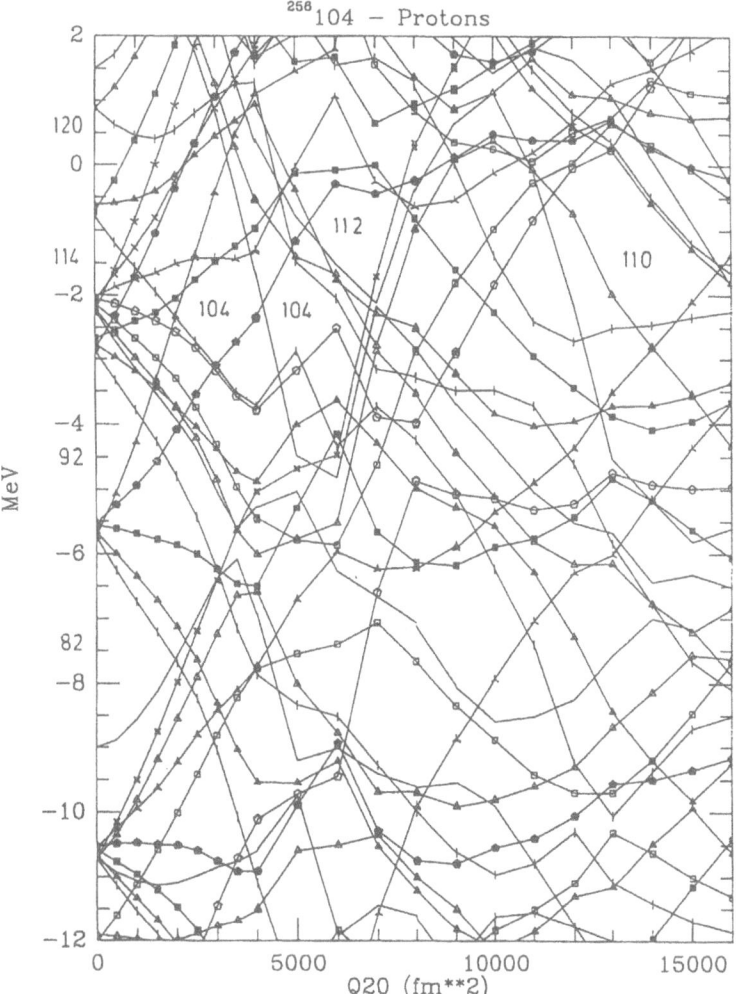

Figure 4. *Proton single particle levels for* $^{256}104_{152}$

approximating the HFB mean-field by a folded-Yukawa potential similar to the one introduced by Nix [3,4]

$$V(\vec{r}) = V_0 \int \rho(\vec{r}\,')v(|\vec{r} - \vec{r}\,'|)d\vec{r}\,', \qquad (3)$$

where the generating densities are obtained from the axial HFB densities by adding non-axial and asymmetric deformation parameters that can be optimized by energy minimization. The free remaining parameters are fitted in order to reproduce experimental GS radii and single-particle spectra. Pairing correlations are introduced within the BCS formalism and adjusted from HFB results. Futhermore a rearrangement energy similar to the one appearing in density-dependent HFB is being tested in order to derive total binding energies and potential energy surfaces directly from the semi-phenomenological hamiltonian.

References

[1] J. Dechargé, D. Gogny, *Phys. Rev. C* **21**, 1568 (1980)
[2] J.F. Berger, M. Girod, D. Gogny, *Nucl. Phys. A* **502**, 85c (1989)
[3] M.Bolsteri, E.O. Fiset, J.R. Nix, J.L. Norton, *Phys. Rev. C* **5**, 1050 (1972)
[4] P. Möller, J.R. Nix, *Nucl. Phys. A* **361**, 117-146 (1981)

ly # IV.

Nuclear Structure

ON THE ORIGIN OF ROTATIONS AND VIBRATIONS IN ATOMIC NUCLEI

J. P. Draayer, C. Bahri and D. Troltenier

Department of Physics and Astronomy, Louisiana State University, Baton Rouge, LA 70803–4001, U.S.A.

1. Introduction

A long-time objective of nuclear structure physics has been to bridge the gap between collective and shell-model descriptions of observed nuclear phenomena. Progress has been slow because of the difficulty in making realistic shell-model calculations, and quite frankly, the success of the much simpler collective models. Algebraic shell-model theories come closest to realizing this objective. There are two basic types of algebraic models: those based on a boson description of the dynamics, such as the Interacting Boson Model (IBM) [1], and those which treat the nucleons as fermions. In this paper we will be focusing on the latter, bosons will only enter when they appear naturally as a consequence of physical circumstances such as low particle or high hole occupancy so Pauli principle constraints are not violated.

The first and most familiar algebraic fermion model is the Elliott SU(3) scheme. It is known to work well for light ($A \leq 28$) nuclei [2]. Another is the Sp(3,R) (denoted Sp(6,R) sometimes) or symplectic model which is a natural multi-$\hbar\omega$ extension of the Elliott scheme [3]. For heavier systems ($A \geq 150$) there are currently two algebraic models being employed: the so-called Fermion Dynamical Symmetry Model (FDSM) which identifies s and d fermion pair operators that form an algebra which closes under commutation (the SO(8) group for the n=4 shell and Sp(6) for n=5 and n=6) and gives a possible microscopic interpretation of the IBM [4], and the pseudo-SU(3) model and its pseudo-symplectic extension which builds on the concept of good pseudo-spin symmetry in heavy nuclei [5,6,7].

In this talk we will focus on the SU(3)/Sp(3,R) models. We will begin by showing how rotational motion can be realized within this framework and then go on to explore the shell-model underpinnings of β and γ vibrational motion as it is normally portrayed in the collective model. The theory yields a deeper understanding of the strengths and limitations of collective-model phenomenology and gives a shell-model interpretation of the potential energy surface concept. It also demonstrates the importance of multi-$\hbar\omega$ shell-model couplings. Possibilities for gaining a better grasp of superdeformed and related phenomena will also be suggested.

2. Rotational motion

A shell-model interpretation of collective motion can be easily given whenever the collective Hamiltonian can be expressed in a frame-independent operator form because that expression can then be directly rewritten in terms of the corresponding microscopic operators. The rotor is a particularly elegant example because expressing the Hamiltonian in a frame-independent form is relatively easy and leads immediately to its shell-model representation. Furthermore, the operators that enter into the expression have historical significance, dating back to Racah's pioneering work on the SU(3) ⊃ SO(3) symmetry group [8]. The argument is so illustrative it bears repeating, but in an abbreviated form. New material which points towards exploiting the connection between the triaxial rotor and SU(3) is discussed in somewhat greater detail.

The triaxial rotor Hamiltonian is given by

$$H_{ROT} = A_1 I_1^2 + A_2 I_2^2 + A_3 I_3^2 \tag{1}$$

where I_α ($\alpha = 1, 2, 3$) is the projection of the total angular momentum on the α-th body-fixed symmetry axis and A_α is the corresponding inertia parameter depending inversely on the moment of inertia about the α-th principal axis. This familiar principal-axis form can be rewritten in a frame-independent representation by introducing three special scalar operators:

$$\begin{aligned}
L^2 &= \sum_\alpha L_\alpha L_\alpha = \sum_\alpha I_\alpha^2, \\
X_3^c &= \sum_{\alpha,\beta} L_\alpha Q_{\alpha\beta}^c L_\beta = \sum_\alpha \lambda_\alpha I_\alpha^2, \\
X_4^c &= \sum_{\alpha,\beta,\gamma} L_\alpha Q_{\alpha\beta}^c Q_{\beta\gamma}^c L_\gamma = \sum_\alpha \lambda_\alpha^2 I_\alpha^2.
\end{aligned} \tag{2}$$

The L_α and $Q_{\alpha\beta}^c$ in this equation are Cartesian forms for the total angular momentum and collective quadrupole operators, respectively. (The superscript c appended to the Q denotes the *collective* quadrupole operator which has non-vanishing matrix elements between major shells ($n' = n, n \pm 2$), in contrast with the *algebraic* quadrupole operators, $Q_{\alpha\beta}^a$, which have non-vanishing matrix elements only within a major shell, $n' = n$.) The last expression given for each scalar in eq.(2) is the form these operators take in the body-fixed, principal-axis system where the eigenvalues of the $Q_{\alpha\beta}^c$ are presumed to be sharp: $\langle Q_{\alpha\beta}^c \rangle = \lambda_\alpha \delta_{\alpha,\beta}$. These equations can be inverted to yield the I_α^2 in terms of L^2, X_3^c, and X_4^c:

$$I_\alpha^2 = \left[(\lambda_1\lambda_2\lambda_3)L^2 + (\lambda_\alpha^2)X_3^c + (\lambda_\alpha)X_4^c\right]/D_\alpha \text{ where } D_\alpha \equiv 2\lambda_\alpha^3 + \lambda_1\lambda_2\lambda_3. \tag{3}$$

Substituting this result for the I_α^2 into eq.(1) yields

$$H_{ROT} = aL^2 + bX_3^c + cX_4^c, \tag{4}$$

where a, b and c depend on the inertia parameters and the eigenvalues of $Q_{\alpha\beta}^c$:

$$a = \sum_\alpha a_\alpha A_\alpha, \quad b = \sum_\alpha b_\alpha A_\alpha, \quad c = \sum_\alpha c_\alpha A_\alpha, \tag{5}$$

$$a_\alpha = \frac{\lambda_\beta \lambda_\gamma}{2\lambda_\alpha^2 + \lambda_\beta \lambda_\gamma}, \quad b_\alpha = \frac{\lambda_\alpha}{2\lambda_\alpha^2 + \lambda_\beta \lambda_\gamma}, \quad c_\alpha = \frac{1}{2\lambda_\alpha^2 + \lambda_\beta \lambda_\gamma}.$$

A shell-model image of the rotor Hamiltonian can be obtained by substituting single-particle forms for the collective L_α and $Q_{\alpha\beta}^c$ operators: $L_\alpha = \sum_i l_\alpha(i)$ and $Q_{\alpha\beta}^c = \sum_i q_{\alpha\beta}^c(i)$. However, this ignores the shell structure and the fermion character of the many-body system. It is important to remember that while the L_α have non-vanishing matrix elements only within a major oscillator shell, the $Q_{\alpha\beta}^c$ couple shells differing by two quanta ($n' = n, n \pm 2$).

Indeed, the off-diagonal ($n' = n \pm 2$) couplings are about equal in magnitude to the diagonal ($n' = n$) ones. It follows from this that operators like $Q^c \cdot Q^c$ and the X_3^c and X_4^c (even if used only as residual interactions) can destroy the shell structure. This catastrophe can be avoided easily by simply setting all off-diagonal couplings between major shells to zero, an action which corresponds to replacing the $Q_{\alpha\beta}^c$ operators by their algebraic counterparts, $Q_{\alpha\beta}^a$. Elliott was the first person to recognize that the $Q_{\alpha\beta}^a$ operators, along with the L_α, generate SU(3), the symmetry algebra of the isotropic harmonic oscillator Hamiltonian. The appropriate shell-model image of the rotor Hamiltonian, eqs.(1) and (4), is thus given by

$$H_{SU3} = H_0 + aL^2 + bX_3^a + cX_4^a, \tag{6}$$

where H_0 is the harmonic oscillator Hamiltonian.

Shell-model values for the λ_α are required to complete the mapping. This follows by equating invariants of the two theories, a very natural thing to do since constants of the motion relate to the important physics, which in turn should be independent of the particular description. Because SU(3) is a rank two group it has two invariants: C_2 with eigenvalue $[\lambda^2 + \lambda\mu + \mu^2 + 3(\lambda + \mu)]$, and C_3 with eigenvalue $[(\lambda - \mu)(\lambda + 2\mu + 3)(2\lambda + \mu + 3)/2]$, where λ and μ are SU(3) representation labels with $(\lambda + \mu)$ and μ, respectively, specifying the number of boxes in the first and second rows in a standard Young tableau labeling of irreducible representations (irreps) of the SU(3) group. Note that C_2 is of degree two in the generators of SU(3) while C_3 is of degree three. The symmetry group of the rotor [$T_5 \wedge$ SO(3)] also has two invariants: traces of the square $\{\text{Trace}[(Q^c)^2]\}$ and cube $\{\text{Trace}[(Q^c)^3]\}$ of the collective quadrupole matrix. The eigenvalues of these two invariant operator forms are $\lambda_1^2 + \lambda_2^2 + \lambda_3^2 \to (k\beta)^2$ and $\lambda_1\lambda_2\lambda_3 \to (k\beta)^3 \cos(3\gamma)$, respectively, where (β, γ) are the shape variables of the collective model. The requirement of a linear correspondence between these two sets of invariants leads to the following relations,

$$\lambda_1 = -(\lambda - \mu)/3, \quad \lambda_2 = -(\lambda + 2\mu + 3)/3, \quad \lambda_3 = (2\lambda + \mu + 3)/3. \tag{7}$$

This correspondence, in turn, sets up a direct relationship between the (β, γ) shape variables of the collective model and the (λ, μ) irrep labels of SU(3),

$$\beta^2 = \frac{4\pi}{5(A\bar{r}^2)^2}\left[\lambda^2 + \lambda\mu + \mu^2 + 3(\lambda + \mu) + 3\right], \quad \gamma = \tan^{-1}\left(\frac{\sqrt{3}(\mu + 1)}{2\lambda + \mu + 3}\right). \tag{8}$$

Since λ and μ are positive integers, this translates into a regular grid when superimposed on a traditional (β, γ) plot, with β the radius vector and γ the azimuthal angle:

$$k\beta_x = k\beta\cos(\gamma) = \frac{2\lambda + \mu + 3}{3}, \quad k\beta_y = k\beta\sin(\gamma) = \frac{\mu + 1}{\sqrt{3}} \tag{9}$$

where $k^2 = \frac{5}{9\pi}(A\bar{r}^2)^2$. Each (λ, μ)-irrep corresponds to a unique value for the (β, γ)-pair. This relationship between (β, γ) and (λ, μ) is illustrated in Figure 1.

A logical consequence of the $(\beta, \gamma) \leftrightarrow (\lambda, \mu)$ mapping, eq.(8), is that constraints should be placed on the (β, γ) values of the collective model, a result that stands in sharp contrast with a liquid drop picture which takes β and γ to be continuous variables. These conditions follow because nucleons obey Fermi-Dirac statistics and the shell structure dictates the allowed set of (λ, μ) irreps for any particular nucleus. On the other hand, the $(\beta, \gamma) \leftrightarrow (\lambda, \mu)$

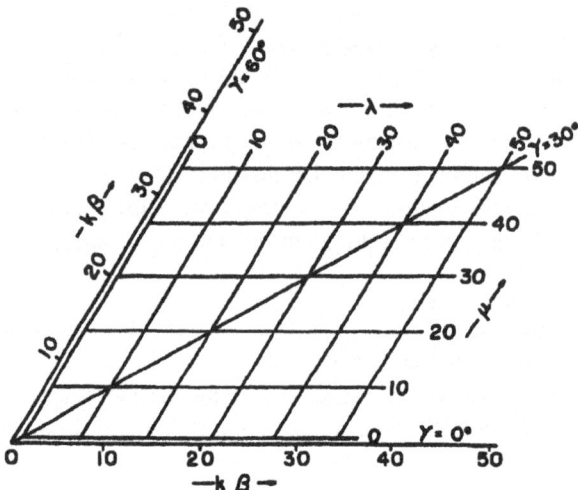

Figure 1. A traditional (β, γ) plot illustrating the mapping between (β, γ) and (λ, μ).

correspondence also teaches us that the collective model is overly restrictive because within the shell-model picture a given (λ, μ) can occur multiple times, the actual number being determined by the embedding of SU(3) in the higher spatial symmetry of the shell-model, for example U(6) for the ds-shell, U(10) for the fp-shell, and so on. Furthermore, whereas the $(\beta, \gamma) \leftrightarrow (\lambda, \mu)$ relationship implies constraints within the (β, γ) plane, it also points to a rich configuration space extending above the plane which is missing in the collective-model picture.

A further condition that follows from the $(\beta, \gamma) \leftrightarrow (\lambda, \mu)$ mapping is band termination. In the collective-model description the rotational bands extend indefinitely whereas because SU(3) is a compact group with finite dimensional representations its bands all terminate. For a prolate geometry the former follows the usual $L(L+1)$ energy rule, while the latter begins to show deviation from this simple behavior at L values greater than about 1/3 the maximum, which for the (λ, μ) irrep is $(\lambda + \mu)$. Whereas the rotor-like characteristics of the bands are extended somewhat in the symplectic extension of the SU(3) theory through the multi-$\hbar\omega$ couplings, they still show definite saturation effects in regions of angular momentum that are experimentally reachable. Accompanying this deviation from the $L(L+1)$ energy rule is a fall-off in predicted B(E2) strengths from the rigid rotor values. This is shown in schematic form in Figure 2 where the square of the SU(3) coupling coefficient which governs the strength of the B(E2) is plotted as a function of various λ values for the $\mu = 0$ case. The rotor limit is clearly seen to correspond to the asymptotic $\lambda \to \infty$ limit of the SU(3) theory.

It is also instructive to push the $(\beta, \gamma) \leftrightarrow (\lambda, \mu)$ connection to a consideration of a coupled double-rotor picture which is commonly used to describe heavy nuclei, with one rotor representing the protons (π) and another the neutrons (ν). For prolate geometries ($\gamma_\pi = 0$ and $\gamma_\nu = 0$) it is convenient to introduce an angle θ which measures the angle between the principal axes of the two distributions. (The scissors mode used to describe the so-called giant B(M1) strengths got its name from this simple double-rotor picture. Also note that the Exclusion Principle is not violated because the two distributions are made up of different

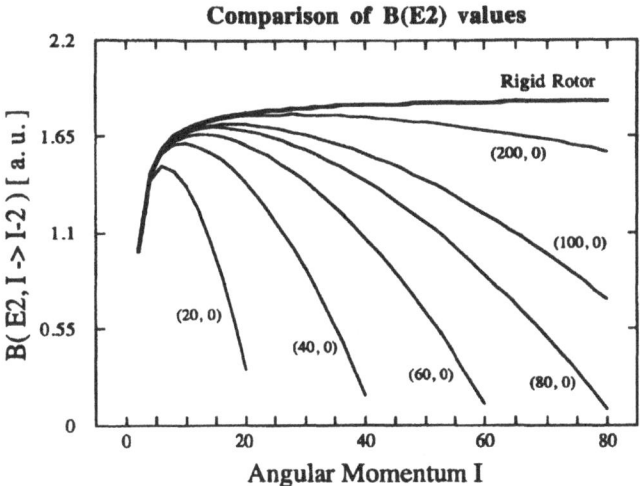

Figure 2. *Comparison of calculated $B(E2: I \to I-2)$ strengths in the $(\lambda, 0)$ irrep of SU(3) with the corresponding values for a prolate rotor. The rotor values are obtained in the $\lambda \to \infty$ limit of the SU(3) theory. Similar results hold for $\mu \neq 0$. Typically $\lambda \sim 100$ (rare earth) and therefore a fall-off from rigid rotor predictions occurs for $L \sim 30\hbar$.*

particle types.) For $\theta = 0°$ the two axially symmetric ellipsoids overlap maximally (aligned principal axes) whereas when $\theta = 90°$ the principal axes are perpendicular to one another and the resulting overlap of the two distributions is a minimum. The (β, γ) value of the joint distribution can be determined once β_π, β_ν and θ are specified. (Recall that β and γ are determined respectively by the trace of the square and cube of the quadrupole matrix and that the quadrupole matrix of the joint distribution is simply the sum of the two, with the second (Q_ν) rotated by an angle θ about an axis which is perpendicular to the symmetry axis of the first (Q_π), $Q = Q_\pi + R Q_\nu R^{-1}$ with rotation matrix $R = \exp(i\theta \cdot \hat{n})$ where $\hat{n}$ is a unit vector that is perpendicular to the symmetry axis of the proton distribution.) Or vice-versa, given β_ν, β_π and (β, γ) one can clearly deduce the relative orientation angle θ. This simple construction corresponds to the $(\lambda_\pi, \mu_\pi = 0) \times (\lambda_\nu, \mu_\nu = 0) \to (\lambda, \mu)$ coupling in the SU(3) picture which is known to be simply reducible, that is, each allowed (λ, μ) irrep in the product $[(\lambda, \mu) = (\lambda_\pi + \lambda_\nu, 0), (\lambda_\pi + \lambda_\nu - 2, 1), (\lambda_\pi + \lambda_\nu - 4, 2), \ldots, (\max(\lambda_\pi, \lambda_\nu) - \min(\lambda_\pi, \lambda_\nu), \min(\lambda_\pi, \lambda_\nu))]$, that occurs once and only once. For prolate SU(3) geometries ($\mu = 0$), a discrete orientation angle given by

$$\sin \theta_n = \sqrt{\frac{n(\lambda_\pi + \lambda_\nu - n)}{\lambda_\pi \lambda_\nu}}$$

can be associated with the $(\lambda_\pi + \lambda_\nu - 2n, n)$ irrep in the product $(\lambda_\pi, \mu_\pi = 0) \times (\lambda_\nu, \mu_\nu = 0)$ where $n = 0$ ($\theta = 0° \to \parallel$), 1, ..., $\min(\lambda_\pi, \lambda_\nu)$ ($\theta = 90° \to \perp$).

In general one must deal with triaxial shapes ($\gamma_\pi \neq 0$ and $\gamma_\nu \neq 0$) and a general product distribution: $(\beta_\pi, \gamma_\pi) \times (\beta_\nu, \gamma_\nu) \to (\beta, \gamma)$. In this case the geometrical interpretation is more complicated because the single variable θ, measuring the angle between the major axes of the ellipsoids, together with (β_π, γ_π) and (β_ν, γ_ν), is no longer sufficient to characterize the final (β, γ) configuration. In general, the three Euler angles (φ, θ, ϕ) of the rotation relating the two sets of principal axis frames must be specified, see Figure 3. For $(\varphi, \theta, \phi) = (0°, 0°, 0°)$

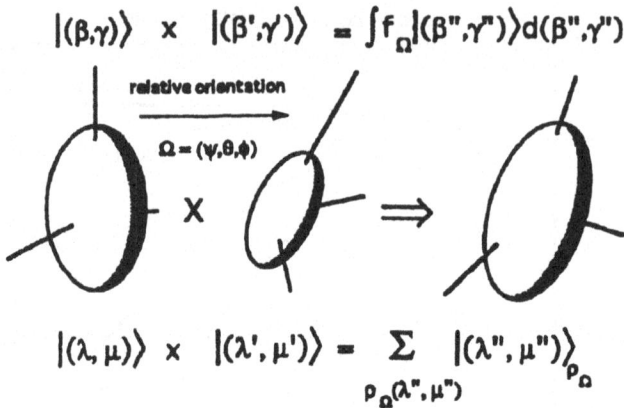

Figure 3. Schematic representation for the expansion of a product of two quadrupole mass distributions in terms of other quadrupole mass distributions. The upper product is for triaxial quantum rotors, which are characterized by the (β, γ) shape variables of the collective model and have a $[T_5 \wedge SO(3)]$ symmetry; the lower coupling is for (λ, μ) irreps of SU(3). The overlap function f_Ω is the inner product $\langle (\beta'', \gamma'')|(\beta, \gamma); (\beta', \gamma')\rangle_\Omega$ where $\Omega = (\psi, \theta, \phi)$ specifies the Euler angles giving the relative orientation of the principal axes of the unprimed ($|(\beta, \gamma)\rangle$) and primed ($|(\beta', \gamma')\rangle$) systems. In the SU(3) case, the decomposition is a sum of SU(3) irreps with integer multiplicity ρ which can be determined by the Littlewood rules for coupling Young diagrams. The multiplicity ρ_Ω, like f_Ω, can be related to the number of distinguishable orientations of the two initial distributions that yield the final one.

the major and minor axes of the sub-distributions coincide (maximum alignment) whereas if $(\varphi, \theta, \phi) = (0°, 90°, 0°)$ the semi-axes (y) remain aligned but the major (z) and minor (x) axes of the two systems are perpendicular to one another, etc. In the SU(3) case the allowed product configurations are determined by the Littlewood rules for the coupling of two three-rowed Young diagrams. There is a need for three $(\varphi, \theta, \phi) \leftrightarrow (m, n, \rho)$ rather than one $(\theta) \leftrightarrow (n)$ quantum label in this case as well: $(\lambda_\pi, \mu_\pi) \times (\lambda_\nu, \mu_\nu) \to (\lambda_\pi + \lambda_\mu + m, \mu_\pi + \mu_\nu + n)^\rho$, where ρ is a non-negative integer index ($\rho = 1, 2, \ldots, \rho_{max}$) labeling distinct occurrences of the same (λ, μ) that can occur in the $(\lambda_\pi, \mu_\pi) \times (\lambda_\nu, \mu_\nu)$ product. Working backwards, it should also be clear that the $(\beta, \gamma) \leftrightarrow (\lambda, \mu)$ correspondence can be used to give a geometrical interpretation to the abstract group theoretical concept of the outer multiplicity at least for the SU(3) case which has up until now escaped a simple physical interpretation. Specifically, the multiplicity ρ, together with m and n, can be considered to be a measure of the relative orientation of the two sub-distributions. In this way the first ($\rho = 1$) occurrence of (λ, μ) corresponds to a parent configuration oriented with one set of angles $(\varphi_1, \theta_1, \phi_1)$ while the second ($\rho = 2$) solution corresponds to another set $(\varphi_2, \theta_2, \phi_2)$, and so on. If $\rho_{max} = 1$, the corresponding (λ, μ) distribution can only be realized in one way. With this interpretation in hand the evaluation of reduced matrix elements and especially SU(3) coupling and recoupling coefficients should be revisited, looking for asymptotic solutions that exploit the geometrical concept of overlapping ellipsoidal mass distributions. This is a study that is currently underway.

It is also instructive to view the relationship between the rotor and SU(3) theories at a more fundamental level. This can be achieved by comparing the algebras of their symmetry groups. The symmetry group of the quantum rotor is the semi-direct product $T_5 \wedge SO(3)$

where T_5 is generated by the five independent components of the (spherical) collective quadrupole operator (Q_μ^c) and SO(3) is generated by the angular momentum operators (L_μ). The generators of SU(3), on the other hand, are the Q_μ^a (see the discussion following eq.(1)) and the L_μ operators. If Q^x denotes a generic quadrupole operator, then the commutation relations of the L_μ and the Q_μ^x are given by

$$\begin{aligned}
[L_\mu, L_\nu] &= -\sqrt{2} <1\mu, 1\nu|1, \mu+\nu> L_{\mu+\nu}, \\
[L_\mu, Q_\nu^x] &= -\sqrt{6} <1\mu, 2\nu|2, \mu+\nu> Q_{\mu+\nu}^x, \\
[Q_\mu^x, Q_\nu^x] &= c <2\mu, 2\nu|1, \mu+\nu> L_{\mu+\nu},
\end{aligned} \qquad (10)$$

where $c = 0$ for $T_5 \wedge$ SO(3), ($Q^x = Q^c$), $c = +3\sqrt{10}$ for SU(3) ($Q^x = Q^a$), and $c = -3\sqrt{10}$ for a heretofore not mentioned group Sl(3,R) [$Q^x = Q^b \sim (x_i p_j + p_j x_i)$] which is associated with shear degrees of freedom. In eq.(10) the $<-,-|->$ symbol denotes an ordinary SO(3) Clebsch-Gordan coefficient. All three of these groups, $T_5 \wedge$ SO(3), SU(3), and Sl(3,R), are subgroups of the symplectic group Sp(3,R), which we will return to in Section 4. From these commutation relations it is easy to see how the SU(3) algebra reduces to that of $T_5 \wedge$ SO(3): if Q^a is divided by the square root of the second order invariant of SU(3) ($Q^a \leftarrow Q^a/\sqrt{C_2}$ where by definition the invariant $C_2 = (Q^a \cdot Q^a + 3L^2)/4$ commutes with the Q_μ^a and L_μ operators), the first and second commutators in eq.(10) remain unchanged, while the $L_{\mu+\nu}$ on the right-hand-side of the third equation goes over into $L_{\mu+\nu}/C_2$ and for low L values in large SU(3) irreps, $L_{\mu+\nu}/C_2 \to 0$. This renormalization of the Q^a operator is a group contraction process and the arguments presented show the SU(3) algebra reduces to the algebra of $T_5 \wedge$ SO(3) in the contraction limit, and consequently, the SU(3) theory reduces to that of the quantum rotor. Differences between observables of the two theories follow because SU(3) is a compact group with finite dimensional representations while $T_5 \wedge$ SO(3) is non-compact with infinite dimensional representations. Band termination and a fall-off in B(E2) strengths are examples.

3. Vibrational motion

The collective model takes vibrational as well as rotational degrees of freedom into account plus coupling between these modes. The coupling enters through the inertia parameters which are functions of the deformation. The macroscopic Hamiltonian can be written as the sum

$$H_{MAC} = H_{VIB} + H_{ROT} \qquad (11)$$

where H_{ROT} is given by eq.(1) with $A_\alpha \to 1/\left(8B_\alpha \beta^2 \sin^2(\gamma - \frac{2\pi}{3}\alpha)\right)$ and H_{VIB} has the form

$$H_{VIB} = \frac{1}{2} B_{\beta\beta} \dot{\beta}^2 + B_{\beta\gamma} \dot{\beta}\beta\dot{\gamma} + \frac{1}{2} B_{\gamma\gamma} \beta^2 \dot{\gamma}^2 + V(\beta, \gamma), \qquad (12)$$

where, in principle, $V(\beta, \gamma)$ can be any rotationally invariant function of β and γ [9]. An early formulation of the theory used a polynomial expansion for $V(\beta, \gamma)$ with the fundamental scalars β^2 and $\beta^3 \cos(3\gamma)$ taken as the variables:

$$V(\beta, \gamma) = \sum_{p,q}^{2p+3q \leq k} C_{p,q}^{mac} \beta^{2p+3q} (\cos(3\gamma))^q . \qquad (13)$$

For this potential it is clear that a second-order theory ($k = 2$) gives simple harmonic motion in the β-degree of freedom ($\omega_\beta = [2C_{1,0}^{mac}/B_{\beta\beta}]^{1/2}$ with $C_{1,0}^{mac} \geq 0$) while a third-order ($k = 3$) term is required for harmonic vibrations in the γ direction ($\omega_\gamma = [-9C_{0,1}^{mac}\beta/B_{\gamma\gamma}]^{1/2}$ with $C_{0,1}^{mac} \leq 0$). The usual theory takes $k = 6$, which for general triaxial geometries ($\gamma \neq 0$) means four additional terms are possible: β^4, $\beta^5 \cos(3\gamma)$, β^6, and $\beta^6 \cos^2(3\gamma)$, which is then a 6-th order polynomial in β (with no linear term) in the prolate ($\gamma = 0$) limit. Recent enhancements to the theory extend this to include general analytic and even numeric forms for the $V(\beta, \gamma)$ potential. The β-vibrational excitations contribute no angular momentum to the system and in the harmonic limit gives excitation energies in multiples of the basic $\hbar\omega_\beta$ unit. In contrast with this, the γ degree of freedom measures departures from axial symmetry and contributes 2 units of angular momentum per basic $\hbar\omega_\gamma$ excitation. The usual hydrodynamic assumption of the collective model picture imposes constraints on H_{MAC} which translate into relations among the inertia parameters of H_{ROT} and the mass parameters of H_{VIB}.

Our shell-model interpretation of rotational motion led to a simple relationship between (β, γ) and (λ, μ), see eqs.(8) and (9). Under this mapping, changes in β and γ correspond to changes in λ and μ. The following linear relations result upon differentiation of eq.(9):

$$\dot{\beta} = \kappa \left(\cos(\tilde{\gamma})\dot{\mu} + \cos(\gamma)\dot{\lambda} \right) \text{ and } \beta\dot{\gamma} = \kappa \left(\sin(\tilde{\gamma})\dot{\mu} - \sin(\gamma)\dot{\lambda} \right) , \qquad (14)$$

where in eq.(14) $\tilde{\gamma} = \frac{\pi}{3} - \gamma$ and $\kappa = 2/(3k)$ where k is defined in eq. (9). (An alternative procedure would, of course, replace time derivatives by momenta conjugate to the coordinate variables.) The $\gamma \leftrightarrow \tilde{\gamma}$ transformation represents a reflection across the $\gamma = \frac{\pi}{6}$ plane and corresponds to a $\lambda \leftrightarrow \mu$ interchange under the $(\beta, \gamma) \leftrightarrow (\lambda, \mu)$ mapping. These expressions for $\dot{\beta}$ and $\beta\dot{\gamma}$ in terms of $\dot{\lambda}$ and $\dot{\mu}$ can be substituted into the kinetic energy part of H_{VIB} to determine its shell-model equivalent. Likewise, the potential energy can be given in terms of λ and μ by expressing it in terms of the invariants of the theories.

For the polynomial form given in eq.(13) this is particularly simple because $\beta^2 \leftrightarrow \kappa^2[C_2(\lambda, \mu) + 3]$ and $\beta^3 \cos(3\gamma) \leftrightarrow \kappa^3 C_3(\lambda, \mu)$. The apparent shell-model image of H_{VIB} – labeled H_{SU2} in what follows, for reasons that will become clear – is then

$$\begin{aligned} H_{SU2} &= \frac{1}{2} B_{\lambda\lambda} \dot{\lambda}^2 + B_{\lambda\mu} \dot{\lambda}\dot{\mu} + \frac{1}{2} B_{\mu\mu} \dot{\mu}^2 \\ &+ \sum_{p,q}^{2p+3q \leq k} C_{p,q}^{mic} [C_2(\lambda, \mu) + 3]^p [C_3(\lambda, \mu)]^q , \end{aligned} \qquad (15)$$

where in this expression the mass parameters are functions of γ which can be written as a sum of symmetric (s) and antisymmetric (a) parts ($B = B^s + B^a$) under the ($\lambda \leftrightarrow \mu$ and $\gamma \leftrightarrow \tilde{\gamma}$) transformation where the antisymmetric parts vanish when $B_{\beta\gamma}$, which if positive (negative) drives the system towards a larger (smaller) prolate or smaller (larger) oblate shape, is zero:

$$B_{\lambda\lambda}^s(\gamma) = \kappa^2 \left(\cos^2(\gamma) B_{\beta\beta} + \sin^2(\gamma) B_{\gamma\gamma} \right) = +B_{\mu\mu}^s(\tilde{\gamma}) ,$$

$$\begin{aligned}
B^a_{\lambda\lambda}(\gamma) &= -\kappa^2 2\cos(\gamma)\sin(\gamma)B_{\beta\gamma} = -B^a_{\mu\mu}(\tilde{\gamma})\,, \\
B^s_{\lambda\mu}(\gamma) &= \kappa^2\left(\cos(\gamma)\cos(\tilde{\gamma})B_{\beta\beta} - \sin(\gamma)\sin(\tilde{\gamma})B_{\gamma\gamma}\right) = +B^s_{\mu\lambda}(\tilde{\gamma})\,, \\
B^a_{\lambda\mu}(\gamma) &= \kappa^2\left(\cos(\gamma)\sin(\tilde{\gamma}) - \sin(\gamma)\cos(\tilde{\gamma})\right)B_{\beta\gamma} = -B^a_{\mu\lambda}(\tilde{\gamma})\,.
\end{aligned} \quad (16)$$

It also is natural to assume that $C^{mic}_{p,q} = \kappa^{2p+3q}C^{mac}_{p,q}$ in eq.(15). Indeed, a polynomial form of this type has been used in some symplectic shell-model calculations [3]. It therefore appears we have discovered the microscopic equivalent of the macroscopic Hamiltonian (time derivatives replaced by conjugate momenta through second quantization, of course):

$$H_{MIC} = H_{SU2} + H_{SU3} \quad (17)$$

However, as is demonstrated below, the theories collide with one another at a fundamental level in the interpretation of $V(\beta,\gamma)$, with differences in the two due to antisymmetrization requirements which the microscopic (shell-model) theory automatically includes, but are absent in the macroscopic (collective-model) approach which deals only with boson modes.

For the special $k = 2$ case, the $V(\lambda,\mu)$ potential in eq.(15) reduces to $C_2(\lambda,\mu) + 3$ and this expression can be brought into a simple quadratic form through a linear transformation:

$$\xi = \frac{\lambda - \mu}{\sqrt{2}},\ \zeta = \frac{\lambda + \mu + 2}{\sqrt{2}} \rightarrow C_2(\lambda,\mu) + 3 = \frac{1}{2}\xi^2 + \frac{3}{2}\zeta^2. \quad (18)$$

It follows that if $C = C^{mic}_{1,0} > 0$ (see discussion below), H_{SU2} reduces to a coupled oscillator in localized regions of configuration space, and this explains our use of the SU(2) subscript ($C = \kappa^2\tilde{C}/2$):

$$\begin{aligned}
H_{SU2} &= \frac{1}{2}B_{\xi\xi}\dot{\xi}^2 + B_{\xi\zeta}\dot{\xi}\dot{\zeta} + \frac{1}{2}B_{\zeta\zeta}\dot{\zeta}^2 + \frac{1}{2}C\xi^2 + \frac{3}{2}C\zeta^2 \\
&= \frac{1}{2}B_{xx}\dot{\beta}_x^2 + B_{xy}\dot{\beta}_x\dot{\beta}_y + \frac{1}{2}B_{yy}\dot{\beta}_y^2 + \frac{1}{2}\tilde{C}\beta_x^2 + \frac{1}{2}\tilde{C}\beta_y^2\,.
\end{aligned} \quad (19)$$

Here the mass parameters are simple functions of those already given in eq.(16):

$$B_{\xi\xi} = \frac{1}{2}(B_{\lambda\lambda} + B_{\mu\mu} - 2B_{\lambda\mu}),\ B_{\zeta\zeta} = \frac{1}{2}(B_{\lambda\lambda} + B_{\mu\mu} + 2B_{\lambda\mu}),\ B_{\xi\zeta} = \frac{1}{2}(B_{\lambda\lambda} - B_{\mu\mu}),$$

$$B_{xx} = \kappa^{-2}B_{\lambda\lambda},\ B_{xy} = \sqrt{\frac{4}{3}}\kappa^{-2}(B_{\lambda\mu} - \frac{1}{2}B_{\lambda\lambda}),\ B_{yy} = \frac{1}{3}\kappa^{-2}(B_{\lambda\lambda} + 4B_{\mu\mu} - 4B_{\lambda\mu}). \quad (20)$$

If $B_{\mu\mu} = B_{\lambda\lambda}$ the first oscillator decouples with frequencies $\omega_\xi = \sqrt{\frac{C}{B_{\xi\xi}}}$ and $\omega_\zeta = \sqrt{\frac{3C}{B_{\zeta\zeta}}}$, and if in addition $B_{\lambda\mu} = 0$, $\omega_\xi = \sqrt{\frac{C}{B}}$ and $\omega_\zeta = \sqrt{\frac{3C}{B}}$ where $B = B_{\lambda\lambda}$.

The two-dimensional harmonic nature of the (β,γ) motion can be seen even more simply and directly in terms of the original (β,γ) variables, starting from eq. (12). For the $k = 2$ case, the $V(\beta,\gamma)$ potential, eq.(13), is proportional to β^2 and as a consequence it is invariant under rotations that take γ into $(\gamma - \theta)$, for arbitrary θ. This γ invariance means one can choose θ so that the kinetic energy terms in eq. (12) decouple. Specifically, starting with the cartesian representation of the dynamics, $(\beta_x,\beta_y) = (\beta\cos(\gamma),\beta\sin(\gamma))$ as introduced in eq.(9), one can transform to primed variables which are related to the unprimed ones in the usual way,

$$\begin{pmatrix} \beta'_x \\ \beta'_y \end{pmatrix} = \begin{pmatrix} \cos(\theta) & \sin(\theta) \\ -\sin(\theta) & \cos(\theta) \end{pmatrix} \begin{pmatrix} \beta_x \\ \beta_y \end{pmatrix}. \qquad (21)$$

When the angle θ is chosen as follows,

$$\tan(2\theta) = \frac{(B_{\beta\beta} - B_{\gamma\gamma})\sin(2\gamma) + 2B_{\beta\gamma}\cos(2\gamma)}{(B_{\beta\beta} - B_{\gamma\gamma})\cos(2\gamma) - 2B_{\beta\gamma}\sin(2\gamma)}, \qquad (22)$$

the Hamiltonian reduces to the decoupled form ($\beta^2 = \beta_x^2 + \beta_y^2 = \beta'^2_x + \beta'^2_y$)

$$H_{VIB} = \frac{1}{2} B'_{xx} \dot{\beta}'^2_x + \frac{1}{2} B'_{yy} \dot{\beta}'^2_y + \frac{1}{2} C \beta^2 \qquad (23)$$

where the primed mass parameters are given in terms of the unprimed ones of eq. (12) by

$$\begin{aligned} B'_{xx} &= B_{\beta\beta}\cos^2(\gamma - \theta) + B_{\gamma\gamma}\sin^2(\gamma - \theta) - B_{\beta\gamma}\sin(2(\gamma - \theta)), \\ B'_{yy} &= B_{\beta\beta}\sin^2(\gamma - \theta) + B_{\gamma\gamma}\cos^2(\gamma - \theta) + B_{\beta\gamma}\sin(2(\gamma - \theta)). \end{aligned} \qquad (24)$$

Hence one obtains

$$H_{SU2} = (n_x + \frac{1}{2})\hbar\omega_x + (n_y + \frac{1}{2})\hbar\omega_y \qquad (25)$$

where as usual the n_i ($i = x$ or y) are non-negative integer numbers and the frequencies are given by the very simple result $\omega_i = \sqrt{\frac{C}{B'_{ii}}}$.

Since λ and μ label SU(3) shell-model representations they are positive integer numbers. The allowed (λ, μ) values are determined by the embedding of SU(3) irreps in the irreps of the higher spatial symmetry, U(6) for the ds shell and U(N) in general for the n-th shell where N= $(n+1)(n+2)/2$. The complementary spin [U(2)] or spin-isospin [U(4)] symmetry group, as appropriate depending upon whether the valence protons and neutrons occupy respectively different major shells or the same one, must be the conjugate of this spatial symmetry to insure overall antisymmetry as required by the Pauli principle for a many-particle fermion system. The reduction of an irrep of U(N) which is labeled by [f] into representations of SU(3) labeled by (λ, μ) is called the $[f] \to (\lambda, \mu)$ plethysm of the U(N) $\supset$ SU(3) structure. These have been tabulated for the N=6, 10, 15 and 21 cases, which covers through the rare earth and actinide regions when these nuclei are described using the pseudo-spin coupling scheme. (A code for calculating any specific case is public domain software [10].) The picture that emerges presents the $0\hbar\omega$ shell-model basis as a collection of SU(3) irreps. As pointed out at the end of the first section, these sequences can be interpreted as incomplete rotational bands due to the fact that SU(3) is a compact group with finite dimensional irreps while $T_5 \wedge SO(3)$ is non-compact with infinite dimensional irreps. In nuclei these sequences are coupled to one another through SU(3) non-conserving interactions, i.e. by operators such as pairing, that lie outside the SU(3) algebra so they cannot be expressed as polynomial functions of SU(3) generators. (Under a $2n\hbar\omega$ Sp(3,R) extension of the $0\hbar\omega$ SU(3) theory, each (λ, μ) irrep is multiplied by tensor operators of the type $[(20) \times (20) \times \ldots \times (20)]$ (m

terms where $m \leq n$) with each of the (20)-factors creating a $2\hbar\omega$ excitation that is symmetric under particle exchange and free of spurious center-of-mass motion.)

The collective and shell-model theories can now be compared at a fundamental level. States of the leading SU(3) irrep, which is the most deformed $0\hbar\omega$ shell-model configuration allowed by the Pauli principle, correspond to members of the $(n_\beta, n_\gamma) = (0,0)$ collective band. States of the first excited SU(3) irrep (also in the $2\hbar\omega$ space) with an $L = 0$ member should correspond to members of the collective band with $(n_\beta, n_\gamma) = (1,0)$, and so on. It can be shown (for example, see below) that the $L = 0$ states of these low-lying SU(3) irreps distribute themselves in a near harmonic fashion. This certainly happens whenever the nucleon-nucleon interaction is dominated by a quadrupole-quadrupole part: $V \to V_{Q \cdot Q}$, where as usually given, $V_{Q \cdot Q} = -\frac{\chi}{2} Q^a \cdot Q^a = -\frac{\chi}{2}(4C_2 - 3L^2) \to -2\chi C_2$ in $L = 0$ states. The negative sign in this expression ensures that: 1) the most symmetric spatial symmetry is favored, and 2) the leading SU(3) irrep within that symmetry lies lowest. Other important but non-leading SU(3) irreps are of the type $(\lambda - \alpha, \mu - \beta)$ where α and β are small integer numbers that are determined by the $[f] \to (\lambda, \mu)$ plethysm. A simple calculation for the change in the total energy of a system that is due solely to the quadrupole-quadrupole interaction shows the expected result:

$$\Delta E = -2\chi[C_2(\lambda - \alpha, \mu - \beta) - C_2(\lambda, \mu)] = 2\chi[(2\lambda + \mu)\alpha + (\lambda + 2\mu)\beta - (\alpha^2 + \alpha\beta + \beta^2)]$$

$$\begin{array}{ccc} (\lambda, \mu) \gg (\alpha, \beta) & & (\alpha, \beta) = (+4\delta, -2\delta) \\ \longrightarrow & 2\chi[(2\lambda + \mu)\alpha + (\lambda + 2\mu)\beta] & \longrightarrow & 12\chi\lambda\delta \end{array} \quad (26)$$

where the last form, which is exactly harmonic since χ and λ are constants and $\delta = 1, 2, \ldots$ applies when blocking due to the Pauli principle does not manifest itself such as for boson systems which yield the SU(3) irrep sequences $(\lambda, \mu) = (2k, 0), (2k - 4, 2), \ldots, (0, k)$ where $k = m, m - 6, m - 12, \ldots$ with m equal to the number of valence particles [11].

Despite this seemingly simple interpretation of the β-vibrational mode, there is a serious problem with the picture it presents. The $(p, q) = (1, 0)$ term in the expansion of the collective-model potential usually enters with a positive sign ($C_{1,0}^{mac} > 0$) whereas in a shell-model theory it is always negative ($C_{1,0}^{mic} < 0$). The first result ($C_{1,0}^{mac} > 0$) is necessary to obtain harmonic motion. (In a fourth order collective theory the β^2 term can enter with a negative sign but then its square must appear with a positive coefficient to keep the motion bounded and harmonic, or at least nearly so: $C_{2,0}^{mac} > 0$ if $C_{1,0}^{mac} < 0$.) The condition ($C_{1,0}^{mic} < 0$) can be appreciated by noting that $Q^a \cdot Q^a = 4C_2 - 3L^2$, which goes over into $4C_2$ in $L = 0$ configurations, and the fact that the residual two-body interactions are normally strongly correlated with $V_{Q \cdot Q} = -\frac{\chi}{2} Q^a \cdot Q^a$, which has the consequence of ensuring that the most deformed configurations lie lowest. How can it be that the $C_{1,0}$ coefficient is positive in one case and negative in the other? The answer to this apparent contradiction lies in the fact that in the microscopic case the Pauli principle restricts the allowed (λ, μ) values to those irreps given by the $[f] \to (\lambda, \mu)$ group plethysm. Non-leading SU(3) irreps by definition lie energetically above the leading one because they display less deformation (lower β values) and are therefore less tightly bound. In the collective-model picture the entire (β, γ)-plane is available with the nature of the solutions of the H_{MAC} eigenvalue problem depending primarily upon the structure of the $V(\beta, \gamma)$ potential. Harmonic solutions arise in the usual way as expansions of the $V(\beta, \gamma)$ potential about the local minimum. So whereas in the collective-model picture the nature of the solution depends on the structure of the $V(\beta, \gamma)$

potential, in the shell-model case it is determined by a combination of the nucleon-nucleon interaction in the nuclear medium and the $[f] \to (\lambda, \mu)$ group plethysm which is governed by the particle statistics.

The scenario just presented focused on β-vibrational motion and stated nothing regarding the γ degree of freedom. The reason for this becomes clearer once one realizes that for most nuclei the leading SU(3) irrep, which by definition is the most deformed intrinsic many-particle configuration allowed by the Pauli principle, has a non-zero value for μ and as a consequence includes within its span of states a rotational sequence with the same structure as the γ-band in a collective model description of low-energy phenomena. The actual L values in the (λ, μ) irrep of SU(3) are given by the rule:

$$\begin{aligned} L &= K, K+1, \ldots, \lambda + \mu + 1 - K, \text{ if } K \neq 0, \\ \text{and } L &= \lambda + \mu, \lambda + \mu - 2, \ldots, 1 \text{ or } 0 \text{ when } K = 0 \\ \text{where } K &= \min(\lambda, \mu), \min(\lambda, \mu) - 2, \ldots, 1 \text{ or } 0. \end{aligned} \qquad (27)$$

Within this framework the $(n_\beta, n_\gamma) = (0, 1)$ bandhead configuration of the collective model is the $L = 2$ state of the $K = 2$ band in the leading SU(3) irrep, provided λ and μ are both even as they must be for a realization of A-type symmetry of the D_2 symmetry group and the $\min(\lambda, \mu) \geq 2$. To fix the position of the $(n_\beta, n_\gamma) = (0, 1)$ mode, interpreted in this way, at an excitation energy $\hbar\omega_\gamma$ above the $(n_\beta, n_\gamma) = (0, 0)$ ground state, requires the $I^2_{a=3}$ operator, which has eigenvalue K^2, to enter in H_{ROT} with the strength $b = \hbar\omega_\gamma/(4 - 3a/2)$ where a is the parameter multiplying the I^2 term in H_{ROT}. However, this simple prescription fails to produce a $(n_\beta, n_\gamma) = (0, 2)$ mode at the second harmonic $2\hbar\omega_\gamma$ level, giving instead a $L = 4$ band-head state at an excitation energy of $4(\hbar\omega_\gamma - a)$. There is a further difference because the $(n_\beta, n_\gamma) = (0, 2)$ configuration of the collective model picture consists of two states, not just one, with angular momenta $L = 0$ and $L = 4$ where these two result from the symmetric coupling of a pair of $L = 2$ phonons. (The $L = 1$ and 3 states are ruled out because harmonic excitations (boson modes) can only couple symmetrically, and furthermore, the experts tell us the intermediate $L = 2$ configuration is ruled out by reflection symmetry that is implicit in the hydrodynamic assumption.) In summary, while the shell-model picture seems to give a relatively simple interpretation of β-vibrational motion of the collective model it only accounts for the first of the γ-vibrational excitations — double γs are not part of the SU(3) shell-model picture.

Differences in the two prescriptions derive from the fact that the collective model adheres to the hydrodynamic assumption which attributes no structure to the nucleus other than its shape. This is clearly an oversimplification — the existence of SU(3) irreps with odd λ and μ values, which correspond to B_α-type D_2 symmetries ($\alpha = 1, 2, 3$), is a clear indication that nuclei have an internal structure not portrayed by a simple collective-model picture. This particular feature is especially important when attempting to interpret the structure of odd-A nuclei. There is also no fixed relationship between "shape" and "moments of inertia" within the shell-model framework — a given SU(3) irrep can display the full range of spectral characteristics: asymmetry parameters $\kappa = -1$ (prolate), $\kappa = 0$ (asymmetric), and $\kappa = +1$ (oblate). Successes of the collective model and its odd-A extension which couples collective excitations to a Nilsson-type single-particle picture for the "unpaired" nucleon, have had a mesmerizing (if not a paralyzing) effect on the nuclear physics community. We believe a deeper understanding of nuclear physics phenomena lies in probing these fundamental differences.

Other boson theories, as elegant and simple as they may be, can be shown to suffer similar types of deficiencies — too few or too many states, depending upon specifics of the theory and application. For example, it is easy to show that the leading SU(3) irrep $(2n, 0)$ in the SU(3) limit of the IBA is spurious when the number of bosons n (equal to 1/2 the number of valence particles) exceeds 1/3 the total number allowed, n_{max}, for the shell under consideration [11]. Whenever $n > n_{max}/3$ some of the $2n$ quanta must be placed in other than the z-direction, and since $\lambda = n_z - n_x$ and $\mu = n_x - n_y$ this means $\lambda < 2n$ and $\mu > 0$. So while it can be argued with good justification that the collective model, and the more sophisticated derivative theories based on boson algebras, can be used to catalog and help understand large bodies of nuclear structure information, a lack of respect for the Pauli principle can lead to the wrong physics, especially for low-lying states of heavy systems where the structure seems to be determined primarily by competition between interactions among the nucleons and the particle antisymmetry requirements of the Pauli principle. The importance of achieving a deeper understanding of the structure of complex nuclei, especially heavy ones, should not be underestimated. For example, if reliable many-particle wavefunctions for special parent and daughter systems [e.g., ^{160}Gd $\to$ ^{160}Dy (87keV) and ^{238}U $\to$ ^{238}Pu (44keV)] can be generated, they can be used to determine structure factors in double-beta decay which in turn can be related to the mass of the neutrino and ultimately to the validity of the standard model of fundamental particles and interactions [12]. In the next section this train of thought is continued — indeed, much of what is known about potential energy surfaces can be traced to strong competition between interaction driven phenomena and particle statistics.

4. Energy surfaces

The $(\beta, \gamma) \leftrightarrow (\lambda, \mu)$ correspondence set down in Section 2 means that the important collective-model concept of a potential energy surface can also be given a shell-model interpretation [13]. Recall that within the framework of the collective model the potential energy of a particular nucleus is given by $V(\beta, \gamma)$, see eq.(13), where the parameters of the corresponding collective-model Hamiltonian have been adjusted to give a best fit to the excitation energies and transition strengths of that system. For this picture to be given a shell-model interpretation it is necessary to generalize the potential energy surface concept to that of a total energy surface. This follows because the potential and kinetic energies of a system are not separately measurable, only their sum (which is the total energy) can be determined experimentally. One possible procedure is to calculate the binding energy of a system by diagonalizing its Hamiltonian which is constrained to act in a region of configuration space where the expectation value of the square and the cube of the quadrupole matrix are β^2 and $\beta^3 \cos(3\gamma)$, respectively. By varying β and γ, this procedure can be used to determine the functional dependence of the total energy on the deformation parameters. Alternatively, one could determine coherent states for the system by requiring sharp expectation values for the square and cube of the quadrupole matrix and then calculate the expectation value of the Hamiltonian with respect to these states. Regardless of the methodology selected, the concept is only meaningful if the observables ($E \sim <H>$, $\beta^2 \sim <Q \cdot Q>$, and $\beta^3 \cos(3\gamma) \sim <Q \cdot (Q \times Q)>$) of the system are, or are approximately, simultaneously diagonalizable.

The simplest and most straightforward procedure, since the $(\beta, \gamma) \leftrightarrow (\lambda, \mu)$ mapping has already been established, is to calculate the expectation of the Hamiltonian in SU(3) shell-model irreps. The expectation value should be evaluated in the $L = 0$ state, which because it is unique is also a Hamiltonian eigenstate for that irrep: $< H >^{(\lambda,\mu)L=0} = E_0\left[(\lambda,\mu)L = 0\right]$. The $L \neq 0$ states in each irrep usually occur more than once and always lie at some positive excitation energy relative to the $L = 0$ configuration. This procedure produces the left-hand portion ($\beta < \beta_0$ where β_0 is the value of β for which $< H >$ is an absolute minimum) of the schematic curve shown in Figure 4. The near inverse quadratic dependence of $< H >$ on β for values of the deformation close to zero follows from the fact that the quadrupole-quadrupole term, with expectation value proportional to β^2 in $L = 0$ states, enters the residual interaction with a negative sign [recall the arguments given immediately preceding and following eq.(26)]. Within the horizontal $0\hbar\omega$ extension of the shell-model the leading SU(3) irrep is the most deformed and therefore lies lowest — the other Pauli allowed configurations correspond to $\beta < \beta_o$ values lying at some positive excitation energy relative to the leading (λ_0, μ_0) irrep, where β_o and γ_o are related to λ_o and μ_o in the usual way through eq.(8).

The $\beta > \beta_o$ portion of the curve shown in Figure 4 cannot be understood within the context of a $0\hbar\omega$ shell-model theory. Larger deformations correspond to radial extensions of the nucleus and these can only be realized by going to a vertical extension of the shell-model, that is, by including $2n\hbar\omega$ configurations ($n = 1, 2, \ldots$) in the set of basis states. From an algebraic perspective this extension (restricting to one-particle excitations, which excludes particle-hole excitations) takes one from the compact SU(3) group to Sp(3,R), a non-compact symplectic group with infinite dimensional irreps. The symplectic model is an extension of the SU(3) scheme which it contains as its $0\hbar\omega$ limit. The fact that irreps of the symplectic group are infinite follows because Q^c, which couples the n and n' oscillator shells where $n' = n$ and $n \pm 2$, is a generator of Sp(3,R). Nonetheless, the underlying oscillator structure and the fact that Sp(3,R) contains SU(3) as a subgroup means the relevant symplectic irreps have a simple structure. To see this, consider the special case where H is the harmonic oscillator Hamiltonian, $H = H_0$. All $2n\hbar\omega$ configurations are then degenerate with excitation energy $2n\hbar\omega$ relative to the $n = 0$ shell. The SU(3) content of these symplectic states is given by the direct product of a starting $0\hbar\omega$ SU(3) irrep, denoted by (λ_s, μ_s), and the SU(3) irreps contained in the symmetric product of the elementary $(2, 0)$ representation taken n times:

$$\begin{aligned}((\lambda,\mu))_n &= (\lambda_s,\mu_s) \times \left(\prod^n [\times(2,0)]\right) \\ &= (\lambda_s,\mu_s) \times \left(\sum_{\substack{n_1+n_2+n_3=2n \\ n_1 \geq n_2 \geq n_3 \geq 0, \text{all } n_i \text{ even}}} [\times(n_1-n_2, n_2-n_3)]\right). \end{aligned} \quad (28)$$

When an interaction is added to the harmonic oscillator Hamiltonian, $H = H_0 - \frac{\chi}{2}(Q^a \cdot Q^a)$ in the simplest case, these representations split at each level in the same way as the $0\hbar\omega$ SU(3) irreps do at the $n = 0$ level. This is indicated schematically in Figure 4 for the maximally deformed $(\lambda_o + 2, \mu_o)$ irrep in the $n = 1$ space which results from the coupling of $(2, 0)$ to the leading $0\hbar\omega$ $n = 0$ shell-model irrep. This structure is repeated for each (λ_s, μ_s) at each of the $n = 2, 3, \ldots$ levels. The most deformed configuration at each level is the $(\lambda_o + 2n, \mu_o)$ irrep.

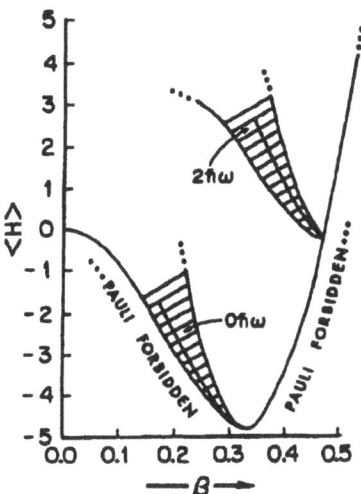

Figure 4. *Shell-model interpretation of the collective-model potential energy surface.*

With no interaction the landscape is evenly terraced (upper levels overhanging flat lower ones) with step sizes $2\hbar\omega$ as one moves from the ground level at $0\hbar\omega$ to the n-th tier $2\hbar\omega$ up. This is modified by the interaction, with the shell structure (terraced landscape) surviving so long as the gain in binding energy ($\langle\Delta H_{BE}\rangle \to$ negative) due to an increase in the deformation is less than the excitation energy added to the system ($\langle\Delta H_{SG}\rangle \to$ positive) to gain that shell-gap configuration energy: $|\langle\Delta H_{BE}\rangle| < |\langle\Delta H_{SG}\rangle|$ or $\Delta BE\left[n(\lambda,\mu)\right] < 2n\hbar\omega$, so the net effect as one moves to larger and larger deformations is an increase in $\langle\Delta H\rangle$. Strongly deformed states, certainly isomeric and especially superdeformed configurations, appear to lie near the base of such cliffs where this condition is just satisfied. Full symplectic calculations for the low-lying properties of ds-shell (see below) as well as rare-earth and actinide nuclei show convergence in the range $8 \leq n \leq 12$ with $(0,2,4) - \hbar\omega$ states typically contributing $(70, 25, 5)\%$, respectively, to the structure of low-lying configurations. So rises in the expectation value of H for $\beta > \beta_o$ result from competition between the increase in the binding due to the availability of a space which includes configurations with larger deformation, and the energy that must be added to the system to reach that level. The binding of nucleons to one another through a common (mean-field) potential allows for moderate deformations but is normally not strong enough to support stable superdeformed shapes. The potential energy surface concept of the collective model appears to be an over-simplification of the interplay between the shell structure which results from particle statistics and a residual effective nucleon-nucleon interaction that distorts the landscape to find an overall minimum energy configuration.

As an illustration of how this scenario plays out in a particular case, consider the ds-shell nucleus ^{24}Mg. The full set of $0\hbar\omega$ SU(3) irreps for ^{24}Mg belong to the $(0s)^4(1p)^{12}(ds)^8$ shell-model configuration. They are determined by the U(6) $\to$ SU(3) group plethysm (U(6) because the spatial degeneracy of the ds-shell is 6: 5 d states and 1 s state). For example, the most symmetric ds-shell spatial symmetry ([f]= [44]) contains the following SU(3) irreps that have an $L = 0$ state in them: $(\lambda,\mu) = (8,4), (4,6), (0,8), (6,2)^2, (2,4)^2, (4,0)^2$, and $(0,2)$ where the superscript is a multiplicity label. Of all the allowed irreps, including those

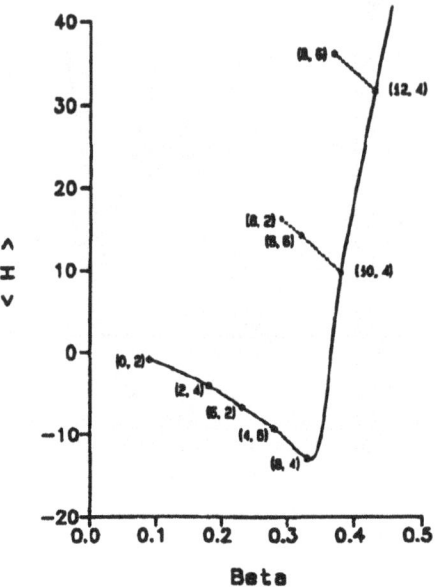

Figure 5. *Expectation of $H = H_0 - \frac{1}{2}\chi Q^c \cdot Q^c$ ($\hbar\omega = 12.6$ MeV and a $\chi = 0.0438$) between basis states of the symplectic model with ds-shell spatial symmetry [f] = [44], which is the leading configuration for ^{24}Mg. (To be compared with Figure 4.)*

in less symmetric spatial symmetries, the $(8,4)$ yields the maximum value for β and for this particular SU(3) irrep $\gamma = 20.6°$. The expectation of a simple $H = H_0 - \frac{1}{2}\chi Q^c \cdot Q^c$ ($\hbar\omega = 12.6$ Mev and a $\chi = 0.0438$) between basis states of the symplectic model are plotted as a function of β in Figure 5. The solid curve traces out the locus of minimum expectation values in the $0\hbar\omega$ space for $\beta < \beta_0$ and connects leading irreps of the $2\hbar\omega$, $4\hbar\omega$, $6\hbar\omega$, etc., spaces in the product $(2,0)^n \times (8,4)$ for $\beta > \beta_0$. The γ values range from $46.1°$ for the $(0,2)$ irrep, to $20.6°$ for the $(8,4)$, and to $15.6°$ for the $(12,4)$. The curve obviously jumps about in the γ direction. As suggested above, the slow rise to the left of β_0 is a $0\hbar\omega$ phenomena while the sharp rise to the right of β_0 is primarily due to the shell structure. If χ were set to zero, all the irreps of the n-th shell would lie at $2n\hbar\omega$. The actual curve will be lower than the one shown because the expectation value is not exact the eigenvalue, but the shift will not be large [14].

5. Conclusion

We have sketched out in very general terms a possible shell-model interpretation for rotational and vibrational modes of excitation in nuclei that seem to be well-described by a simple collective-model phenomenology. The equivalence of the shell-model SU(3) and the collective-model $T_5 \wedge$ SO(3) symmetries in the limit of angular momentum values that are small compared to the maximum $\lambda + \mu$ value that can occur in a (λ, μ) irrep, explains the appearance of rotational-like behavior in nuclei. The $(\beta, \gamma) \leftrightarrow (\lambda, \mu)$ mapping discovered in establishing the shell-model image of rotational motion was shown to lead to a simple shell-model interpretation of the collective-model vibrational kinetic energy. However, there are irreconcilable differences in the collective and shell-model interpretations of the vibrational potential. This can be traced to the fact that in the collective model all regions of configuration space are considered to be equally accessible. The potential is the only term in the Hamiltonian giving preference to specific (β,γ) shapes. On the other hand, the fact that nucleons are fermions and therefore must satisfy the Exclusion Principle imposes a structure that even in the absence of residual interactions blocks out some regions of configuration space. At the same time the shell model picture adds a third dimension to the problem because there can be multiple independent solutions for any set of (λ,μ) and therefore (β,γ) values. The macroscopic-microscopic comparison is indicated schematically as follows:

$$H_{MAC} = T_{VIB} + V_{VIB} + H_{ROT}$$

$$\updownarrow \qquad \downarrow \qquad \updownarrow \qquad \downarrow$$

$$H_{MIC} = T_{SU2} + V_{SU2} + H_{SU3}$$

Two of three terms in the macroscopic (collective-model) Hamiltonian have unambiguous microscopic (shell-model) counterparts. The macroscopic (vibrational) and microscopic potentials are however quite different. Paradoxically, the macroscopic form is both more general and more restrictive than the microscopic one: more general because solutions are continuous functions of (β,γ) and not constrained by other than the form of the potential, and more restrictive because it gives but one solution per (β,γ) value. In the microscopic case the [f] $\to (\lambda,\mu)$ plethysm specifies allowed configurations and multiple independent configurations for any specific (λ,μ) exist, especially if one includes the multi-$\hbar\omega$ configurations of the symplectic model. SU(3) symmetry breaking interactions couple the various SU(3) irreps. This corresponds to a softening of the grid-moving the shell-model picture in line with the collective-model dynamics of continuous (β,γ) variables. The collective model will remain a purely phenomenological theory unless a means is found for incorporating into the potential restrictions imposed by antisymmetrization requirements. Likewise, a shell-model theory that does not include multi-$\hbar\omega$ configurations should not be expected to give a proper account of the structure of strongly deformed systems. Since knowing the limitations of existing models is a necessary first step towards the development of more fundamental theories, we believe the future of nuclear physics is challenging and bright.

Acknowledgements

The work reported here was supported in part by the U.S. National Science Foundation (Grant Number PHY89-22550). It was completed while one of the authors (J. P. D.) enjoyed a guest professorship with Professor Amand Faessler at the Institut für Theoretische Physik, Universität Tübingen, D-72076 Tübingen, FRG. Discussions with Andrey Blokhin and Jutta Escher are gratefully acknowledged.

References

[1] Iachello F and Arima A, *Ann. Phys.* **99**, 253 (1976); **111**, 201 (1978); **123**, 468 (1979)
[2] Elliott J P, *Proc. Roy. Soc. London A* **245**, 128, 562 (1958)
[3] Rowe D J, *Rep. Prog. Phys.* **48**, 1419 (1985)
[4] Hecht K T, McGrory J B and Draayer J P, *Nucl. Phys. A* **197**, 369 (1972); Wu C-L, Feng D H, Chen X-G, Chen J-Q and Guidry M W, *Phys. Rev. C* **36**, 1157 (1987)

[5] Arima A, Harvey M and Shimizu K, *Phys. Lett.* **30B**, 517 (1969)
[6] Hecht K T and Adler A, *Nucl. Phys. A* **137**, 129 (1969)
[7] Bahri C, Draayer J P and Moszkowski S A, *Phys. Rev. Lett.* **68**, 1419 (1992)
[8] Racah G, *Group Theoretical Concepts*, edited by Gürsey F (Gordon and Breach, New York, 1964)
[9] Troltenier D, Maruhn J A, Greiner W and Hess P O, *Z. Phys. A* **343**, 25 (1992)
[10] Draayer J P, Leschber Y, Park S C and Lopez R, *Compt. Phys. Com.* **56**, 279 (1989)
[11] J.P. Draayer in: Casten R F, Lipas P O, Warner D D, Otskuka T, Heyde K, Draayer J P, *Algebraic Approaches to Nuclear Structure*, edited by Casten R F (Harwood, Singapore, 1993)
[12] Castanos O, Hirsch J G, Civitarese O, and Hess P O (1993 preprint, UNAM, Mexico)
[13] Draayer J P, Park S C, and Castanos O, 1989 Phys. Rev. Lett. **62** 20
[14] Castanos O and Draayer J P, *Nucl.Phys. A* **491**, 349 (1989)

PARTICLE-ROTOR MODEL DESCRIPTION OF DEFORMED NUCLEI

Aldo Covello, Angela Gargano and Nunzio Itaco

Dipartimento di Scienze Fisiche, Università di Napoli Federico II, and Istituto Nazionale di Fisica Nucleare, Mostra d'Oltremare, Pad. 20, 80125 Napoli, Italy

1. Introduction

An appropriate theoretical basis for the study of strongly deformed nuclei is provided by the many-particles plus rotor (MPR) model.[1,2] The main difficulty with this model is the description of the correlated motion of the valence nucleons in the intrinsic deformed field. Although the residual interaction is usually assumed to be a simple pairing force, the large number of particles and levels which have to be taken into account make a complete-basis diagonalization very impractical, if not unfeasible. This has caused the BCS theory to become the most widely used approach to this problem. The errors inherent in this approximation may well result, however, in a poor description of the intrinsic structure, thus making it difficult to assess the real scope of the MPR model.

We are currently handling this problem by making use of a new method for treating pairing correlations in deformed nuclei. This method, which we call chain-calculation method (CCM), provides the natural framework for a sequence of number-conserving approximations. We have shown[3-5] that it has the power to reduce the complexity of multi-level, multi-particle pairing problems drastically while yielding practically exact results.

In this paper, we apply the MPR model to the nucleus ^{163}Er considering twenty-one valence neutrons distributed over eighteen twofold degenerate single-particle levels. We first treat the pairing Hamiltonian by means of the CCM and then diagonalize the recoil term within the set of seniority-one states. As a final step the Coriolis term is taken into account. Along the same lines we have also performed a calculation where the pairing correlations are treated in the BCS approximation.

In Sec. 2 we give an outline of the essentials of the MPR model. In Sec. 3 we describe the salient features of the CCM and evidence its practical value. In Sec. 4 we discuss the recoil term and give the relevant matrix elements. The Coriolis coupling is then discussed in Sec. 5. In Sec. 6 we present some results of our study of ^{163}Er. It is shown that, in contrast to the results of the BCS approximation, we obtain a good agreement with experiment without any ad-hoc attenuation factor of the Coriolis term. A brief summary is given in Sec. 7.

2. Many-Particles plus Rotor Model

In the strong coupling representation the model Hamiltonian describing a system of N valence particles coupled to an axially-symmetric rotor is written

$$H_{MPR} = H_I + H_{intr} + H_c, \tag{1}$$

where
$$H_I = A(\mathbf{I}^2 - I_3^2), \tag{2}$$

$$H_{intr} = H_0 + H_{pair} + H_{rec}, \tag{3}$$

$$H_{rec} = A(\mathbf{J}^2 - J_3^2), \tag{4}$$

$$H_c = -A(I_+J_- + I_-J_+), \tag{5}$$

with standard notation. The recoil term H_{rec} is of particular relevance in the MPR model,[2,6] as it contains both one-body and two-body terms. In fact, the angular momentum $\mathbf{J}$ due to the valence particles has the form

$$\mathbf{J} = \sum_{i=1}^{N} \mathbf{j}(i), \tag{6}$$

which implies that H_{rec} becomes

$$H_{rec} = A\left[\sum_{i=1}^{N} \mathbf{j}^2(i) - \left(\sum_{i=1}^{N} j_3(i)\right)^2 + 2\sum_{i<k} \mathbf{j}(i) \cdot \mathbf{j}(k)\right]. \tag{7}$$

The intrinsic deformed field H_0 is described by a nonspheroidal axial and reflection symmetric Woods-Saxon potential.[7] The eigenstates of $H_I + H_{intr}$ can be written in the form

$$\Psi^I_{M\Omega\tau} = \sqrt{\frac{2I+1}{16\pi^2}}\left[D^I_{M\Omega}(\omega)\chi_{\Omega\tau} + (-)^{I+\Omega}D^I_{M-\Omega}(\omega)\chi_{\overline{\Omega}\tau}\right], \tag{8}$$

where the intrinsic wave functions $\chi_{\Omega\tau}$ are solutions to the eigenvalue equation

$$(H_0 + H_{pair} + H_{rec})\chi_{\Omega\tau} = \mathcal{E}_{\Omega\tau}\chi_{\Omega\tau}. \tag{9}$$

Once Eq. (9) is solved the Coriolis term can be diagonalized in this representation. In the next three sections we shall discuss in some detail our treatment of the intrinsic Hamiltonian and of the Coriolis coupling.

3. Treatment of Pairing Correlations by means of the CCM

The Hamiltonian $H = H_0 + H_{pair}$ is written as

$$H = \sum_{\nu}\epsilon_\nu \hat{N}_\nu - \sum_{\nu\nu'} G_{\nu\nu'} A^\dagger_\nu A_{\nu'}, \tag{10}$$

where
$$\hat{N}_\nu = a^\dagger_\nu a_\nu + a^\dagger_{\overline{\nu}} a_{\overline{\nu}}, \tag{11}$$

$$A^\dagger_\nu = a^\dagger_\nu a^\dagger_{\overline{\nu}}. \tag{12}$$

The index ν stands for all quantum numbers specifying the single-particle states while $\overline{\nu}$ denotes the time reversal partner. In cases where Ω is essential, ν will represent only the asymptotic quantum numbers $[Nn_3\Lambda]$.

In our approach the wave function for a system with N identical particles is given the form

$$|N,\beta,L\rangle = \sum_{\nu\gamma} c^L_{\nu\beta\gamma}(N) A^\dagger_\nu |N-2,\gamma,L\rangle, \tag{13}$$

where the index L stands for the quantum numbers of the blocked levels and ν runs only over the unblocked ones. The indices β and γ specify the states with N and $N-2$ particles, respectively. It is precisely the use of (13) which leads to solving the N-particle problem through a chain calculation across systems differing by two in nucleon number.

It should be noted that inherent in this formalism is the use of an overcomplete set of basis vectors $A^\dagger_\nu |N-2,\gamma,L\rangle$. To single out a linearly independent set of states we analyze the metric matrix, whose elements are defined as

$$d^L_{\nu\gamma\nu'\gamma'}(N-2) = \langle N-2,\gamma,L | A_\nu A^\dagger_{\nu'} | N-2,\gamma',L\rangle, \tag{14}$$

through the Cholesky decomposition of symmetric positive definite matrices. An account of our procedure for removing the redundant states may be found in Ref. 8. It is therefore to be understood that the sum in (13) is restricted to pairs of indices ($\nu\gamma$) corresponding to linearly independent basis vectors. Making use of expansion (13) in the Schrödinger equation,

$$H|N,\beta,L\rangle = E_{L\beta}(N)|N,\beta,L\rangle, \tag{15}$$

leads straightforwardly to the formulation of the eigenvalue problem in the form

$$\sum_{\nu'\gamma'}\Big\{\sum_{\nu''\gamma''}[\bar{d}^L(N-2)]^{-1}_{\nu\gamma\nu''\gamma''} H^L_{\nu''\gamma''\nu'\gamma'}\Big\} c^L_{\nu'\beta\gamma'}(N) = E_{L\beta}(N)\, c^L_{\nu\beta\gamma}(N), \tag{16}$$

where

$$H^L_{\nu''\gamma''\nu'\gamma'} = \langle N-2,\gamma'',L|A_{\nu''}H A^\dagger_{\nu'}|N-2,\gamma',L\rangle \tag{17}$$

and $\bar{d}^L$ denotes the metric matrix obtained from the original matrix by removing rows and columns corresponding to redundant states according to the above mentioned procedure.

The elements of the metric matrix (14) as well as those of the Hamiltonian matrix (17) are calculated by making use of the explicit expression of the wave function $|N-2,\gamma,L\rangle$ in the occupation-number representation,

$$|N-2,\gamma,L\rangle = \sum_{n_1\ldots n_l} K^L_{n_1\ldots n_l\gamma}(N-2)(A^\dagger_1)^{n_1}\ldots(A^\dagger_l)^{n_l}|N_0,L\rangle, \tag{18}$$

where $n_i = 0$ or 1, and $n_1 + n_2 + \cdots + n_l = (N-2-N_0)/2$, l being the number of doubly degenerate single-particle levels and N_0 the number of unpaired particles (of course, for blocked levels $n_i = 0$). The explicit expression of the elements of the metric matrix as well as those of the Hamiltonian matrix may be found in Ref. 5.

The coefficients $K^L_{n_1\ldots n_l\beta}(N)$ are related to the c^L's through

$$K^L_{n_1\ldots n_l\beta}(N) = \sum_{\nu\gamma} c^L_{\nu\beta\gamma}(N) K^L_{n_1\ldots n_{\nu-1} 0 n_{\nu+1}\ldots n_l\gamma}(N-2)\delta_{n_\nu 1}. \tag{19}$$

The solution of the N-particle problem can therefore be obtained by a chain calculation starting from N_0 and proceeding by adding pairs of particles up to the desired value of N.

The formalism given above is exact to the extent that all the existing core states $|N - 2, \gamma, L\rangle$ are taken into account. Its practical value, however, resides in the fact that it provides the framework for a sequence of number-conserving approximations which are obtained by truncating the expansion (13). We call $k - th$ order theory the approximation scheme in which the core states are restricted to the lowest k states. We have verified[3-5] that the use of very low orders of approximation (typically fourth or fifth order) produces extremely accurate results for at least the three or four lowest states. Clearly, this brings about an enormous reduction in the size of the matrices to be diagonalized. In fact, they are at most of the order $k \times l$. It is possible, however, to further simplify our chain-calculation approach by restricting the sums on the r.h.s. of (18). Of course, the main question is how to perform this truncation without spoiling the accuracy of the method. We shall not touch on this point here, but refer the reader to Ref. 5 for a detailed discussion. To evidence the degree of accuracy that can be attained by our approach we report, however, some representative results of an application to a system of sixteen particles in as many doubly degenerate single-particle levels (the single-particle level scheme may be found in Ref. 5). The pairing strength G (we use the usual constant pairing force) is 0.20 MeV and we consider states in which all the particles are distributed in time-reversed pairs over the available orbitals (in the following we shall denote these seniority-zero type states as $v = 0$ states).

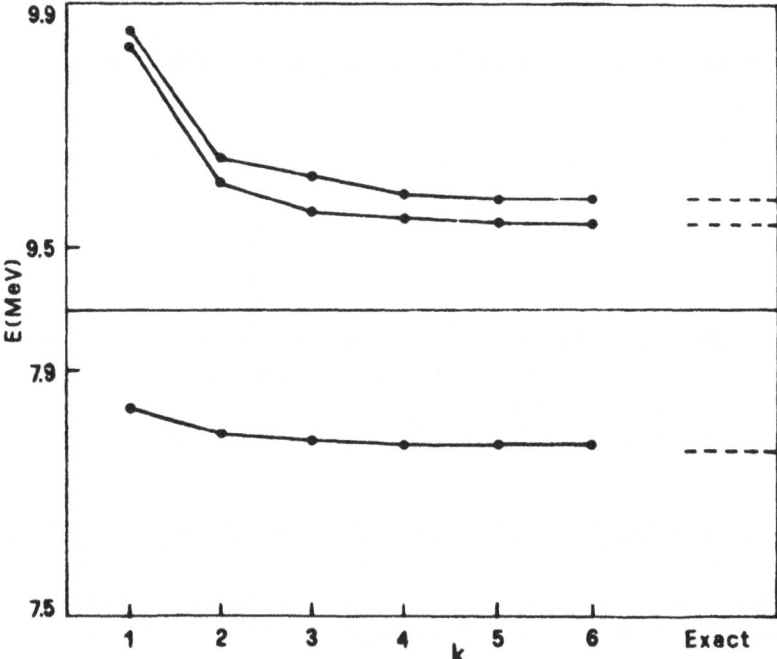

Figure 1. *Energies of the three lowest $v = 0$ states for a system of 16 particles distributed among as many single-particle levels. The results are plotted as a function of the order of approximation k. The number of configurations included in the calculation is 4000. See text for details.*

Fig. 1 shows the energy of the three lowest $v = 0$ states obtained at orders of approximation ranging from 1 to 6 with inclusion of 4000 configurations (they are 11 440 in all) in (18). We see that the fifth-order theory yields results which are quite close to the exact ones. In fact, the ground-state energy is only 10 keV higher than the exact one while for both the two excited states the difference is 3 keV. Regarding the reduction in the size of the energy matrices, the fifth order calculation involves the diagonalization of an 80×80 matrix, whereas the total number of $v = 0$ states is 12870.

The above example shows that our chain-calculation approach makes it possible to treat multi-level pairing-force problems in a simple but extremely accurate way.

4. The Recoil Term

The recoil term (7) in second quantization reads

$$H_{rec} = A\left[\sum_{\nu\nu'} F_{\nu\nu'} a_\nu^\dagger a_{\nu'} + \sum_{\nu_1\nu_2\nu_3\nu_4} R_{\nu_1\nu_3} R_{\nu_4\nu_2} a_{\nu_1}^\dagger a_{\nu_2}^\dagger a_{\nu_4} a_{\nu_3}\right], \tag{20}$$

where

$$F_{\nu\nu'} = \langle \nu | \boldsymbol{j}^2 - j_3^2 | \nu' \rangle, \tag{21}$$

$$R_{\nu'\nu'} = \begin{cases} \langle \nu | j_- | \nu' \rangle & \text{if } \Omega_\nu < \Omega_{\nu'}, \\ \langle \nu' | j_- | \nu \rangle & \text{if } \Omega_\nu > \Omega_{\nu'}, \\ \langle \overline{\nu} | j_- | \nu' \rangle & \text{if } \Omega_\nu = \Omega_{\nu'} = \frac{1}{2}. \end{cases} \tag{22}$$

Here we focus attention on the treatment of odd-A nuclei. In this case we diagonalize H_{rec} within the set of $v = 1$ states obtained by treating the Hamiltonian (10) as described in the previous section. The intrinsic wave function has therefore the form

$$\chi_{\Omega\tau} = \sum_{\beta\mu} f_{\beta\tau\mu}^\Omega |\beta, \mu\Omega\rangle. \tag{23}$$

Here and in the following the label μ refers to the quantum numbers of the blocked level.

The matrix elements of H_{rec} between $v = 1$ states are given by

$$\langle \beta, \mu\Omega | H_{rec} | \beta', \mu'\Omega \rangle = A\left\{\delta_{\mu\mu'}\left[2\sum_{\nu>0} F_{\nu\nu} T_{\beta\beta'\nu}^{\mu\mu'} - \sum_{\nu,\nu'>0} R_{\nu\nu'}^2 T_{\beta\beta'\nu\nu'}^{\mu\mu'}\right] + \right.$$

$$\left. [1 - (1 - \delta_{\beta\beta'})\delta_{\mu\mu'}] F_{\mu\mu'} S_{\beta\beta'}^{\mu\mu'} - \sum_{\nu>0} R_{\nu\mu'} R_{\nu\mu} S_{\beta\beta'\nu}^{\mu\mu'}\right\}, \tag{24}$$

where the quantities S and T are expressed in terms of the coefficients (19). Their explicit expressions may be found in Ref. 9.

Let us now discuss briefly the effects produced by the recoil term. We first consider the diagonal matrix elements. The first two sums in Eq. (24) represent the so-called core contribution.[6,10] In fact, they are just the expectation value of the recoil term with respect to the even-mass nucleus, whose $N - 1$ valence particles are distributed over all the available levels except the blocked one. This contribution, being almost independent of the level blocked by the unpaired particle, does not significantly affect the spectrum of the odd nucleus. The remaining two terms in Eq. (24) are the contribution of the unpaired particle. The third one reduces to the single-particle matrix element $F_{\mu\mu}$ ($S_{\beta\beta}^{\mu\mu} = 1$) while the quantity

$\sum_{\nu>0} R^2_{\nu\mu} S^{\mu\mu}_{\beta\beta\nu}$ represents the correction arising from the other valence nucleons. The latter is strongly dependent both on the nature of the level μ and on the pairing correlations.

As regards the off-diagonal matrix elements, they may have $\beta \neq \beta'$ and/or $\mu \neq \mu'$. We have found, in agreement with the results of other authors,[2] that the matrix elements with $\mu \neq \mu'$ are very small. Through the matrix elements with $\beta \neq \beta'$ different eigenstates of the pairing Hamiltonian may be brought into the intrinsic wave function. In this way the recoil term affects the pair distribution of particles.[11,12] This effect may be particularly important when the involved levels originate from the so-called intruder states (in the case considered in this paper this is the spherical $i_{13/2}$ state).

5. The Coriolis Coupling

The matrix elements of the Coriolis term (5) are given by

$$\langle \Psi^I_{M\Omega\tau}|H_c|\Psi^I_{M\Omega'\tau'}\rangle = -A\Bigg\{\sum_{\mu\mu'}\left[(I+\Omega')(I-\Omega'+1)\right]^{\frac{1}{2}} R_{\mu\mu'}\delta_{\Omega'\Omega+1}$$

$$+\left[(I-\Omega')(I+\Omega'+1)\right]^{\frac{1}{2}} R_{\mu'\mu}\delta_{\Omega'\Omega-1} + (-)^{I+\frac{1}{2}}(I+\frac{1}{2})R_{\mu\mu'}\delta_{\Omega\frac{1}{2}}\delta_{\Omega'\frac{1}{2}}\Bigg\}P^{\Omega\Omega'}_{\tau\mu\tau'\mu'}, \quad (25)$$

where

$$P^{\Omega\Omega'}_{\tau\mu\tau'\mu'} = \sum_{\beta\beta'} f^{\Omega}_{\beta\tau\mu} f^{\Omega'}_{\beta'\tau'\mu'} S^{\mu\mu'}_{\beta\beta'}. \quad (26)$$

The matrix elements (25) are written as the product of two factors. The first one corresponds to the contribution of the odd particle while the second one takes into account the many-particle correlations induced from both the pairing and the recoil interaction. Since the quantities $P^{\Omega\Omega'}_{\tau\mu\tau'\mu'}$ are all ≤ 1, they produce an attenuation of the Coriolis coupling.

6. MPR Model Study of ^{163}Er

We report here some results of a study of the nucleus ^{163}Er within the framework of the MPR model. A similar study by Engeland[13] is based on an exact diagonalization of the intrinsic Hamiltonian in a model space including only a rather limited number of Nilsson levels. We have considered twenty-one valence neutrons distributed over eighteen single-particle levels. As already mentioned in Sec. 2, the intrinsic deformed field has been described by a nonspheroidal axial and reflection symmetric Woods-Saxon potential. The parameters of the potential and the single-particle level scheme are given in Ref. 9.

As for the pairing strength G we have used a value of 0.191 MeV which gives an odd-even mass difference P_n for neutrons,

$$P_n = \frac{1}{2}\left[B\left(^{164}Er\right) + B\left(^{162}Er\right) - 2B\left(^{163}Er\right)\right], \quad (27)$$

in agreement with the experimental value 0.970 MeV. It should be noted that the recoil term contributes to P_n about 240 keV. For the rotational parameter A we take the value 14 keV which comes close to that corresponding to the $11/2^-[505]$ band, which is essentially rotational in character.

In Fig. 2 we show the spectrum of the lowest positive-parity band (originating from the $i_{13/2}$ shell-model state) obtained by using our treatment of pairing correlations, which,

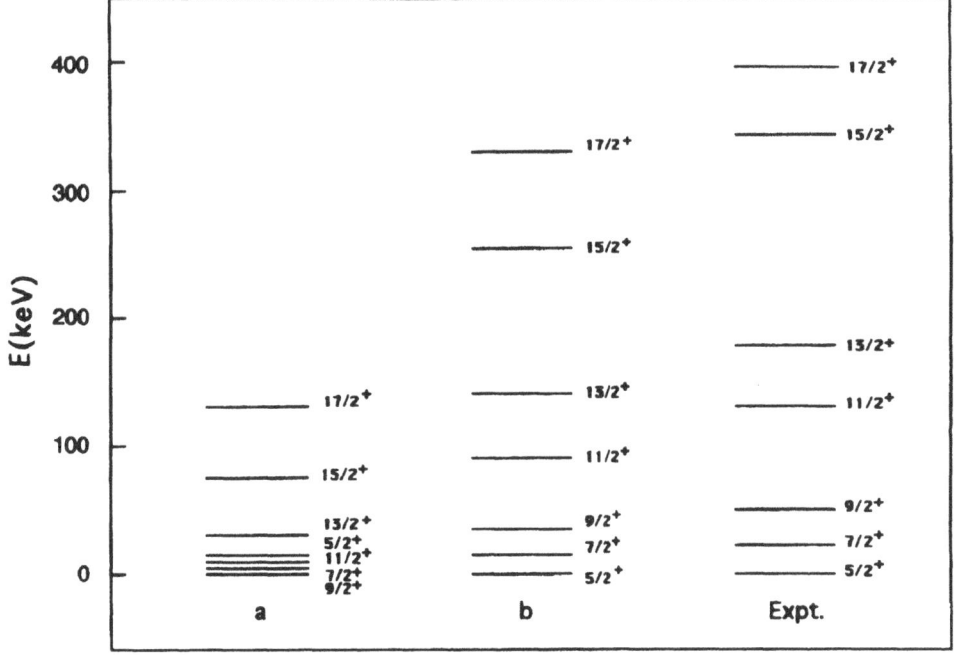

Figure 2. *The spectrum of ^{163}Er obtained by using the CCM: a) without recoil, b) with recoil.*

as pointed out in Sec. 3, is practically exact. We see that the right level ordering and an overall satisfactory agreement with experiment is obtained without any attenuation factor. The comparison between case a) and b) shows the relevance of the recoil term.

By using the same single-particle spectrum and the same rotational parameter we have also performed a calculation where the pairing correlations have been treated in the BCS approximation. As in the calculation based on the CCM, the value of the pairing strength, $G = 0.197$ MeV, has been determined by reproducing the experimental value of P_n while the recoil term has been diagonalized within the set of one quasi-particle states. In Fig. 3 we show the spectrum obtained in this approximation. The comparison between case a) and b) evidences again the role of the recoil term in the MPR model. However, as we see from the spectrum c), a satisfactory agreement with experiment can only be obtained by introducing an ad-hoc attenuation factor.

As regards the spectra of the negative-parity bands $3/2^-[521], 5/2^-[523]$ and $11/2^-[505]$, which are practically purely rotational, a very good agreement with experiment is obtained from both calculations.

Let us now discuss briefly the origin of the difference in the spectra of the positive-parity band obtained from our method and the BCS approximation. As already mentioned, in the final step of the MPR-model calculations the Coriolis interaction is diagonalized in the set of basis states (8). Therefore the difference between the CCM and the BCS results may arise from the differences in the values of the intrinsic energies which appear in the diagonal matrix elements as well as from the attenuation factors $P^{\Omega\Omega'}$ in the off-diagonal ones (25). For ^{163}Er the latter do not differ significantly in the two cases, while rather large differences show up in the intrinsic energies, as it appears from Table 1 (the intrinsic states are characterized

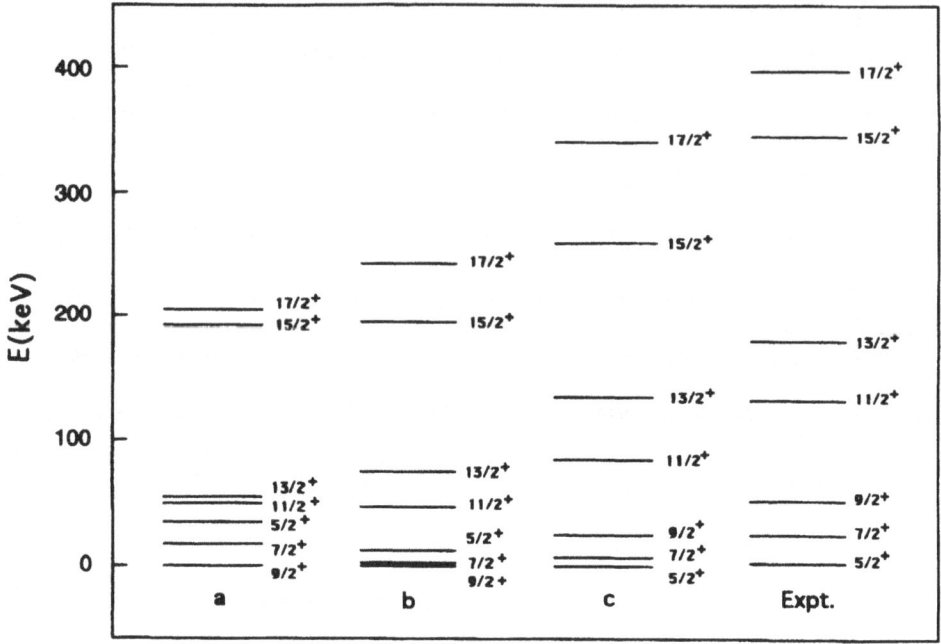

Figure 3. *The spectrum of* ^{163}Er *obtained in the BCS approximation: a) without recoil, b) with recoil, c) with recoil, Coriolis attenuation of 0.9.*

by $[Nn_3\Lambda]$, since, as already mentioned in Sec. 4, there is no appreciable mixing of states corresponding to different blocked levels). Indeed, from this table we see an increase in the intrinsic energy spacings $\mathcal{E}_{[633]7/2} - \mathcal{E}_{[642]5/2}$ and $\mathcal{E}_{[651]3/2} - \mathcal{E}_{[642]5/2}$ of 334 keV and 254 keV, respectively, from the BCS to our treatment of pairing correlations. This reduces the Coriolis coupling bringing the final energy spectrum of the decoupled band in agreement with experiment.

Table 1. *Values of the intrinsic energies* $\mathcal{E}_{\mu\Omega} - \mathcal{E}_{[642]5/2^+}$ *(keV).*

Odd particle state $\mu\Omega$	ε^{CCM}	ε^{BCS}
$[660]\frac{1}{2}^+$	1470	1187
$[651]\frac{3}{2}^+$	946	692
$[642]\frac{5}{2}^+$	0	0
$[633]\frac{7}{2}^+$	905	571
$[624]\frac{9}{2}^+$	2488	2138

Finally, in Table 2 we compare the band-head energies with experiment. It appears that while a very good agreement is obtained from the calculation based on the CCM, the BCS results show rather large discrepancies. Since the Coriolis coupling has little influence on the

Table 2. *Values of the band-head energies (keV) obtained from the CCM and from the BCS method.*

Band head	E^{CCM}	E^{BCS}	E_{expt}
$\frac{5}{2}-$	0	0	0
$\frac{5}{2}+$	85	214	69
$\frac{3}{2}-$	87	86	112
$\frac{11}{2}-$	435	338	444

band-head energies (all the considered bands have $\Omega \neq \frac{1}{2}$), the results of Table 2 are again a consequence of the differences in the values of the intrinsic energies.

7. Summary

In this paper we have applied a new method for treating pairing correlations in deformed nuclei to the study of the nucleus ^{163}Er within the framework of the many-particles plus rotor model. While involving a limited amount of computational work this method yields extremely accurate results.

The results of our calculation have turned out to be in good agreement with experiment without any ad-hoc attenuation factor of the Coriolis term. This is not the case when the pairing correlations are treated by the usual BCS method. In fact, we have found out that use of this approximation makes it necessary to introduce an attenuation factor of 0.9. The primary reason for this is the inaccuracy in the values of the intrinsic energies.

In conclusion, the results of our study of ^{163}Er point to the fact that, once the intrinsic structure is properly treated, the MPR model provides an appropriate framework for the description of deformed nuclei.

Acknowledgements

This work was supported in part by the Italian Ministero dell'Università e della Ricerca Scientifica e Tecnologica (MURST).

References

[1] P. Ring and P. Schuck, *The Nuclear Many-Body Problem*, Chap. 3, and references therein, (Springer-Verlag, New York, 1980)

[2] T. Engeland, *The particle-rotor model*, in: *International Review of Nuclear Physics*, Vol. 2, and references therein, edited by T. Engeland, J. Rekstad and J. S. Vaagen, (World Scientific, Singapore, 1984)

[3] A. Covello, F. Andreozzi, A. Gargano and A. Porrino, *A chain-calculation approach to nuclear structure studies*, in: *Proceedings of the Third International Spring Seminar on Nuclear Physics*, edited by A. Covello, (World Scientific, Singapore, 1991)

[4] A. Covello, F. Andreozzi, A. Gargano and A. Porrino, *A new truncation scheme for nuclear structure calculations*, in: *Proceedings of the Predeal International Summer School*, edited by A. A. Raduta, D. S. Delion and I. I. Ursu, (World Scientific, Singapore, 1992)

[5] A. Covello, F. Andreozzi, A. Gargano and A. Porrino, *Treatment of pairing correlations in deformed nuclei*, in: *Proceedings of the Fourth International Spring Seminar on Nuclear Physics*, edited by A. Covello, (World Scientific, Singapore, 1993)

[6] E. Osnes, J. Rekstad and O. Gjøtterud, *Nucl. Phys. A* **253**, 45 (1975)
[7] F. A. Gareev, S. P. Ivanova, V. G. Soloviev, S. I. Fedotov, *Sov. J. Particles Nucl.* **4**, 148 (1973)
[8] F. Andreozzi, A. Covello, A. Gargano and A. Porrino, *Equations-of-motion approach to shell-model calculations*, in: *Proceedings of the International Symposium on Nuclear Shell Models*, edited by M. Vallieres and B. H. Wildenthal, (World Scientific, Singapore, 1985)
[9] F. Andreozzi, A. Covello, A. Gargano, N. Itaco and A. Porrino, to be published.
[10] G. Ehrling and S. Wahlborn, *Physica Scripta* **6**, 94 (1972)
[11] T. Engeland and J. Rekstad, *Phys. Lett. B* **89**, 8 (1979)
[12] J. Rekstad and T. Engeland, *Phys. Lett. B* **89**, 316 (1980)
[13] T. Engeland, *Physica Scripta* **25**, 467 (1982)

SPIN-DEPENDENT GENERALIZED COLLECTIVE MODEL

Martin Greiner[1], Dirk Heumann[1], Werner Scheid[1], Günter Braunss[2] and Peter Hess[3]

[1] Institut für Theoretische Physik der Justus-Liebig-Universität, Heinrich-Buff-Ring 16, D-35392 Gießen, Germany
[2] Mathematisches Institut der Justus-Liebig-Universität, Arndtstraße 2, D-35392 Gießen, Germany
[3] Instituto de Ciencias Nucleares, UNAM, Circuito Exterior, C.U., A.P. 70-543, 04510 México D.F., Mexico

In an advanced course on quantum mechanics a student learns that the Dirac equation is constructed from a linearization of the Klein-Gordon equation. Whereas the latter describes a particle with no spin the former includes a spin $\frac{1}{2}$ degree of freedom. This appears to be a consequence of the linearization procedure and not of the theory of relativity, because the same result also follows from the linearization of the Schrödinger equation[1], which represents the nonrelativistic limit of the Klein-Gordon equation.

We now apply the same linearization procedure to the collective Schrödinger equation describing quadrupole surface vibrations of a nucleus. The quadrupole coordinates $\alpha_{2\mu}$ are defined by an expansion of the nuclear surface in terms of spherical harmonics; the corresponding five canonically conjugate spherical momenta $\pi_{2\mu}$ can be transformed into Euclidean momenta p_i with $i=1,\ldots,5$. The relevant "free" Schrödinger equation for quadrupole degrees is of first order in the energy operator and of second order in the momentum operators p_i, whereas the linearized Schrödinger equation is linear in both energy and momenta; it follows:

$$i\hbar\frac{\partial}{\partial t}\Psi = \frac{1}{2B_2}\sum_{i=1}^{5}p_i^2\Psi \quad \stackrel{linearization}{\Longrightarrow}$$
$$\begin{pmatrix} I & 0 \\ 0 & 0 \end{pmatrix} i\hbar\frac{\partial}{\partial t}\Psi = \left[-\begin{pmatrix} 0 & 0 \\ 0 & I \end{pmatrix}2B_2 + \sum_{j=1}^{5}\begin{pmatrix} 0 & -i\gamma_j \\ i\gamma_j & 0 \end{pmatrix}p_j\right]\Psi \quad . \tag{1}$$

B_2 is a mass parameter and the 4×4 dimensional γ_j-matrices fulfill the Clifford algebra $\gamma_i\gamma_j + \gamma_j\gamma_i = 2\delta_{ij}I$. The structure of this linearized Schrödinger equation can be understood most easily, if the nonrelativistic limit $E_{rel} = B_2 + E_{nrel}$ of a five-dimensional Dirac equation is taken.

The linearized Schrödinger equation (1) incorporates a new spin degree of freedom[2]; the diagonalisation of the appropriate spin operator $S_{1\mu} = \frac{\sqrt{10}\hbar}{4}\left[\gamma^{[2]}\otimes\gamma^{[2]}\right]^{[1]}_{\mu}$ leads to a spin $\frac{3}{2}$. This can also be understood from the group chain $SO(5)\supset SO(3)$, where the lowest half-integer representation $(\frac{1}{2},\frac{1}{2})$ of SO(5) reduces to the representation $(\frac{3}{2})$ of SO(3).

With the appearance of this new spin degree of freedom, we are led to the assumption, that the linearized Schrödinger equation for quadrupole degrees of freedom may describe some aspects of spin $\frac{3}{2}$ even-odd nuclei. Therefore, coordinate-, momentum- and spin-dependent potentials are introduced in the most general form and added to the linearized Hamiltonian, which is given by the square bracket of eq.(1). Because the lower components χ of the spinor $\Psi = (\varphi, \chi) \exp(-\frac{i}{\hbar} E t)$ are redundant and completely expressable in terms of the upper components φ, the linearized Schrödinger equation including general potentials can be transformed into an effective Schrödinger equation for the upper components φ; it reads[3]:

$$E\varphi = \left[\frac{\sqrt{5}}{2B_2} \left[\pi^{[2]} \otimes \pi^{[2]} \right]^{[0]} + T_{corr}(\alpha_{2\mu}, \pi_{2\mu}) + V(\alpha_{2\mu}) + V_{spin}(\gamma_{2\mu}, \alpha_{2\mu}, \pi_{2\mu}) \right] \varphi. \quad (2)$$

The effective Hamiltonian of this Spin-dependent Generalized Collective Model (SGCM) consists of four parts, namely a kinetic energy, a correction to the kinetic energy, a potential energy and a spin-dependent potential. T_{corr} and V are considered up to third and sixth order, respectively, in $\alpha_{2\mu}$ and $\pi_{2\mu}$; their parameters together with B_2 are all taken from the neighbouring even-even nucleus and are not refitted. The spin-dependent potential decomposes into

$$V_{spin}(\gamma_{2\mu}, \alpha_{2\mu}, \pi_{2\mu}) = \left[\gamma^{[2]} \otimes A^{[2]}(\alpha_{2\mu}, \pi_{2\mu}) \right]^{[0]} + \sum_{k=1,3} \left[\left[\gamma^{[2]} \otimes \gamma^{[2]} \right]^{[k]} \otimes F^{[k]}(\alpha_{2\mu}, \pi_{2\mu}) \right]^{[0]}. \quad (3)$$

The vector and tensor potentials $A^{[2]}$ and $F^{[k]}$ are chosen to be up to second order in $\alpha_{2\mu}$ and $\pi_{2\mu}$, which renders them to five independent terms; the corresponding five parameters are determined from a fit to the five lowest excited states of the considered spin $\frac{3}{2}$ even-odd nucleus. – The Hamiltonian in eq.(2) is diagonalized within the basis states of a five-dimensional harmonic oscillator coupled to spinor states with spin $\frac{3}{2}$.

Figure 1 shows the energy spectrum obtained for the spin $\frac{3}{2}$ even-odd nucleus ^{191}Ir. It

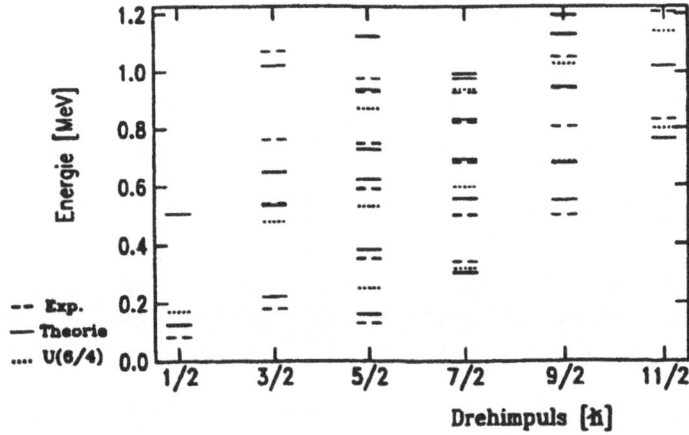

Figure 1. *Calculated energy spectrum (solid) for $^{191}_{77}$Ir compared with experimental results (dashed) and calculations within the U(6/4) IBFM (dotted).*

agrees well with the experimental data[4] and is comparable to a U(6/4) Interacting Boson Fermion Model calculation[5]. – With the quadrupole operator

$$Q_{2\mu} = \rho_0 R_0^5 \left\{ \alpha_{2\mu} - \frac{10}{\sqrt{70\pi}} \left[\alpha^{[2]} \otimes \alpha^{[2]} \right]_{2\mu} \right\} \quad , \tag{4}$$

where $\rho_0 = \frac{3Ze}{4\pi R_0^3}$ is the constant nuclear charge density and R_0 is the nuclear radius, also electromagnetic transition probabilities have been calculated. Table 1 shows the results for the B(E2) values; be aware that no new parameter has been introduced. The results are in

Table 1. *Calculated B(E2) values for $^{191}_{77}$Ir compared with experimental results and calculations within the U(6/4) IBFM.*

transition	exp.	SGCM	U(6/4)
$1/2_1 \to 3/2_1$	0.166	0.119	0.400
$3/2_2 \to 3/2_1$	0.110	0.156	0.000
$5/2_1 \to 3/2_1$	0.593	0.647	0.400
$7/2_1 \to 3/2_1$	0.250	0.327	0.400
$7/2_1 \to 5/2_1$	0.196	0.347	-
$9/2_1 \to 5/2_1$	0.653	0.480	-
$11/2_1 \to 7/2_1$	0.457	0.596	-
$3/2_2 \to 1/2_1$	0.379	0.332	0.188
$5/2_2 \to 1/2_1$	0.359	0.417	0.323
$5/2_2 \to 3/2_1$	0.011	0.0083	0.000

excellent agreement with the experimental data. – Similiar results also hold for other spin $\frac{3}{2}$ nuclei, as for example $^{187}_{77}$Ir, $^{189}_{77}$Ir, $^{193}_{77}$Ir and $^{193}_{79}$Au, $^{195}_{79}$Au, $^{197}_{79}$Au, $^{199}_{79}$Au[6,7].

In order to extend this approach to describe also even-odd nuclei with a different spin we present the following consideration: The operator

$$_{3/2}A^{[2]}_{\mu} = \left[a^{+[3/2]} \otimes \tilde{a}^{[3/2]} \right]^{[2]}_{\mu} \tag{5}$$

is constructed from a spin $\frac{3}{2}$ fermion creation and annihilation operator and will be presented in the vector basis $a^{+[3/2]}_{\mu} | 0 >$. We then find

$$\left\{ 2 \left(_{3/2}A^{[2]}_{\mu} \right)_V , 2 \left(_{3/2}A^{[2]}_{\nu} \right)_V \right\} = 2(-1)^{\mu} \delta_{\mu - \nu} \quad , \tag{6}$$

which is identical to the Clifford algebra of the spherical $\gamma^{[2]}_{\mu}$-matrices. In addition, the spin-dependent parts of the spin-dependent potentials (3) can be represented as combinations of one fermion creation and annihilation operators in the vector basis.

With this consideration we have achieved two things: First of all, the spin-dependent potentials (3) can now directly be translated into the boson-fermion potentials of the U(6/4) Interacting Boson Fermion Model[5], so that as a consequence the SGCM has a very similar structure compared with the IBFM, although both approaches start off from very different grounds. – From the identification $\gamma^{[2]}_{\mu} = 2(_{3/2}A^{[2]}_{\mu})_V$ it is evident to introduce the replacement $_{3/2}A^{[J]}_{\mu} \to {}_jA^{[J]}_{\mu} = \left[a^{+[j]} \otimes \tilde{a}^{[j]} \right]^{[J]}_{\mu}$ in order to describe even-odd nuclei with a spin different from $\frac{3}{2}$. Calculations have been done for $^{103,105}_{45}$Rh, $^{107,109}_{47}$Ag as spin $\frac{1}{2}$ nuclei and for $^{89}_{38}$Sr, $^{93}_{42}$Mo as spin $\frac{5}{2}$ nuclei; the results for the energy spectra and E2-transitions are of similiar quality as for the afore mentioned spin $\frac{3}{2}$ nuclei[7].

The linearization of the collective Schrödinger equation for quadrupole degrees of freedom has lead to a new spin $\frac{3}{2}$. The resulting Spin-dependent Generalized Collective Model is able to describe energy spectra and E2-transitions of various spin $\frac{3}{2}$ even-odd nuclei quantitatively. This model can also be extended to describe other even-odd nuclei with a different spin; it also has a very similar structure to the Interacting Boson Model. – Besides this application of the linearization scheme in nuclear structure, it might be worth looking also on other fields in physics as for example on cosmology, where a gravitino with spin $\frac{3}{2}$ could be introduced as the counterpart of the graviton with spin 2.

Acknowledgement

One of us (M.G.) wants to thank the organizers of this NASI summer school on "Frontier Topics in Nuclear Physics" for their invitation to experience the friendly and inspiring atmosphere of this meeting.

References

[1] J. M. Levy-Leblond, *Comm. Math. Phys.* **6**, 286 (1967)
[2] M. Greiner, W. Scheid and R. Herrmann, *Mod. Phys. Lett. A* **3**, 859 (1988)
[3] M. Greiner, D. Heumann and W. Scheid, *Z. Phys. A - Atomic Nuclei* **336**, 139 (1990)
[4] E. Browne, *Nucl. Data Sheets* **56**, 709 (1989)
[5] A. Balantekin, I. Bars and F. Iachello, *Nucl.Phys. A* **370**, 284 (1981)
[6] D. Heumann, M. Greiner and W. Scheid, *Mod. Phys. Lett. A*, **6**, 3653 (1991)
[7] D. Heumann, *Spinabhängige Hamiltonoperatoren nach Linearisierung in der Kernphysik und der Gravitationstheorie*, PhD-thesis, Justus-Liebig-Universität, Gießen (1992)

TWO PHONON EXCITATIONS IN HEAVY NUCLEI STUDIED IN PHOTON SCATTERING EXPERIMENTS

P. von Brentano[1], A. Zilges[1], R.-D. Herzberg[1], U. Kneissl[2] and H. H. Pitz[2]

[1] Institut für Kernphysik, Universität zu Köln, D-50937 Köln, Germany
[2] Institut für Strahlenphysik, Universität Stuttgart, D-70569 Stuttgart, Germany

Introduction

The description of nuclei near closed shells can be simplified if one uses a phonon model to describe the low lying excitations. In this picture the lowest collective excitations of spherical nuclei, i.e. the quadrupole and octupole surface oscillations are treated as 2^+ quadrupole phonons and 3^- octupole phonons, respectively. One property of the phonon model is the existence of Multi Phonon Excitations near the sum energy of the single phonons due to the coupling of two or more phonons. Much experimental effort has been spent throughout the last years to investigate such states. But due to the admixture of quasiparticle states and anharmonicities in the coupling mechanism the identification of Multi Phonon States is rather difficult.

The best studied two phonon excitations are the quadrupole–quadrupole $2^+ \otimes 2^+$ states in the Cd-isotopes [1]. One finds a multiplet of states with $J^\pi = 0^+, 2^+$ and 4^+ at around twice the energy of the 2^+ quadrupole vibration. Another possibility is the coupling of the quadrupole with an octupole vibration creating a multiplet of five states with $J^\pi = 1^-, 2^-, ..., 5^-$. The 1^- member of this multiplet has been identified in several nuclei but the information on the other states is still rather sparse [2, 3, 4].

In this contribution we will start with a presentation of results of our photon scattering experiments on various spherical even–even nuclei in the N=82 region. We will show that it is possible to observe the 1^-–member of the quadrupole–octupole multiplet in all examined nuclei and give absolute B(E1) values. The second part deals with the coupling of an additional particle to the $2^+ \otimes 3^-$ states. The examination of the nuclei ^{139}La, ^{141}Pr and ^{143}Nd in (γ, γ') experiments gave for the first time evidence for the existence of two phonon particle states. The experimental findings could be reproduced in a harmonic approximation by using a quadrupole-quadrupole interaction.

Dipole Strength in even-even Nuclei

Collective electric dipole transitions are forbidden if one assumes a homogeneous charge distribution and reflection symmetry in the nucleus [5]. And indeed, most of the observed E1–transitions are strongly hindered and exhibit B(E1)–values around 10^{-4}–10^{-5} Weisskopf units [6, 7]. Nevertheless a small number of states shows E1 ground state transitions which are several orders of magnitude larger [8]. Photon scattering experiments are especially well suited to populate such levels because the electromagnetic excitation mechanism from the ground state is selective on the strength. Therefore one excites in these experiments all 1^- states which have E1 transition strengths above a certain detection limit. Enhanced E1 transitions have been found in well deformed [9, 10] as well as in spherical nuclei [11]. The E1 transitions in well deformed nuclei have been described in different theoretical approaches [12, 13, 14, 15, 16].

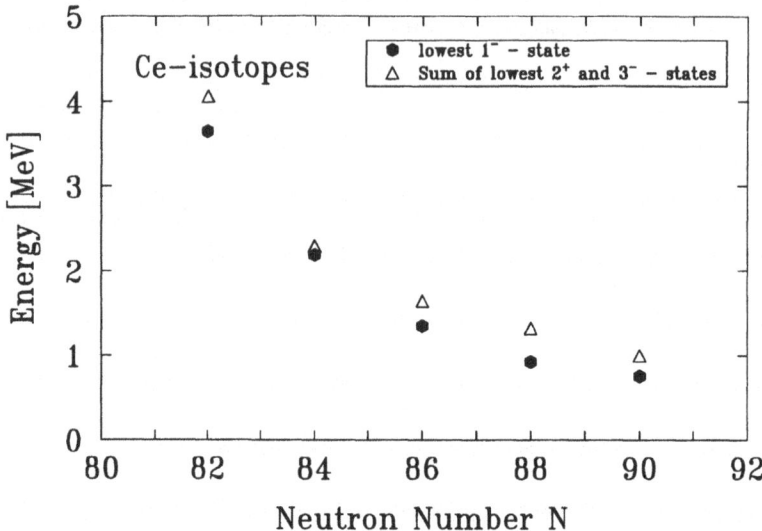

Figure 1. *Energies of the lowest 1^- state versus the sum energy of the 2_1^+ and 3_1^- state.*

It has been pointed out that the lowest 1^- state can be described in terms of octupole correlations [12]. In fig. 1 we show the energy of the lowest 1^- state observed in various Ce-isotopes versus the sum energy of the lowest 3^--state (the octupole vibration) and the lowest 2^+-state (the quadrupole vibration). The Ce-isotopes cover the whole range of nuclear deformations from the spherical ^{140}Ce to the well deformed ^{148}Ce. The strong correlation between the energies is a clear signature for the existence of octupole excitations in spherical as in well deformed nuclei. The difference is the coupling of the 2^+ state: In spherical nuclei the octupole vibration couples to the quadrupole vibration with an energy of approximately 2 MeV whereas in deformed nuclei the octupole vibration couples to the quadrupole deformed ground state of the nucleus. Octupole excitations in deformed nuclei have been discussed in detail in ref. [9]. In the following we will discuss the octupole excitations in spherical nuclei.

Figure 2 shows the photon scattering spectra of the N=82 nuclei ^{142}Nd, ^{140}Ce, and ^{138}Ba in the energy range between 3 and 4 MeV. The most prominent features are very strong lines at 3.425, 3.643, and 4.027 MeV, respectively. These lines results from the E1 ground state transitions from the 1^- state of the $2^+ \otimes 3^-$ multiplet. The absolute transition strengths or the lifetimes can be determined unambiguously from the photon scattering experiments due to the well known electromagnetic excitation mechanism. The nuclei are irradiated by

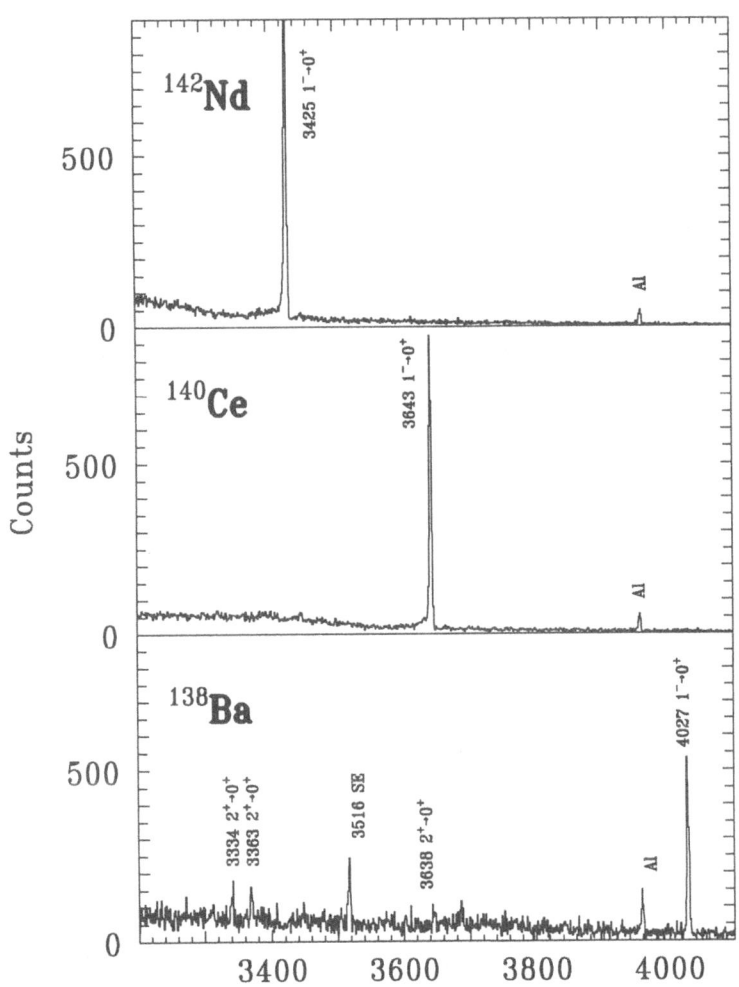

Figure 2. *Photon scattering spectra of the nuclei* ^{142}*Nd,* ^{140}*Ce, and* ^{138}*Ba.*

a continuous bremsstrahlung spectrum and the resonant excitation obeys the Breit–Wigner resonance law

$$\frac{d\sigma^0_{abs}(E)}{d\Omega} = \pi \lambdabar^2 \frac{2J+1}{2(2J_0+1)} \cdot \frac{\Gamma_0 \Gamma_f}{(E-E_r)^2 + \frac{1}{4}\Gamma^2} \cdot \frac{W(\Theta)}{4\pi}, \qquad (1)$$

where $\lambdabar = \hbar c/E_r$, J_0 denotes the ground state spin, J is the spin of the excited state n, $\Gamma_0 = \Gamma(n, J \to J_0)$ is the ground state transition width, $\Gamma_f = \Gamma(n, J \to J_f)$ the transition width to the final state (in case of elastic scattering $\Gamma_f = \Gamma_0$, otherwise $\Gamma_f = \Gamma_1$), and $W(\Theta)$ is the angular correlation function which depends on the spin of the excited state and the observation angle. Integration over the energy yields the integrated cross section which is compared to the well known cross sections of simultaneously measured ^{27}Al levels to obtain the absolute transition strength. For the three nuclei mentioned above we calculated B(E1)↑ values of $\simeq 20 \times 10^{-3} e^2 fm^2$ which correspond to some mWu and are large compared to the usual E1 strengths in this mass region.

To reproduce the data in a model approach the interacting boson model [17] was chosen due to its ability to calculate low lying collective excitations in a simple way. To calculate negative parity states one has to extend the sd–IBA by p–bosons with $J^\pi = 1^-$ and by f–bosons with $J^\pi = 3^-$ [18, 19, 20, 21]. The improved T(E1) operator contains a one–body and a two–body term. It can be shown that it is impossible to reproduce the experimental data without the two body operator which supports again the interpretation of the 1^- state as a $2^+ \otimes 3^-$ excitation [22].

A recent study of the level structure in ^{140}Ce by Grinberg et al. finds that the 1^- state at 3.643 MeV consists to 94% of a coupling of the 2^+ and 3^- states described in a Random Phase Approximation [23].

To summarize the arguments favouring the two phonon structure for these 1^- states:

- The sum energy of the octupole vibration and the lowest quadrupole state is strongly correlated to the energy of the first 1^- state in deformed as well as in spherical nuclei.
- There is only one single dipole excitation in the energy region where one expects the $2^+ \otimes 3^-$ state.
- The E1 transition strength to the ground state is two to three orders of magnitude larger than the usual E1 strength in this mass region.
- Various theoretical works support the idea of a two phonon structure.

Dipole Excitations in odd A Nuclei

What will happen if one couples an additional particle to the excitations described above? At low energies the single particle will couple to the collective quadrupole and octupole vibrations of the core creating one phonon particle multiplets. The real situation is a bit more complicated due to the admixture of close lying single particle levels. Nevertheless these states have been observed and theoretically described in a number of nuclei in the N=82 region, see e.g. ref.[24, 25].

As discussed above, another group of collective excitations with a $2^+ \otimes 3^-$ structure appears in the core nucleus due to the coupling of the quadrupole and octupole phonon. In a simple picture one would expect a fragmentation of the E1 strength to the 1^- state at 3.425 MeV in the core nucleus ^{142}Nd into three states with a $2^+ \otimes 3^- \otimes$ particle structure in ^{143}Nd. The additional neutron is in the $f_{7/2}$ shell, therefore we expected to excite three

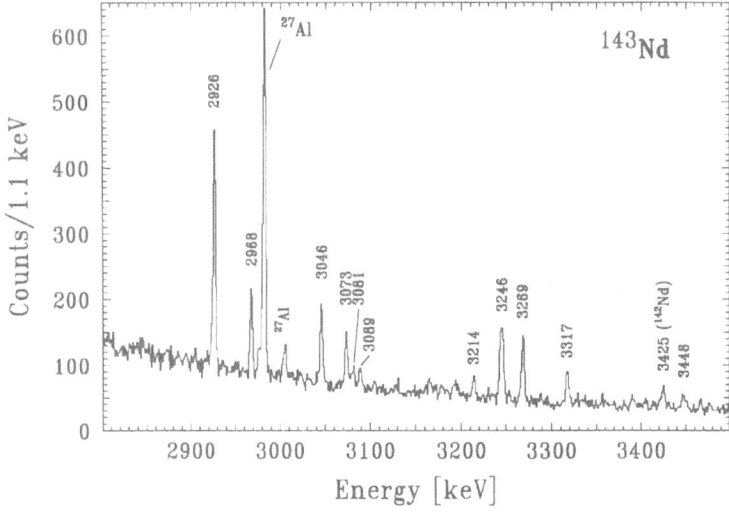

Figure 3. *Photon scattering spectrum of ^{143}Nd between 2.8 and 3.5 MeV.*

states in our photon scattering experiment with $J^\pi = 5/2^+$, $7/2^+$, and $9/2^+$ in the energy region around 3.4 MeV.

Figure 3 shows the photon scattering spectrum between 2.8 and 3.5 MeV. One observes

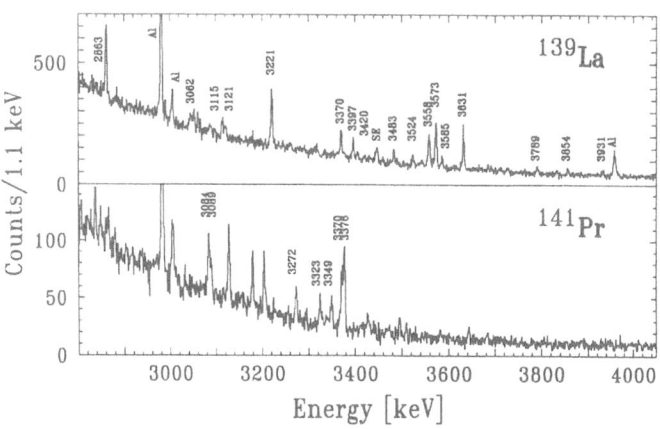

Figure 4. *Photon scattering spectra of ^{141}Pr and ^{139}La.*

a large number of relatively strong dipole transitions to the ground state. To explain the fragmentation into a fairly large number of states one should take into consideration that the $2^+ \otimes 3^-$ multiplet consists of five states with $J^\pi = 1^-$ to 5^-. The additional particle couples to each of these excitations generating a total of 31 levels. Therefore we can find five levels which have spin J=5/2, five with J=7/2 and five with J=9/2. Due to mixing with the $1^- \otimes f_{7/2}$ state, the E1 strength fragments over these fifteen levels.

One can compare the measured strength in ^{143}Nd with the strength in ^{142}Nd using the sum rule for B(E1)↓ for weak particle coupling [26],

$$\sum_n B(E1; n, J \to J_0) = B(E1; 1^- \to 0^+) \,. \tag{2}$$

As mentioned above the pure configurations are mixed, therefore the total sum of the B(E1)↓ strength for each spin J in the nucleus ^{143}Nd corresponds to the B(E1)↓ value in the core ^{142}Nd. In our photon scattering experiment we are at present unable to determine the spin of the excited states in odd A nuclei. Nevertheless it is possible to compare the measured strength with the strength in ^{142}Nd if one assumes that we only excite states with J=5/2, 7/2 and 9/2. If one adds up the B(E1)↓ strength in the energy region around 3.2 MeV the total strength should be 3 times as large as the dipole strength of the $1^- \to 0^+$ transition in the core nucleus.

Thus one should compare the measured summed strength for ^{143}Nd

$$\sum_{J,n} \frac{2J+1}{8} B(E1; n, J) \downarrow = (12.8 \pm 1.6) \times 10^{-3} \, e^2 fm^2 \,, \tag{3}$$

with

$$3 \cdot B(E1) \downarrow (^{142}Nd) = (16.3 \pm 2.4) \times 10^{-3} \, e^2 fm^2 \tag{4}$$

for the E1 transition from the 3425 keV state in ^{142}Nd. The fact that the sum rule is almost completely exhausted is an impressive proof for the correctness of a two phonon particle structure for the excitations in ^{143}Nd.

In addition we performed model calculations in a harmonic approximation with quadrupole-quadrupole interaction. The core is described by the sdf–IBM Hamiltonian mentioned above. Then we couple a single $f_{7/2}$ neutron to the collective excitations of the core. One observes a very good agreement between the experimental energies and the model prediction [27].

We have also performed (γ, γ') experiments on the odd Z nuclei ^{139}La and ^{141}Pr. Figure 4 shows the photon scattering spectra in the energy region where we expect the two phonon particle states. Again a number of strong dipole transitions to the ground state is observed. Due to the relatively low lying first excited state in these nuclei the theoretical description is more complicated than for ^{143}Nd. However, the preliminary data evaluation points to a $2^+ \otimes 3^- \otimes$ particle structure in ^{139}La and ^{141}Pr.

We thank our colleagues from the NRF–collaboration in Stuttgart and Giessen, especially, H. Friedrichs, J. Margraf, S. Lindenstruth, B. Schlitt, and C. Wesselborg for their outstanding engagement to obtain and discuss the experimental data on dipole excitations.

The authors gratefully acknowledge valuable discussions with S. Albers, T. Belgya, R. F. Casten, F. Dönau, A. Faessler, A. Gelberg, R. V. Jolos, G. Molnar, B. R. Mottelson, N. Pietralla, A. Oros, A. Raduta, A. Richter, V.G. Soloviev, L. Trache, S.W. Yates and N. V. Zamfir.

Two of us (A.Z. and R.-D.H.) want to thank the Institut für Strahlenphysik for the kind hospitality during our numerous stays in Stuttgart.

This work was partially supported by the Deutsche Forschungsgemeinschaft (DFG) under contracts Br 799-33/34 and Kn 154-21 and by the BMFT under contract number 06OK143.

References

[1] R. F. Casten, *Nuclear Structure from a Simple Perspective*, (Oxford University Press, Oxford, 1990), and references therein

[2] R. A. Gatenby, J. R. Vanhoy, E. M. Baum, E. L. Johnson, S. W. Yates, T. Belgya, B. Fazekas, A. Veres and G. Molnar, *Fast E1 transitions and evidence for octupole–octupole and quadrupole–octupole excitations in ^{144}Sm*, Phys. Rev. C **41**, R414 (1990)

[3] D. Hofer, M. Bisenberger, R. Hertenberger, H. Kader, H. J. Maier, E. Müller–Zanotti, R. Schiemenz, G. Graw, P. Maier–Komor, G. Molnar, W. Unkelbach, J. Hebenstreit, H. Ohm, D. Paul, P. von Rossen, and M. Fujiwara, *Direct and Multiple Excitations in ^{96}Zr from Inelastic-Scattering Experiments*, Nucl. Phys. A **551**, 173 (1993)

[4] S. Albers, *private communication*, Cologne, 1992

[5] S. G. Rohozinski, *Octupole Vibrations in Nuclei*, Rep. Prog. Phys. **51**, 541 (1988)

[6] P. M. Endt, *Strengths of Gamma–Ray Transitions in A=91–150 Nuclei*, At. Dat. Nucl. Dat. Tab. **26**, 47 (1981)

[7] T. Lönnroth, *Evidence for Three Microscopically Different Kinds of E1 Transitions in Lead-Region Nuclei*, Z. Phys. A - Atomic Nuclei **331**, 11 (1988)

[8] F. R. Metzger, *Low–lying E1–transitions in the stable even Sm isotopes*, Phys. Rev. C **14**, 543 (1976)

[9] A. Zilges, P. von Brentano, H. Friedrichs, R. D. Heil, U. Kneissl, S. Lindenstruth, H. H. Pitz, and C. Wesselborg, *A survey of $\Delta K=0$ dipole transitions from low lying J=1 states in rare earth nuclei*, Z. Phys. A – Hadrons and Nuclei **340**, 155 (1991)

[10] H. Friedrichs, B. Schlitt, J. Margraf, S. Lindenstruth, C. Wesselborg, R. D. Heil, H. H. Pitz, U. Kneissl, P. von Brentano, R.-D. Herzberg, A. Zilges, D. Häger, G. Müller, M. Schumacher, *Evidence for Enhanced Electric Dipole Excitations in Deformed Rare Earth Nuclei Near 2.5 MeV*, Phys. Rev. C **45**, R892 (1992)

[11] H. H. Pitz, R. D. Heil, U. Kneissl, S. Lindenstruth, U. Seemann, R. Stock, C. Wesselborg, A. Zilges, P. von Brentano, S. D. Hoblit, and A. M. Nathan, *Low energy photon scattering off $^{142,146,148,150}Nd$: An investigation in the mass region of a nuclear shape transition*, Nucl. Phys. A **509**, 587 (1990)

[12] W. Donner and W. Greiner, *Octupole vibrations of deformed nuclei*, Z. Phys. **197**, 440 (1966)

[13] F. Iachello, *Local versus Global Isospin Symmetry in Nuclei*, Phys. Lett. B **160**, 1 (1985)

[14] V. G. Soloviev and V. A. Sushkov, *Electric–dipole transitions in doubly even deformed nuclei*, Phys. Lett. B **262**, 189 (1991)

[15] A. A. Raduta, I. I. Ursu, and N. Lo Iudice, *Low–Lying Bands as Alpha-like Dipole Excitations of a Coherent Quadrupole Boson State*, Nuovo Cimento A **105**, 663 (1992)

[16] P. von Brentano, N. V. Zamfir, and A. Zilges, *E1 operator in the sdf-Interacting Boson Model from an Alaga rule constraint*, Phys. Lett. B **278**, 221 (1992)

[17] A. Arima and F. Iachello, *Interacting boson model of collective states*, Ann. of Phys. **99**, 253 (1976)

[18] J. Engel and F. Iachello, *Interacting boson model of collective octupole states*, Nucl. Phys. A **472**, 61 (1987)

[19] D. Kusnezov and F. Iachello, *A study of collective octupole states in barium in the interacting boson model*, Phys. Lett. B **209**, 420 (1988)

[20] T. Otsuka and M. Sugita, *Unified description of quadrupole–octupole collective states in nuclei*, Phys. Lett. B **209**, 140 (1988)

[21] F. Iachello and A. Arima, *The interacting boson model*, (Cambridge University press, Cambridge, 1987)

[22] A. F. Barfield, P. von Brentano, A. Dewald, K. O. Zell, N. V. Zamfir, D. Bucurescu, M. Ivascu, and O. Scholten, *Evidence for the two-body nature of the E1-transition operator in the sdf-interacting boson model*, Z. Phys. A - Atomic Nuclei **332**, 29 (1989)

[23] M. Grinberg, Thai Khac Dinh, Ch. Protochristov, I. Penev, C. Stoyanov, and W. Andrejtscheff, *Level structure and transition probabilities in ^{140}Ce*, J. Phys. G: Nucl. Part. Phys. **19**, 1179 (1993)

[24] J. Wrzesinski, A. Clauberg, C. Wesselborg, R. Reinhardt, A. Dewald, K. O. Zell, and P. von Brentano, *Medium- and Low-Spin States in ^{143}Nd studied in the (α,n) Reaction*, Nucl. Phys. A **515**, 297 (1990)

[25] L. Trache, K. Heyde, and P. von Brentano, *Particle-core coupling in the N=83 nucleus ^{143}Nd: Extended unified-model calculations*, Nucl. Phys. A **554**, 118 (1993)

[26] D. J. Rowe, *Nuclear Collective Motion*, (Methuen, London, 1970)

[27] A. Zilges, R.-D. Herzberg, P. von Brentano, F. Dönau, R. D. Heil, R. V. Jolos, U. Kneissl, J. Margraf, H. H. Pitz, and C. Wesselborg, *First Identification of Dipole Excitations to a $2^+ \otimes 3^- \otimes$ Particle Multiplet in an Odd-A Nucleus*, Phys. Rev. Lett. **70**, 2880 (1993)

TOWARD A COMPLETE UNDERSTANDING OF THE SCISSORS MODE

N. Lo Iudice

Dipartimento di Scienze Fisiche, Università di Napoli
and
Istituto Nazionale di Fisica Nucleare, Sezione di Napoli, Napoli, Italy

Introduction

The understanding of the nature of the low lying collective magnetic dipole excitation, known as scissors mode, discovered in ^{156}Gd by high-resolution inelastic electron scattering[1] and observed by now in all deformed nuclei, has been the goal of extensive experimental as well as theoretical investigations[2].

An important, maybe conclusive step forward has been made recently with the discovery that the summed orbital $M1$ strength in Sm isotopes increases quadratically with deformation[3] and is closely correlated with the strength of the $E2$ transition to the lowest 2^+ state[4]. The same deformation law has been observed also in Nd isotopes[5].

This discovery has induced several investigators to re-examine the many theoretical models adopted to describe the mode in order to check their consistency with these new properties[6-14]. The result of these investigations was that after appropriate manipulations, practically all approaches give a M1 strength quadratic in the deformation parameter δ, differing to more or less extent from the experimental data in magnitude and slope.

The list of the model descriptions includes a phenomenological approach based on the use of the M1 strength in the form derived within the two rotor model (TRM), which first predicted the mode and inspired its geometrical picture (an out of phase rotational oscillation between deformed proton and neutron systems)[15]. It has been found that if the physical constants entering into the strength are estimated by an empirical, model independent method, the observed δ^2 behaviour of the M1 strength is reproduced quite satisfactorily and its saturation properties accounted for[14].

This is a very important result since the TRM expression of the M1 strength is of general validity and can be assumed to define the scissors M1 strength. Its full consistency with the observed deformation properties can therefore be considered as a model independent proof of the scissors character of the observed M1 excitations. Moreover, the M1 strengths obtained in all phenomenological or schematic models can be cast into such a TRM form[16]. The different model descriptions of the mode seem therefore to offer different recipes for computing the physical constants entering into the scissors M1 strength.

In order to illustrate the above aspects we will present first a summary of the analysis which shows the consistency of the scissors M1 strength with the deformation law. We will then discuss a semiclassical description from which it will emerge that a new scissors-like mode may exist at high energy and that strongly collective scissors-like modes may occur in superdeformed nuclei. We will finally illustrate how the TRM relate to other models. We will discuss in particular the relation to the random-phase approximation (RPA)[7], the generalized coherent state model (GCSM)[13,17] and the proton-neutron interacting boson model (IBM2)[11]. In the final section the implications of the present analysis regarding the nature of the observed low lying M1 excitations and of the other possible modes are discussed.

Two-Rotor Model and the Scissors M1 Strength

The basic assumption of the model[15] is that protons and neutrons form two axially symmetric rotors free to rotate around a common axis orthogonal to their symmetry axes and interacting via a potential $V(\vartheta)$ dependent on the angle 2ϑ between the symmetry axes. The total Hamiltonian has then the form

$$H = \frac{1}{2\Im_p}\vec{J}_p{}^2 + \frac{1}{2\Im_n}\vec{J}_n{}^2 + V(\vartheta), \tag{1}$$

where $\Im_p$ and $\Im_n$ are the proton and neutron moments of inertia, $\vec{J}_p$ and $\vec{J}_n$ their angular momenta. Expressed in terms of the total and relative angular momenta

$$\vec{J} = \vec{J}_p + \vec{J}_n, \qquad \vec{S} = \vec{J}_p - \vec{J}_n, \tag{2}$$

the Hamiltonian decouples into a rotational and an intrinsic part describing the relative motion. For small values of the angle ϑ the intrinsic Hamiltonian takes the form of a two dimensional harmonic Hamiltonian with ϑ playing the role of a radial variable

$$H = H_{int} = \frac{1}{2\Im_{sc}}(S_1^2 + S_2^2) + \frac{1}{2}C\vartheta^2, \tag{3}$$

where

$$S_i = -i\frac{d}{d\vartheta_i}, \qquad \vartheta^2 = \vartheta_1^2 + \vartheta_2^2 \tag{4}$$

and

$$\Im_{sc} = 4\frac{\Im_p \Im_n}{\Im_p + \Im_n}. \tag{5}$$

For $\Im_p = (Z/A)\Im$ and $\Im_n = (N/A)\Im$, $\Im_{sc}$ practically coincides with the nuclear mass parameter $\Im$.

The scissors mode is the first excited state with quantum numbers $\{n\ K^\pi\} = \{0\ 1^+\}$ and energy $\omega = (C/\Im_{sc})^{1/2}$. It is excited mainly through a magnetic dipole operator of the form

$$\begin{aligned}\mathcal{M}(M1,\mu) &= \sqrt{\frac{3}{4\pi}}(g_p J_\mu^{(p)} + g_n J_\mu^{(n)})\mu_N \\ &= \sqrt{\frac{3}{4\pi}}(g_R J_\mu + \frac{1}{2}g_r S_\mu)\mu_N, \end{aligned} \tag{6}$$

where g_R and g_r are respectively the rotational and scissors gyromagnetic factors

$$g_R = \frac{1}{2}(g_p + g_n), \qquad g_r = g_p - g_n . \tag{7}$$

The corresponding strength is

$$\begin{aligned} B(M1) \uparrow &= B(M1, 0^+ \to K^\pi = 1^+, J^\pi = 1^+) \\ &= \frac{3}{8\pi} |< K^\pi = 1^+ |S_{+1}| 0 >|^2 g_r^2 \mu_N^2 \simeq \frac{3}{16\pi} \Im_{sc} \, \omega g_r^2 \, \mu_N^2 . \end{aligned} \tag{8}$$

The above expression is valid beyond the TRM and can actually be assumed to define the scissors M1 strength. We may indeed exploit the harmonic nature of the Hamiltonian (eq. (3)) and write

$$\Im_{sc} = <0|S^2|0> = \frac{1}{\omega} \sum_{\mu=\pm 1} |<\mu|S_\mu|0>|^2 \tag{9}$$

$$C = \Im \omega^2 = \omega \sum_{\mu=\pm 1} |<\mu|S_\mu|0>|^2 . \tag{10}$$

By a natural generalization we obtain the following defining relations

$$\Im_{sc} = \sum_{n\mu} \frac{1}{\omega_n} |<n\mu|S_\mu|0>|^2 = <0|S_1 \frac{1}{H-E_0} S_1|0> \tag{11}$$

$$C = \sum_{n\mu} \omega_n |<n\mu|S_\mu|0>|^2 = <0|[S_1,[H,S_1]]|0> , \tag{12}$$

where now operators and wave functions can be either macroscopic or microscopic. It is now a simple matter to derive the following scissors sum rule

$$\begin{aligned} \sum_n \omega_n B_n(M1) \uparrow &= \frac{3}{16\pi} \sum_{n,\mu} \omega_n |<n\mu|S_\mu|0>|^2 g_r^2 \mu_N^2 \\ &= \frac{3}{16\pi} <0|[S_1,[H,S_1]]|0> g_r^2 \mu_N^2 \\ &\simeq \frac{3}{16\pi} \Im \omega^2 g_r^2 \mu_N^2 . \end{aligned} \tag{13}$$

This is based on the only assumption that the M1 transition is promoted by the scissors operator S. It is otherwise quite general. The last equation is valid under the (experimentally supported) assumption of small fragmentation of the mode. In this limit eq. (8) defining the scissors strength follows immediately.

Scissors M1 Strength and Nuclear Deformation

The M1 transition probability is linearly related to the strength of the $E2$ transition to the lowest 2^+ state[14]. The starting point is the relation between the E2 strength and the classical energy weighted sum rule $S(E2)$ [18]

$$\omega_2 B(E2) \uparrow = (E_2 - E_0) B(E2, 0^+ \to 2^+) = \frac{2}{5} \frac{Z}{A} \chi_D S(E2) , \tag{14}$$

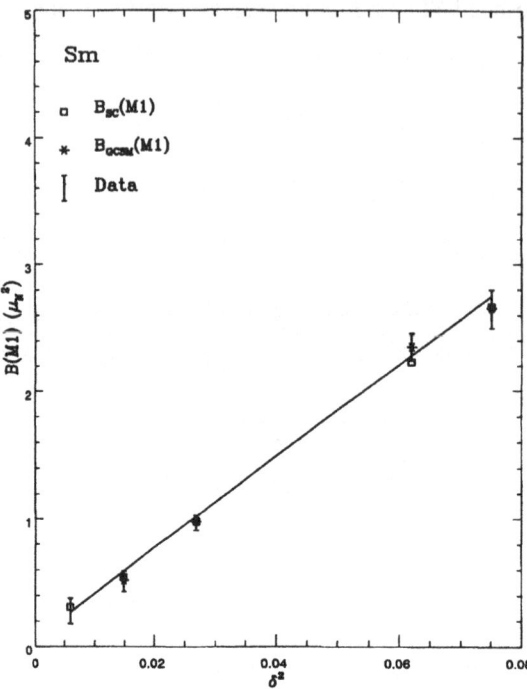

Figure 1. *Summed M1 strength versus δ^2 in Sm isotopes. The lines connect the experimental points. The data are taken from ref.3.*

where $\chi_D = D_2(irr)/D_{rot} \simeq 1/5$ is the ratio between the irrotational and rotational mass parameters and

$$S(E2) = \frac{15}{4\pi}\frac{1}{m}R^2 Z e^2, \tag{15}$$

with m the nucleon mass and $R = 1.2 A^{1/3}\, fm$. The above two equations yield the relation

$$\Im \simeq \frac{3}{\omega_2} = \frac{\pi}{28.8\chi_D}\frac{A^{1/3}}{Z^2}B(E2, 0^+ \to 2^+)\frac{(MeV^{-1})}{e^2 fm^4}. \tag{16}$$

Inserting this expression into eq. (8) we obtain

$$B(M1)\uparrow \simeq 0.0065\frac{A^{1/3}}{\chi_D Z^2}\omega B(E2, 0^+ \to 2^+)\frac{\mu_N^2}{e^2 fm^4}(g_p - g_n)^2. \tag{17}$$

This $M1 - E2$ relation implies a quadratic dependence of the M1 transition probability on deformation. To this purpose we adopt the expression valid to lowest order in the deformation parameter δ used in ref.3

$$B(E2)\uparrow = \frac{5}{16\pi}Q_0^2 \simeq \frac{1}{5\pi}Z^2 R^4 \delta^2 e^2, \tag{18}$$

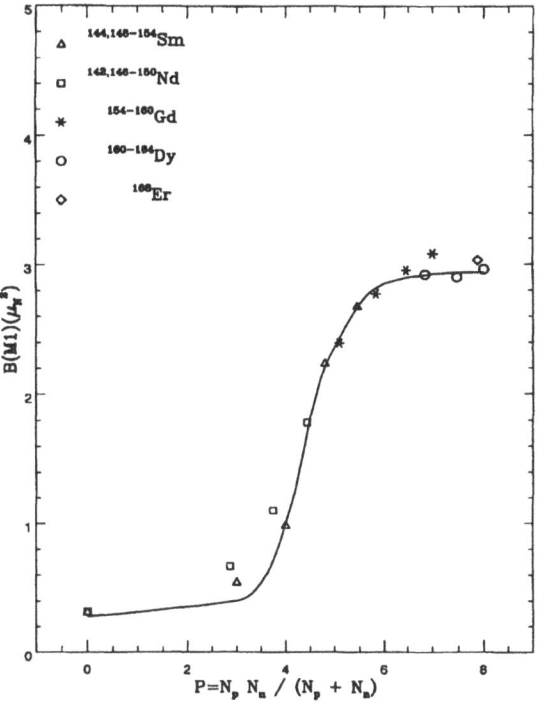

Figure 2. *Saturation plot of the M1 strength (see ref.4). The solid line is drawn to guide the eye.*

where $Q_0 \simeq \frac{4}{5} Z e R^2 \delta$ is the nuclear quadrupole moment. Upon substitution into eqs. (16)-(17), we obtain[14]

$$\Im \simeq \frac{1.44}{20} A^{5/3} \delta^2 \ (MeV^{-1}), \tag{19}$$

$$B_{sc}(M1) \uparrow \simeq 0.004 \ \omega A^{5/3} \ \delta^2 \ g_r^2 \mu_N^2 \ . \tag{20}$$

For a quantitative estimate we put $g_R = Z/A$ and $g_n = 0$, so that from eq. (7) $g_r = g_p = (2Z)/A$. The M1 strength so determined follows the observed δ^2 law (fig.1) and saturates with the fractional number $P = (N_p N_n)/(N_p + N_n)$ of valence protons and neutrons in agreement with experiments[14] (fig.2).

The consistency of the scissors $M1$ strength computed in such a model-independent way, with the new observed properties enforces the scissors character of the mode. The mass parameters used in the calculations are given by eq. (19), obtained after having neglected higher order terms in δ. They are therefore somewhat smaller than the nuclear mass parameters. The difference is in general small so that we may infer from the results obtained that protons and neutrons behave approximately as deformed "superfluids" in their relative rotational oscillation. It is natural to ask at this stage if similar results and conclusions can be obtained by using entirely theoretical tools.

A Low and a High Energy Scissors Mode in the TRM

The basic ingredients of the M1 strength are the restoring force constant and the mass parameter. In the TRM the computation of the restoring force constant is based on the observation that a rotational oscillation between protons and neutrons induces a density variation of the form

$$\delta\varrho = \delta\varrho_p(\vartheta) - \delta\varrho_n(-\vartheta) \simeq k_\varrho \varrho_0 \sum_{\nu=1,-1} a_{2\nu} r^2 Y_{2\nu}^*, \tag{21}$$

where $k_\varrho = 7/R^2$ is fixed by the normalization condition

$$\int \varrho(\vec{r}) r^\lambda Y_{\lambda\mu}(\vec{r}) d\vec{r} = \alpha_{\lambda\mu} \varrho_0 R^{\lambda+3} = \frac{3}{4\pi} A R^\lambda \alpha_{\lambda\mu}. \tag{22}$$

The shape variable can be easily shown to be related to the angle ϑ by

$$a_{21} = a_{2-1} = -i\sqrt{\frac{3}{2}} \beta \vartheta. \tag{23}$$

This is the key relation which will be exploited to link the TRM to other approaches.

The restoring force constant can be determined from the symmetry energy mass formula

$$\Delta V = \frac{1}{2} b_s \int \frac{(\delta\varrho)^2}{\varrho_0} d\vec{r} = \frac{1}{2} C \vartheta^2, \tag{24}$$

where $b_s \simeq 50 MeV$ is the symmetry energy coefficient. Solving the integral we obtain

$$C \simeq \frac{28}{5} b_s A \delta^2. \tag{25}$$

High Energy Mode

If the mode is assumed to arise from a relative motion between two irrotational fluids we get a two-rotor Hamiltonian with an irrotational moment of inertia

$$\Im = \Im_{irr} = \delta^2 \Im_{rig}, \qquad \Im_{rig} \simeq \frac{2}{5} m A R^2. \tag{26}$$

The energy of the mode and the corresponding strength are given by

$$\omega \simeq 139.4 A^{-1/3} \ MeV, \qquad B(M1)\uparrow \simeq 0.12 \delta^2 A^{4/3} g_r^2 \mu_N^2. \tag{27}$$

The strength is quadratic in the deformation parameter, the energy however is far higher than the observed one. We obtain indeed for ^{154}Sm

$$\omega \simeq 26 \ MeV, \qquad B(M1)\uparrow \simeq 4.7 \ \mu_N^2, \tag{28}$$

having put $g_n = 0$ and $g_p = 2g_R = 2Z/A$. Although the $M1$ strength is reasonably close to the experimental value ($\simeq 2.7 \mu_N^2$), the energy lies in the region of the giant isovector quadrupole resonance. It is indeed the energy of the $K^\pi = 1^+$ mode of the isovector quadrupole resonance as it emerges from decomposing the restoring force constant into a kinetic and potential part

$$C = C_0 + \mathcal{K} \tag{29}$$

by putting $b_s = b_0 + b_1 \simeq 50 MeV$. Since $b_1 = V_1/4 \simeq 32.5 MeV$, it follows that $b_0 \simeq 17.3 MeV$. Hence

$$\sqrt{\frac{C_0}{\Im}} \simeq 2\omega_0 , \qquad \omega_0 \simeq 41 A^{-1/3} , \tag{30}$$

which is the unperturbed energy adopted for such a resonance within the unified theory[18].

Low Energy Mode

In order to get a low energy mode we may adopt the definitions (9-10) and require in analogy with eq. (30) that

$$\frac{1}{\Im_{sc}} <S^2> = \frac{1}{\Im_{sc}} \sum_{n\mu} |<n|S_\mu|0>|^2 \simeq \delta\omega_0 . \tag{31}$$

This relation, treated in the anisotropic harmonic oscillator basis, yields a rigid body moment of inertia

$$\Im_{sc} = \Im_{rig} = \frac{2}{\delta\omega_0} \sum |(S_1)_{\alpha\beta}|^2 \simeq \frac{2}{5} m A R^2 . \tag{32}$$

From eqs. (31) and (32) we obtain for the kinetic component of b_s $b_0 \simeq 17.3/4$. We fix the potential component from the ratio between the nuclear average potentials $b_1/b_0 = -V_1/(4V_0) \simeq 0.6$ obtaining

$$\omega = \sqrt{\frac{C}{\Im_{rig}}} \simeq 53\delta A^{-1/3} MeV , \qquad B(M1) \uparrow \simeq 0.045 \delta A^{4/3} (g_p - g_n)^2 \mu_N^2 . \tag{33}$$

We may alternatively impose for the low energy mode

$$\frac{1}{\Im_{sc}} <S^2> = E(\epsilon) + E(\epsilon + \delta\omega_0) \simeq 2E , \tag{34}$$

where $E(\epsilon) = \sqrt{(\epsilon/2)^2 + \Delta^2}$ are the quasi-particle energies with $\epsilon = \epsilon_{sp} - \lambda$. We obtained the last relation by choosing the chemical potential to be $\lambda = \delta\omega_0/2$ so that $(\epsilon_{sp} = 0)$

$$E(\epsilon) = E(\epsilon + \delta\omega_0) = \sqrt{(\frac{\delta\omega_0}{2})^2 + (\Delta)^2} . \tag{35}$$

The mass parameter results then to be

$$\Im_{sc} \simeq \frac{\delta\omega_0}{(E(\epsilon) + E(\epsilon + \delta\omega_0))} (u(\epsilon)v(\epsilon+\delta\omega_0) - v(\epsilon)u(\epsilon+\delta\omega_0))^2 \Im_{rig} \simeq (\frac{\delta\omega_0}{2E})^3 \Im_{rig} . \tag{36}$$

Eqs. (35) and (36) yield for the energy and the strength

$$\omega \simeq (2E)(1+\frac{b_1}{b_0})^{1/2}, \qquad B(M1)\uparrow \simeq \frac{3}{16\pi}\omega\Im_{rig}(\frac{\delta\omega_0}{E})^3 g_r^2\mu_N^2, \qquad (37)$$

or more quantitatively

$$\omega \simeq 1.26(2\Delta)\sqrt{1+x^2}, \quad B(M1)\uparrow \simeq 0.001(2\Delta)A^{5/3}\frac{x^3}{1+x^2}\mu_N^2, \qquad (38)$$

where $x = \delta\omega_0/(2\Delta)$. The strength goes like δ^3 for small deformations ($x \ll 1$) and becomes linear for large deformations ($x \gg 1$). In the range of the observed deformations the above M1 strength is approximately quadratic in δ.

If we had averaged the two quasi-particle energies and the moment of inertia with respect to λ as suggested in ref. 19, we would have obtained[7,18]

$$\Im \simeq \Im_{rig}(1-g(x)), \qquad (39)$$

where

$$g(x) = \frac{\ln(x+(1+x^2)^{1/2})}{x(1+x^2)^{1/2}} \qquad (40)$$

and

$$2E \simeq 2\Delta(\sqrt{1+4x^2} + \frac{1}{2x}\ln\left|\sqrt{1+4x^2}+2x\right|). \qquad (41)$$

For weak deformations the above relations give

$$\omega \simeq 2\Delta(1+\frac{b_1}{b_0})^{1/2}, \qquad \Im \simeq \frac{(\delta\omega_0)^2}{6\Delta^2}\Im_{rig} \qquad (42)$$

and therefore

$$B(M1)\uparrow \simeq \frac{0.58}{\Delta}\delta^2 A\mu_N^2, \qquad (43)$$

which exhibits a δ^2 behaviour. Such a behaviour is however spoiled for stronger deformations ($x \simeq 1$). We have gained here the result already obtained by Hamamoto and Magnusson within a schematic RPA context[7]. The complete equivalence between semiclassical and schematic RPA approaches will be indeed proved explicitly.

Scissors Modes in Superdeformed Nuclei?

We have seen that the semiclassical model predicts a low and a high energy mode. In the first mode proton and neutron behave approximately as superfluid systems, in the second as irrotational fluids. Being the modes switched by deformation, the model predictions should apply in principle to superdeformed nuclei as suggested recently in a RPA calculation[20].

Let us assume that K is still a good quantum number. The $M1$ operator couples the $|IMK>$ to the $|I'M'K+1>$ states with $I' = I-1, I, I+1$. Using the TRM intrinsic wave function and assuming $I \gg K$ we obtain for the summed strength

$$\sum_{I'} B(M1, IK \to I'K+1) \simeq \frac{3}{16\pi} \Im_{sc} \omega \frac{1}{K+1} g_r^2 \mu_N^2. \tag{44}$$

For $K = 0$ we gain eq. (8) which gives the standard scissors strength.

Numerically we may use for the high energy mode eqs. (27) with $g_r = 2g_R = 2Z/A$ obtaining for the superdeformed ^{152}Dy ($\delta \simeq 0.62$)

$$\omega \simeq 26 \; MeV, \qquad B(M1) \uparrow \simeq 26.12 \; \mu_N^2, \tag{45}$$

For the low energy mode we may assume rigid rotors and use eqs. (33) with $g_r = 1$ obtaining for the same superdeformed nucleus

$$\omega \simeq 6.15 \; MeV, \qquad B(M1) \uparrow \simeq 22.63 \; \mu_N^2. \tag{46}$$

The choices made for the gyromagnetic ratios were dictated by the analogy with schematic RPA to be discussed below.

The two modes in superdeformed nuclei should therefore be strongly collective. Proton and neutron systems however, while remaining irrotational at high energy, become rigid in the low energy mode.

Two-Rotor Model and other Approaches

Using the key relation between ϑ and a_{21} (eq. (23)) we can turn the TRM Hamiltonian in θ into an equivalent Hamiltonian in $a_{2\mu}$.

$$H = \frac{1}{2}\Im_{sc}\dot{\vartheta}^2 + \frac{1}{2}C\vartheta^2 = \frac{1}{2}B \sum_{\mu=\pm 1} |\dot{a}_{2\mu}|^2 + \frac{1}{2}C_\alpha \sum_{\mu=\pm 1} |a_{2\mu}|^2, \tag{47}$$

where

$$\Im_{sc} = 3\beta^2 B, \qquad C = 3\beta^2 C_\alpha. \tag{48}$$

This is just the unified theory Hamiltonian of Bohr and Mottelson[18] which allow us to make a bridge between TRM and several other models.

Two-rotor Model and Schematic RPA

In order to show the complete equivalence with schematic RPA we first switch from the normalization given by eq. (22) to the new one

$$\int \varrho(\vec{r}) r^\lambda Y_{\lambda\mu}(\vec{r}) d\vec{r} = \alpha_{\lambda\mu}. \tag{49}$$

This induces a rescaling of the mass and restoring force parameters entering into the new TRM Hamiltonian

$$B = \frac{\Im}{3\beta^2(\frac{3}{4\pi}AR^2)^2} = \frac{2\pi}{3}\frac{m}{AR^2}, \qquad (50)$$

$$C_\alpha = \frac{C}{3\beta^2(\frac{3}{4\pi}AR^2)^2} = C_0^\alpha + \mathcal{K}_\alpha, \qquad (51)$$

where

$$C_0^\alpha \simeq \frac{8\pi}{3}\frac{m\omega_0^2}{AR^2}, \qquad \mathcal{K}_\alpha \simeq \frac{7}{3}\frac{\pi V_1}{AR^4}. \qquad (52)$$

The above quantities are just the mass parameters and coupling strengths derived within the unified theory approach[18]. A complete equivalence between the semiclassical TRM and the unified theory description has therefore been established. The link with schematic RPA is now obtained simply by imposing the condition

$$\sum_\mu |a_{2\mu}|^2 = \sum_\mu \sum_{ph} |(Q_\mu)_{ph}|^2, \qquad Q_\mu = \tau_3 r^2 Y_{2\mu}. \qquad (53)$$

Also in schematic RPA one obtains a low and a high energy mode in deformed as well as superdeformed nuclei[20]. The energies are obviously exactly the same as the one obtained by the semiclassical approach. The M1 strength has the TRM form with $g_r = 1$ for the low energy mode and $g_r \simeq (1 + \mathcal{K}_\alpha/C_\alpha)/2 \simeq 2g_R$ for the high energy mode, a confirm that the two procedures are interchangeable.

Schematic RPA does not describe correctly the deformation properties of the mode. This is mainly because the spin degree of freedom is ignored. The above properties are described satisfactorily in more realistic RPA calculations[7,9,24], where the spin is properly taken into account.

Two-Rotor Model and GCSM

In the GCSM approach[17] the scissors state Ψ_1^1 is obtained by projecting the $J = 1$ component out of an intrinsic state of the form

$$\Phi_1 = (b_n^+ \otimes b_p^+)_{K=1}^{J=1}\Phi_0, \qquad (54)$$

where

$$\Phi_0 = \Phi_{\alpha=K=0} = exp[d_p(b_{p0}^+ - b_{p0}) + d_n((b_{n0}^+ - b_{n0})]|0> \qquad (55)$$

is a coherent intrinsic ground state and $d_p = d_n = \rho/\sqrt{2}$ is a function of the deformation parameter $\beta = (16\pi/45)^{1/2}\delta$.

This is one of the six states that render diagonal an interacting boson Hamiltonian of the form

$$H = A_1(N_p + N_n) + A_2(N_{pn} + N_{np}) + \sqrt{5}\frac{A_1 + A_2}{2}(\Omega_{pn}^+ + \Omega_{np})$$
$$+ A_3(\Omega_p^+\Omega_n + \Omega_n^+\Omega_p - 2\Omega_{np}^+\Omega_{np}) + A_4J^2, \qquad (56)$$

where A_i are free parameters and

$$N_{\tau\tau'} = \sum_\mu b^+_{\tau\mu} b_{\tau'\mu}, \qquad \Omega^\dagger_{\tau\tau'} = (b^\dagger_\tau \otimes b^\dagger_{\tau'})_0 - \frac{2}{\sqrt{5}}\rho^2. \tag{57}$$

The $M1$ strength is given by

$$\begin{aligned} B(M1)\uparrow &= \frac{3}{16\pi}|<\Psi^{(0)}_0\|S\|\Psi^{(1)}_1>|^2 g_r^2 \mu_N^2 \\ &= \frac{9}{40\pi}\rho^4\left(1+\frac{1}{10}\left(\frac{N_{00}}{N02}\right)^2\right)^2\left(\frac{N_{11}}{N_{00}}\right)^2 g_r^2 \mu_N^2, \end{aligned} \tag{58}$$

where the N_{ik} are normalization factors. For strong deformation (rotational limit) we have $\rho \simeq k_p\beta$ where k_p is a scale parameter which in the harmonic limit is just $k_p = (B_p C_p)^{1/4}$. It is easy to derive in this limit the $M1 - E2$ relation[13]

$$B_{rot}(M1)\uparrow \simeq 2\pi \frac{k_p^2}{Z^2 R^4} B_{rot}(E2; 0^+ \to 2^+) \frac{g_r^2 \mu_N^2}{e^2 fm^4} \tag{59}$$

as well as the deformation law

$$B_{rot}(M1)\uparrow \simeq \frac{9}{8\pi}\rho^2 g_r^2 \mu_N^2 \simeq \frac{9}{8\pi} k_p^2 \beta^2 g_r^2 \mu_N^2. \tag{60}$$

Using the relations $k_p^2 = (B_p C_p)^{1/2} = B_p\omega$ and $3B_p\beta^2 = \Im^{(p)} \simeq \Im/2$ it is straightforward to write the above strength in the TRM form (eq. (8)). In this limit even the M1 GCSM state assumes the TRM form[13]

$$\Phi_1 \simeq \vartheta e^{-\frac{1}{2}\Im\omega\vartheta^2} = \vartheta e^{-\frac{1}{2}(\frac{\vartheta}{\vartheta_0})^2}, \qquad \vartheta_0^2 = \frac{1}{\Im\omega}. \tag{61}$$

A quadratic law but with a different coefficient is obtained also for weak deformations

$$B_{vib}(M1)\uparrow \simeq \frac{9}{40\pi} g_r^2 \rho^4 \mu_N^2 \simeq \frac{9}{2\pi} k_p^2 \beta^2 g_r^2 \mu_N^2. \tag{62}$$

We have computed the exact GCSM M1 strength given by eq. (58) having fixed the parameters by a fit of some selected levels of the ground beta and gamma bands. We have further put $g_p = 1$ and $g_n = 0$ as in other interacting boson models. The agreement with experiments is quite satisfactory (fig.1). They enforce the scissors nature of the mode and suggest once again that protons and neutrons behave as superfluids. Indeed, the computation of the GCSM nuclear moment of inertia

$$\Im \simeq \frac{3\rho^2}{A_1+A_2} \simeq \frac{3k_p}{A_1+A_2}\beta^2 \simeq \Im_{exp} \tag{63}$$

yields values very close to the experimental ones. This result implies also that pairing correlations, though not explicitly present, are effectively accounted for.

Two-rotor Model and IBM

It has been proved[21] that in the classical limit the IBM-2 Hamiltonian can be turned into a TRM Hamiltonian of the form given by eq. (1). The only main difference with respect to the rigid TRM is that in the IBM case only valence nucleons contribute to the TRM physical constant. We have then to make the following substitutions in the TRM Hamiltonian

$$\Im_p \simeq \frac{Z}{A}\Im \rightarrow \Im_\pi \simeq \frac{N_\pi}{N_\pi + N_\nu}\Im, \quad \Im_n \simeq \frac{N}{A}\Im \rightarrow \Im_\nu \simeq \frac{N_\nu}{N_\pi + N_\nu}\Im, \quad (64)$$

where N_π and N_ν are the number of valence proton and neutron pairs. The above correspondence allows to write the M1 strength into the form

$$B(M1)\uparrow = \frac{3}{16\pi}\Im_{sc}\omega g_r^2 \mu_n^2 \simeq \frac{3}{4\pi} P \frac{\Im\omega}{2(N_\pi + N_\nu)} g_r^2, \quad (65)$$

where $P = 2N_\pi N_\nu/(N_\pi + N_\nu)$. This is quite close to the general IBM-2 M1 strength[11]

$$B(M1)\uparrow = \frac{3}{16\pi} <0|S^2|0> g_r^2 \mu_N^2 \simeq \frac{9}{4\pi} P \frac{<N_d>}{N_\pi + N_\nu - 1} g_r^2, \quad (66)$$

where $<N_d>$ is the average number of quadrupole bosons in the ground state. This term is essential for the saturation properties of the IBM M1 strength. The two expressions actually coincide in the limit of large N if

$$\frac{1}{2}\Im\omega \simeq 3 <N_d> . \quad (67)$$

Such a relation may indeed be derived by making use of coherent states[16]. If computed phenomenologically, the strength given by eq. (65) reproduces fairly well the experimental data, an indication that the scissors mode is mainly promoted by valence nucleons.

The IBM M1 strength cast into the TRM form displays its δ^2 dependence. On the other hand, casting the TRM M1 strength into the IBM form allows to underline the essential contribution of monopole and quadrupole pairing to the saturation properties of the M1 strength since these correlations are built into the IBM basis states.

Concluding Remarks

The present analysis has shown that the TRM M1 strength represents the common root of several schematic or phenomenological descriptions of the scissors mode. We saw the cases of schematic RPA and of two interacting boson models, the GCSM and the IBM-2. The same close relation can be stated[16] with the neutron-proton deformation model[6], the projected Hartree-Bogoliubov[8] and the truncated shell-model[12]. The possibility of casting the M1 strengths of so many model descriptions into the TRM form reflects the general validity of such an expression. Its success in describing the observed deformation properties of the mode in a phenomenological model-independent way proves that the observed low lying M1 excitations correspond to a scissors-like mode.

All models converging to the TRM yield a M1 strength quadratic in the deformation parameter more or less in agreement with experiments. In all of them pairing correlations play a crucial role in enforcing the quadratic deformation law, a further indication that protons and neutrons behave in their relative motion approximately as deformed superfluids. The deviations from experiments are to be ascribed to the simplifying assumptions or the approximations peculiar to each model. The δ^2 law allows in this way a preliminary selection of different theoretical descriptions or at least can suggest how to improve them. A more complete and effective selection can be made by combining such a law with the analysis of the M1 form factor. This operator in fact has a more complicated structure, which differs from model to model, and explores more in detail the model wave functions. The responses are consequently different in different approaches.

The schematic RPA description appears to be not very satisfactory. The main reason is to be found in the fact that spin is ignored. Realistic RPA calculations which account for spin are needed. For a correct RPA treatment of the mode however, several conditions are to be fulfilled[22,23]. Most of these requirements are satisfied in the latest RPA calculations[7,9,24] with satisfactory results regarding the deformation properties.

The same δ^2 law suggests that the mode is collective. We observe in this respect that the collectivity of the mode should not be measured in terms of single particle units[22]. The mode in fact appears only with deformation and its M1 strength increases and saturates with it. These two properties, observed in Sm and Nd isotopes[3-5], suggest that the mode can be considered collective in well deformed nuclei to the extent that the observed strengths exhaust the scissors sum rule. This is the case for the observed low lying M1 excitations. We can then conclude that these low lying excitations correspond to a true collective scissors-like mode.

According to the semiclassical as well as schematic RPA analysis there is in principle room also for a high energy scissors mode, in which protons and neutrons behave as irrotational normal fluids. The estimate of its M1 strength appears reliable since the pairing correlations play a negligible role at high energy. The observation of such a high energy M1 mode would therefore give precious informations about the isovector E2 giant resonance in deformed nuclei.

The semiclassical and schematic RPA approaches predict also strongly collective low and high scissors-like excitations in superdeformed nuclei. While the high energy mode should be promoted by a relative rotational oscillation between irrotational normal fluids, the low one should be generated by the motion of rigid rotors as in the original model[15]. The exciting perspective of the possible occurrence of these excitations deserves further theoretical and especially experimental investigations.

References

[1] D. Bohle, A. Richter, W. Steffen, A. E. L. Dieperink, N. Lo Iudice, F. Palumbo, and O. Scholten, *Phys. Lett. B* **137**, 27 (1984)

[2] For a summary see A. Richter, *Nucl. Phys. A* **507**, 99c (1990); **522**, 139c (1991)

[3] W. Ziegler, C. Rangacharyulu, A. Richter and C. Spieler, *Phys. Rev. Lett.* **65**, 2515 (1990)

[4] C. Rangacharyulu, A. Richter, H.-J. Wörtche, W. Ziegler and R. F. Casten, *Phys. Rev. C* **43**, R949 (1991)

[5] J. Margraf et al., *Phys. Rev. C* **47**, 1474 (1993)

[6] S. G. Rohozinski and W. Greiner, *Z. Phys. A* **322**, 271 (1985)

[7] I. Hamamoto and C. Magnusson, *Phys. Lett. B* **260**, 6 (1991)

[8] E. Garrido, E. Moya de Guerra, P. Sarriguren and J. M. Udias, *Phys. Rev. C* **44**, R1250 (1991)

[9] C. De Coster and K. Heyde, *Phys. Rev. C* **44**, R2262 (1991); K. Heyde, C. De Coster, A. Richter and H.-J. Wörtche, *Nucl. Phys. A* **549**, 103 (1992)

[10] T. Mizusaki, T. Otsuka and M. Sugita, *Phys. Rev. C* **44**, R1277 (1991)

[11] J. N. Ginocchio, *Phys. Lett. B* **265**, 6 (1995)

[12] L. Zamick and D. C. Zheng, *Phys. Rev. C* **44**, 2522 (1991); **46**, 2106 (1992)

[13] N. Lo Iudice, A. A. Raduta and D. S. Delion, *Phys. Lett. B* **300**, 195 (1993); submitted to *Phys. Rev C* for publication

[14] N. Lo Iudice and A. Richter, *Phys. Lett. B* **304**, 193 (1993)
[15] N. Lo Iudice and F. Palumbo, *Phys. Rev. Lett.* **41**, 1532 (1978); G. De Franceschi, F. Palumbo and N. Lo Iudice, *Phys. Rev. C* **29**, 1496 (1984)
[16] N. Lo Iudice to be submitted for publication.
[17] A. A. Raduta, Amand Faessler and V. Ceausescu, *Phys. Rev. C* **36**, 2111 (1987)
[18] A. Bohr and B. R. Mottelson, *Nuclear Structure*, Vol. II., (Benjamin, N.Y., 1975)
[19] N. Lo Iudice in: *Understanding the variety of nuclear excitations*, edited by A. Covello, (World Scientific, Singapore, 1991)
[20] I. Hamamoto and W. Nazarewicz, *Phys. Lett. B* **297**, 25 (1992)
[21] H. R. Walet, P. J. Brussard and A. E. L. Dieperink, *Phys. Lett. B* **163**, 4 (1985)
[22] N. Lo Iudice and A. Richter, *Phys. Lett. B* **228**, 291 (1989)
[23] N. Lo Iudice in: *New trends in theoretical and experimental nuclear physics*, edited by A. A. Raduta et al., (World Scientific, Singapore, 1992)
[24] P. Sarriguren, Moya de Guerra, R. Nojarov and A. Faessler, to be published.

ELECTRIC EXCITATIONS IN BACKWARD ELECTRON SCATTERING

M. Dingfelder, R. Nojarov and Amand Faessler

Institut für Theoretische Physik, Universität Tübingen, Auf der Morgenstelle 14, D-72076 Tübingen, Germany

The low lying magnetic dipole excitations, found in deformed nuclei through inelastic electron scattering in 1984 by A. Richter and coworkers[1], have attracted since then a considerable experimental and theoretical interest. One possible microscopic description is the quasiparticle random-phase approximation (QRPA) with a model Hamiltonian[2]

$$H = H_0 + H_{SCQ} + H_{SS} + H_{RV}. \tag{1}$$

The mean field H_0 is described by an axially symmetric deformed Woods-Saxon potential in cylindrical coordinates. It includes pairing correlations in the BCS approximation. Such a phenomenological mean field violates the rotational invariance, but provides realistic single-particle energies and wave functions. The separable residual interactions consist of a selfconsistent quadrupole-quadrupole force H_{SCQ}, a spin-spin force H_{SS} and a rotational-vibrational coupling H_{RV} which restores the rotational invariance of the RPA-Hamiltonian violated by the deformation.

The experimental M1 strength distribution is well reproduced, but the (e,e') cross sections provide a more stringent test for the theory, because they are very sensitive to details of the nuclear structure. The (e,e') cross sections have been measured at backward scattering angle $\theta = 165°$, which is more favourable for $M1$ transition. The experimental cross sections in light, e.g. Titanium, nuclei are described well in RPA under the assumption of purely $M1$ excitations[3]. However, a rotational band develops in heavier deformed (e.g. rare earth) nuclei, giving rise to an accompanying $E2$ transition. Its excitation energy is only 30 - 50 keV larger than that of the corresponding $M1$ transition, because of the large moment of inertia. This energy separation is comparable with the energy resolution of the present high precision (e,e') experiments. The theoretical (e,e') form factors of rare earth nuclei[4] deviate systematically from experiment at higher momentum transfer. Therefore, the contribution of the accompanying $E2$ excitations to the backward (e,e') cross sections in heavy deformed nuclei are taken here into account.

We obtain the cross sections in DWBA, which must be used in heavier nuclei, because the incoming electron wave is distorted by the strong Coulomb field of the nucleus. The DWBA calculations require the reduced transition densities,

$$\rho_{21}(r) = \int \rho(\vec{r}) Y_{21}(\Omega) \, d\Omega, \tag{2}$$

$$\mathcal{J}_{LL'1}(r) = \int \vec{\mathcal{J}}(\vec{r}) \cdot \vec{Y}_{LL'1}(\Omega) \, d\Omega, \tag{3}$$

$$\vec{\mathcal{J}}(\vec{r}) = \vec{j}^C(\vec{r}) + \vec{\nabla} \times \vec{\mu}^S(\vec{r}), \tag{4}$$

which are calculated microscopically with the RPA wave functions. The $M1$ transition density $\mathcal{J}_{111}(r)$ is purely transverse, while both transverse $\mathcal{J}_{211}(r)$, $\mathcal{J}_{231}(r)$ and longitudinal $\rho_{21}(r)$ densities contribute to the $E2$ transition.

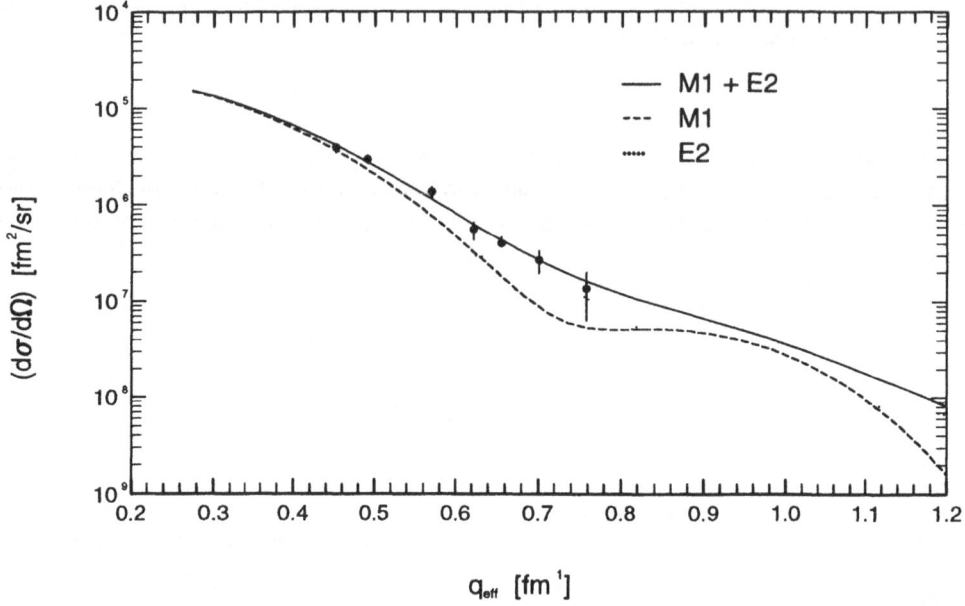

Figure 1. *Total DWBA-cross sections ($\theta = 165°$) for $M1$, $E2$ and $M1 + E2$ transitions versus the momentum transfer, calculated for the strongest M1 excitation in ^{156}Gd and compared to experiment[1].*

The calculated $M1$ and $E2$ cross sections for the strongest M1 excitation in ^{156}Gd are shown in figure 1. It is seen that the $E2$ cross section is two orders of magnitude smaller at low incident energies, but becomes comparable with the $M1$ cross section already at $q_{eff} = 0.6 fm^{-1}$. The higher multipolarity shifts the theoretical cross sections to higher transferred momenta. In this way the discrepancies between the experimental and theoretical $M1$ cross sections, which appear in heavy nuclei at higher incident energies even below the first diffraction minimum, are removed after adding the $E2$ contributions to the cross sections.

References

[1] D. Bohle et al., *Phys. Lett. B* **137**, 27 (1984)
[2] A. Faessler, R. Nojarov, *Phys. Rev. C* **41**, 1243 (1990)
[3] R. Nojarov, A. Faessler, P. O. Lipas, *Nucl. Phys. A* **533**, 381 (1991)
[4] A. Faessler, R. Nojarov, F. G. Scholtz, *Nucl. Phys. A* **515**, 237 (1990)

OBSERVATION OF NON YRAST STATES IN BARIUM NUCLEI FOLLOWING ISOMERIC β-DECAY WITH THE OSIRIS CUBE DETECTOR ARRAY

P. von Brentano, K. Kirch, U. Neuneyer, G. Siems and I. Wiedenhöver

Institut für Kernphysik, Universität zu Köln, D-50937 Köln, Germany

Introduction

There is a large variety of different quadrupole shapes being observed in the investigation of nuclear deformation. In Köln we have worked on a number of nuclei with particularly interesting shapes. We will mention just a few recent results. In the doubly magic ^{146}Gd nucleus spherical shell model states up to 16 MeV have been observed [1]. In the same nucleus two superdeformed bands, which extend to an excitation energy of 32 MeV, were found by the OSIRIS collaboration [2] [3]. This is an extreme case of shape coexistence. A similar phenomenon is observed in the "doubly magic" nucleus ^{114}Sn in which we found a new prolate band of negative parity which is related to the deformed Amsterdam positive parity band and coexists with spherical shell model states and vibrations [4]. J. Eberth has observed with the OSIRIS spectrometer the coexistence of oblate and prolate bands in ^{69}Se. These oblate bands are well developed and a negative deformation of $\beta = 0.3$ was determined [5].

Although most nuclei are axially symmetric there are interesting exceptions which are triaxial [6] [7] [8]. The nuclei in the xenon - barium region with mass values A between 126 and 134 are a particularly large region of triaxial nuclei. An important question is if either there is a rigid γ-deformation or the potential is γ-soft [9][10]. The difference of the spectra of two nuclei with a rigid and a soft γ-potential which have the same average value of the γ-deformation is actually not that obvious as one might expect. Rather this difference is a second order effect. A decision on the two alternatives can be obtained by measuring non-yrast collective excitations in these nuclei.

Population and observation of low lying collective states beside the ground state and γ-band, require a careful selection of reactions and equipment. Heavy ion reactions are not suitable because even in rather excellent high spin data observed with the OSIRIS 12 spectrometer in Köln and Berlin and with the 8π-spectrometer in Chalk River these states were not observed [11] [12]. For example a beautiful experiment on ^{126}Ba with this Chalk River spectrometer recently revealed 15 side bands, but in fact the 984 keV 0_2^+ state in ^{126}Ba [13] was not observed. Concerning the xenon nuclei 126 and 128 we obtained very good results using (α,xn)-reactions among others [14][15][16][17]. The use of these reactions in the investigation of the low spin structure of barium isotopes requires xenon gas targets.

Gamma-coincidence spectroscopy following β-decay

We alternatively used the reactions $Sn(^{14}N, xn)La \xrightarrow{\beta^+} Ba$ to populate excited levels in barium nuclei. Many of the odd-odd lanthanum nuclei possess medium spin isomers with lifetimes in the region of about a few minutes. So it is possible to observe medium spin levels in the barium nuclei. We applied cyclic measuring procedures with alternating periods, one to produce lanthanum parent nuclei and one other to detect the γ-rays following the β-decay. Of course this is not a new method in principle, but by using arrays of Compton suppressed detectors one can go beyond the experimental border reached so far. The use of the Cologne OSIRIS cube spectrometer [18] leads to an excellent quality of the $\gamma\gamma$-coincidence data. The spectrometer was equipped with six anti Compton shielded HPGe-detectors. The application of the BGO Anti Compton Shields yields an increase of the peak-to-background ratio by a factor of about twenty in $\gamma\gamma$-coincidence spectra measured following β-decay [19]. So it is possible to do new physics with an old method and we are able to detect very weak and hitherto unobserved transitions. Despite their small intensities such transitions can still have large B(E2)-values and can thus be of great interest for the nuclear models. The big advantage of using the Anti Compton Shields is shown in figs. 1 and 2. For the transitions $2_1^+ \to 0_1^+$ (465 keV in ^{132}Ba, 256 keV in ^{126}Ba) the peak-to-background ratio is about 13 times better in the measurement with BGO-shields. The 465 keV transition is the biggest peak in fig. 2 and cut off (original height: 3.55E06 counts). The peak at 108 keV is the transition $(3/2)^+ \to (1/2)^+$

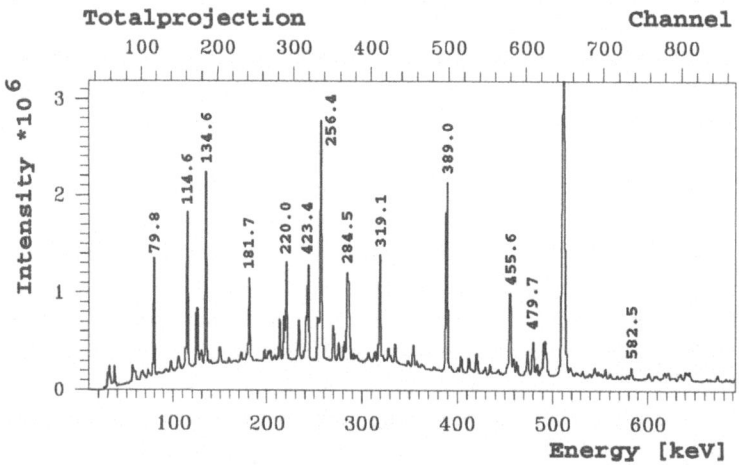

Figure 1. *Total projection of the ^{126}Ba coincidence matrix. The BGO-anti-Compton-shields were not available in this measurement.*

to the ground state of ^{131}Ba. By taking the efficiency corrected ratio of this two transitions we get the rough estimation that our spectra are contaminated only by about two percent by the ^{131}Ba channel. Other secondary reactions are even smaller.

Low lying collective states in the barium nuclei

The observation of additional low lying collective bands gives us important structural information in respect to the concerned nuclei. Idrissi et al. found a "K=0" band in ^{124}Ba using the β-decay of ^{124}La [20]. In ^{126,128}Ba we observed the same "K=0" band (see

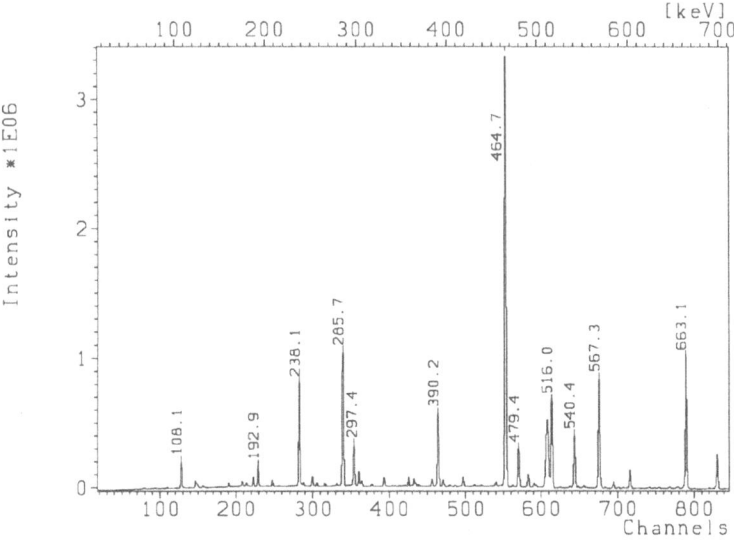

Figure 2. *Total projection of the ^{132}Ba coincidence matrix. The Compton suppression reduces the background by an order of magnitude.*

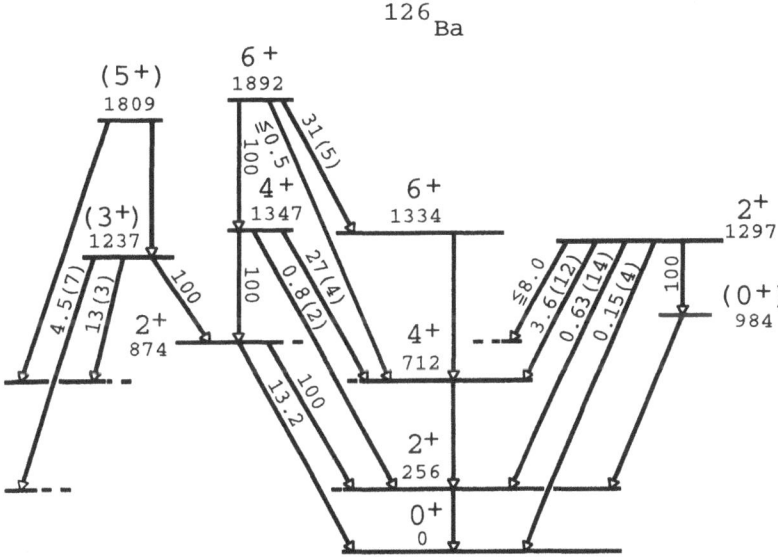

Figure 3. *Partial level scheme of ^{126}Ba with the ground state band, the quasi gamma band and two members of the "K=0" band. The values given at the transitions are E_γ^5 corrected branching ratios.*

figs. 3,4). Unfortunately there is no experimental branching ratio for the decays of the 0_2^+ level because the transition to the 2_2^+ was not observed. This is due to the small difference of the energies of these levels. Taking into account an E_γ^5 correction the unobserved transition might still be favoured and have the largest relative BE(2)-value. In the case of ^{130,132}Ba these important branchings were found using the same β-decay method [19][21]. Another feature in the level schemes which we should mention is the typical staggering of the levels in the

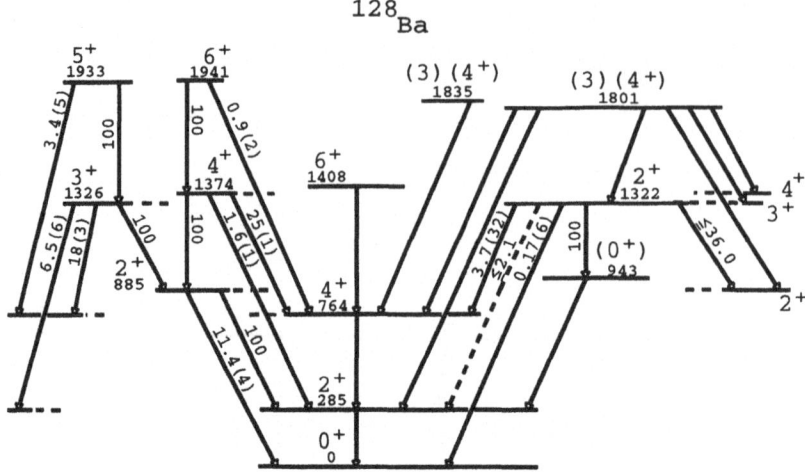

Figure 4. Partial levelscheme of ^{128}Ba.

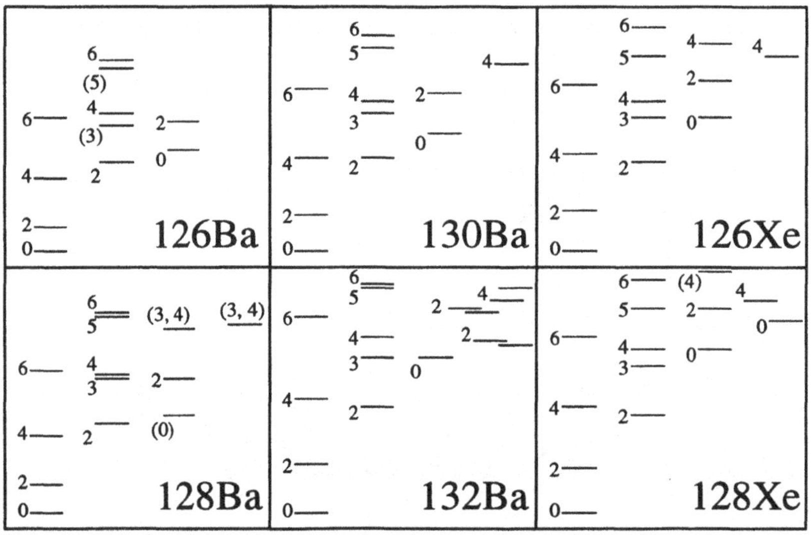

Figure 5. Low lying levels of some barium and xenon nuclei [16][21] [22] [23]

quasi gamma band, i.e. the 3_1^+ is nearer to the 4_2^+ than to the 2_2^+ level and the 5_1^+ is nearer to the 6_2^+ than to the 4_2^+. This staggering is a characteristic feature of a γ-soft rotor whereas in the case of a γ-rigid rotor it is just the other way around [8][27]. To get quantitative results from this argument one introduces the staggering index S for the 4^+ state as follows:

$$S(4,3,2) = \frac{[E(4) - E(3)] - [E(3) - E(2)]}{E(2)}. \tag{1}$$

It is used in the same way for the γ band triplett (6,5,4). We will give the values in the next section. A particularly nice measurement was done for ^{130}Ba. In this case we were able to evaluate the angular correlations and could therefore determine some of the spins of the

"K=0" band. Also the 4_3^+ bandhead of the "K=4" band was found as it is known from ^{126}Xe which belongs to the same F-spin multiplet as ^{130}Ba. In another recent barium measurement a similar correspondence between ^{132}Ba and ^{128}Xe was found. In fig. 5 the systematics of the energies of the lowest positive parity levels of the mentioned nuclei is shown.

Table 1. Comparison of experimental and IBA-1 reduced branching ratios [24] [13]. The values for ^{130}Ba are deduced from real B(E2)-values because the E2/M1 mixing is known [21]. The other transitions might be mixed but the transition probabilities are only corrected by E_γ^5, supposing that there are only small M1 admixtures.

I_i	I_f	$|\Delta\tau|$	^{126}Ba	O(6)$_\chi$	^{128}Ba	O(6)$_\chi$	^{130}Ba	^{126}Xe	O(6)$_\chi$	^{132}Ba	^{128}Xe	O(6)$_\chi$
0_2^+	2_2^+	1	100	100	100	100	100	100	100	100	100	100
	2_1^+	2					3.3(2)	7.7(22)	2.1	0.20(5)	7.1(4)	1.8
2_2^+	2_1^+	1	100	100	100	100	100	100	100	100	100	100
	0_1^+	2	13.2(20)	2.9	11.4(4)	2.1	6.2(7)	1.5(4)	1.5	2.8(5)	1.3(3)	1.3
3_1^+	2_2^+	1	100	100	100	100	100	100	100	100	100	100
	2_1^+	2	4.5(7)	3.9	6.5(6)	2.9	4.5(6)	2.4(7)	2.0	3.5(4)	2.1(1)	1.8
	4_1^+	1	13.0(25)	40	18.0(32)	40	22(3)	34(10)	40	30(6)	39(1)	40
4_2^+	2_2^+	1	100	100	100	100	100	100	100	100	100	100
	2_1^+	2	0.8(2)	3.9	1.55(8)	2.9	2.3(4)	0.4(1)	2.0	1.8(3)	1.3(1)	1.8
	4_1^+	1	27.3(42)	91	25.4(11)	91	54(10)	83(23)	91	82(15)	76(2)	110

Comparison of the data with the dynamical O(6) symmetry of the Interacting Boson Model (IBA-1)

In table 1 we compare reduced experimental branching ratios with calculations in the IBA-1 O(6) limit [10][25][26]. In order to reproduce nonvanishing $\Delta\tau = 2$ transitions we left the pure O(6) and took χ different from zero in the transition operator. Of great interest are the branchings of the 0_2^+ decays, where available. It is seen that the bandhead of the "K=0" band decays much stronger to the quasi gamma band than to the ground state band. This implies that the "K=0" band is built up on a quasi double γ-excitation rather than a β-excitation where of course a decay to the ground state band would be preferred. The good correspondence of the experimental and the calculated values is obvious. It is remarkable that the experimental values are always rather small when we look at $\Delta\tau = 2$ transitions which are forbidden in the pure O(6). The O(6) symmetry is equivalent to γ-instability, since the hamiltonian of the Wilets-Jeans γ-unstable rotor [9] is reduced to the O(6) Casimir operator if β is set to be constant [28]. By this we get a corroboration for the γ-softness of the potential or – if you want to – of the nuclear shape. In table 2 we show the experimental staggering indices compared to the calculated ones. The most important fact is that the index is always negative. The rigid Assymetric Rotor Model [6] gives a positive staggering. It is remarkable that the only nuclei for which the staggering index is significantly different from zero are the xenon, barium and platinum nuclei and these nuclei seem to be γ-soft and not γ-rigid.

Table 2. *Experimental energy ratios S in Xe- and Ba-nuclei compared to the predicted values. S measures the staggering in the quasi γ-band.* [15][16]

	^{126}Ba	^{128}Ba	^{130}Ba	^{126}Xe	^{132}Ba	^{128}Xe	$O(6)_\chi$
S(4,3,2)	-0.99	-1.38	-0.94	-0.69	-0.25	-0.65	-1.41
S(6,5,4)	-1.49	-1.94	-1.25	-0.27	-0.28	-0.25	-1.79

Conclusions

We have reported that "quasi double γ" "K=0" and "K=4" bands have been observed in several barium isotopes in the β-decay of medium spin isomeres using the OSIRIS cube spectrometer in Cologne. From their γ-decay one can assign these bands to be "quasi double γ-excitations". Furthermore one finds that these triaxial nuclei are all γ-soft and not γ-rigid in their low lying collective bands. The same conclusion is reached from the energy staggering in the γ-band. The fact that such nuclei are γ-soft means they have no definite shape even in the intrinsic system. These nuclei have thus a rather interesting geometry and it would be of great interest to investigate whether Hartree-Fock type calculations and electron scattering data will corroborate these indirect findings on the shape of these nuclei.

Acknowledgement

We would like to thank A. Dewald, T. Fricke, A. Gelberg, R. Kühn, T. Otsuka, K. Schiffer, O. Vogel and K. O. Zell for discussions. This work was partly funded by the DFG under contract No. Br. 799/5-1.

References

[1] H. Wolters, E. Ott, R. Wirowski, A. Dewald, J. Theuerkauf, J. Eberth, K.O. Zell, P.von Brentano, W. Gast, G. Hebbinghaus, P. Kleinheinz, A. Krämer-Flecken, R.M. Lieder, T. Morek, T. Rzaca-Urban, H. Schnare, C. Senff, W. Urban, H. Grawe, H. Kluge, K.H. Maier, S. Heppner, H. Hübel, A.P. Byrne and W. Schmitz, Z. Phys. A 333, 413 (1989)

[2] T. Rzaca-Urban, K. Strähle, G. Hebbinghaus, D. Balabanski, W. Gast, R.M. Lieder, H. Schnare, W. Urban, P. von Brentano, A. Dewald, J. Eberth, E. Ott, J. Theuerkauf, H. Wolters, K.O. Zell, D. Alber, K.H. Maier, E.M. Beck, H. Hübel and W. Schmitz, Z. Phys. A 339, 421 (1991)

[3] G. Hebbinghaus, K. Strähle, T. Rzaca-Urban, D. Balabanski, W. Gast, R.M. Lieder, H. Schnare, W. Urban, H. Wolters, E. Ott, J. Theuerkauf, K.O. Zell, J. Eberth, P. von Brentano, D. Alber, K.H. Maier, W. Schmitz, E.M. Beck, H. Hübel, T. Bengtsson, J. Ragnarsson, S. Aberg, *Phys. Lett. B* **240**, 311 (1990)

[4] M. Schimmer, S. Albers, A. Dewald, A. Gelberg R. Wirowski and P. von Brentano *Nucl. Phys.* A **539**, 527 (1992)

[5] M. Wiosna, J. Busch, J. Eberth, M. Liebchen, T. Mylaeus, N. Schmal, R. Sefzig, S. Skoda, W. Teichert, *Phys. Lett.* B **200**, 255 (1988)

[6] A. S. Davydov and G. F. Filippov, *Nucl. Phys.* **8**, 237 (1958); A. S. Davydov and A. A. Chaban, *Nucl. Phys.* **20**, 499 (1960)

[7] A. Bohr and B.R. Mottelson, *Nuclear Structure*, W.A. Benjamin Reading (1975)

[8] R.F. Casten *Nuclear Structure from a Simple Perspective* (Oxford University Press, Oxford, New York, 1990)

[9] L. Wilets and M. Jean, *Phys. Rev.* **102**, 788 (1956)

[10] F. Iachello and A. Arima, *The Interacting Boson Model* (Cambridge University Press, Cambridge, 1987)
[11] U. Neuneyer, H. Wolters, A. Dewald, W. Lieberz, A. Gelberg, et al., *Z. Phys. A* **336**, 245 (1990)
[12] D. Ward et al.*Nucl. Phys. A* **529**, 315 (1991)
[13] G. Siems, T.Fricke, U. Neuneyer, K. Schiffer, D.Lieberz, M. Eschenauer, I. Wiedenhöver and P. von Brentano, *Proc. 6th Int. Conf. on Nuclei far from Stability + 9th Int. Conf. on Atomic Masses and Fundamental Constants*, Bernkastel-Kues (Germany), 1992
[14] R. Reinhardt, A. Dewald, A. Gelberg, W. Lieberz,K. Schiffer, K. P. Schmittgen, K. O. Zell and P. von Brentano, *Z. Phys. A* **329**, 207 (1988)
[15] F. Seiffert, W. Lieberz, A. Dewald, S. Freund, A. Gelberg, A. Granderath, D. Lieberz, R. Wirowski and P. von Brentano. *Nucl. Phys. A***554**, 287 (1993)
[16] W. Lieberz, A. Dewald, W. Frank, A. Gelberg, W. Krips, D. Lieberz, R. Wirowski, P. von Brentano *Physics Lett. B* **240**, 38 (1990)
[17] P. von Brentano, A. Dewald, A. Gelberg, A. Granderath, D. Lieberz and W. Lieberz. in: *High spin physics and gamma-soft nuclei*, eds. J. X. Saladin, R. A. Sorensen and C. M. Vincent (World Scientific, Singapore, 1991)
[18] R. Wirowski, *Ph.D. thesis*, Universität zu Köln, 1993
[19] K. Kirch, G. Siems, I. Wiedenhöver, U. Neuneyer, R. Wirowski and P. von Brentano, *Proc. 8th ISCGRSART*, Fribourg 1993, to be published
[20] N. Idrissi et al., *Z. Phys. A* **341**, 427 (1992)
[21] G. Siems, *Ph.D. thesis*, Universität zu Köln, 1993
[22] E. S. Paul, D. B. Fossan, Y. Liang, R. Ma and N. Xu, *Phys. Rev. C* **40**, 1255 (1989)
[23] R. Reinhardt et al., *Z. Phys. A* **329**, 507 (1988)
[24] W. Krips, W. Frank, W. Lieberz and P. von Brentano, *Nucl. Phys. A* **529**, 485 (1991)
[25] A. Arima and F. Iachello, *Phys. Rev. Lett.* **35**, 1069 (1975); **40**, 385 (1978); *Ann. Phys. (NY)* **99**, 253 (1976); **111**, 201 (1978); **123**, 468 (1979)
[26] R.F. Casten and P. von Brentano, *Phys. Lett. B* **152**, 22 (1985)
[27] N.V. Zamfir and R.F. Casten, *Phys. Lett. B* **260**, 265 (1991)
[28] T. Otsuka and M. Sugita, *Phys. Rev. Lett.* **59**, 1541 (1987)

IDENTIFICATION OF THE TWO PAIRING-VIBRATION PHONON⊗PARTICLE MODE IN ^{145}Sm AND NEIGHBOURS

A. M. Oros[1,2], L. Trache[1], G. Cata-Danil[1], P. von Brentano[2], K. O. Zell[2], G. Graw[3], D. Hofer[3], E. Müller-Zanotti[3]

[1] *Institute of Atomic Physics, Bucharest, Romania*
[2] *Institute of Nuclear Physics, University of Cologne, Germany*
[3] *Sektion Physik der Universität München, Germany*

The pairing vibration mode is associated to pairing correlations in nuclei [1] and there is rich information about its existence in even-even nuclei. It is connected to fluctuations of the pairing gap parameter Δ around its ground state value. The collectivity of this excitation mode is reflected in the cross section for the population of some 0^+ and 2^+ states in two like nucleon transfer reactions, which are strongly enhanced upon the two-particle estimate. To learn more about the underlying fermionic structure of a collective mode, one should study both its coupling with single-particle (fermionic) degrees of freedom, and the multiphonon states and their anharmonicities.

We report here on the comparison of the only two cases where the neutron pairing vibration is studied in odd-neutron nuclei and about the boson-fermion couplings extracted from these studies. This comparison is possible due to a recent experiment ^{147}Sm(p,t)^{145}Sm, carried out at the Tandem accelerator and Q3D spectrometer in Munich. The first case reported concerned also an N=83 isotone, ^{143}Nd [2]. The even-even closed-shell N=82 isotones in the rare earth region have been previously studied through the two-neutron transfer reactions and the Monopole Pairing Vibration (MPV) was identified [3]. The comparison of the L=0 excitations in (p,t) reactions leading to the even-even and the corresponding odd nuclei ^{142}Nd, ^{143}Nd, ^{144}Sm and ^{145}Sm is shown in Fig. 1 (top to bottom). The simplest multiphonon state, consisting of one pair removal and one pair addition quanta and belonging to the closed shell nucleus, is a 0^+ state usually referred to as the Monopole Pairing Vibration (MPV) state; in the odd nucleus the corresponding state has the spin of the lowest single particle orbital which the odd nucleon occupies. In the specific case of the N=83 isotones, this is the $\nu 2f_{7/2}$ orbital.

Besides the g.s., the states having mainly MPV and MPV⊗neutron configuration are identified around E^*=3 MeV through their high cross section relative to that for other configurations, and by the characteristic angular distributions showing L=0 angular momentum transfer. For both even-even nuclei, the strength of the MPV was found in two excited 0^+ states, due to the mixing of the MPV state with another configuration. The unperturbed positions of the two states were calculated in a two-state-mixing model (open bars). In the Nd case, the correspondence between the 0^+ states in the core and the states with MPV⊗$\nu f_{7/2}$ component in the odd nucleus is nearly one-to-one.

The energy shifts are opposite in sign: positive for the centroid of the two states corresponding to the stronger 0^+ in the core (mainly MPV, labeled B) and negative for the state

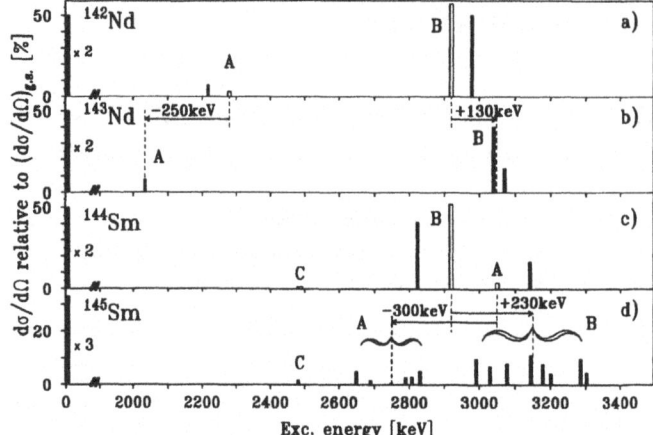

Figure 1. *Comparison between the features of MPV in Nd and Sm as seen in the (p,t) reactions.*

corresponding to the weaker 0^+ in the core (labeled A). The two states of the group B in the odd nucleus are only 30 keV apart, which demonstrates the very weak interaction between MPV and other configurations at the same energy.

In ^{145}Sm the MPV has the same strength and is located at roughly the same excitation energy as in the ^{144}Sm core, but is very much fragmented. The strength is divided in two narrow groups of 8+5=13 $J^\pi = 7/2^-$ states. The strength division for the two groups follows that in the core, but the opposite energy shifts of their centroids lead to inversion (weaker group lower, stronger group higher). We note that these energy shifts are similar in sign and size with the corresponding ones (A and B) from the Nd case, pointing to similar conclusions about the configuration of the 0^+ states and about the particle-MPV coupling. The small energy spreading within the groups allows to estimate the interaction of MPV with neighbouring configurations as very weak: < 50 keV. It follows that all interactions are in fact identical in Nd and Sm, and the observed different fragmentation is due to an accidentally high density of $7/2^-$ states in ^{145}Sm.

In conclusion, the coupling of the odd neutron to the two-pairing phonon mode was found to be weak: the collectivity is not affected, the energy shift of the MPV $\otimes$ neutron configuration is small and directed upwards due to the effect of the Pauli principle. The interaction of MPV with other configurations was found very weak. The study of such configurations through the (p,t) reaction proved to be a powerful tool for an easy identification of an important number of states of same spin and parity.

References

[1] A. Bohr in: *Nuclear Structure* (Dubna Symposium), IAEA, Vienna, p. 179 (Benjamin, Reading, MA, 1968); A. Bohr and B. Mottelson, *Nuclear Structure*, Vol. 2 (Benjamin, Reading, MA, 1975)

[2] L. Trache, C. Wesselborg, P. von Brentano, J. Wrzesinski and G. P. A. Berg, *Nucl. Phys. A* **540**, 66 (1992)

[3] L. Trache, A. M. Oros, Gh. Cata-Danil, K. O. Zell, P. von Brentano, G. Graw, D. Hofer, E. Zanotti-Müller, submitted to *Phys. Rev. C*

TRIAXIAL DEFORMATION IN PROTON-ODD Cs NUCLEI

O. Vogel, A. Gelberg and P. von Brentano

Institut für Kernphysik, Universität zu Köln, D-50937 Köln, Germany

The aim of this work is to show some results of Triaxial Rotor plus Particle (TRP) calculations for odd Cs. Since we have used the TRP model successfully for the neutron odd Xe and Ba nuclei [1, 2], it was natural to extend it to proton odd nuclei in this mass region. The Hamiltonian of the TRP model is given by

$$H_{TRP} = H_{rot} + H_{sp} + H_{pair}, \qquad (1)$$

where we have used hydrodynamical moments of inertia for H_{rot} and an axially asymmetric oscillator potential model for H_{sp}. The model is described in more detail in ref. [2, 3, 4]. Our main interest lies in the determination of the deformation parameter γ. One feature of the energy spectra which is very sensitive to γ is the staggering between favored and unfavored states. A convenient description of this is given by [5]

$$S(I) = \frac{E(I) - E(I-1)}{E(I) - E(I-2)} \cdot \frac{I(I+1) - (I-2)(I-1)}{I(I+1) - I(I-1)} - 1. \qquad (2)$$

In fig. 1 we compare the experimental values of $S(I)$ for ^{129}Cs [6] with the calculated ones for three selected values of γ. One can see that in the case of axial symmetry ($\gamma = 0°$) no agreement could be obtained. The value $\gamma = 27°$, which is the γ-value of the core ^{128}Xe [7], does not agree with the data either. The best fit could be reached for $\gamma = 22°$, which is below the γ-value of the core. Since we have derived γ for the neighboring odd and even

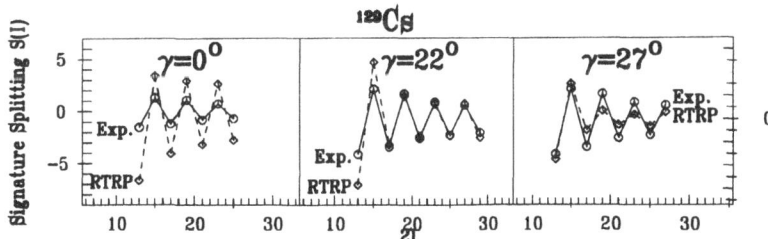

Figure 1. *Comparison of the experimental signature splitting $S(I)$ with the TRP calculated one for several values of γ for ^{129}Cs.*

Xe in earlier works [1, 2, 7], it is interesting to compare them to our γ values for Cs, in order to discuss the core polarization in γ due to the odd particle. This is done in fig. 2, where we plotted γ vs. the core neutron number N_C, which is N for the even Xe and proton odd Cs and

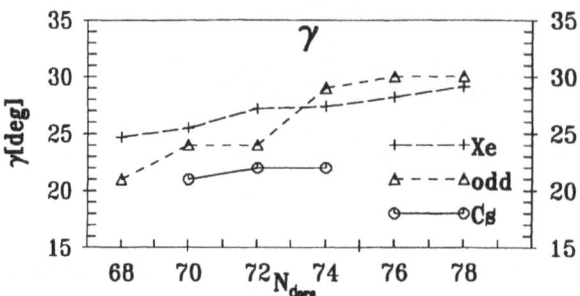

Figure 2. *Deformation parameter γ versus core neutron number N_c for Xe and and odd Cs nuclei.*

$N - 1$ for the neutron odd Xe. The existing data allow us to deduce some systematic trends of γ–deformation in odd nuclei. For the proton odd Cs nuclei the γ value lies quite below the value of the core, which can be understood by the location of the Fermi level below the $h_{11/2}$ subshell, resulting in an prolate driving force from the odd proton. For corresponding neutron odd Xe the situation is quite different since the Fermi level is located in the upper part of the $h_{11/2}$ subshell for larger N and diminishes with decreasing N. This trend is also reproduced by the γ values of Xe, where we see an oblate driving polarization for larger N and an opposite trend for smaller N. So we can draw the conclusion that the odd Cs nuclei can be well described with Triaxial Rotor plus Particle model. Although the proton odd Cs nuclei are definitely triaxial, their triaxiality is smaller than in the corresponding core, due to the polarization by the odd particle.

The authors would like to thank I. Ragnarsson and P.B. Semmes for providing the TRP–code and for fruitful discussions. We also want to thank R.V. Jolos and I. Wiedenhöver for stimulating discussions. This work was supported by the BMFT under contract 06OK143.

References

[1] D. Lieberz et al., *Nucl. Phys. A* **529**, 1 (1991)
[2] A. Gelberg et al., *Nucl. Phys. A* **557**, 439c (1993)
[3] S. E. Larsson et al., *Nucl. Phys. A* **307**, 189 (1978)
[4] H. Toki and A. Faessler, *Nucl. Phys. A* **253**, 23 (1975)
[5] N. V. Zamfir and R. Casten, *Phys. Lett. B* **260**, 265 (1991)
[6] Y. Liang et. al., *Phys. Rev. C* **42**, 890 (1990); L. Hildingsson et al., *Z. Phys. A* **340**, 33 (1991); J. R. Hughes et al., *Phys. Rev. C* **44**, 2390 (1991)
[7] J. Yan et al., *Phys. Rev. C* **48**, 1046 (1993)

SHAPE COEXISTENCE IN NEUTRON-DEFICIENT ^{111}Sb

V. E. Iacob, C. Stan-Sion, M. Parlog, A. Berinde, N. Scintei, N. Nica and L. Trache

Institute of Atomic Physics, Bucharest, MG-6, Romania

Nuclei near closed shells are considered of special interest for detailed experimental and theoretical studies because of their rather simple and transparent behavior. Nuclei with one particle outside a closed shell provide valuable information relative to the interplay between the single particle and the collective degrees of freedom.

In this respect, nuclei near the closed shell Z=50 were subject to a series of experimental investigations. The systematical appearance of a quasi-rotational band was previously reported and is discussed in the review by K.Heyde et al. [1] and in the paper by L.M.Yang et al.[2]. It was found that nuclei in this region manifest shape coexistence due to proton excitations across the major shell closure Z=50. For the cases in which protons and neutrons lie (a) in orbitals with same j or (b) in orbitals with large n-p overlap (such as $\pi 1g$-$\nu 1g$ and $\pi 1g$-$\nu 2d$) the n-p interaction is strong enough to drive a core deformation.

So far, identification of deformed bands in odd-mass antimony nuclei (Z=51) was limited to masses $A \geq 113$ [3]. In the present paper we report the identification of the band-head of a rotational-like band in the neutron-deficient nucleus ^{111}Sb.

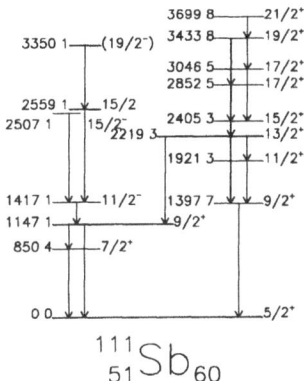

Figure 1. *Partial level scheme of ^{111}Sb*

We complete here the data reported in ref. [4] on ^{111}Sb. We add new states in the level scheme and complete the spin and parity assignments (fig.1). New experiments ^{95}Mo+^{19}F with a beam energy E_{max}=72 MeV were done in order to improve the statistics of the γ-γ coincidences and to measure neutron-gated angular distributions. The neutrons were detected

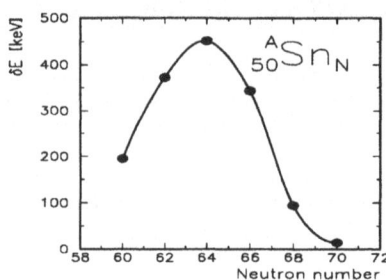

Figure 2. *Differences between experimental and estimated positions of the 0_2^+ levels in tin nuclei*

with a NE213 (Φ 110*100 mm) liquid scintillator. The pulse-shape discrimination technique was used in order to obtain the signal that triggered the acquired gamma spectra.

The main result of this work is the identification of a low excitation deformed band based on the $9/2^+$ state at E^*=1397.7 keV in ^{111}Sb, which coexists with the spherical single particle excitations from the main cascade. The arguments we use in the identification of the deformed band are the following: the band contains ΔJ=1 transitions, with constant sign of the mixing ratio $\delta(E2/M1) > 0$, suggesting a prolate deformation; the energy spacing of the levels in the band; the absence of gammas linking the deformed band and the spherical band; the systematics of the $\pi 1g_{9/2}$ band head in heavier odd mass Sb isotopes ($113 \leq A \leq 123$).

In order to give a stronger support for the position of the deformed band-head in ^{111}Sb, in fig. 2 we analyzed the discrepancies between the experimental and estimated positions of the first 0_2^+ level in the even mass tin nuclei. For the estimates we used $E_{est}(0_2^+)_{Sn} = E(9/2^+)_{Sb} + E(5/2^+)_{In}$ for fixed even neutron number. This comes out from the hypothesis that the first excited 0_2^+ state in even tin nuclei is mainly a two-proton excitation across the major shell closure Z=50 [5]. The 2p-2h excitation involves 5 steps: (i) proton pair breaking, (ii) first particle excitation, (iii) second particle excitation, (iv) pairing of excited particles and (v) pairing of the holes. One observes that the deformed $9/2^+$ band-head in odd-mass Sb nuclei involves steps (i),(ii) and (iv) whereas the deformed $5/2^+$ band head in odd-mass In nuclei involves steps (iii) and (v). One must notice from this figure as well the largest experiment-estimate discrepancy at neutron number N=64, suggesting that this neutron number is quasi-magic when the proton number is Z=50 (mutual support of magicity).

In conclusion, we have identified a low-lying quasi-rotational band in neutron deficient $^{111}_{51}$Sb$_{60}$. It is based on the $9/2^+$ state at E^*=1398 keV and we give arguments that this is a 2p-1h state analogue to those observed in heavier odd Sb isotopes, arising from the promotion of a $1g_{9/2}$ proton across the shell gap. Its interaction with $1g_{9/2}$ neutrons strongly polarizes the core.

References

[1] K. Heyde, P. Van Isacker et al., *Phys. Reports* **102**, 291 (1983)
[2] L. M. Yang, T. Song and X. H. Wang, *Phys. Letters B* **175**, 6 (1986)
[3] R. E. Shroy, A. K. Gaigalas, G. Schatz and D. B. Fossan, *Phys. Rev. C* **19**, 1324 (1979)
[4] V. E. Iacob, C. Stan-Sion et al., *Z. Phys. A* **329**, 381 (1988)
[5] K. Heyde, J. Jolie, J. Moreau et al., *Nucl. Phys. A* **466**, 189 (1987)

$\mathcal{J}^{(2)}$ ANOMALIES IN THE YRAST SUPERDEFORMED BAND OF ^{149}Gd

G. Duchêne and the EUROGAM collaboration

CRN CNRS-IN2P3, Université Louis Pasteur, BP 20 Cro, F-67037 Strasbourg Cedex 2, France

The yrast superdeformed (SD) band of ^{149}Gd has been studied using the Eurogam[1,2] multidetector array which consisted of 44 large-volume HP Ge detectors. The ^{149}Gd residual nucleus was populated via the ^{124}Sn(^{30}Si,5n) reaction at a bombarding energy of 158 MeV. The target consisted of a stack of two thin Sn foils (500 μg/cm^2). At least six unsuppressed Ge detectors were required in coincidence to trigger the electronics which leads to about 3×10^9 four-fold coincidences after unpacking the higher-fold events.

The spectrum obtained by triple gating on all non-contaminated SD transitions shows a large peak-to-background ratio. (see figure 1 in reference 3). Therefore the γ-ray energies of this SD band are established with a large accuracy, within $\pm$ 0.1 keV. Regular rotational bands correspond in γ-ray spectra to regularly spaced γ-lines. The energy spacing ΔE_γ between two consecutive transitions of the collective structure is inversely proportional to the dynamical moment of inertia $\mathcal{J}^{(2)}$ ($\mathcal{J}^{(2)} = 4\hbar^2/\Delta E_\gamma$) of the deformed nucleus. Figure 1 shows the $\mathcal{J}^{(2)}$ of ^{149}Gd yrast SD band for rotational frequencies $\hbar\omega$ ($\hbar\omega$ = E$_\gamma$/2) larger than 0.5 MeV. Small oscillations are observed with amplitudes which are much larger than the experimental error bars. The regularity of these oscillations over the 0.5-0.75 MeV frequency range is more in favour of a perturbation of the SD energy levels than of accidental $\mathcal{J}^{(2)}$ irregularities.

We compare the experimental $\Delta E_\gamma(I)$ energy spacing value at a given spin I to a reference value $\Delta E_\gamma(I)$ corresponding to a perfect (unperturbed) rotational band. $\Delta E_\gamma^{ref}(I)$ is calculated as follows :

$$\Delta E_\gamma^{ref}(I) = \frac{1}{4}[\Delta E_\gamma(I+2) + 2\Delta E_\gamma(I) + \Delta E_\gamma(I-2)]. \tag{1}$$

From the difference $\delta = \Delta E_\gamma(I) - \Delta E_\gamma^{ref}(I)$ emerge two sets of experimental points corresponding on average to $\delta = \pm 230$ eV. Assuming that the perturbation affects the excitation energy of each state of the band by the same quantity e, the $\mathcal{J}^{(2)}$ oscillations can be reproduced by shifting the states of spin I, I + 4, I + 8, ... and I + 2, I + 6, I + 10, ... by respectively plus and minus e. In our case e equals 58 $\pm$ 11 eV. The presence of the two regular $\Delta I = 4$ families in a rotational band suggests an explanation based on a fourfold rotational symmetry which corresponds to the invariance of the intrinsic Hamiltonian with respect to a rotation of $\pi/2$ around the rotation x axis. Due to the β_4-softness of ^{149}Gd SD nucleus[4], a weak hexadecapole vibration coupled to the rotation of the superdeformed nucleus may explain the experimental data.

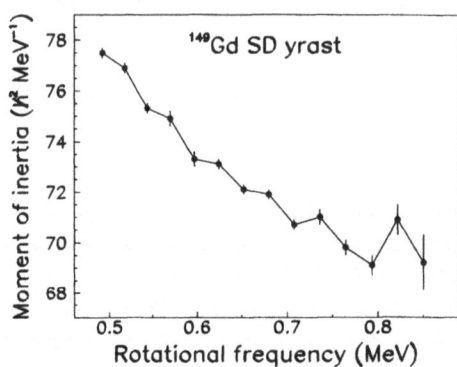

Figure 1. *Dynamical moment of inertia of the yrast superdeformed band in ^{149}Gd for rotational frequencies higher than 0.5 MeV. The solid line joining the points is intended to guide the eye.*

References

[1] F. A. Beck, *EUROBALL: Large gamma-ray spectrometers through European collaborations*, Prog. Part. Nucl. Phys. **28**, 443 (1992)

[2] C. W. Beausang et al., *Measurements on prototype Ge and BGO detectors for the Eurogam array*, Nucl. Instr. and Meth. A **313**, 37 (1992)

[3] S. Flibotte, H. R. Andrews, G. C. Ball, C. W. Beausang, F. A. Beck, G. Bélier, T. Byrski, D. Curien, P. J. Dagnall, G. de France, D. Disdier, G. Duchêne, C. Finck, B. Haas, G. Hackman, D. S. Haslip, V. P. Janzen, B. Kharraja, J. C. Lisle, J. C. Merdinger, S. M. Mullins, W. Nazarewicz, D. C. Radford, V. Rauch, H. Savajols, J. Styczen, C. Theisen, P. J. Twin, J. P. Vivien, J. C. Waddington, D. Ward, K. Zuber and S. Åberg, $\Delta I=4$ *bifurcation in a superdeformed band: evidence for a C_4 symmetry*, to be published

[4] W. Nazarewicz, R. Wyss and A. Johnson, *Structure of superdeformed bands in the A ≈ 150 mass region*, Nucl. Phys. A **503**, 285 (1989)

LIFETIME MEASUREMENTS WITH THE GAMMA RAY INDUCED DOPPLER (GRID) BROADENING METHOD

A. Jungclaus[1,2], H. G. Börner[1], J. Jolie[1,3], K. P. Lieb[2], and S. Ulbig[1]

[1] *Institut Laue-Langevin, F-38042 Grenoble*
[2] *II. Physikalisches Institut, Universität Göttingen, D-37073 Göttingen*
[3] *Institut de Physique, Université de Fribourg, CH-1700 Fribourg (present address)*

In recent years the gamma ray induced Doppler (GRID) broadening technique has been established and shown to be a suitable method to determine nuclear lifetimes in the range below several picoseconds in nuclei populated via thermal neutron capture[1]. It is based on the simple fact that the energy of a γ ray emitted from a nucleus which is moving with a certain velocity is shifted relative to a γ ray emitted from a nucleus at rest. In contrast to the traditional Doppler techniques (e.g., DSA and RDDS), the recoil in GRID is not induced by the reaction process but by the electromagnetic decay of the nucleus of interest produced by thermal neutron capture. The recoil velocity and hence the Doppler shift of a subsequently emitted secondary γ ray is therefore smaller by some orders of magnitude and only the unique resolving power of the ultra high resolution crystal spectrometer GAMS4, installed at the Institut Laue-Langevin (Grenoble), allows the detection of the small line broadenings. The Doppler profile is essentially determined by three parameters: the lifetime of the state under investigation, the recoil velocity distribution of the recoiling atoms, i. e. the history of decay processes which contribute to the recoil at the moment of the γ emission, and the atomic collisions which govern the slowing-down process of the recoiling atoms in the target. Whenever two of these three quantities are known, the third can be determined.

The situation is most favorable in light nuclei, where the levels under investigation are very often strongly populated directly from the capture state and often even details of the remaining feeding branches are well known. Therefore a series of GRID measurements in the light nuclei ^{36}Cl, ^{49}Ti, ^{54}Cr, ^{57}Fe and ^{59,61}Ni has been performed[2] to study the goodness and consistency of different slowing-down descriptions. A systematic comparison between the values obtained from the analysis of the Doppler broadened line profiles, using molecular dynamics simulations to describe the stopping of the recoiling atoms in the bulk of the target material, and from literature lifetime values from DSA measurements proved an excellent agreement[3].

Once the method was established by means of the results in light nuclei it has been used to attack some open questions in nuclear structure in heavy nuclei. The main problem one has to deal with in these nuclei is the in most cases incomplete knowledge of the feeding into the state of interest. However, in a very conservative approach, model-independent lifetime limits can be deduced from the experimental line-shape considering extreme assumptions for the unknown part of the population. The GRID studies in ^{168}Er (ref.4) and ^{196}Pt (ref.5) have shown that in some cases the restriction to such a lifetime window is already sufficient

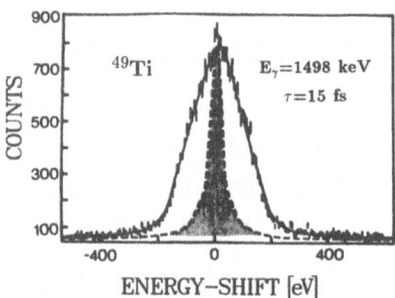

Figure 1. *Doppler broadened line profile measured with the double flat crystal spectrometer GAMS4. The 1498 keV transition depopulates an excited state in ^{49}Ti with a lifetime of 15 fs. The dashed line refers to the instrumental response function.*

to clarify important nuclear structure questions: The measurement of the absolute transition rate for the decay of the double-γ $K^{\pi} = 4^+$ vibration in ^{168}Er using the GRID method gave for the first time definite evidence for the existence of two-phonon collective states in deformed nuclei. In the O(6) nucleus ^{196}Pt the deduced absolute E2 transition rate for the decay from the 0_3^+ (σ = N-2) to the 2_1^+ state allowed a quantitative test of the goodness of the σ quantum number, which labels the fundamental irreducible representation of the O(6) group, and thereby provided definite and unassailable signature of O(6) character extending up to very high excitation energies. An alternative way to tackle the problem of incomplete knowledge of the population pattern of the state under investigation is the simulation of the decay scheme with the statistical model of the nucleus. This approach has been used in the analysis of GRID line-shapes in ^{148,150,152}Sm (ref.6) and ^{156}Gd (ref.7) and the deduced lifetimes have been compared[8] to results obtained in nuclear resonance fluorescence and DSAM experiments.

References

[1] H. G. Börner and J. Jolie, *J. Phys. G* **19**, 217 (1993) and references therein
[2] H. G. Börner et al., *Phys. Lett. B* **215**, 45 (1988); K. P. Lieb et al., *Phys. Lett. B* **215**, 50 (1988); S. Ulbig et al., *Nucl. Phys. A* **505**, 193 (1989); S. Ulbig et al., *Z. Phys. A* **338**, 397 (1991)
[3] A. Kuronen et al., *Nucl. Phys. A* **549**, 59 (1992)
[4] H. G. Börner et al., *Phys. Rev. Lett.* **60**, 691 (1991)
[5] H. G. Börner et al., *Phys. Rev. C* **42**, R2271 (1990)
[6] A. Jungclaus et al., *Phys. Rev. C* **47**, 1020 (1993)
[7] J. Klora et al., *Nucl. Phys. A* **561**, 1 (1993)
[8] A. Jungclaus et al., *Phys. Rev. C* **48**, 1005 (1993)

ON LINE STUDY OF NEUTRON DEFICIENT HAFNIUM ISOTOPES USING FAST RADIOCHEMICAL SEPARATIONS

D. Trubert[1], M. Hussonnois[1], J. F. Le Du[1], L. Brillard[1], V. Barci[2], G. Ardisson[2], Z. Szeglowski[3], O. Constantinescu[3], Yu.Ts. Oganessian[3]

[1] *Institute of Nuclear Physics, BP 1, 91406 Orsay cédex, FRANCE*
[2] *Laboratory of Radiochemistry, University of Nice, 06034 Nice cédex, FRANCE*
[3] *Joint Institute for Nuclear Research (JINR), 141980 Dubna, RUSSIA*

Introduction

We have developed and implanted at the accelerator Tandem MP of Orsay, a device for continuous and "on line" studies of short-life isotopes, named RACHEL (Rapid Aqueous Chemistry apparatus for Heavy ELements). This setup was originally concepted at the Laboratory of Nuclear Reactions in Dubna and was previously used for some chemical studies of element 104[1,2].

The RACHEL apparatus could be separated in two main distinct parts:

- the production of a selected isotope through nuclear reactions and the transportation of reaction products with a helium jet system to a chemistry laboratory,
- the isolation and purification of the selected element from all other reaction products using fast radiochemical techniques mostly based on the use of ion exchangers.

Experimental Setup and Results

Reaction products recoiling from a thin target during irradiation with heavy ions are adsorbed on KCl aerosols in helium gas, which fill continuously the irradiation chamber. Reaction products could be so transported far from the irradiation device in a very short time. A very important care was given to the aerosol-gas system in order to improve a high transportation yield, and more over to obtain a constant yield during the whole time of irradiation. In a chemistry room, aerosols together with reaction products are solved continuously with an appropriated solvent (which depends of the studied element), in a dissolver-degazer system which allowed a complete dissolution of reaction products and a separation of the solution from the gas phase in a very short time. Then, fast radiochemical separations could be done in order to isolate the selected element. Those separations are mostly based on the principle of ion exchanger resins or liquid-liquid chromatography methods. Concerning the hafnium isotopes we produced, the cross sections are "very large" (few hundred millibarns) and so the detection of these isotopes could be performed directly using gamma and X spectroscopies. In the case of transactinide elements, cross sections are very low (nanobarn or less), and these isotopes are alpha emitters. The originality of this technique for heavy elements relies on the fact that the measurements are not performed on the element itself, but on the daughter nuclides. Indeed, elements 104 or 105 is sorbed continuously on an anionic exchanger, daughter elements (actinides) are not retained and then are fixed on a cationic exchanger following the anionic one. The daughter elements are desorbed "off-line", a thin source is then prepared using the electrospray method, and measured with barrier surface detectors.

A general scheme of the RACHEL facility, modified for the study of hafnium isotopes, is given in figure 1.

In order to improve our setup, we chose to produce Hf isotopes, which are the more close chemical homologue of element 104[1,5]. During our first tests, we saw that the gamma spectroscopy of some of the neutron deficient hafnium isotopes is mostly unknown. For this purpose, we have produced, using ^{16}O ion beam (energy ranging from 70 to 100 MeV and intensities around 700 nA), and with monoisotopic targets of gadolinium (masses 154, 155, and 156), neutron deficient hafnium isotopes with masses from 169 to 164. In this cases, the solvent of reaction products was hydrofluoric acid (concentration ranging from 0.2 to 1.5 M), and the radiochemical separations were based on the difference of the chemical behaviour of hafnium and lanthanides on ion exchange resins. As a matter of fact, in hydrofluoric media, Hf ions are under negatively charged complexes, sorbed on anionic exchangers with a very large retention coefficient, while lanthanides, which composed the most important part of transfer reaction products, remain under cationic forms[1,2].

With the help of continuous purification of hafnium isotopes, and in selecting targets and energies, we were able to separate and purify each isotope ranging from mass 169 to 164. Then, we could registrate practically pure gamma and X spectra of all these isotopes. Therefore, direct and coincidence measurements were performed on each of these isotopes and are until now, still in progress. As an example, for ^{168}Hf, only 20 transitions were known (2 through gamma measurements and 18 with the help of electron conversion spectroscopy[3,4]). In our work, more than 120 new gamma lines ranging from 6 keV up to 1 MeV, with intensities higher than 1%, were identified and registred. The scheme level of ^{168}Lu fed through the β^+-decay of ^{168}Hf is completely unknown. The ground state level was assumed to be a 6^- or a 5^- level (with 5.5 minutes half-life), while an excited state of $J^\pi = 3^+$ was predicted at 220 ± 140 keV (with 6.7 minutes half-life)[3]. The half-life was measured using five of the major gamma rays, and with these data, we derived a half-life of 26.0 ± 0.5 minutes, which is in perfect agreement with the previous values (25.95 minutes)[3,4].

The isotope ^{168}Lu, daughter of ^{168}Hf through β^+-decay, was also studied. This nucleus was already investigated, but we found, according to the high level of purification of our sources, again more than 50 new gamma lines. For all the other hafnium isotopes studied, coincidence data are also still in progress, and show clearly the presence of many new gamma rays.

Conclusion

These first experiments show without ambiguity the performances and the feasability of our setup. This technique has been initially developed for the study of chemical and physical properties of transactinide elements 104, 105, and 106. Experiments done with Hf, considered as the homologue of element 104, has proved that this technique could be also applied to the study of chemical as well as nuclear properties of any short-life radionuclides ($T > 10$ seconds) and is particularly suitable for transactinide elements. Therefore, in the next future such studies will be performed with ^{261}104, ^{262}105, and ^{263}105 isotopes.

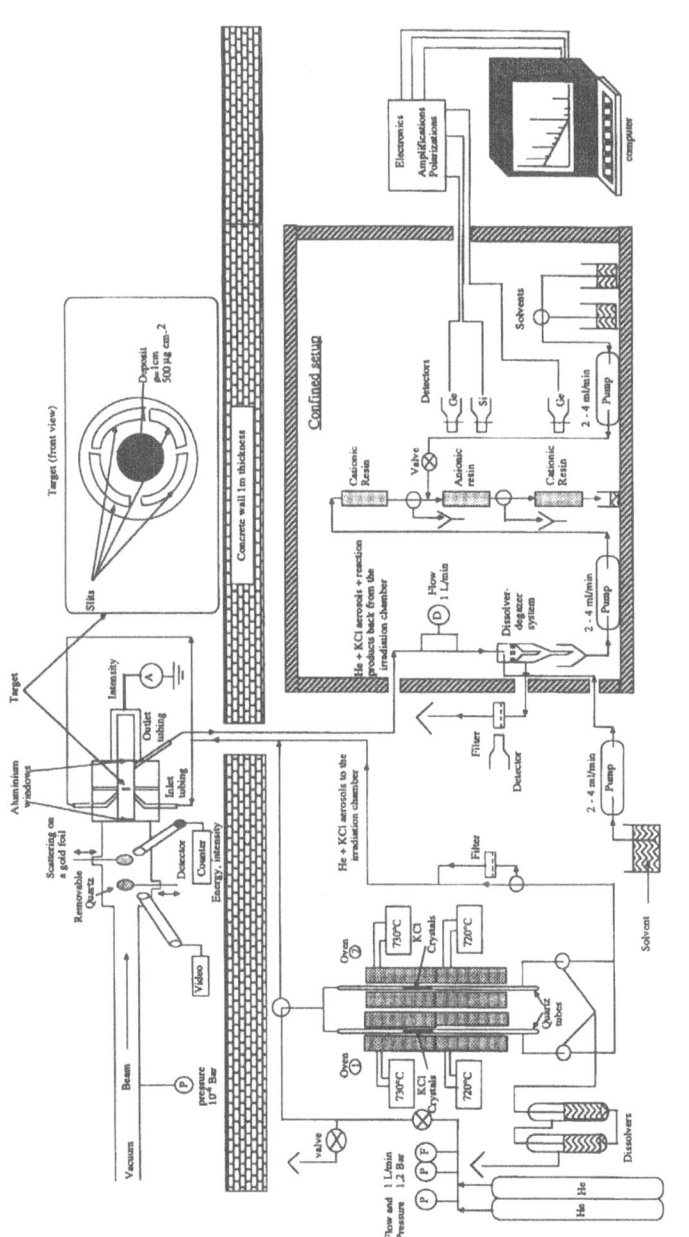

Figure 1. *General scheme of the RACHEL apparatus.*

References

[1] Z. Sglelowski, H. Bruchertseifer, V. P. Domanov, B. Gleisberg, L. J. Guseva, M. Hussonnois, G. S. Tikhomirova, I. Zvara, Y. T. Oganessian, *Radiochimica Acta* **51**, 71-76 (1990)
[2] H. Bruchertseifer, B. Eichler, J. Estevez, I. Zvara, *Radiochim. Acta* **47**, 41-46 (1989)
[3] B. Harmatz and T. H. Handley, *Nucl. Phys.* **81**, 481-522 (1966)
[4] J. K. Tuli (editor), *Nuclear Data Sheets, "mass 168"* **53**, 223 (1988)
[5] D. C. Hoffman, *Chemistry of the Transactinide Elements*, October 1990, Report LBL - 29815, University of California, CA 94720, Berkeley, USA

TOROIDAL MULTIPOLE TRANSITIONS IN COLLECTIVE EXCITATIONS OF ATOMIC NUCLEI

Ş. Mişicu

Center for Earth Physics, Institute of Atomic Physics, PO Box MG-6, RO-76900 Bucharest, Romania

Introduction

Recently, in the frame of nuclear fluid dynamics, a new isoscalar $\lambda^\pi = 1^-$ state, called the dipole torus mode was proposed[1]. Calculations of the electromagnetic form factors entering into the photoabsorbtion or electron scattering cross sections show that only the transverse electric form factor contributes to the excitation of the predicted mode. Following the ideas outlined by Dubovik et al.[2], the matrix elements of the transverse electric form factor $t_{E\lambda}(q)$ can be separated in two parts: one depending on the low momentum transfer q limit of the longitudinal (Coulomb) multipoles $t_{C\lambda}(0)$ and the other on the so-called toroidal multipoles $t_{T\lambda}(q)$:

$$t_{E\lambda}(q) = -\frac{\omega}{q}\sqrt{\frac{\lambda+1}{\lambda}}t_{C\lambda}(0) + q^2 t_{T\lambda}(q). \tag{1}$$

The $t_{C\lambda}$ and $t_{E\lambda}$ multipoles were worked out for various nuclear models, and can be found in the literature[3,4]. I stress the fact that the toroidal multipoles are important at high-momentum transfer processes where the validity of the Siegert theorem breaks down[5]. In the case of the dipole torus mode, the dipole longitudinal $t_{C1}(0)$ vanishes and thus, according to eq.(1) the electromagnetic nature of the excitation is purely toroidal.

Toroidal Quadrupole Transitions in the Collective Vibrational Model

The toroidal multipoles turn out to be important also in the case of exciting the ground-state rotational band of an even-even nucleus ($0^+, 2^+, 4^+, ...$) by real-photon electron scattering reactions. For such excitations the t_{C2} were intensively investigated in the past. Information on t_{E2} may be obtained from the interference term ($t_{E2}t_{C2}$) by measuring the angular distribution of decay photons in coincidence with the scattered electron[4,6].

In the framework of an oscillating incompressible liquid-drop model, for the transition to the one-surfon 2^+ state, the transverse electric quadrupole t_{E2} is related to t_{C2} through the relation

$$t_{E2}(q) = -\frac{\omega_2}{q}\sqrt{\frac{3}{2}}t_{C2}(q) \tag{2}$$

with $\omega_2 = [C_2/B_2]^{1/2}$, the energy of the $\lambda^\pi = 2^+$ one-surfon state. Making use of the decomposition (1) for the t_{E2} one gets

$$t_{E2}(q) = -\frac{\omega_2}{q}\sqrt{\frac{3}{2}}t_{C2}(0)[1 + \eta_2(q)] \tag{3}$$

where $\eta_2(q) = q^2 t_{T2}(q)/t_{C2}(0)$. This form of t_{E2} is in particular useful for obtaining some information on the amount of toroidal quadrupole in t_{E2} at various q values. In the low-q limit the ratio η_2 vanishes and the Siegert theorem is recovered, whereas in the high-q region ($q \geq 400 MeV/c$), $\eta_2 \gg 1$ and the toroidal quadrupole transition dominates the excitation of the one-surfon state. Therefore, $\eta_2(q)$ can be also viewed as a quantity which measures how large are the deviations from the usual low-q limit form, governed by the Siegert theorem. It can be related to the ratio between t_{E2} and t_{C2}, at arbitrary q, introduced in ref.[6]

$$\eta_2(q) = -\left[1 + \sqrt{\frac{2}{3}}\frac{q}{\omega_2}\xi_2(q)\frac{t_{C2}(q)}{t_{C2}(0)}\right]. \qquad (4)$$

Numerical calculations performed for ^{16}O and ^{90}Zr show that the toroidal quadrupole is shifted toward smaller q with increasing mass number. The introduction of the magnetization density by means of a crude model[3] does not change the main conclusion on the relative importance of toroidal effects at high momentum transfer.

References

[1] S. I. Bastrukov, Ş. Mişicu and A. V. Sushkov, *Dipole torus mode in nuclear fluid-dynamics*, JINR preprint E4-92-521, Dubna (1992)
[2] V. M. Dubovik and V. V. Tugushev, *Toroid moments in electrodynamics and solid-state physics*, Phys. Rep. **187**, 147 (1990)
[3] J. de Forest Jr. and J. D. Walecka, *Electron scattering and nuclear structure*, Adv.in Phys. **15**, 1 (1966)
[4] E. Moya De Guerra, *Rotational nuclear models and electron scattering*, Phys. Rep. **138**, 293 (1986)
[5] J. L. Friar and S. Fallieros, *Extended Siegert theorem*, Phys. Rev. C **29**, 1645 (1984)
[6] C. Garcia-Recio, T. W. Donnelly and E. Moya De Guerra, *Studies of electron coincidence reactions and electromagnetic currents in rotational nuclei*, Nucl.Phys. A **509**, 221 (1990)

HOW IMPORTANT IS THE PROPER TREATMENT OF TRANSLATIONAL INVARIANCE IN THE ANALYSIS OF ELECTRON SCATTERING FROM NUCLEI ?[1]

K. W. Schmid

Institut für Theoretische Physik, Universität Tübingen, FRG

Introduction

We shall consider the nucleus as a non-relativistic, finite quantum many-body system. Obviously its physics should be the same all over the world and even in a space-lab orbiting around it. This is what we call *translational* and *Galilean* invariance. It has the consequence that the corresponding Hamiltonian, however complicated it may be, cannot depend on the center of mass coordinate of the constituents and only in a trivial way on the total linear momentum:

$$\hat{H} = \hat{H}_{int}(\vec{r}_i - \vec{R}, \hat{p}_i - \frac{1}{A}\hat{P}, \tau_i, \sigma_i) + \frac{\hat{P}^2}{2MA} \quad (1)$$

Here i runs from 1 to the number of nucleons A, M is the nucleon mass, $\hat{P} \equiv \sum_{i=1}^{A} \hat{p}_i$ the operator of the total linear momentum, $\vec{R} \equiv \sum_{i=1}^{A} \vec{r}_i / A$ the center of mass coordinate, and τ_i, σ_i denote the isospin and spin quantum numbers of the individual constituents. The eigenfunctions of (1)

$$|\Psi\rangle = |\Psi_{int}\rangle|\vec{P}\rangle \quad (2)$$

can thus be factorized into an internal part and a trivial plane wave describing the motion of the system as a whole. Obviously we are only interested in the former and hence we have to solve the Schrödinger equation

$$\hat{H}_{int}|\Psi_{int}\rangle = E_{int}|\Psi_{int}\rangle \quad (3)$$

for the bound (or the corresponding coupled channel equations for the scattering) states of the system. This is precisely what is done for few nucleon problems: one writes the Hamiltonian in Jacobi coordinates and solves the corresponding Schrödinger (or Fadeev) equations. However, nucleons are Fermions and thus the internal wave function has to be antisymmetric. Since the Jacobi coordinates are functions of all the particle coordinates, the antisymmetrisation has to be done explicitly. Because the number of terms needed is equal to the factorial of the number of identical particles involved, in many-body physics such an explicit antisymmetrisation is impossible. Therefore here one usually sacrifices translational and Galilean invariance for the sake of Pauli principle and expands the total wave function in

---

[1] This work was supported by the Graduiertenkolleg Tübingen (DFG, Mu 705/3).

terms of (generalized) Slater-determinants

$$|\Psi\rangle = \sum_D |D\rangle\langle D|\Psi\rangle \tag{4}$$

being constructed from the eigenstates of some basis creating potential. As has been demonstrated long ago by Elliot and Skyrme [1], at least for a pure harmonic oscillator basis, internal and center of momentum motion in the expansion (4) can still be separated with group theoretical or diagonalisation techniques provided "complete $n - \hbar\omega$" spaces are taken into account and only discrete states are considered. For more general wave functions (and especially in case that a continuum nucleon is involved) these methods fail and we have to get rid of the so called "spurious" components in (4) due the motion of the system as a whole in another way.

This is done by *projecting* into the center of momentum (COM) rest frame via the operator [2]

$$\hat{C}(0) \equiv \hat{C}(\vec{P} = 0) = \int d^3\vec{a} \exp\{i\vec{a} \cdot \hat{P}\}. \tag{5}$$

Provided that (4) describes a localized state of finite extension the application of this operator yields the desired internal component and, as we shall see below, with some additional care about the recoil effects this concept can also be used if continuum nucleons are involved. Note, however, that determining (4) in some standard way (e.g., in the Hartree-Fock or some shell-model configuration-mixing approach) and then applying (5) is *not* sufficient. The projection has to be done *before* determining (4) via variational or diagonalisation techniques. Only then ambiguities in the weight function can be avoided and full Galilean invariance be achieved [3].

Obviously up to here the discussion has been purely academic. Does the violation of translational and Galilean invariance really matter in nuclear physics ? Isn't it just a $1/A$ effect which can be safely neglected as soon as nuclei above about ^{16}O are considered ? If we only look for energies the last argument is obviously true. Indeed, even in the light ^{4}He Hartree-Fock with COM projection before the variation [4] yields only about half an MeV energy gain with respect to standard Hartree-Fock provided the COM term $\hat{P}^2/2MA$ is removed from the Hamiltonian (1). However, things change if we look into the wave functions themselves as it is done for example by scattering experiments. As we shall see, here the proper treatment of the COM effects does make quite some difference. I shall demonstrate this with the help of two simple examples.

First Example: Elastic Scattering

In the first example we consider elastic electron scattering from a doubly even target [5,6]. For simplicity let us start with a system of Z unit point charges at positions $\vec{r}_i$. In momentum space the charge density operator for such a system has the simple form

$$\hat{\rho}(\vec{q}) = \sum_{i=1}^{Z} \exp\{i\vec{q} \cdot \vec{r}_i\} \tag{6}$$

where $\vec{q}$ denotes the 3-momentum transfer. If we don't worry about any COM effects, the charge form factor for elastic scattering from the target ground state $|\Psi_{gs}\rangle$ nucleus can then be written as

$$F_{normal}(\vec{q}) \equiv \langle \Psi_{gs}|\hat{\rho}(\vec{q})|\Psi_{gs}\rangle. \tag{7}$$

This expression is, however, not correct. First of all, the operator (6) is not translationally invariant. Instead of (6)

$$\hat{\rho}_{inv}(\vec{q}) \equiv \hat{\rho}(\vec{q})\hat{M}(\vec{q}) = \hat{\rho}(\vec{q})\exp\{-i\vec{q}\cdot\vec{R}\} \quad (8)$$

should be taken in the calculations. Furthermore, as discussed in the introduction, in general the ground state wave function will be of the form (4) and hence has to be projected into the COM rest frame via (5). Thus instead of (7) we obtain for the translational invariant form factor

$$F_{proj}(\vec{q}) \equiv \frac{\langle\Psi_{gs}|\hat{\rho}(\vec{q})\hat{M}(\vec{q})\hat{C}(\vec{P}=0)|\Psi_{gs}\rangle}{\langle\Psi_{gs}|\hat{C}(\vec{P}=0)|\Psi_{gs}\rangle}. \quad (9)$$

Obviously (9) is hard to calculate and one may be tempted to try some approximations. As has been demonstrated in Ref.[6], by assuming Gaussian shape for the many-body matrix elements of the shift operator in (5) we arrive at an "intrinsic" form factor

$$F_{intr}(\vec{q}) \equiv \langle\Psi_{gs}|\hat{\rho}(\vec{q})\hat{M}(\vec{q})|\Psi_{gs}\rangle. \quad (10)$$

An additional Gaussian overlap approximation for the Gartenhaus-Schwartz operator $\hat{M}(\vec{q})$ yields then

$$F_{osc}(\vec{q}) \equiv \exp\left\{\frac{3}{8}\frac{\vec{q}^2}{\langle\Psi_{gs}|\hat{P}^2|\Psi_{gs}\rangle}\right\} F_{normal}(\vec{q}) \quad (11)$$

which we call the "oscillator" form factor since for a "non-spurious" oscillator configuration $|\Psi_{gs}\rangle$ the folding exponential in front of the "normal" form factor in eq.(11) reduces to the famous Tassie-Barker [7] factor $\exp\{(qb)^2/(4A)\}$ (b being the oscillator length) which is the usually applied "COM-correction" in the analysis of elastic electron scattering from nuclei.

How these different approaches compare with each other is demonstrated in Fig. 1 for the case of elastic electron scattering from the nucleus ^{16}O, the ground state of which has been obtained here by a Skyrme Hartree-Fock calculation. Obviously here the finite extension of the nucleons has been taken into account by using the appropriate free nucleon electromagnetic form factors. As can be seen the oscillator form factor (11) does reproduce the projected result rather well up to about 2 fm^{-1}. For higher momentum transfers, however, drastic differences do occur. Here, and this is exactly the region where we hope to learn something about correlations in nuclei, the full projection (9) has to be done to obtain reliable results. The same is by the way true also for the heavier nuclei ^{40}Ca and ^{48}Ca, which have been studied in Ref.[6]. These results indicate that at least at high momentum transfers all nuclei become "light" in the $1/A$ sense and hence the proper treatment of COM plays here an important role. Thus in the next example we shall study a situation were we have high momentum tranfer right from the beginning.

Second Example: Inclusive Scattering

"Inclusive" scattering means here that only the scattering angle and energy loss of the outgoing electron is measured, i.e. one sums over all possible final states of the excited A nucleon system. The cross section for such a process can be written as

$$\frac{d\sigma}{d\Omega} = \sigma_{Mott}\{v_L R_L + v_T R_T\} \quad (12)$$

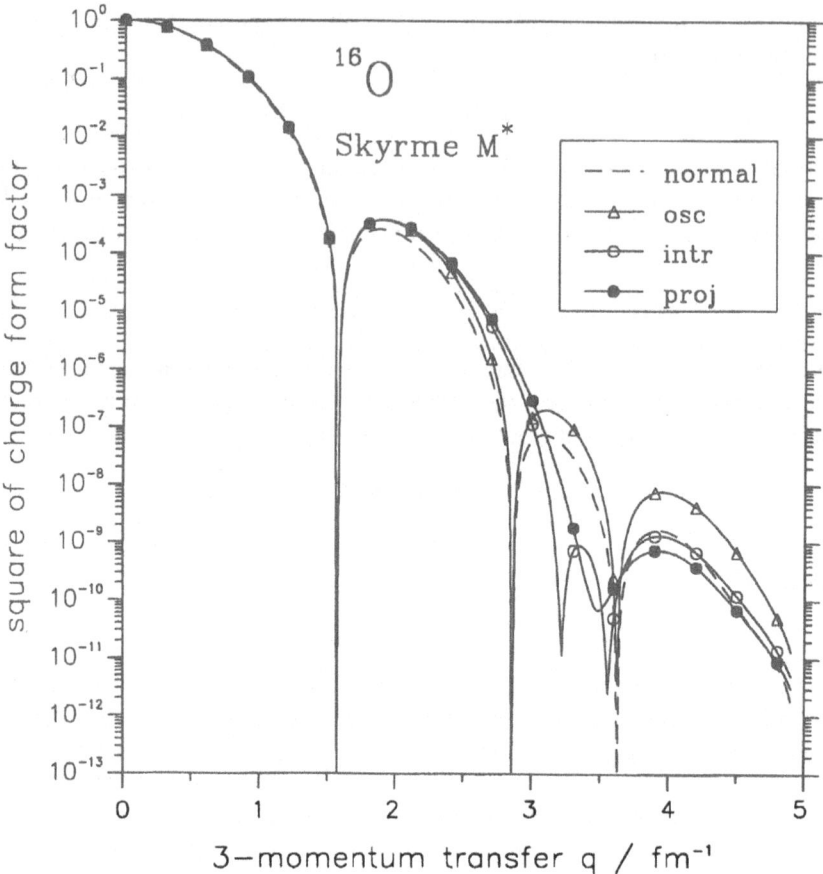

Figure 1. The square of the charge formfactor for elastic electron scattering from a Skyrme-Hartree-Fock ground state of ^{16}O is presented. Compared are the translational invariant "projected" formfactor (9), the "intrinsic" result (10), the usual "oscillator" correction (11) and, finally the "normal" formfactor (7) with no COM correction whatsoever. The figure has been taken from Ref.[6].

where σ_{Mott} is the Mott cross section and the v_L and v_T are kinematic factors depending only on the energy loss and 3-momentum transfer of the electron. The interesting physics is in the "longitudinal" and "transverse" response functions R_L and R_T, respectively, which can be obtained from the experimental cross section via a so called "Rosenbluth decomposition" which makes use of the different angular dependence of the kinematic factors v_L and v_T.

Calculations of these response functions for inclusive electron scattering from complex nuclei usually give fair agreement with experiment for the transverse strength but overestimate the data for the longitudinal strength considerably, no matter whether they use the Fermi gas or the plane wave impulse approximation (PWIA) or even take into account part of the final state interaction (FSI) via some optical potential (see, e.g., Ref.[8] for a recent review). This problem also occurs in ^{4}He, where the PWIA calculations of the Rome group [9] clearly fail to reproduce the Bates data [10] for the longitudinal strength, especially at low momentum transfers. On the other hand the "mathematical" Coulomb sum rules extracted from the ^{4}He data, as in other few body systems, agree rather well with those obtained via the closure approximation from the correlated ground state derived with conventional two- and three-nucleon forces [10,11], and recent Green function Monte Carlo calculations for the longitudinal response in ^{4}He by Carlson and Schiavelli [12] yield an excellent agreement with the data, too. Thus at least in light nuclei there is no indication for "exotic" contributions as they have been widely discussed in the past (see, e.g., Ref.[13] for a recent review).

In this chapter we report the results of recent calculations [14] which are "conventional" in the sense that no sub-nucleonic degrees of freedom are used for the wave functions but "unconventional" in the special care to treat the reaction mechanism correctly and, especially, to preserve translational as well as Galilean invariance. The approach fulfills the following requirements:

(i) Bound and scattering states are calculated from the same Hamiltonian. Only this makes sum rule arguments meaningful and ensures orthogonality.

(ii) Only purely *internal* states not contaminated by admixtures due to motions of the system as a whole are admitted. This implies either the use of Jacobi coordinates and consequently *explicit* antisymmetrisation (as usual in the $A = 2, 3$ systems) or *projection* into the center of momentum (COM) rest frame via the operator (5). Again we prefer the second method since it can be used in heavier systems, too. Here, furthermore, the projection is done *before* solving the Schrödinger equation in order to avoid ambiguities in the weight functions [3].

(iii) Calculating the *relative* wave functions for a nucleon in ^{4}He ensures that bound as well as scattering states display the correct asymptotic behaviour. This implies the exact treatment of the recoil as well as of the long range Coulomb interaction.

(iv) The Pauli principle is strictly respected. The individual nucleon-nucleon interactions are not replaced by a local mean field for the FSI between the outgoing nucleon and the residual nucleus. From sum rule arguments [15] it is obvious that such potentials tend to give a considerably smaller average energy loss for the electron in light nuclei than observed. Instead we use a microscopic many body Hamiltonian

$$\hat{H} \equiv \hat{T} + \hat{V}_c + \hat{V}_s \tag{13}$$

with $\hat{T}$ denoting the kinetic energy, $\hat{V}_c$ the 2-body Coulomb interaction between the protons and $\hat{V}_s$ being an effective short range 2-body nucleon-nucleon interaction. Since we project into the COM rest frame *before* applying the Hamiltonian (13), it is not necessary to substract the kinetic energy for the motion of the total system.

The requirements are met by a resonating group type ansatz for the total wave function (for simplicity we skip the tedious but straightforward angular momentum coupling):

$$|\Psi_c\rangle = \sum_{p,h} \int d^3\vec{k} \mathcal{A}\big[|\vec{k},p\rangle \hat{C}(0)|h\rangle\big] f_c(\vec{k},p,h) \tag{14}$$

where $\mathcal{A}$ is the antisymmetrizer, $|\vec{k},p\rangle$ a plane wave with momentum $\vec{k}$ in the relative coordinate of the last nucleon with respect to the center of mass of the residual system carrying the spin-isospin quantum numbers $p = \sigma, \tau$ and $\hat{C}(0)|h\rangle$ denotes the (properly orthonormalized) internal states of the residual $A-1$ nucleon systems. Using the separability of the plane wave and the fact that $\hat{C}(0)$ and the COM coordinate $\vec{R}$ of the A nucleon system are symmetric under the antisymmetrizer, we obtain

$$|\vec{k},p,h\rangle \equiv \mathcal{A}\big[|\vec{k},p\rangle \hat{C}(0)|h\rangle\big] = \hat{C}(0)\exp\big[-i\frac{A}{A-1}\vec{k}\cdot\vec{R}\big]a^\dagger(\frac{A}{A-1}\vec{k},p)|h\rangle \qquad (15)$$

where $a^\dagger(\frac{A}{A-1}\vec{k},p)$ creates a particle with quantum numbers p in a plane wave with momentum $\frac{A}{A-1}\vec{k}$. Applied to $|h\rangle$ this yields a (non orthogonal) determinantal structure which can be treated with standard methods of many body theory.

In order to obtain the expansion coefficients f_c out of eq.(14) we apply the recently developed separable approach to multi-non-orthogonal channel problems including the Coulomb interaction [16] using the fact that a Hamiltonian of the type (13) can always be splitted into a long range part (relative kinetic energy plus, if present, the Coulomb interaction in each channel) and a residual short range part which includes the channel to channel coupling. The former is treated exactly with the help of the corresponding Green's operator, the latter expanded in a discrete, finite basis of bound Coulomb Sturmian functions. Details of the calculations will be presented in a forthcoming paper [17]. The procedure may be summarized as follows:

(i) We calculate the Hamiltonian and overlap kernels within the configurations (15) and expand them in bound Coulomb Sturmian functions.

(ii) We orthonormalize according to the prescription given in Ref.[18].

(iii) We remove the long range part and solve the resulting coupled channel equations with the proper boundary conditions for the bound as well as the scattering states as prescribed in Ref. [16].

(iv) We calculate the kernels for the charge density and the current density operators in the configurations (15) and expand them in the Sturmian basis. We orthonormalize with the transformation from step (ii) and calculate the response functions. For the charge and current densities the usual operators resulting from a Foldy-Wouthuysen transformation of the free nucleon Dirac Hamiltonian coupled to the electron field up to order $1/M^2$ in the nucleon mass neglecting the spin-orbit contributions are used. Obviously the nucleon coordinates and momenta entering these expressions are internal coordinates of the type $\vec{r}_i - \vec{R}$ and $\hat{p}_i - (1/A)\hat{P}$, respectively. For the Sachs form factors of the nucleons the standard dipole parametrisation [19] is taken.

For ^{4}He two additional approximations were made:

(1) We took $(0s)^3$ oscillator determinants projected into their COM rest frames for the description of the $A=3$ systems. This introduces an oscillator parameter b into the model and limits the ground state wave function of a nucleon in ^{4}He to relative s.

(2) We parametrized the effective interaction by linear combinations of Gaussians for the central, spin-orbit and tensor components using the Eikemeier-Hackenbroich parametrisation [20] with an additional r^2 in front of the latter. Together with the oscillator assumption for the residual nuclei this allows the analytic evaluation of the non-local overlap and Hamiltonian kernels. Detailed formulas will be given in Ref.[17]. Four different forces were used. (a) The purely central Brink-Boeker force B1 [21] being composed out of two Gaussians with ranges of 0.7 and 1.4 fm in each channel. This force has only Wigner and Majorana components. (b) A modified version of B1 obtained by keeping the even triplet and singlet components and adjusting the difference of Bartlett and Heisenberg components (using the B1 ranges) to the singlet odd channel of the Eikemeier-Hackenbroich potential [20] by a least square fit. With respect to B1 the MB1 force is more attractive in the triplet odd, more repulsive in the

singlet odd channel. (c) We took the Bonn A potential [22] and calculated a G-matrix for ^{4}He via the methods presented in Ref.[23] using a fixed starting energy of -40 MeV. This yields relative matrix elements in an oscillator basis for the various channels. We took again two Gaussians with the B1 ranges for each of the various force components and adjusted the strength parameters by least square fits to these matrix elements for low l-values. This force is called GBA. (d) A modified version MGBA of the GBA was obtained by using the same tensor, spin-orbit and even central components but making again the central triplet odd components more attractive with respect to the singlet odd ones. The modification was here only half as large as in the case of the MB1.

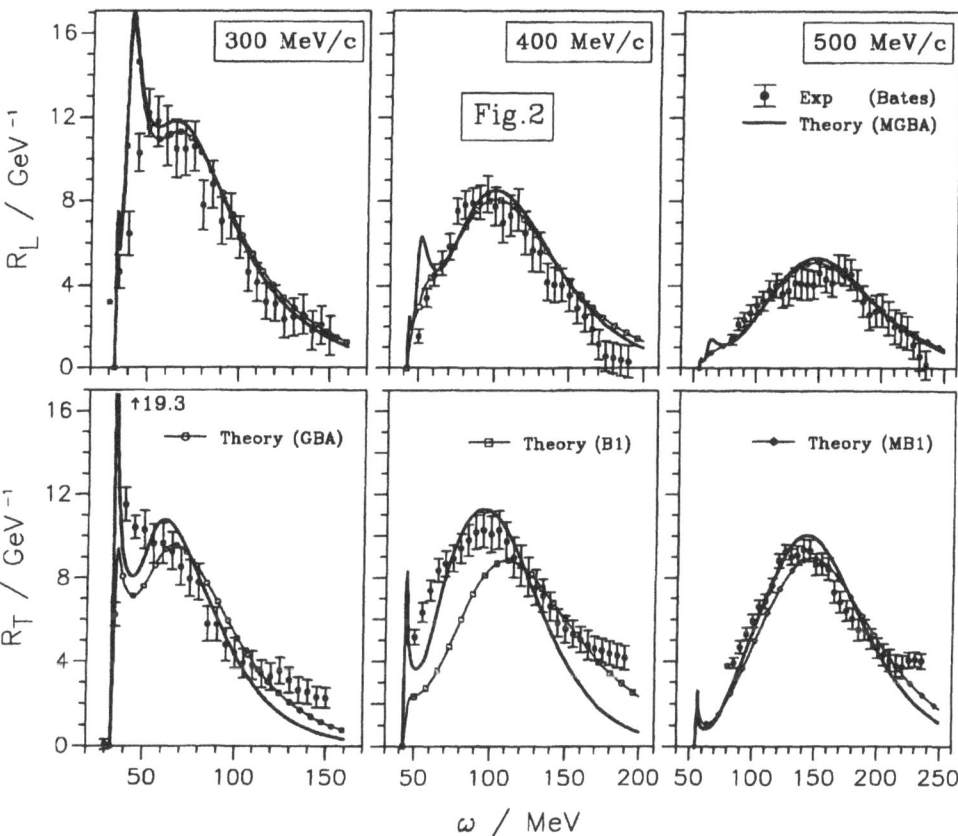

Figure 2. *The theoretical results for the longitudinal (upper half) and transverse (lower half) response functions in ^{4}He are compared to the experimental data obtained by the Bates group [10] at three different values of the 3-momentum transfer. ω denotes the energy loss of the electron in the lab-system. In each channel partial waves up to $l = 10$ have been taken into account. The full lines result from the modified G-Matrix (MGBA). For 300 MeV/c in addition the results obtained with the bare G-matrix (GBA), for 400 Mev/c those calculated with the Brink-Boeker force (B1), and for 500 MeV/c those obtained with the modified Brink-Boeker force (MB1) are presented. The figure has been taken from Ref.[14].*

In Fig. 2 the results are compared to the experimental data [10] at the three different 3-momentum transfers 300, 400 and 500 MeV/c. Apart from the high energy tails of the transverse response functions, where meson exchange currents, the onset of pion production and the influence of the Delta resonance being not included in the calculations show up in the data, the agreement of the MGBA results with experiment is striking. Strengths, positions, widths and shapes of both the longitudinal as well as the transverse response functions are well reproduced.

Using one of the three other forces has little effect on the longitudinal responses. Only the resonance structures near the threshold are somewhat modified due to the tensor and spin-orbit components which are not present in the forces B1 and MB1. For the transverse response functions, MGBA and MB1 yield again almost the same results. The peak positions, however, are shifted to higher energy losses if the force B1 and also, though considerably less, if the GBA interaction is used.

The only free parameter in the model is the oscillator length b. We have chosen $b = 1.56$ fm, where the energy of the internal $(0s)^3$ states is minimal. Using instead $b = 1.252$ fm, where the energy of the coupled channel ground state for ^{4}He is minimal, yields an energy gain of only about half an MeV for the latter, an only marginally improved elastic form factor (which by the way reproduces the data up to about 2.7 fm^{-1} rather nicely), and has on the response functions effects of about the same order as switching from the MGBA to the MB1 results in the transverse response function at 500 MeV/c. Offering the system a few more degrees of freedom for the residual nuclei would make the results completely independent of b. A detailled discussion will be given elsewhere [17].

There can be no doubt that the proper treatment of recoil and COM effects is extremely important in light nuclei like ^{4}He. It is, however, common folklore that these effects should decrease like $1/A$. Now, the missing strength is largest in ^{56}Fe. How could that be attributed to translational invariance breaking? And why just in the longitudinal but not in the transverse response functions? We propose a surprisingly simple answer: obviously the $1/A$ argument is true for the energies. The large effects which recoil and COM motion have on the position of the quasielastic peak in ^{4}He (they will be discussed in Ref.[15]) will therefore decrease rapidly with increasing A. The missing strength, however, is not a problem of the peak position, but of the wave functions. It is common knowledge, too, that spurious components in the wave functions due to the COM motion do persist with increasing A. They occur in ^{208}Pb as well as in ^{4}He and usually do carry a considerable amount of electromagnetic transition strength. Conventional scattering states do not live in their COM rest frame and thus collect this spurious strength. Even via the usual random phase approximation only the $L^\pi = 1^-$ components of them can be eliminated. All the other multipolarities remain untouched. Now these admixtures are all "non-spin-flip" components and thus contribute (apart from the spuriousity in the ground state itself) only to the longitudinal and to the rather small convection current part of the transverse response functions. The latter are dominated by the spin currents and thus the total transverse transition strength is affected only slightly.

This argument is also consistent with the observation that the "mathematical" Coulomb sum rules extracted from experiment [10] for ^{4}He agree much better with the theoretical values obtained from correlated ground state wave functions [10,11] than the "physical" ones with their explicitly calculated counterparts. The closure approximation leads to the expectation value of an operator depending only on *relative* coordinates and thus here only part of the spurious admixtures in the ground state and none of those in the excited states do contribute.

Obviously the above suggestion remains to be proven quantitatively. Thus translational and Galilean invariance preserving calculations have to be performed also in heavier nuclei like ^{16}O and ^{40}Ca. Furthermore exclusive data should be looked at, too. We are working on computer codes allowing such more general calculations.

Conclusions

Though still a lot of work remains to be done, we have to consider the above examples at least as a rather strong hint: the proper treatment of center of momentum and recoil effects seems to play indeed a very important role in the analysis of electron scattering. The $1/A$ argument, which is definitely true for the energy, is of no use here: form factors and response functions explicitly test the momentum distributions in the wave functions and thus are rather sensitive to spurious admixtures resulting from motions of the system as a whole. The latter occur in all nuclei and can only be eliminated by projection into the center of momentum rest frame, which as already mentioned, has to be done *before* the (approximate) solution of the many body Schrödinger (or Lippmann-Schwinger) equation in order to avoid ambiguities in the weight functions. Admittedly, this is a technically rather involved procedure. However, only if the "trivial" correlations between the nucleons resulting from the restoration of translational and Galilean invariance are taken into account correctly, we may, as we all hope, learn something about the "real" correlations between the nucleons in the nucleus from electron scattering experiments.

ACKNOWLEDGMENTS

I would like to thank my collaborators: the work on elastic scattering has been done together with F. Grümmer (Bochum/Jülich) and P.-G. Reinhard (Erlangen). The study of inclusive scattering was a collaboration with G. Schmidt (Tübingen) and will be part of his PhD-thesis. Here furthermore G. DoDang (Orsay), L. Egido (Madrid), H. Müther (Tübingen) and Z. Papp (Debrecen) contributed by very helpful discussions. In particular Z. Papp's coupled channel formalism was essential for these investigations and deserves a special credit.

References

[1] J. P. Elliot and T. H. R. Skyrme, *Proc. Phys. Soc. A (London)* **232**, 561 (1955)
[2] R. E. Peierls and J. Yoccoz, *Proc. Phys. Soc. A (London)* **70**, 381 (1957)
[3] J. Yoccoz, *Varenna Lectures* **36**, 474 (1966)
[4] K. W. Schmid and F. Grümmer, *Z. Phys. A* **336**, 5 (1990)
[5] K. W. Schmid and F. Grümmer, *Z. Phys. A* **337**, 267 (1990)
[6] K. W. Schmid and P.-G. Reinhard, *Nucl.Phys. A* **530**, 283 (1991)
[7] L. J. Tassie and C. F. Barker, *Phys.Rev.* **111**, 940 (1958)
[8] G. Orlandini and M. Traini, *Rep. Prog. Phys.* **54**, 257 (1991)
[9] C. Ciofi degli Atti, *Nucl.Phys. A* **463**, 127c (1987); G. Salmé and E. Pace, unpublished
[10] K. F. von Reden et al., *Phys.Rev. C* **41**, 41 (1990)
[11] R. Schiavilla et al., *Nucl.Phys. A* **473**, 267 (1987)
[12] J. Carlson and R. Schiavilla, *Phys. Rev. Lett.* **68**, 3682 (1992)
[13] P. Mulders, *Phys. Rep.* **185**, 83 (1990)

[14] K. W. Schmid and G. Schmidt, submitted for publ.
[15] K. W. Schmid, L. Egido and G. Schmidt, in preparation
[16] Z. Papp, *Phys. Rev. A* **46**, 4437 (1992); ibid **38**, 2457 (1988); *Comp. Phys. Comm.* **71**, 426 and 435 (1992)
[17] K. W. Schmid and G. Schmidt, in preparation
[18] E. W. Schmid and G. Spitz, *Z. Phys. A* **321**, 581 (1985)
[19] see, e.g., G. DoDang et al., *Phys. Rev. C* **35**, 1637 (1987)
[20] H. Eikemeier and H. H. Hackenbroich, *Nucl.Phys. A* **169**, 407 (1971)
[21] D. M. Brink and E. Boeker, *Nucl.Phys.* **91**, 1 (1967)
[22] R. Machleidt, *Adv. Nucl. Phys.* **19**, 189 (1989)
[23] H. Müther and P. U. Sauer, preprint 1992, to be published in *Computational nuclear physics. Vol II*, edited by K. Langanke, J. A. Maruhn and S. E. Koonin (Springer, Berlin)

VARIATIONAL APPROACH TO COMPLEX NUCLEAR STRUCTURE PROBLEMS[1]

K. W. Schmid[1] and A. Petrovici[2]

[1] Institut für Theoretische Physik, Universität Tübingen, FRG
[2] Institute of Atomic Physics, Bucharest, Romania

Many nuclear structure problems require the use of single particle basis systems which are far too large to allow for the complete diagonalisation of a suitably chosen effective many-nucleon Hamiltonian as it is done in the shell-model configuration-mixing (SCM) approach. Obvious examples are the spectra of medium-heavy and heavy deformed nuclei but also even such comparatively simple tasks as the description of negative parity states in light doubly even nuclei. Being interested in such problems one is therefore forced to truncate the complete SCM-expansion of the nuclear wave functions to a numerically feasible number of configurations, and the problem how this can be done without loosing the essential degrees of freedom being relevant for the particular states under consideration is one of the central issues of nuclear structure physics.

A couple of years ago we have proposed a whole hierarchy of such truncation schemes. They all work with symmetry-projected Hartree-Fock-Bogoliubov (HFB)-type quasi-particle determinants as basic building blocks, differ, however, in the degree of sophistication of the variational procedures used to determine the underlying mean-fields as well as the configuration-mixing, up to finally a level is reached on which the selection of the configurations itself is entirely left to the dynamics of the considered system and determined by a chain of variational calculations. In this way successively an "optimal" many-nucleon basis is created and arbitrary drastic changes of the structure of the nuclear excitations with angular momentum as well as excitation energy can be described. All these approaches allow furthermore to use rather large single particle basis systems as well as general two-body interactions, and thus for the first time provide the possibility of really "large-scale" nuclear structure studies going far beyond the possibilities of the SCM-approach. See Ref. 1-15 for more details.

References

General aspects:

[1] K. W.Schmid, F. Grümmer, *Rep. Progr. Phys.* **50**, 731 (1987)

[2] K. W.Schmid, F. Grümmer, A. Faessler, *Ann.Phys.(N.Y.)* **180**, 1 (1987)

[3] Zheng Ren-Rong, K. W. Schmid, F. Grümmer, A. Faessler, *Nucl. Phys.* A **494**, 214 (1989)

[4] K. W. Schmid, Zheng Ren-Rong, F. Grümmer, A. Faessler, *Nucl. Phys.* A **499**, 63 (1989)

---

[1] This work was supported by the Internationales Büro des KfK Karlsruhe and the DFG.

[5] K. W. Schmid, M. Kyotoku, F. Grümmer, A. Faessler, *Ann. Phys. (N.Y.)* **190**, 182 (1989)

[6] A. Petrovici, K. W. Schmid, A. Faessler, *Sov. Jour. Part. Nucl.* **23**, 402 (1992)

[7] F. Dönau, K. W. Schmid, A. Faessler, *Nucl. Phys. A* **539**, 403 (1992)

[8] K. W. Schmid in *Nuclear Structure Models*, p. 333, edited by R. Bengtsson, J. Draayer, W. Nazarewicz, (World Scientific, Singapore, 1992)

Special applications in the A $\sim$ 70 region:

[9] A. Petrovici, K. W. Schmid, F. Grümmer, A. Faessler, T. Horibata, *Nucl. Phys. A* **483**, 317 (1988)

[10] A. Petrovici, K. W. Schmid, F. Grümmer, A. Faessler, *Nucl. Phys. A* **504**, 277 (1989)

[11] A. Petrovici, K. W. Schmid, F. Grümmer, A. Faessler, *Nucl. Phys. A* **517**, 108 (1990)

[12] L. Chaturvedi et al., *Phys. Rev. C* **43**, 2541 (1991)

[13] A. Petrovici, K. W. Schmid, F. Grümmer, A. Faessler, *Z. Phys. A* **339**, 71 (1991)

[14] A. Petrovici, E. Hammarén, K. W. Schmid, F. Grümmer, A. Faessler, *Nucl. Phys. A* **549**, 352 (1992)

[15] A. Petrovici, K. W. Schmid, A. Faessler, 1 paper accepted by *Nucl. Phys. A*; 2 papers accepted by *Z. Phys. A*

DESCRIPTION OF THE CONTINUUM IN CALCULATING PARTIAL DECAY WIDTHS OF GIANT RESONANCES

T. Vertse[1], P. Lind[2], R. J. Liotta[3] and E. Maglione[4]

[1] *Institute of Nuclear Research of the Hungarian Academy of Sciences, Debrecen, P. O. Box 51, H–4001, Hungary*
[2] *Department of Mathematical Physics, Lund Institute of Technology, P. O. Box 118, S–221 00 Lund, Sweden*
[3] *Royal Institute of Technology, Frescativägen 24, S–104 05 Stockholm, Sweden*
[4] *Dipartimento di Fisica "Galileo Galilei" Via Marzolo 8, I–35 100 Padua, Italy*

Since giant resonances are collective excitations, lying in general, above the threshold of the particle emission one has to take into account the possibility of the nucleon escape in the theoretical description. The usual treatment is in the frame of the *continuum random phase approximation*[1] (CRPA), in which the effect of the continuum is taken into account exactly. For heavy nuclei CRPA calculations are time consuming since a large number of p-h configurations form the giant multipole resonance. Dealing with a realistic single particle potential one has to integrate the Schrödinger equation numerically in order to get the solutions which determine the Green's function. Therefore the major part of the computation time in the CRPA is spent on calculating the Green's function at different energies. In order to speed up the CRPA calculations we introduced[2] a pole expansion of the Green's function:

$$g(r,r';k) = \sum_n \frac{w_n(r,k_n)w_n(r',k_n)}{2k_n(k-k_n)}, \tag{1}$$

in which the normalized pole solutions w_n of the Schrödinger equation belonging to complex wave numbers k_n are used and the sum runs over all classes of poles, i. e. bound and antibound states, decaying and capturing resonances. This form is also given by the Mittag-Leffler expansion of g. But it was shown only recently by Berggren and Lind[3] that eq.(1) is a reasonable approximation even if the potential has a Coulomb tail (i.e. it does not vanish beyond the range of the nuclear forces). The w_n "states" have to be determined only once for a given potential and the Green's function can be calculated for any complex value of the wave number k quickly by using eq. (1).

Our approach is applied for the calculation of the total and partial neutron widths of the giant multipole resonances in ^{208}Pb using a separable multipole-multipole form of the residual interaction. For the isoscalar giant monopole resonance (GMR) where the number of p-h states are not too large we carried out also an exact CRPA calculation and used its result for checking the accuracy of our method. We found that the correlated energies agreed up to 3-4 decimal digits with the CRPA results and the values of the partial widths are also reproduced up to 5-15% accuracy with a basis which consisted of 5-7 resonant states besides the bound and antibound states in each partial wave[2].

Our calculated values[4] for the partial neutron decay widths are in reasonable agreement with both the calculated values of ref. 5 and with the experimental values[5,6] too, with the exception of the $p_{1/2} + i_{13/2}$ unresolved doublet. We calculated the giant multipole resonances also for $J = 1, 2, 3, 4, 5$ and found that all the $J \neq 2$ giant resonances were considerably fragmented.

In order to understand the structure of the giant resonances we used the *resonant RPA*[7,8] (RRPA). In the RRPA the basis is composed of a subset of the pole functions used above. The truncated form of the completeness relation of Berggren[9] is used, which includes only bound states and decaying resonances. The wave functions of the correlated states are expanded in terms of p-h configurations with complex energies. For a separable interaction the RRPA gives the forward amplitudes as

$$X(k, i; n) = N_n \frac{<k||f||i>}{\omega_n - (\mathcal{E}_k - \mathcal{E}_i)}, \qquad (2)$$

where ω_n, $\mathcal{E}_k$ and $\mathcal{E}_i$ are the energies of the correlated, the particle and the hole states, respectively. N_n is the normalization factor and $<k||f||i>$ the matrix element of the separable interaction. A common property of the giant resonances is that the wave function is dominated by those p-h configurations in which the particle is in bound or quasi-bound states. Therefore the escape width of these resonances is small.

The attractive feature of our approach is that our code runs 2-3 orders of magnitude faster than the CRPA and this is promising indeed for the calculation of the spreading width where the number of configurations is excessive.

This work was supported in part by the OTKA Foundation Hungary (contract numbers 3017 and 3010).

References

[1] S. Shlomo and G. Bertsch, *Nucl. Phys. A* **243**, 507 (1975)
[2] P. Lind, R. J. Liotta, E. Maglione, T. Vertse, submitted to *Z. Phys. A*
[3] T. Berggren and P. Lind, *Phys. Rev. C* **47**, 768 (1993)
[4] E. Maglione, R. J. Liotta and T. Vertse, *Phys. Lett. B* **298**, 1 (1993)
[5] A. Bracco, J. R. Beene, N. Van Giai, P. F. Bortignon, F. Zardi and R. A. Broglia, *Phys. Rev. Lett.* **60**, 2603 (1988)
[6] S. Brandenburg, W. T. A. Borghols, A. G. Drentje, A. van der Woude, M. N. Harakeh, L. P. Ekström, A. Hakanson, L. Nilsson, N. Olsson and R. De Leo, *Phys. Rev. C* **39**, 2448 (1989)
[7] T. Vertse, P. Curutchet, O. Civitarese, L. S. Ferreira and R. J. Liotta, *Phys. Rev. C* **37**, 876 (1988)
[8] P. Curutchet, T. Vertse and R. J. Liotta, *Phys. Rev. C* **39**, 1020 (1989)
[9] T. Berggren, *Nucl. Phys. A* **109**, 265 (1968)

HIGH-QUASIPARTICLE CALCULATION IN SPHERICAL NUCLEI

J. Blomqvist[1], A. Insolia[2], R. J. Liotta[1] and N. Sandulescu[3]

[1] *Manne Siegbahn Institute of Physics, S-10405 Stockholm*
[2] *University of Catania and INFN, I-9529 Catania*
[3] *Institute of Atomic Physics, PO Box MG-6, RO-76900 Bucharest, Romania*

The complex high spin spectra measured in the last years renewed the interest for high-quasiparticle approximations. One alternative is offered by high-QRPA, but in this approximation, besides quite complicated equations which have to be managed, the control of the spurious state, coming from particle number non-conservation, is a difficult task[1]. In this communication we present another alternative,so called QMSM(quasiparticle multistep shell model method)[2-5].

Within this framework, starting with a given two- body interaction, one calculates first the BCS one quasiparticle (1qp) states. Then, by QRPA, 2qp states are calculated. At this level the spurious BCS state can be easily identified since it has zero energy. In the next step one calculates the 3qp states in a basis defined as the tensorial product of the one and two quasiparticle states, previously computed:

$$|\alpha_3> = \sum_{i\alpha_2} X(i\alpha_2;\alpha_3)(\alpha_i^+\Gamma^+(\alpha_2))_{\alpha_3}|0> \qquad (1)$$

where $\Gamma^+(\alpha_2)$ are the 2qp operators. By α_i^+ we denote the one quasiparticle creation operators and by α_k the quantum numbers of k quasiparticle states. One advantage is that one can control the BCS spuriousity excluding the state $\Gamma^+(\alpha_2 = 0_1^+)$ from the summation (eq(1)). Another advantage is that, by neglecting the backward going amplitude in the two quasiparticle operator, the dynamical equation can be writen in a compact way, only in terms of previously calculated energy and wave functions[4,5]:

$$\omega(\alpha_3)F(i\alpha_2;\alpha_3) = (E_i + \omega(\alpha_2))F(i\alpha_2;\alpha_3)$$
$$+ \sum_{j\beta_2}\sum_k (\omega(\beta_2) - E_i - E_k)A(i\alpha_2,j\beta_2;k)F(j\beta_2;\alpha_3) \qquad (2)$$

where, with standard notation,

$$A(i\alpha_2,j\beta_2;k) = \hat{\alpha_2}\hat{\beta_2} \begin{Bmatrix} i & k & \beta_2 \\ j & \alpha_3 & \alpha_2 \end{Bmatrix} x(kj;\alpha_2)x(ki;\beta_2) \qquad (3)$$

and $x(ij;\alpha_2) = \sqrt{1+\delta_{ij}}X(ij;\alpha_2)$. The amplitudes F are the projections of the basis vectors on the physical state $|\alpha_3>$:

$$F(i\alpha_2;\alpha_3) = <\alpha_3|(\alpha_i^+\Gamma^+(\alpha_2))_{\alpha_3}|0> . \qquad (4)$$

This formulation is well suited for the truncation of the model space, which can be controlled in a self-consistent way by the overlap matrix between the basis vectors:

$$< 0|(\alpha_j^+\Gamma^+(\beta_2))_{\alpha_3}^\dagger (\alpha_i^+\Gamma^+(\alpha_2))_{\alpha_3}|0> = \delta_{ij}h\delta_{\alpha_2\beta_2} + \sum_k A(i\alpha_2;j\beta_2;k). \quad (5)$$

One notes that in eq.(2) the same matrix A appears as in eq.(5), which simplifies very much the numerical calculation.

One can further calculate four quasiparticle excitations in a basis defined as a tensorial product of two times two quasiparticle states:

$$|\alpha_4> = \sum_{\alpha_2 \leq \beta_2} X(\alpha_2\beta_2;\alpha_4)(\Gamma^+(\alpha_2)\Gamma^+(\beta_2))_{\alpha_4}|0>. \quad (6)$$

Again the dynamical equation can be expresed in a compact way [2,3]:

$$\omega(\alpha_4)F(\alpha_2\beta_2;\alpha_4) = (\omega(\alpha_2) + \omega(\beta_2))F(\alpha_2\beta_2;\alpha_4)$$
$$+ \sum_{\pi_2,\tau_2}\sum_{ijkl}(\omega(\pi_2) - E_j - E_k)A(ijkl;\alpha_2\beta_2\pi_2\tau_2)F(\pi_2\tau_2;\alpha_4) \quad (7)$$

where

$$A(ijkl;\alpha_2\beta_2\pi_2\tau_2) = \hat{\alpha}_2\hat{\beta}_2\hat{\pi}_2\hat{\tau}_2 \begin{Bmatrix} i & j & \alpha_2 \\ k & & \beta_2 \\ \pi_2 & \tau_2 & \alpha_4 \end{Bmatrix} x(ij;\alpha_2)x(kl;\beta_2)x(ik;\pi_2)x(jl;\tau_2). \quad (8)$$

The overlap matrix can be also written in terms of the matrix A (eq.(8)). In a similar way one can calculate five quasiparticle excitations, in a basis defined as a tensorial product of two times three quasiparticle states, with the same simple connection between the dynamical matrix and overlap matrix as before.

This procedure has been applied in lead[2,5] and tin regions[3] where, by using the same interaction and single-particle energies for the same chain of isotopes, a unitary description of 1qp,2qp,3qp and 4qp excitations has been proposed. The truncation procedure gave us the opportunity to obtain simple interpretations of the spectra. For instance the one vector approximation of weak coupling model has been analysed and generalised from a microscopical point of view[5]. Also the formalism has been applied to analyse the complex excitation spectrum of the isotope ^{194}Pb, where the coexistence of a neutron spherical degree of freedom part has been identified[4]. Presently, within the same framework, we investigate the region around ^{100}Sn.

References

[1] L. Zhao and A. Sustich, *Ann. of Phys.* **213**, 378 (1992)
[2] C. Pomar, J. Blomqvist, R. J. Liotta and A. Insolia, *Nucl.Phys.* A **515**, 381 (1990)
[3] A. Insolia, N. Sandulescu, J. Blomqvist and R. J. Liotta, *Nucl.Phys.* A **550**, 34 (1992)
[4] N. Sandulescu, A. Insolia, B. Fant, J. Blomqvist, R. J. Liotta, *Phys.Lett.* B **288**, 235 (1992)
[5] N. Sandulescu, A. Insolia, J. Blomqvist and R. J. Liotta, *Phys.Rev.* C **47**, 554 (1993)

CLUSTER APPROACH TO ATOMIC NUCLEI

G. S. Anagnostatos

Institute of Nuclear Physics, GR-15310 Aghia Paraskevi, Attiki, Greece

Introduction

Cluster approach has recently become an attractive way of research for atomic nuclei due to the following main reasons.

- The very fact that nucleons are not point particles but rather extensive particles of about 1 fm radius which is well comparable with that of the nucleus itself.[1]
- The success of the liquid drop model despite the fact that the independent particle assumption of the shell model is well supported by many experimental facts.
- The apparent success of the α-cluster model.[2]
- The many similarities found between atomic clusters and nuclei in such a way that both could be thought as members of a new science that of many (in comparison to few body physics) but not too many (in comparison to solid state physics) particles physics.[3]
- The fact that recently semi-classical mechanics has experienced a renaissance as a relevant theory for regular and chaotic systems of the microworld.[4]

The present study employs the Isomorphic Shell Model,[5,6] as cluster approach to atomic nuclei, which is applicable to atomic clusters as well.[3]

The isomorphic shell model

The isomorphic shell model is a microscopic nuclear-structure model that incorporates into a hybrid model the prominent features of single-particle and collective approaches in conjunction with the nucleon finite size.[5,6]

The single-particle component of the model is along the lines of the conventional shell model with the *only* difference that in the model the nucleons creating the central potential are the nucleons of each particular nuclear shell alone, instead of all nucleons in the nucleus as assumed in the conventional shell model.[6] That is, our Hamiltonian is analyzed into partial state-dependent Hamiltonians for neutrons (N) and for protons (Z) as follows, where crossing terms between partial Hamiltonians of different shells, H_{ij}, have been omitted.

$$
\begin{aligned}
H &= {}_N H + {}_Z H \\
  &= {}_N H_{1s} + {}_N H_{1p} + {}_N H_{1d2s} + \cdots \\
  &\quad + {}_Z H_{1s} + {}_Z H_{1p} + {}_Z H_{1ds} + \cdots
\end{aligned}
\qquad (1)
$$

While a finite square-well or Woods-Saxon potential would be a more realistic choice of the potential, for reasons of simplicity, we take the harmonic oscillator (HO) potential without spin-orbit coupling, where the expressions of the mean square radius and of the energy eigenvalues, necessary in demonstrating the model, are exceptionally simple and have closed mathematical forms. In addition, the appearance of the finite negative constants $-_N V_i$ and $-_Z V_i$ in the neutron and the proton harmonic oscillator potentials below, reduces the boggling impression given when an infinite potential is used for determining total-binding energies.

Thus, for each partial neutron or proton Hamiltonian we take

$$_N H_i = {_N V_i} + {_N T_i} = -_N \bar{V}_i + \frac{1}{2} m (_N \omega_i^2) r^2 + {_N T_i} \quad (2)$$

$$_Z H_i = {_Z V_i} + {_Z T_i} = -_Z \bar{V}_i + \frac{1}{2} m (_Z \omega_i^2) r^2 + {_Z T_i} \quad (3)$$

That is, each harmonic oscillator potential has its own state-dependent frequency ω. These ω are *not* taken as adjustable parameters, but all are determined from the harmonic oscillator relation[7]

$$\hbar \omega = \left(\frac{\hbar^2}{m < r^2 >} \right) \left(n + \frac{3}{2} \right), \quad (4)$$

where n is the harmonic oscillator quantum number and $< r_i^2 >^{1/2}$ is the average radius of the relevant high fluximal shell determined by the semiclassical part of the model specified below.

The solution of the Schrödinger equation with Hamiltonian (1), in spherical coordinates, is

$$\Psi_{nlm}(r, \theta, \phi) = R_{nl}(r) Y_l^m(\theta, \phi), \quad (5)$$

where $Y_l^m(\theta, \phi)$ are the familiar spherical harmonics and the expressions for the $R_{nl}(r)$ are given in several books of quantum mechanics and nuclear physics, for example see Table 4–1 of Ref. 7.

The only difference between our wave functions and those in these books is the different ω's as stated in (2) – (3) above. Those of our wave functions, however, which have equal ℓ value, because of the different $\hbar \omega$, are not orthogonal, since in these cases the orthogonality of Legendre polynomials does not suffice. Orthogonality, of course, can be obtained by applying established procedures, e.g., the Gram-Schmidt process.

According to Hamiltonian (1), the binding energy of a nucleus with A nucleons in the case of orthogonal wave functions takes the simple form given by

$$BE = 1/2 (V \cdot A) - 3/4 \left[\sum_{i=1}^{A} \hbar \omega_i (n_i + 3/2) \right], \quad (6)$$

where V is the average potential depth. The coefficients 1/2 and 3/4 take care of the double counting of nucleon pairs in determining the potential energy.

Applications and details of the quantum mechanical part of the model are given in Ref. 6. Here an application of the semiclassical part (see Refs. 5 and 8–13) in the place of the quantum mechanical part of the model is considered in the spirit of the Ehrenfest's theorem,[14] which for the observables of position (R) and momentum (P) takes the form

$$\frac{d}{dt}<R> = \frac{1}{m}<P> \quad \text{and} \tag{7}$$

$$\frac{d}{dt}<P> = -<\nabla V(R)> \tag{8}$$

The quantity $<R>$ represents a set of 3 time-dependent numbers $\{<X>,<Y>,<Z>\}$ and the point $<R>(t)$ is the centre of the wave function at the instant t. The set of those points which correspond to the various values of t constitutes the trajectory followed by the centre of the wave function.

From (7) and (8) we get

$$m\frac{d^2}{dt^2}<R> = -<\nabla V(R)> \tag{9}$$

Furthermore, it is known that, for the *special* case of the harmonic oscillator potential assumed by the isomorphic shell model in (3), the following relationship is valid

$$<\nabla V(R)> = [\nabla V(r)]_{r=<R>}, \quad \text{where} \tag{10}$$

$$[-\nabla V(r)]_{r=<R>} = F \tag{11}$$

That is, for this potential the average of the force over the whole wave function is rigorously equal to the classical force F at the point where the centre of the wave function is situated. Thus, for the special case (harmonic oscillator) considered, the motion of the centre of the wave function precisely obeys the laws of classical mechanics. Any difference between the quantum and the classical description of the nucleon motion exclusively depends on the degree the wave function may be approximated by its centre. Such differences will contribute to the magnitude of deviations between the experimental data and the predictions of the semiclassical part of the model employed here.

Now, in the semiclassical treatment the nuclear problem is reduced into that of studying the centres of the wave functions presenting the constituent nucleons or, in other words, of studying the average positions of these nucleons. For this study the following two assumptions are employed by the isomorphic shell model.

i. The neutrons (protons) of a closed neutron (proton) shell, considered at their *average* positions, are in *dynamic equilibrium* on the sphere presenting the average size of that shell.

ii. The average sizes of the shells are determined by the *close-packing* of the shells themselves, provided that a neutron and a proton are represented by *hard spheres* of definite sizes (i.e., $r_n = 0.974$ fm and $r_p = 0.860$ fm).

It is apparent that assumption (i) is along the lines of the conventional shell model, while assumption (ii) is along the lines of the liquid-drop model.

The model employs a specific equilibrium of nucleons, considered at their average positions on concentric spherical cells, which is valid whatever the law of nuclear force may be: assumption (i). This equilibrium leads uniquely to Leech[15] (equilibrium) polyhedra as average forms of nuclear shells. All such nested polyhedra are closed-packed, thus taking their minimum size: assumption (ii). The cumulative number of vertices of these polyhedra, counted successively from the innermost to the outermost, reproduce the magic numbers each time a polyhedral shell is completed[5] (see the numbers in the brackets in Fig. 1 there and in this paper).

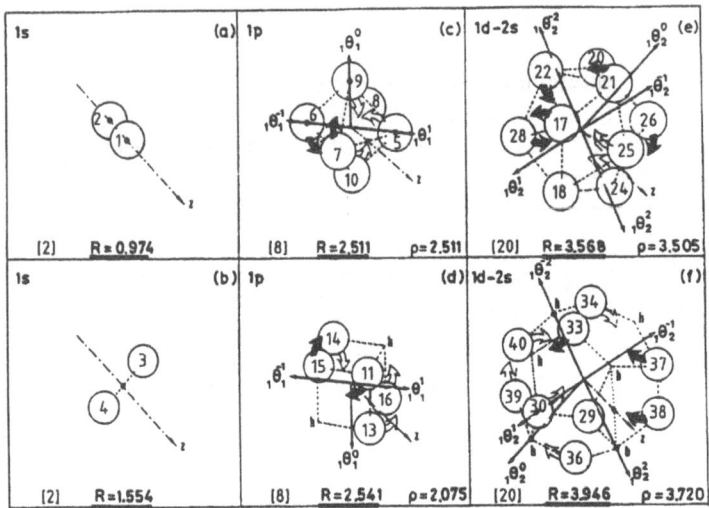

Figure 1. *The isomorphic shell model for the nuclei up to $N = 20$ and $Z = 20$. The high-symmetry polyhedra in row 1 (i.e. the zerohedron, the octahedron and the icosahedron) stand for the average forms for neutrons of (a) the $1s$, (c) the $1p$ and (e) the $1d2s$ shells, while the high-symmetry polyhedra in row 2 (i.e. the zerohedron, the hexahedron (cube) and the dodecahedron) stand for the average forms of (b) the $1s$, (d) the $1p$ and (f) the $1s2s$ shells for protons. The vertices of polyhedra stand for the average positions of nucleons in definite quantum states (τ, n, l, m, s). The letters h stand for the empty vertices (holes). The z axis is common for all polyhedra when these are superimposed with a common centre and with relative orientations as shown. At the bottom of each block the radius R of the sphere assigned to the relevant polyhedron and the radius ρ of the relevant classical orbit equal to the maximum distance of the vertex-state (τ, n, l, m, s) from the axis $_n\theta_l^m$ precisely representing the orbital angular momentum axis with definite n, l and m values are given. Curved arrows shown help the reader to visualise for each nucleon round what axis is rotated where close (open) arrows show rotations directed up (down) the plane of the paper. All polyhedra vertices are numbered as shown. The backside (hidden) vertices of the polyhedra and the related numbers are not shown in the figure.*

For one to conceptualize the isomorphic shell model, he should first relate this model to the conventional shell model. Specifically, the main assumption of the simple shell model, i.e. that each nucleon in a nucleus moves (in an average potential due to all nucleons) independently of the motion of the other nucleons, may be understood here in terms of a *dynamic equilibrium* in the following sense.[5] Each nucleon in a nucleus is *on average* in a *dynamic equilibrium* with the other nucleons and, as a *consequence*, its motion may be described independently of the motions of the other nucleons. From this one realizes that dynamic equilibrium and independent particle motion are *consistent* concepts in the framework of the isomorphic shell model.

In other words, the model implies that at some instant in time (reached *periodically*) all nucleons could be thought of as residing at their individual average positions, which coincide with the vertices of an equilibrium polyhedron for each shell. This system of particles evolves in time according to each independent particle motion. This is possible, since axes standing for the angular-momenta quantization of directions are *identically* described by the rotational symmetries of the polyhedra employed.[16-19] For example, see Ref. 18, where one can find a complete interpretation of the independent particle model in relation to the symmetries of these polyhedra. Such vectors are shown in Fig. 1 for the orbital angular-momentum quantization of directions involved in all nuclei up to $N = 20$ and $Z = 20$.

Since the radial and angular parts of the polyhedral shells in Fig. 1 are well defined, the coordinates of the polyhedral vertices (nucleon average positions) can be easily computed. These coordinates up to $N = Z = 20$, needed here for the application of the model on ^{11}Li (see next section), are already published in footnote 14 of Ref. 8, and in footnote 15 of Ref. 9. These coordinates correspond to the relevant R values of the assigned polyhedral spheres given in Fig. 1 (see bottom line at each block).

According to the isomorphic shell model, the nucleon average positions of a nucleus are distributed at the vertices of the polyhedral shells as shown, for example, in Fig. 1. The specific vertices occupied, for a given (closed- or open-shell) nucleus at the ground state, form a vertex configuration (corresponding to a state configuration) that possesses a maximum binding energy (BE) in relation to any other possible vertex configuration. This maximum BE vertex configuration defines the average form and structure of the ground state of this nucleus. All bulk (static) ground-state properties of this nucleus (e.g. BE, rms radii, etc.) are derived as properties of this structure, as has been fully explained in Ref. 5 and will become apparent below.

The quantities estimated by the model in the framework of this paper (see the next section) are potential energy V_{ij}, Coulomb energy $(E_c)_{ij}$; average kinetic energy $<T>_{nlm}$; odd-even energy E_δ; binding energy E_{BE}; collective rotational energy E_{rot}; and rms charge, mass and effective radii $<r^2>^{1/2}$ by using (12)–(21), respectively.[5,8,10]

$$V_{ij} = 1.7 \cdot 10^{17} \cdot \frac{e^{-(31.8538)r_{ij}}}{r_{ij}} - 187 \frac{e^{-(1.3538)r_{ij}}}{r_{ij}}, \tag{12}$$

where the internucleon distances r_{ij} are estimated following Fig. 1 or (the same) the corresponding coordinates of polyhedral vertices.[8-9]

$$(E_C)_{ij} = \frac{e^2}{r_{ij}}, \tag{13}$$

where distances r_{ij} are computed as explained above.

$$<T>_{nlm} = \frac{\hbar^2}{2M} \left[\frac{1}{R_{\max}^2} + \frac{l(l+1)}{\rho_{nlm}^2} \right], \tag{14}$$

where $R_{\max}$ is the outermost polyhedral radius (R) plus the relevant nucleon radius (i.e., $r_n = 0.974$ fm or $r_p = 0.860$ fm), i.e., the radius of the nuclear volume in which the nucleons are confined, M is the nucleon mass, ρ_{nlm} is the distance of the vertex (n, l, m) from the axis $_n\theta_l^m$ (see Fig. 1 and Ref. 10).

$$E_{BE} = -\sum_{\substack{\text{all nucleon} \\ \text{pairs}}} V_{ij} - \sum_{\substack{\text{all proton} \\ \text{pairs}}} \frac{e^2}{r_{ij}} - \sum_{\text{all nucleons}} <T>_{nlm} - E_\delta + E_{\text{rot}}, \tag{15}$$

where distances r_{ij} are estimated as above and E_δ is a correction "odd-even" term familiar from the liquid drop model. Here the value of E_δ is equal to zero for even-Z even-N nuclei for which the potential in (12) is exclusively derived[8] and thus no correction is needed, while for odd-A nuclei its value is taken equal[7] to 80/A MeV, i.e.

$$E_\delta = \frac{80}{A} \tag{16}$$

$$E_{\rm rot} = \frac{\hbar^2 I(I+1)}{2J}, \tag{17}$$

where J is the moment of inertia of the rotating part of the nucleus given by (18)

$$J = \sum_i^{N_{\rm rot}} m\rho_i^2 = m \sum_i^{N_{\rm rot}} \rho_i^2 = mN_{\rm rot} <r^2>_{\rm rot}, \tag{18}$$

where $N_{\rm rot}$ is the number of nucleons participating in the collective rotation and $<r^2>_{\rm rot}$ is the rms radius of these nuclei.

The term $E_{\rm rot}$ in (15) is meaningful *only* for the cases where the angular speed ω due to independent particle motion is comparable (about equal) to that due to collective motion in such a way that these two motions are coupled even at the ground state, i.e., for these cases the adiabatic approximation is not valid.

$$<r^2>_m^{1/2} = \left[\frac{\sum_{i=1}^Z R_i^2 + \sum_{i=1}^N R_I^2 + Z(0.8)^2 + N(0.91)^2}{Z+N} \right]^{1/2}, \tag{19}$$

$$<r^2>_{ch}^{1/2} = \left[\frac{\sum_{i=1}^Z R_i^2}{Z} + (0.8)^2 - (0.116)\frac{N}{Z} \right]^{1/2}, \tag{20}$$

where the subscripts ch and m refer to charge and mass, R_i is the radius of the ith proton or neutron average position from Fig. 1, Z and N are the proton and the neutron numbers of the nucleus, 0.8 and 0.91 fm are the rms radii of a proton and of a neutron, and -0.116 fm^2 is the ms charge radius of a neutron.[20] The 0.91 fm value for a neutron is taken from the 0.8 fm value for a proton by considering proportionality according to the sizes of their bags 0.974 and 0.860 fm, respectively, i.e. $0.91 = 0.8(0.974/0.860)$.

$$<r^2>_{\rm eff}^{1/2} = \left[<r^2>_m + <r^2>_{\rm rot} \right]^{1/2} \tag{21}$$

Applications of the model

Since the average radial sizes R_i of all nuclear shells up to ^{208}Pb, in the framework of the isomorphic shell model, have been specified,[5] the values of all radii (proton, neutron, and mass radii) can be determined. For example, by assuming that the filling of subshells follows the simple shell model and by applying (20) the charge radii of all nuclei from H to Pb can be determined as listed in Table 1. The closeness of the model predictions to the experimental values, also listed in the table, is apparent. A specific application of the model to exotic nucleus ^{11}Li follows.

In Fig. 2 the neutron and the proton average positions for ^{11}Li, in the framework of the isomorphic shell model, are shown. Numbering of positions in this figure follows that of

Table 1. *Charge root mean square radii* [1] *in units Fermi.*

NUCL.	MOD.	EXP.	NUCL.	MOD.	EXP.
H		0.8	^{98}Mo	4.40	4.391(26)
^{4}He	1.71	1.71 (4)	^{98}Tc	4.43	
^{7}Li	2.06	2.39(3)	^{102}Ru	4.46	4.480(22)[4]
^{9}Be	2.22	2.50(9)	^{103}Rh	4.49	4.510(44)
^{11}B	2.31	2.37	^{106}Pd	4.52	4.541(33)
^{12}C	2.37	2.40(56)[2]	^{107}Ag	4.55	4.542(10)[4]
^{14}N	2.54	2.540(20)	^{114}Cd	4.57	4.624(8)
^{16}O	2.70	2.710(15)[3]	^{115}In	4.60	4.611(10)[4]
^{19}F	2.84	2.85(9)[2]	^{120}Sn	4.63	4.630(7)
^{20}Ne	2.98	3.00(3)	^{121}Sb	4.65	4.63(9)
^{23}Na	2.95	2.94(4)[2]	^{130}Te	4.67	4.721(6)
^{24}Mg	3.06	3.08(5)	^{127}I	4.72	4.737(7)
^{27}Al	3.14	3.06(9)	^{132}Xe	4.77	4.790(22)[4]
^{28}Si	3.21	3.15(5)	^{135}Cs	4.82	4.801(11)[4]
^{31}P	3.27	3.24	^{138}Ba	4.85	4.839(8)[4]
^{32}S	3.33	3.263(20)	^{139}La	4.91	4.861(8)
^{35}Cl	3.37	3.335(18)	^{140}Ce	4.95	4.883(9)
^{40}Ar	3.40	3.42(4)	^{141}Pr	4.99	4.881(9)
^{39}K	3.44	3.436(3)[3]	^{142}Nd	5.03	4.993(35)
^{40}Ca	3.47	3.482(25)	^{146}Pm	5.06	
^{45}Sc	3.51	3.550(5)[3]	^{152}Sm	5.10	5.095(30)[4]
^{48}Ti	3.55	3.59(4)	^{153}Eu	5.13	5.150(22)[4]
^{51}V	3.59	3.58(4)	^{158}Gd	5.16	5.194(22)[4]
^{52}Cr	3.62	3.645(5)[3]	^{159}Tb	5.19	
^{55}Mn	3.65	3.68(11)	^{164}Dy	5.22	5.222(30)[4]
^{56}Fe	3.68	3.737(10)	^{165}Ho	5.25	5.210(70)[4]
^{59}Co	3.71	3.77(7)	^{166}Er	5.28	5.243(30)[4]
^{58}Ni	3.73	3.760(10)	^{169}Tm	5.30	5.226(4)[5]
^{63}Cu	3.81	3.888(5)[3]	^{174}Yb	5.32	5.312(60)[4]
^{64}Zn	3.87	3.918(11)	^{175}Lu	5.35	5.378(30)[5]
^{69}Ga	3.93		^{180}Hf	5.37	5.339(22)[4]
^{72}Ge	3.99	4.050(32)[4]	^{181}Ta	5.40	5.500(200)[4]
^{75}As	4.04	4.102(9)[4]	^{184}W	5.42	5.42(7)
^{80}Se	4.08	4.142(3)[5]	^{187}Re	5.44	
^{79}Br	4.13	4.163(79)[5]	^{192}Os	5.46	5.412(22)[4]
^{86}Kr	4.17	4.160[3]	^{193}Ir	5.48	
^{87}Rb	4.21	4.180[3]	^{195}Pt	5.50	5.366(22)[4]
^{88}Sr	4.25	4.26(1)	^{197}Au	5.52	5.434(2)
^{89}Y	5.29	4.27(2)	^{202}Hg	5.54	5.499(17)[4]
^{90}Zr	4.32	4.28(2)	^{205}Tl	5.56	5.484(6)
^{93}Nb	4.36	4.317(8)[4]	^{208}Pb	5.58	5.521(29)

[1] Most of the experimental radii come from Ref. 20 except as noted below in 2–5.
[2] See Ref. 24
[3] See Ref. 25
[4] See Ref. 26
[5] See Ref. 27

Fig. 1. Thus, the numbers appearing in Fig. 2 are not in order, but they depict those positions from Fig. 1 which lead to the maximum binding energy for the system of three protons and eight neutrons appearing in the nucleus ^{11}Li.

The estimation of the rms mass radius of ^{11}Li is based on the average-nucleon-position configuration shown in Fig. 2 and the corresponding coordinates from Refs. 8 and 9. Application of (12) – (14) and (16) leads to the energy components given below.

$$-\sum_{\substack{\text{all nucleon} \\ \text{pairs}}} V_{ij} = 145.66 \text{ MeV} \tag{22}$$

$$\sum_{\substack{\text{all proton} \\ \text{pairs}}} \frac{e^2}{r_{ij}} = 1.46 \text{ MeV} \tag{23}$$

$$\sum_{\text{all nucleons}} <T>_{nlm} = 67.25 \text{ MeV} \tag{24}$$

$$E_\delta = 7.27 \text{ MeV} \tag{25}$$

That is, from (15) and (22) – (25),

$$BE - E_{\text{rot}} = 69.68 \text{ MeV} . \tag{26}$$

We set now

$$BE = BE_{\text{exp}} = 45.54 \text{ MeV} . \tag{27}$$

The rotational energy from (26), which is valid here since $\omega_{i.p.} \approx \omega_{\text{rot}}$ for ^{11}Li (that is, the adiabatic approximation is not valid[21]), is equal to

$$E_{\text{rot}} = 24.14 \text{ MeV}, \tag{28}$$

which from (17) and (18) for $N_{\text{rot}} = 3$ (that is, the number of nucleons in incomplete subshells, i.e. of the $1p_{1/2}$ proton and of the two $1d_{5/2}$ neutrons) leads to

$$<r^2>_{\text{rot}}^{1/2} = 1.31 \text{ fm} . \tag{29}$$

Since the rms mass radius of ^{11}Li, based on the nucleon average positions shown in Fig. 2 and also on (19) and the relevant coordinates from Refs. 8 and 9, is

$$<r^2>_m^{1/2} = 2.84 \text{ fm}, \tag{30}$$

the effective radius for ^{11}Li is now from (21)

$$<r^2>_{\text{eff}}^{1/2} = 3.13 \text{ fm} . \tag{31}$$

This model radius is in very good agreement with the experimental radius[22]

$$<r^2>_{\text{exp}}^{1/2} = 3.11 \pm 0.8 \text{ fm} . \tag{32}$$

The results of both applications of the model, i.e. that to the rms charge radii of all nuclei from H to Pb and that to ^{11}Li give support to the applicability of the isomorphic shell model and further to the applicability of the cluster approach in atomic nuclei.

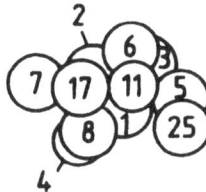

Figure 2. *Average form and structure of ^{11}Li. Spheres shown stand for the average positions of nucleons, where larger spheres depict neutrons ($r_p = 974$ fm) and smaller protons ($r = 0.860$ fm). Numbering of positions in this figure follows that of Fig.1 and depicts those positions from Fig.1 which lead to the maximum binding energy for the nucleus of three protons and eight neutrons. Specifically, Nos. 1–2 stand for the two neutrons in the state $1s_{1/2}$, while Nos. 3–4 for protons in the state $1s_{1/2}$, Nos. 5–8 for the four neutrons in the state $1p_{3/2}$, No. 11 for the one proton in the state $1p_{3/2}$, Nos. 17 and 25 for the two neutrons in the state $1d_{5/2}$. The last proton and the last two neutrons belong to incomplete subshells and constitute the rotating (in a way which is not apparent) part of the nucleus.*

Further, a very sensitive application of the model to the variation of the charge radius for the string of isotopes ^{40}Ca – ^{48}Ca may be found in Ref. 13, while an application of the model concerning the properties of ^{6}He and ^{8}He exotic nuclei, and the possible existence of neutron nuclei (e.g. ^{20}n) may be found in Refs. 21 and 23, respectively.

Conclusions

In the present and previous studies the cluster approach, in the framework of the isomorphic shell model, has been proved useful in predicting properties of nuclei all the way from the usual nuclei up to the most exotic nuclei, that is, the neutron nuclei.

This approach has the unique advantage of providing for each nucleus the nuclear shape simultaneously with other nuclear properties. This shape is not limited to the average nuclear shape as usually in nuclear physics, but it is a detailed shape made up of the average positions of the individual nucleons. In addition, the symmetries of this shape provide the possible axes of rotation and the corresponding moments of inertia in a unique way much superior to those provided by the average shape.

Coming from the above, a privilege of the present approach is the determination of the angular rotational speed of the nuclear collective excitation and thus the identification of those cases where a coupling of intrinsic and collective motions occurs, as e.g. in the very neutron rich nuclei.

Finally, it may be said that the way is open for many other applications of the approach in nuclear physics.

References

[1] S. Théberge and A.W. Thomas, Magnetic moments of the nucleon octet calculated in the cloudy

bag model, *Nucl. Phys A* **393**, 252 (1982).
[2] D.M. Brink, H. Friedrich, A. Weiguny, and C.W. Wong, Investigation of the alpha-particle model for light nuclei, *Phys. Lett.* **33B**, 143 (1970).
[3] G.S. Anagnostatos, Fermion/boson classification in microclusters, *Phys. Lett. A* **157**, 65 (1991).
[4] H. Friedrich, Paths to chaos, *Phys. World* **5**, 32 (1992).
[5] G.S. Anagnostatos, Isomorphic shell model for closed-shell nuclei, *Int. J. Theor. Phys.* **24**, 579 (1985).
[6] G.S. Anagnostatos, Multiharmonic nuclear Hamiltonian, *Can. J. Phys.* **70**, 361 (1992).
[7] W.F. Hornyak. "Nuclear Structure", Academic, New York (1975).
[8] G.S. Anagnostatos and C.N. Panos, Effective two-nucleon potential for high-energy heavy-ion collisions, *Phys. Rev. C* **26**, 260 (1982).
[9] G.S. Anagnostatos and C.N. Panos, Simple static central potentials as effective nucleon-nucleon interactions, *Lett. Nuovo Cimento* **41**, 409 (1984).
[10] C.N. Panos and G.S. Anagnostatos, Comments on 'A relation between average kinetic energy and mean-square radius in nuclei', *J. Phys. G* **8**, 1651 (1982).
[11] G.S. Anagnostatos, Classical equations-of-motion model for high-energy heavy-ion collisions, *Phys. Rev. C* **39**, 877 (1989).
[12] G.S. Anagnostatos and C.N. Panos, Semiclassical simulation of finite nuclei, *Phys. Rev. C* **42**, 961 (1990).
[13] G.S. Anagnostatos, T.S. Kosmas, E.F. Hefter, and C.N. Panos, Semiclassical approximation for the isotopic shift of charge radii in Ca isotopes, *Can. J. Phys.* **69**, 114 (1991).
[14] E. Merzbacher. "Quantum Mechanics", John Wiley and Sons, Inc., New York (1975).
[15] J. Leech, Equilibrium of particles on a sphere, *Math. Gaz.* **41**, 81 (1957).
[16] G.S. Anagnostatos, Angular structure of unique spherical shells, *Lett. Nuovo Cimento* **22**, 507 (1978).
[17] G.S. Anagnostatos, The geometry of the quantization of angular momenta (l, s, j) in fields of central symmetry, *Lett. Nuovo Cimento* **28**, 573 (1980).
[18] G.S. Anagnostatos, Symmetry description of the independent particle model, *Lett. Nuovo Cimento* **29**, 188 (1980).
[19] G.S. Anagnostatos, J. Yapitzakis, and A. Kyritsis, Rotational invariance of orbital-angular-momentum quantization of direction for degenerate states, *Lett. Nuovo Cimento* **32**, 332 (1981).
[20] C.W. de Jager, H. de Vries, and C. de Vries, Nuclear charge and magnetization-density-distribution parameters from elastic electron scattering, *At. Data Nucl. Data Tables* **14**, 479 (1974).
[21] G.S. Anagnostatos, Intrinsic-collective coupling in ^{6}He and ^{8}He, in: "Exotic Nuclei", Y. Penionzevic, ed., Springer, Berlin (1992).
[22] I. Tanihata, Structure of neutron-rich nuclei studied by radioactive beams, *Nucl. Phys. A* **522**, 275c (1991).
[23] G.S. Anagnostatos, Cluster approach to exotic nuclei: neutron nuclei, Invited talk at the "International School on Nuclear Physics". Dubna, Russia, May 1993 (in press).
[24] R. Engfer, H. Schneuwly, J.L. Vuilleumier, H.K. Valter, and A. Zehnder, Charge-distribution parameters, isotope shifts, isomer shifts, and magnetic hyperfine constants from muonic atoms, *At. Data Nucl. Data Tables* **14**, 509 (1974).
[25] B.A. Brown, C.R. Bronk, and P.E. Hodgson, Systematics of nuclear rms charge radii, *J. Phys. G* **10**, 1683 (1984).
[26] E. Wesolowski, The rms radii of nuclear proton distributions, *J. Phys. G* **10**, 321 (1984).
[27] I. Angeli, Effect of valence nucleons on rms charge radii and surface thickness, *J. Phys. G* **17**, 439 (1991).

DEFORMATION, CLUSTERIZATION, FISSION AND SU(3)

J. Cseh[1], G. Lévai[1], K. Varga[1], R. K. Gupta[2], and W. Scheid[3]

[1] Institute of Nuclear Research of the Hungarian Academy of Sciences, Debrecen, P. O. Box 51, H–4001, Hungary
[2] Physics Department, Panjab University, Chandrigarh-160014, India
[3] Institut für Theoretische Physik, Justus-Liebig-Universität, Heinrich-Buff-Ring 16, D-35392 Giessen, Germany

Introduction

The SU(3) group, being the symmetry group of the harmonic oscillator problem, entered the nuclear physics through the shell model of light nuclei[1]. In addition to playing an important role in the shell model, it helped to establish the close relation between the quadrupole deformation and clusterization in some transparent cases[2,3]. The purpose of this lecture is to draw attention to the fact, that similar symmetry considserations can be very interesting and of practical use in a couple of new territories as well. We consider here examples for the clusterization of light nuclei in their ground-state region, as well as in highly excited states (molecular resonances), interrelation between the superdeformed and cluster configurations, binary fission modes of light nuclei, and heavy cluster radioactivity. The SU(3) symmetry appears in these investigations as the dynamical symmetry of the semimicroscopic algebraic cluster model[4,5].

In what follows first we mention some historical preliminaries, and the changes of circumstances against and for the extended applications of SU(3)-based analysis. Then we introduce the semimicroscopic cluster model, and present some of its applications. Finally a brief discussion is given.

1958

In 1958 Elliott deduced the collective motion, in particular the rotation, of the nucleus from the shell model picture, giving this way a microscopic foundation of a macroscopic (or phenomenologic) model[1]. The rotational bands of light nuclei are obtained in his model as a set of shell model states with a specific and common symmetry feature. The harmonic oscillator shell model with quadrupole residual interaction proved to be a good approximation to light nuclei, consequently the truncation is done according to the SU(3) symmetry.

In the same year Wildermuth and Kanellopoulos published another important analysis about the interrelation of nuclear models. They have shown, that in the harmonic oscillator approximation the Hamiltonian of the shell model and that of the cluster model can be rewritten into each other exactly[2]. (In the harmonic oscillator cluster model the clusters are described in terms of harmonic oscillator shell models, and they interact via harmonic oscillator potentials.

All oscillators have the same excitation quantum.) Due to the equivalence of the Hamiltonians, there is, of course, a close relation between the basis states of the two descriptions.

Based on these two important results, still in 1958 Bayman and Bohr showed, that some quadrupole states are identical with certain cluster configurations[3]. The connection is built up through the shell model content, more specifically via their SU(3) basis. In spite of the fact that we associate seemingly different pictures to the deformation and clusterization, the corresponding wavefunctions can be very similar, even identical with each other, due to the antisymmetrization effects. The same statement holds for different cluster configurations as well.

The analyses mentioned before enlightened the relation between the basic nuclear structure models, and provided us with some textbook examples[6]. E.g. the ground-state-band of ^{20}Ne consists of deformed states, which, at the same time can be considered as a fairly pure shell model configuration, and good cluster model states of ^{16}O + α structure.

Against and for the SU(3) Symmetry

Following the beautiful results of 1958, most of the calculations were subsequently carried out within the framework of one of the basic structure models, and much less attention was paid to the question of the interrelation between their states. The reasons are numerous; and here we mention only some of them. *i)* The validity of the SU(3) shell model is rather limited. Being an $L - S$ coupled model, it breaks down beyond the sd shell, due to the spin-orbit coupling and other interactions. *ii)* The shell model picture gets more complicated not only when we go to heavier nuclei, but also when we go to the region of high deformation. *iii)* The symmetry analysis of Refs.[2,3] gave qualitative results, but not quantitative. *iv)* The fission-like processes, which most probably have to do with cluster configurations, seemed to be too difficult for a microscopic treatment.

More recently, however, some changes of the circumstances have been realized, which seem to act in favour of the symmetry analysis. *i)* A pseudo SU(3) shell model scheme seems to work for heavy nuclei (due to the very special fine tuning of the effective two-nucleon forces)[7]. *ii)* When we go to extreme high deformations (with ratios of main axes of the spheroidal shape of 2:1:1, or 3:1:1), then the shell model picture becomes simpler again, like in the spherical case[8]. *iii)* Algebraic models proved to be successful in quantitative analysis of the experimental data based on the concept of dynamical symmetry[9]. *iv)* Harvey succeeded to formulate a simple heuristic prescription, based on the harmonic oscillator shell model, which seems to explain the structural aspects of the fusion or fission of light nuclei[10].

In the next section the semimicroscopic algebraic cluster model is introduced, in which the favourite features of the last paragraph are put together in a coherent way, and some of its applications are shown in the subsequent section in order to illustrate its capibilities.

The Semimicroscopic Algebraic Cluster Model

In the semimicroscopic algebraic cluster model[4,5] the internal structure of a cluster is described in terms of the $SU(3)$ shell model[1], therefore its wavefunction is characterised by the $U_C^{ST}(4) \otimes U_C(3)$ symmetry, where C refers to cluster, and $U^{ST}(4)$ is Wigner's spin-isospin group[11]. The relative motion of the clusters is accounted for by the vibron model with $U_R(4)$ group structure[12]. The representation labels of the group-chain

$$U^{ST}_{C_1}(4) \otimes U_{C_1}(3) \otimes U^{ST}_{C_2}(4) \otimes U_{C_2}(3) \otimes U_R(4)$$
$$\supset U^{ST}_C(4) \otimes U_C(3) \otimes U_R(3) \supset U^S_C(2) \otimes U(3)$$
$$\supset U^S_C(2) \otimes O(3) \supset U(2) \supset O(2) \qquad (1)$$

provide us with the quantum numbers for the basis states of a two-cluster system. From this set we have to skip those states, which are Pauli forbidden, or which correspond to spurious excitations of the center of mass. To this end we can take into account the norm kernel eigenvalues of the microscopic $SU(3)$ cluster model[13], or one can apply a matching requirement between the quantum numbers of the shell model basis of the whole nucleus and its cluster model basis (within the harmonic oscillator approximation)[2,4,5]. For the low-lying major shells, where the influence of the Pauli blocking is the strongest, the latter calculation is usually simpler.

When the internal structure of each cluster is described by a single $U^{ST}_C(4) \otimes U_C(3)$ representation, then the physical operators of the system can be obtained in terms of the generators of the $U^{ST}_{C_1}(4) \otimes U_{C_1}(3) \otimes U^{ST}_{C_2}(4) \otimes U_{C_2}(3) \otimes U_R(4)$ group. In such a case the description is algebraically closed, i.e. the matrix elements can be deduced by means of group theoretical techniques. In the limiting case when the Hamiltonian is given by the invariant operators of (1), then the eigenvalue problem has an analytical solution, and a $U(3)$ dynamical symmetry is said to hold.

If one or both of the clusters are $U^{ST}_C(4)$ and/or $U_C(3)$ scalars, then the problem becomes simpler. In Ref.[5] the formalism is presented in detail for the $U_{C_1}(3) \otimes U_R(4)$ and $U_{C_1}(3) \otimes U_{C_2}(3) \otimes U_R(4)$ models, as well as for the restricted $U^{ST}_{C_1}(4) \otimes U_{C_1}(3) \otimes U_R(4)$ model. In this latter case the restriction implies that only spin and isospin free interactions and a single $U^{ST}_C(4)$ representation are considered. If both of the clusters are $U^{ST}_C(4)$ and $U_C(3)$ scalars, the model reduces to that of the simple vibron model with a basis truncation corresponding to the Wildermuth condition[14,15].

Here we give a brief account of the $U_C(3) \otimes U_R(4)$ model, which is able to describe two-cluster systems in which one of the clusters is a closed-shell nucleus (e.g. ^4He, ^{16}O, or ^{40}Ca), while the other one is an even-even nucleus. In this simple case the basis states can be labeled without explicit reference to the $U^{ST}(4)$ group, (unless some higher excitations of the non-closed-shell nucleus are also considered), and the cluster model basis states are characterized by the representation labels of the group chain:

$$U_C(3) \otimes U_R(4) \supset U_C(3) \otimes U_R(3) \supset SU_C(3) \otimes SU_R(3) \supset SU(3) \supset O(3) \supset O(2)$$
$$|[n^C_1, n^C_2, n^C_3], [N, 0, 0, 0], \quad [n_\pi, 0, 0,], (\lambda_C, \mu_C), (n_\pi, 0), (\lambda, \mu), K_L, L \quad , M\,\rangle.$$
$$(2)$$

The irreducible representations (λ, μ) of $SU(3)$ are obtained by taking the outer product of $(\lambda_C, \mu_C) \otimes (n_\pi, 0)$. N stands for the maximal number of the π-bosons, i.e. that of the excitation quanta assigned to the relative motion, and it determines the size of the model space. The angular momentum content of a (λ, μ) representation is given by the usual relations of the Elliott model[1]. For technical reasons, however, it is more convenient to use the orthonormal $SU(3)$ basis of Draayer and Akiyama[16], rather than the Elliott basis, which is not orthogonal. The parity of the basis states is determined by the parity assigned to the relative motion: $P_R = (-1)^{n_\pi}$. (The internal states of the non–$U(3)$–scalar cluster carry positive parity $P_C = (-1)^{n_1^C + n_2^C + n_3^C}$, unless major shell excitations of the clusters are also considered.)

The wave function can be written as:

$$|(\lambda_C, \mu_C), N(n_\pi, 0); (\lambda, \mu)\chi LM\rangle$$
$$= \sum_{\chi_C L_C M_C} \sum_{L_R M_R} \langle(\lambda_C, \mu_C)\chi_C L_C M_C; N(n_\pi, 0)L_R M_R|(\lambda, \mu)\chi LM\rangle$$
$$\times |(\lambda_C, \mu_C)\chi_C L_C M_C\rangle |N(n_\pi, 0)L_R M_R\rangle. \tag{3}$$

The Racah factorization lemma[17] can be used to factor the coupling coefficient into two terms:

$$\langle(\lambda_C, \mu_C)\chi_C L_C M_C; (n_\pi, 0)L_R M_R|(\lambda, \mu)\chi LM\rangle$$
$$= \langle(\lambda_C, \mu_C)\chi_C L_C; (n_\pi, 0)L_R\|(\lambda, \mu)\chi L\rangle$$
$$\times \langle L_C M_C L_R M_R|LM\rangle. \tag{4}$$

Here the double barred symbols are $SU(3) \supset O(3)$ isoscalar factors or Wigner coefficients.

The physical operators can be constructed from the generators of the groups present in group chain (2). In particular, the most general form of the Hamiltonian can be obtained in terms of a series expansion of these generators. In the simplest case, however, when we use the $SU(3)$ dynamical symmetry approximation, and consider only one $U_C(3)$ representation to describe the structure of the non–closed–shell even–even cluster, the energy eigenvalues can be obtained in a closed form:

$$E = \epsilon + \gamma n_\pi + \delta n_\pi^2 + \eta C_2(\lambda, \mu) + \beta L(L+1), \tag{5}$$

The electromagnetic transition operators are also constructed from the group generators (which automatically implies selection rules in the dynamical symmetry approximation). The electric quadrupole transition operator, for example, is written as the sum of the rank–2 generators of the $U_C(3)$ and the $U_R(3)$ groups:

$$T^{(E2)} = q_R Q_R^{(2)} + q_C Q_C^{(2)}. \tag{6}$$

The matrix elements of the operators with the basis states (3) can be calculated using tensor algebraic techniques[17].

The formulation of the $U_{C_1}(3) \otimes U_{C_2}(3) \otimes U_R(4)$ and $U_C^{ST}(4) \otimes U_C(3) \otimes U_R(4)$ models can be done via a straightforward generalization of the results presented here. These models can also be used away from the $SU(3)$ dynamical symmetry limit: in this case the diagonalization of the Hamiltonian becomes necessary.

Although the interactions applied in this approach are phenomenological ones, they can be related to the effective two-nucleon forces, due to the use of the microscopic $SU(3)$ cluster model basis. In Ref.[18] the vibron model Hamiltonian of the $U(3)$ dynamical symmetry was chosen as

$$E = \beta L(L+1) + \gamma n_\pi + \delta n_\pi^2 + \epsilon \tag{7}$$

with parameters obtained from the most often used effective two-nucleon forces (Table 1). It turns out that this simple phenomenological Hamiltonian can reproduce the matrix elements of the microscopic interactions (in a range of ca 100 MeV) reasonably well.

Table 1. *Parameters of the phenomenological cluster–cluster interactions (in MeV) obtained from effective two-nucleon forces. (The abbreviations stand for the Volkov No. 2, Hasegawa-Nagata No. 1, No. 2, and modified forces, respectively.) The last line gives the corresponding values obtained from a fit to the experimental spectrum of ^{20}Ne. See Ref.[18] for more details.*

Force	β	γ	δ	ϵ
V2	0.0493	12.289	-0.119	-91.329
HN1	0.0720	13.776	-0.149	-123.28
HN2	0.0687	13.042	-0.136	-106.01
MHN	0.0562	12.415	-0.121	-93.145
Exp.	0.161	13.601	-0.571	-71.040

Applications

The applications of the semimicroscopic algebraic cluster model have been carried out so far within the dynamical symmetry approximation.

The $U(3)$ dynamical symmetry of this approach has an interesting new feature: it is usually related to several cluster configurations. The consequences have been studied in two cases.

i) The link between the superdeformed and cluster states of alpha–like nuclei ($N = Z = even$) was explored[19]. A spheroidal shape is called superdeformed here, when the ratio of its main axes can be expressed as ratios of natural numbers. They were obtained from Nilsson-Strutinsky calculations[20], and we addressed the question, what kind of alpha-like cluster configurations correspond to these states. It turns out, that seemingly different clusterizations have the same wavefunction (within the approximation of the leading SU(3) representation), and in addition, the same cluster structure appears at different excitation energies of the same nucleus, depending on the relative orientations of the deformed clusters (Fig. 1).

ii) The allowed and forbidden binary fission modes of ground–state–like configurations in sd shell nuclei were determined[21] (Table 2).

Detailed quantitative analyses have been performed for a few cluster systems so far. In these calculations the model spectra were obtained with interactions containing not only linear and quadratic functions of the quantum numbers (like in Eq. (5)), but also some higher order terms.

In Ref.[23] the $U_{C_1}(3) \otimes U_{C_2}(3) \otimes U_R(4)$ model is used to treat the low–lying $T = 0$ states and the molecular resonances of the ^{24}Mg nucleus in a uniform $^{12}C + ^{12}C$ cluster description. We have analysed about 150 experimental levels in the energy range of 0 to 40 MeV (see Fig. 3.), and nearly 100 electric quadrupole transition probability data in our study, which is a more complete account of the energy spectrum and $E2$ transitions of the ^{24}Mg nucleus than any previous model calculation. We have displayed the $B(E2)$ values

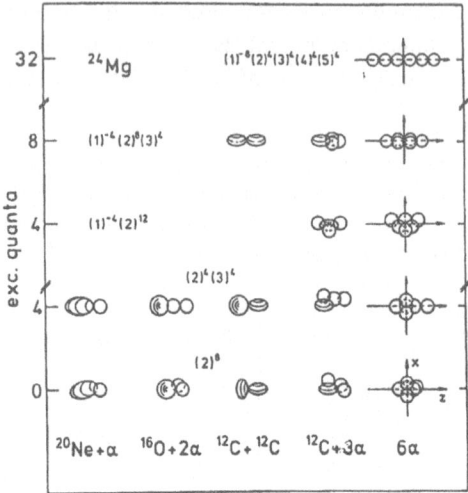

Figure 1. *Possible cluster configurations[19] of some highly deformed states of the ^{24}Mg nucleus. Only the directions of the symmetry axes and the directions of amalgamations of the clusters carry physical content, while the relative distances do not. The y axis is perpendicular and coming out of the plain.*

for the in–band transitions in Table 3. Our results for interband $E2$ transitions are also satisfactory. The fact that most of the transitions forbidden by the selection rules due to the $SU(3)$ dynamical symmetry have very weak experimental counterparts seems to indicate that the $SU(3)$ dynamical symmetry approach is a realistic approximation of the actual physical situation here. The model was able to describe the general features of the moleular resonance spectrum as well. $E2$ transition probabilities calculated for in–band transitions within this region were significantly smaller than most of the corresponding results of other models. The example of the $^{12}C + ^{12}C$ system demonstrated that a large number of experimental data, including the ground–state region as well as the highly excited molecular resonances can be reconciled in terms of relatively straightforward calculations, which is one of the major advantages of the semimicroscopic algebraic cluster model.

The $T = 1$ states of the ^{18}O nucleus have been described in terms of core–plus–alpha–particle configurations[22] by means of the restricted $U_C^{ST}(4) \otimes U_C(3) \otimes U_R(4)$ model. 34 energy levels of the experimental spectrum were identified with model states, and were assigned to 11 bands, some of which were new (Fig. 2). This spectrum is more complete, than those of the previous calculations. Reduced $E2$ and $E1$ transition probabilities were also determined. Their agreement with the experimental data is similar to those of the best microscopic or semimicroscopic calculations. Our results also showed strong correlation with those of fully microscopic calculations, which seems to indicate that the semimicroscopic algebraic cluster model approximates certain microscopic effects reasonably well.

Preliminary results indicate, that for the $^{24}Mg + ^{24}Mg$ and $^{28}Si + ^{28}Si$ systems the nuclear resonances can also be described simultaneously with the low-lying spectra of the corresponding compound nuclei[24].

The extension of the model to heavy nuclei seems to be possible based on the pseudo $SU(3)$ scheme[25]. The $^{210}Pb + ^{14}C$ clusterization was found to be present in the ground

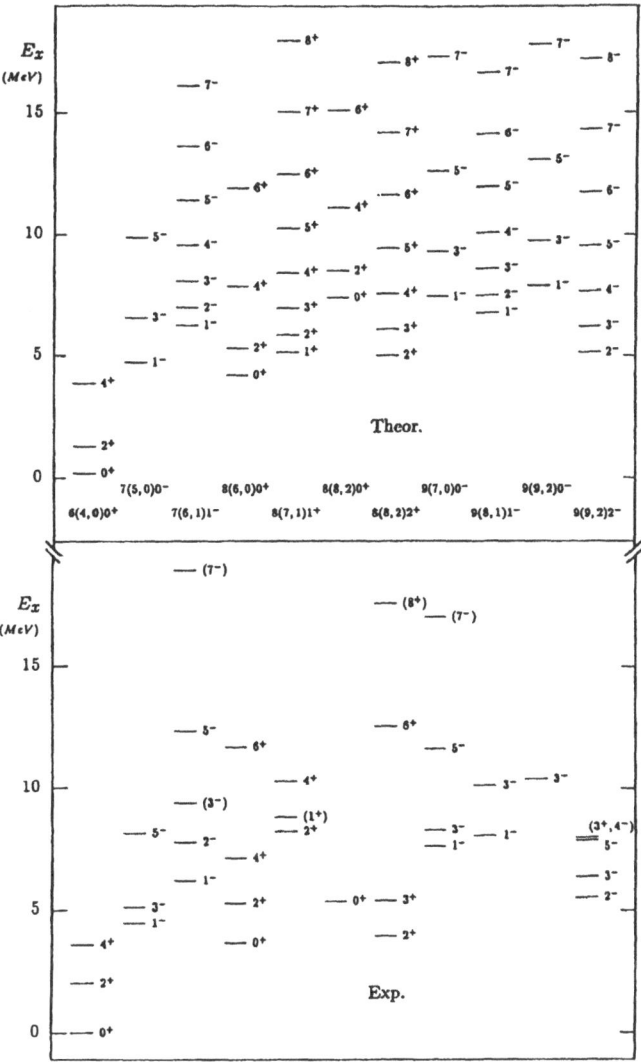

Figure 2. *Experimental and model spectra*[22] *of* ^{18}O.

state wavefunction of the ^{224}Ra nucleus. In addition, the low-energy spectrum could also be reproduced in terms of this cluster configuration (Fig. 4).

Discussion

In this lecture we have presented a semimicroscopic algebraic cluster model, and some of its applications. In this approach the model space is constructed microscopically, and the interactions we apply are phenomenological ones. Nevertheless, their relation to effective two-nucleon-forces can be established. The model has been developed so far for two-cluster systems, and the applications have been carried out within the dynamical symmetry approach.

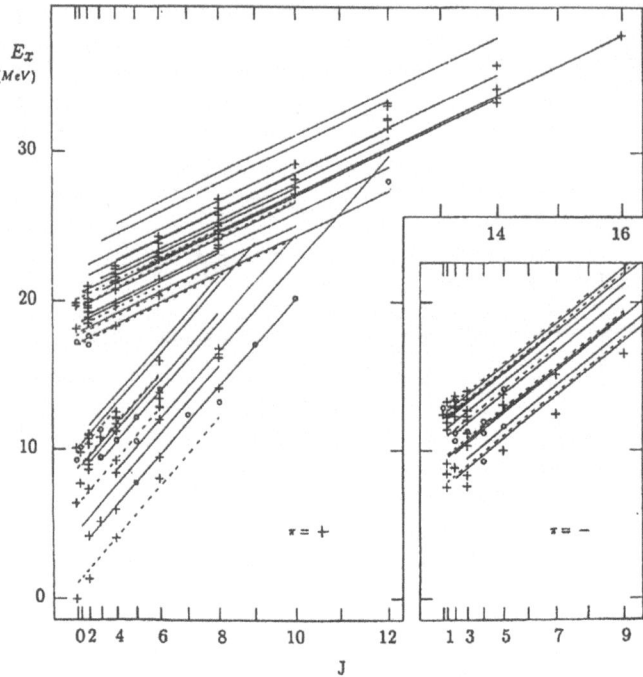

Figure 3. Positive- and negative-parity $T = 0$ energy levels of the ^{24}Mg nucleus displayed separately in rotational diagram form[23]. Circles (○) stand for states with uncertain J^π assignment. The lines denote the position of the calculated model bands. (Dashed lines indicate bands with $K = 0$, which contain only every second possible J value.)

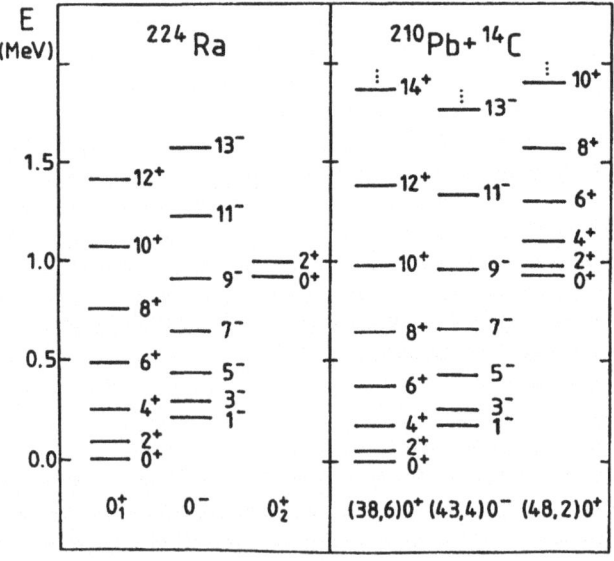

Figure 4. Low–lying bands in the energy spectrum of ^{224}Ra in comparison with the results of a $^{210}Pb+^{14}C$ model calculation[25].

Table 3. In–band transitions for the ^{24}Mg nucleus. See Ref.[23] for the sources of the experimental data. The quantun numbers $n_\pi(\lambda,\mu)\chi$ assigned to the bands are also displayed.

$J_i^\pi(E_{xi})$	$J_f^\pi(E_xf)$	$B(E2)_{Exp}$	$B(E2)_{Th}$	$n_\pi(\lambda,\mu)\chi$
$2^+(1.37)$	$0^+(0.0)$	21.0 ± 0.4^a	21.0^a	$12(8,4)0$
$4^+(4.12)$	$2^+(1.37)$	37.8 ± 3.0	28.0	
$6^+(8.11)$	$4^+(4.12)$	38 ± 13	27.1	
$8^+(13.21)$	$6^+(8.11)$	30 ± 14	23.1	
$3^+(5.24)$	$2^+(4.24)$	38.0 ± 5.5	37.5	$12(8,4)2$
$4^+(6.01)$	$2^+(4.24)$	18.7 ± 2.4	11.4	
$5^+(7.81)$	$3^+(5.24)$	35.0 ± 4.9	17.5	
$5^+(7.81)$	$4^+(6.01)$	24 ± 10	19.5	
$6^+(9.53)$	$4^+(6.01)$	18 ± 8	18.0	
$7^+(12.35)$	$5^+(7.81)$	21 ± 14	19.7	
$8^+(14.15)$	$6^+(9.53)$	9.1 ± 2.4	13.7	
$2^+(8.65)$	$0^+(6.43)$	14.0 ± 4.3	12.4	$12(6,2)0$
$6^+(12.86)$	$4^+(9.30)$	11.2 ± 2.1	12.2	
$5^-(10.03)$	$3^-(8.36)$	20^{+8}_{-5}	34.7	$13(9,4)0$
$7^-(12.44)$	$5^-(10.03)$	51 ± 10	32.3	
$5^-(13.06)$	$3^-(10.33)$	22 ± 4	28.1	$13(8,3)1$
$4^-(9.30)$	$3^-(7.62)$	29 ± 6	35.1	$13(8,3)3$
$5^-(11.60)$	$3^-(7.62)$	4.6 ± 1.4	7.3	
$5^-(11.60)$	$4^-(9.30)$	37 ± 11	31.8	

[a] Used to fit model parameters.

Table 2. Allowed and forbidden binary fission modes of ground–state–like configurations in sd shell nuclei according to the $U(3)$ selection rule.[21] (For nuclei in the first half of the sd shell all the binary fragmentations into stable isotopes are allowed).

Nucleus	Allowed	Forbidden
^{28}Si		$^{14}N + ^{14}N$ $^{16}O + ^{12}C$
^{29}Si	$^{17}O + ^{12}C$ $^{19}F + ^{10}B$ $^{20}Ne + ^{9}Be$ $^{23}Na + ^{6}Li$	$^{15}N + ^{14}N$ $^{16}O + ^{13}C$
^{30}Si	$^{21}Ne + ^{9}Be$ $^{23}Na + ^{7}Li$	$^{15}N + ^{15}N$ $^{17}O + ^{13}C$ $^{18}O + ^{12}C$ $^{19}F + ^{11}B$
^{31}P	$^{19}F + ^{12}C$ $^{21}Ne + ^{10}B$ $^{20}Ne + ^{11}B$ $^{23}Na + ^{8}Be$ $^{24}Mg + ^{7}Li$ $^{25}Mg + ^{6}Li$	$^{16}O + ^{15}N$ $^{17}O + ^{14}N$
^{32}S	$^{20}Ne + ^{12}C$	$^{16}O + ^{16}O$
^{33}S	$^{21}Ne + ^{12}C$ $^{23}Na + ^{10}B$ $^{24}Mg + ^{9}Be$ $^{27}Al + ^{6}Li$	$^{17}O + ^{16}O$ $^{19}F + ^{14}N$ $^{20}Ne + ^{13}C$
^{34}S	$^{22}Ne + ^{12}C$ $^{23}Na + ^{11}B$ $^{25}Mg + ^{9}Be$ $^{27}Al + ^{7}Li$	$^{18}O + ^{16}O$ $^{17}O + ^{17}O$ $^{19}F + ^{15}N$ $^{21}Ne + ^{13}C$
^{36}S		$^{18}O + ^{18}O$
^{35}Cl	$^{23}Na + ^{12}C$ $^{24}Mg + ^{11}B$ $^{25}Mg + ^{10}B$ $^{29}Si + ^{6}Li$ $^{28}Si + ^{7}Li$	$^{19}F + ^{16}O$ $^{20}Ne + ^{15}N$ $^{21}Ne + ^{14}N$
^{37}Cl	$^{30}Si + ^{7}Li$	$^{19}F + ^{18}O$ $^{22}Ne + ^{15}N$ $^{26}Mg + ^{11}B$
^{36}Ar	$^{24}Mg + ^{12}C$	$^{20}Ne + ^{16}O$
^{38}Ar	$^{31}P + ^{7}Li$	$^{19}F + ^{19}F$ $^{20}Ne + ^{18}O$ $^{21}Ne + ^{17}O$ $^{22}Ne + ^{16}O$ $^{23}Na + ^{15}N$ $^{25}Mg + ^{13}C$ $^{26}Mg + ^{12}C$ $^{27}Al + ^{11}B$ $^{29}Si + ^{9}Be$
^{39}K	$^{28}Si + ^{11}B$ $^{32}S + ^{7}Li$ $^{33}S + ^{6}Li$	$^{20}Ne + ^{19}F$ $^{23}Na + ^{16}O$ $^{24}Mg + ^{15}N$ $^{25}Mg + ^{14}N$ $^{27}Al + ^{12}C$ $^{29}Si + ^{10}B$
^{40}Ca	$^{28}Si + ^{12}C$	$^{20}Ne + ^{20}Ne$ $^{24}Mg + ^{16}O$

The algebraic structure of the model is given by a direct product of compact unitary algebras. We have performed a simple model calculation in order to study the effects of the more general q-deformed algebras, i.e. the effects of a kinematic symmetry breaking. In particular, the quantum deformed $U_q(3) \supset O_q(3) \supset O_q(2)$ dynamical symmetry has been applied for the description of the $^{16}O + \alpha$ system[26]. Although the quantum algebraic treatment gave a better fit to the experimental data than the Lie algebraic one, the improvement was not significant.

Extensions of this model to several directions are possible and desirable. E.g. symmetry breaking interactions could be applied, spin- and isospin-dependent forces can be introduced, multicluster systems should be investigated, etc. The relative advantages and disadvantages of the present approach can be deduced only after more systematic applications. Nevertheless, based on the first studies mentioned here, it seems to be promising for the description of complex cluster systems.

This work was supported in part by OTKA, Hungary (Grant Nos. 3008, 3010), by the Alexander von Humboldt-Stiftung, and by the DAAD (Germany).

References

[1] J. P. Elliott, *Proc. Roy. Soc. A* **245**, 128 and 562 (1958)
[2] K. Wildermuth and Th. Kanellopoulos, *Nucl. Phys.* **7**, 150 (1958)
[3] B. F. Bayman and A. Bohr, *Nucl. Phys.* **9**, 596 (1958/59)
[4] J. Cseh in: *Proc. Int. Conf. on Nuclear and Atomic Clusters*, Turku, 1991, edited by M. Brenner, T. Lönnroth, F. B. Malik, (Springer, Berlin); J. Cseh, *Phys. Lett. B* **281**, 173 (1992)
[5] J. Cseh and G. Levai, *Ann. Phys. (N.Y.)*, in press
[6] K. Wildermuth and Y. C. Tang, *A Unified Theory of the Nucleus*, (Acad. Press, New York, 1977)
[7] J. P. Draayer, *Nucl. Phys. A* **520**, 259c (1990)
[8] C. Y. Wong *Phys. Lett. B* **32**, 668 (1970)
[9] F. Iachello and A. Arima, *The Interacting Boson Model*, (Cambridge Univ. Press, Cambridge, 1987)
[10] M. Harvey in: *Proc. 2nd Conf. on Clustering Phenomena in Nuclei*, College Park, 1975, USDERA report ORO-4856-26 (1975)
[11] E. P. Wigner, *Phys. Rev.* **51**, 106 (1937)
[12] F. Iachello and R. D. Levine, *J. Chem. Phys.* **77**, 3046 (1982)
[13] H. Horiuchi and K. Ikeda, *Cluster Models and Other Topics*, (World Scientific, Singapore, 1986)
[14] J. Cseh and G. Levai, *Phys. Rev. C* **38**, 972 (1988)
[15] J. Cseh, *J. Phys. Soc. Jpn. Suppl.* **58**, 604 (1989)
[16] J. P. Draayer and Y. Akiyama, *J. Math. Phys.* **14**, 1904 (1973)
[17] B. G. Wybourne, *Classical Groups for Physicists*, (Wiley, New York, 1974)
[18] K. Varga and J. Cseh, *Phys. Rev. C* **48**, 602 (1993)
[19] J. Cseh and W. Scheid, *J. Phys. G* **18**, 1419 (1992)
[20] G. Leander and S. E. Larsson, *Nucl. Phys. A* **239**, 93 (1975)
[21] J. Cseh, *J. Phys. G* **19**, L97 (1993)
[22] G. Levai, J. Cseh and W. Scheid, *Phys. Rev. C* **46**, 548 (1992)
[23] J. Cseh, G. Levai and W. Scheid, *Phys. Rev. C* **48**, 1724 (1993)
[24] J. Cseh, Z. Hornyak and Sz. Vattamany, unpublished
[25] J. Cseh, R.K. Gupta and W. Scheid, *Phys. Lett. B* **299**, 205 (1993)
[26] J. Cseh, *J. Phys. G* **19**, L63 (1993)

NUCLEAR SHAPES AT VERY HIGH ANGULAR MOMENTA

Raj K. Gupta[1,2], J. S. Batra[1], S. S. Malik[1], P. O. Hess[3] and W. Scheid[2]

[1] *Physics Department, Panjab University, Chandigarh-160014, India*
[2] *Institut für Theoretische Physik, Justus-Liebig-Universität, Giessen, Germany*
[3] *Instituto de Ciencias Nucleares, UNAM, Mexico D.F., Mexico*

The knowledge of nuclear shapes from ground-state up to scission is of interest from the points of views of the not-yet measured quadrupole deformations or B(E2) transitions for very high spin states, the variations of moments of inertia with angular momenta, the limiting angular momentum at scission and for distinguishing between the processes of fission and cluster preformation in nuclei. Some of these questions are answered here in terms of a model that uses the two centre shell model (TCSM) parametrization of the nuclear shape and the measured ground-state (g.s.) yrast band energies and quadrupole deformations of these states. This work is an extension of the one for quasi-molecular resonance states in heavy-ion collisions[1].

We assume that the g.s. yrast band energies of nuclei are given by

$$E_J = \frac{\hbar^2}{2\mathcal{I}_J}J(J+1). \tag{1}$$

By knowing E_J from experiment, this relation allows us to compute the empirical $\mathcal{I}_J^{expt}$, which is fitted to the calculated moment of inertia for different nuclear shapes arising from TCSM parametrization.

The moment of inertia for TCSM nuclear shape is given in Ref. 2, in terms of the major- and minor-axes a_i, b_i, respectively, the c.m. positions z_i (i=1,2) and the neck parameter ϵ:

$$\mathcal{I}_{TCSM} = f(a_i, b_i, z_i, \epsilon). \tag{2}$$

This relation has the interesting feature of giving in one limit the g.s., spheroidal nucleus, moment of inertia for $z_1 = z_2 = 0$ and $\epsilon = 1$ and, in the other limit, the sticking moment of inertia $\mathcal{I}_s$ of two touching spheres (scission configuration).

For symmetric division, the four parameters of TCSM nuclear shape are obtained by solving eq. (1) simultaneously, with the volume of TCSM shape made equal to the volume of an equivalent spherical nucleus, minimizing the surface area of the TCSM shape, and letting the deformation of each fragment be equal to that of the rotating nucleus at each angular momentum. This resulted in unique nuclear shapes, variations of moments of inertia $\mathcal{I}$, the necking-in parameter ρ_0 (related to ϵ), and the calculated deformations β as a function of angular momenta J. Extrapolating $\rho_0(J)$ to zero, gives the limiting J-value for the touching configuration, called J_s. Finally, the role of deformation energy in the potential energy surface is studied, which modifies the moment of inertia $\mathcal{I}_s$ to, say, $\mathcal{I}_{lim}$.

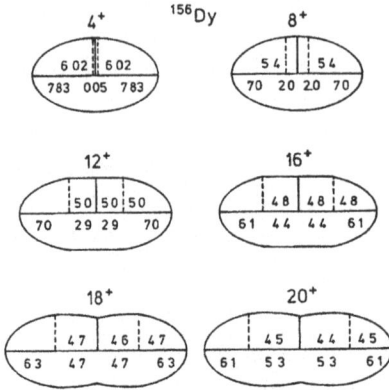

Figure 1. *The derived nuclear shapes for the g.s. yrast band of* ^{156}Dy.

We have applied the above model to ^{156}Dy, ^{158}Er and ^{164}Hf nuclei. The necked nuclear shapes appear already at $J^\pi = 14^+ - 18^+$. This is illustrated in Fig. 1 for ^{156}Dy. For touching configuration ($\rho_0 = 0$), the limiting angular momentum is $J_s^\pi = 60^+ - 118^+$. The interesting result is that the empirically determined moments of inertia $\mathcal{I}_J^{expt}$ and the ($\mathcal{I}_s, J_s$) point fit nicely the nuclear softness (NS) model[3] expression (in terms of various orders of nuclear softness parameters, $\sigma_n = \frac{1}{\mathcal{I}_0 n!}\frac{\partial^n \mathcal{I}_0}{\partial J^n}$)

$$\mathcal{I}_J = \mathcal{I}_0(1 + \sigma_1 J + \sigma_2 J^2 + \sigma_3 J^3), \tag{3}$$

as well as an empirical expression[4] in terms of a ratio of two polynomials

$$\mathcal{I}_J = \frac{a + bJ^2 + cJ^4 + dJ^6}{1 + sJ^2}. \tag{4}$$

Eq. (3) represents the gradual softening of a rigid rotator and eq. (4) gives both the backward- and forward-bending for $J^\pi \leq 30^+$. Also, the empirical variation of the quadrupole deformation of the rotating nucleus $\beta_A^{expt}(J)$, is given, on the average, by the calculated $\beta(J)$. The deformation energy is shown to decrease the moment of inertia $\mathcal{I}_s$, which in (1) is included via the J-dependence of $\mathcal{I}$.

The model developed is useful for predicting both the energies (using the intrapolated $\mathcal{I}_J$) and quadrupole deformations or the B(E2) transitions (from intrapolated nuclear shapes or the extrapolated calculated $\beta(J)$) for high spin states of a nucleus before the fission stage is reached.

Acknowledgements: This work is supported in parts by the UGC (India), the AvH-Stiftung and DAAD (Germany) and the DGAPA (Mexico).

References

[1] H. S. Khosla, S. S. Malik and R. K. Gupta, *Nucl. Phys. A* **513**, 115 (1990)
[2] R. Aroumougame, N. Malhotra, S. S. Malik and R. K. Gupta, *Phys. Rev. C* **35**, 994 (1987)
[3] R. K. Gupta, *Phys. Lett.* **36B**, 173 (1971)
[4] S. Wahlborn and R. K. Gupta, *Phys. Lett.* **40B**, 27 (1972)

V.

Weak Interaction

and Double Beta Decay

DOUBLE BETA DECAY AND NEUTRINO MASS. THE HEIDELBERG–MOSCOW EXPERIMENT

H. V. Klapdor-Kleingrothaus[1]

Max-Planck-Institut für Kernphysik, Heidelberg, Germany

Introduction

The neutrino is one of the best examples for the merging of the different disciplines of micro- and macrophysics. The neutrino plays, by its nature (Majorana or Dirac particle) and its mass, a key role for the structure of modern particle physics theories (GUTs, SUSYs, SUGRAs, ...) [1]–[3]. At the same time it is candidate for non-baryonic dark matter in the universe, and the neutrino mass is connected, by the sphaleron effect, to the matter – antimatter asymmetry of the early universe [4].

In the search for the neutrino mass we can differentiate between different approaches, which may be classified as terrestrial and extra-terrestrial, among the latter the most spectacular being the investigation of solar neutrinos and neutrinos from supernovae. In principle also the up to now unobserved cosmic neutrino background radiation contains a mass information. Since the neutrino oscillation experiments measure the difference of the (squared) mass eigenvalues of different neutrino flavors $\Delta m^2 = |m_1^2 - m_2^2|$, they primarily yield information on the heavier neutrino flavors and thus are *complementary* to those non-accelerator experiments like β and $\beta\beta$ decay measuring directly the electron neutrino mass. The sharpest limits on the electron neutrino come at present from non-accelerator experiments: tritium decay and double beta decay. Both of these experiments again are complementary: while the effect of a neutrino mass in the β decay spectrum of tritium is independent of the nature of the neutrino (Dirac or Majorana particle), double beta decay can measure only an effective Majorana mass of the neutrino. So only study of both processes together gives the required information on the Dirac and Majorana mass terms in the neutrino mass matrix. Double beta decay can also probe in a very sensitive way the hypothesis of a 17 keV neutrino, if this is a Majorana particle, by its influence on the effective mass observed in $\beta\beta$ decay.

There are several decay modes discussed in $\beta\beta$ decay, the main ones being two neutrino (2ν) and neutrinoless (0ν) decay (Fig. 1)

$$^A_Z X \rightarrow {}^A_{Z+2} X + 2e^- + 2\bar{\nu}_e \tag{1}$$

$$^A_Z X \rightarrow {}^A_{Z+2} X + 2e^- . \tag{2}$$

---

[1] Speaker of the HEIDELBERG–MOSCOW collaboration

In addition there could occur 0ν decay accompanied by emission of one or two majorons (Fig. 1)

$$\begin{align}
{}^A_Z X &\rightarrow {}^A_{Z+2} X + 2e^- + \chi \tag{3} \\
{}^A_Z X &\rightarrow {}^A_{Z+2} X + 2e^- + \chi + \chi \tag{4}
\end{align}$$

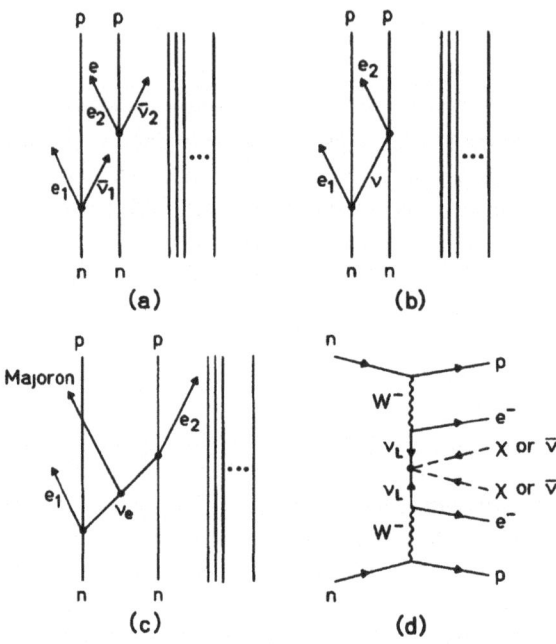

Figure 1. Graphs for 2ν (a), 0ν (b), $0\nu\chi$ (c) and $0\nu\chi\chi$ (d) double beta decay

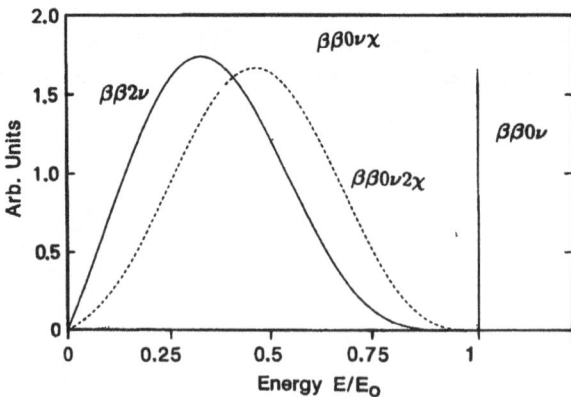

Figure 2. Shapes of the electron sum spectra of different $\beta\beta$ decay modes

Fig. 2 shows the corresponding spectral shapes of the electron sum spectra. The first process is a usual effect of second order in the classical weak interaction, the second requires as a prerequisite a non-vanishing Majorana neutrino mass or a right-handed weak interaction. In the framework of GUTs, or more general gauge theories, both of these conditions cannot be seen independently, but in any case a non-vanishing mass is required. In left-right symmetric models more exotic processes then [1]–[4] might contribute to the $0\nu\beta\beta$ amplitude like exchange of very heavy neutrinos or Higgs particles (see e.g. [3]). In SUSY models modes connected with gaugino exchange might dominate $0\nu\beta\beta$ decay [5] (see Fig. 3).

There are three possibilities to break B-L (baryon number minus lepton number) in theory [3], which is required to produce a Majorana mass: (1) explicit B-L breaking, i.e. the Lagrangian contains B-L violating expressions; (2), (3) spontaneous breaking of a local or global B-L symmetry. The existence of the majoron χ is associated with the third possibility. From the different majoron models characterized by their weak isospin, connected with different possibilities to generate Majorana mass terms in extensions of the standard model two are ruled out by the recent LEP measurements [6]: the triplet majoron [7], which would contribute an effect equivalent to two neutrino flavors to the Z^0 width, and the doublet majoron [3], which should contribute half a neutrino width. However, the existence of singlet majorons or a mixture of singlet- and doublet majorons is possible and it might show up in $\beta\beta$ decay (see e.g. [8]). Investigation of the doublet majoron, which would be associated with a double beta decay mode with emission of two majorons, could allow to make conclusions on the Zino mass [3]. The model of singlet majorons has recently experienced large interest in connection with attempts to construct neutrino mass hierarchies involving a 17 keV neutrino [9, 10]. It is therefore important to search for this decay mode.

It has to be noted that if the electron neutrino is a mixed state (mass matrix is not diagonal in the flavor space):

$$|\nu_e> = \sum_i U_{ei} |\nu_i>, \tag{5}$$

then the measured quantity in double beta decay is an effective mass

$$<m_\nu> = \sum_i m_i U_{ei}^2 \tag{6}$$

which for appropriate CP phases of the mixing coefficients U_{ei} could be smaller than m_i for all i. Neutrino oscillation experiments yield some restrictions of this possibility (see [2]). In general not too pathological models seem to yield $m_{\nu_e} = <m_{\nu_e}>$ (see [1]).

We can differentiate between two classes of direct (non-geochemical) $\beta\beta$ decay experiments (Fig. 4):
a) active source experiments (source = detectors)
b) passive source experiments.

In the first class of experiments the $\beta\beta$ process usually is identified only on the basis of the distribution of the total energy of the electrons. The second class of experiments yields more complete information on the $\beta\beta$ events by measuring time coincidence, tracks and vertices of the electrons and their energy distribution. But also time projection chambers (TPCs) using $\beta\beta$ active counting gas are belonging to the first class - such as the Gotthard ^{136}Xe experiment.

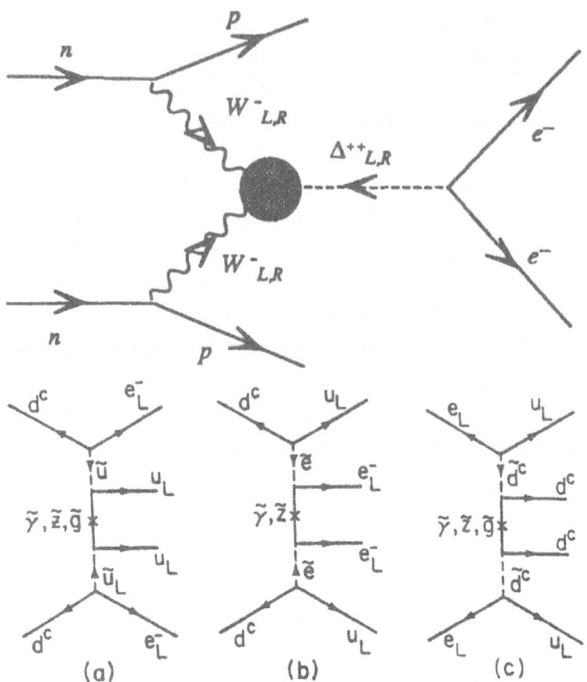

Figure 3. *Graphs contributing to $0\nu\beta\beta$ decay in left-right symmetric models by Higgs exchange, and in some SUSY models by gaugino exchange (from [3,5])*

It is obvious from Fig. 5, which shows an overview over measured $0\nu\beta\beta$ half-life limits and deduced mass limits, that the largest sensitivity for $0\nu\beta\beta$ decay is obtained at present by active source experiments, in particular ^{76}Ge [11] and ^{136}Xe [12] (only the geochemical experiment with ^{128}Te reaches a similar limit). The main reason is that large source strengths can be used (simultaneously with high energy resolution), in particular when enriched $\beta\beta$ emitter materials are used.

Other criteria for the "quality" of a $\beta\beta$ emitter are:

- a small product $T^{0\nu}_{1/2} \cdot <m_\nu>^2$, i.e. a large matrix element $M^{0\nu}$
- a $Q_{\beta\beta}$ value beyond the limit of natural radioactivity (2.614 MeV).

The future (next 5 years) of $\beta\beta$ experiments will be dominated by use of enriched detectors, ^{76}Ge playing a particularly favourable role here [11, 13], and enriched source material such as ^{136}Xe (see [12]), ^{100}Mo [14], ^{116}Cd [15] (Fig. 6). Such experiments can probe the neutrino mass in the next years down to about 0.1 eV.

Calculation of $\beta\beta$ matrix elements, OEM

For deduction of an (effective) neutrino mass (limit) from a measured 0ν decay rate (limit), the calculation of nuclear matrix elements is required. The half-life for $0\nu\beta\beta$ decay is given by

$$[T^{0\nu}_{1/2}(0^+_i \to 0^+_f)]^{-1} =$$

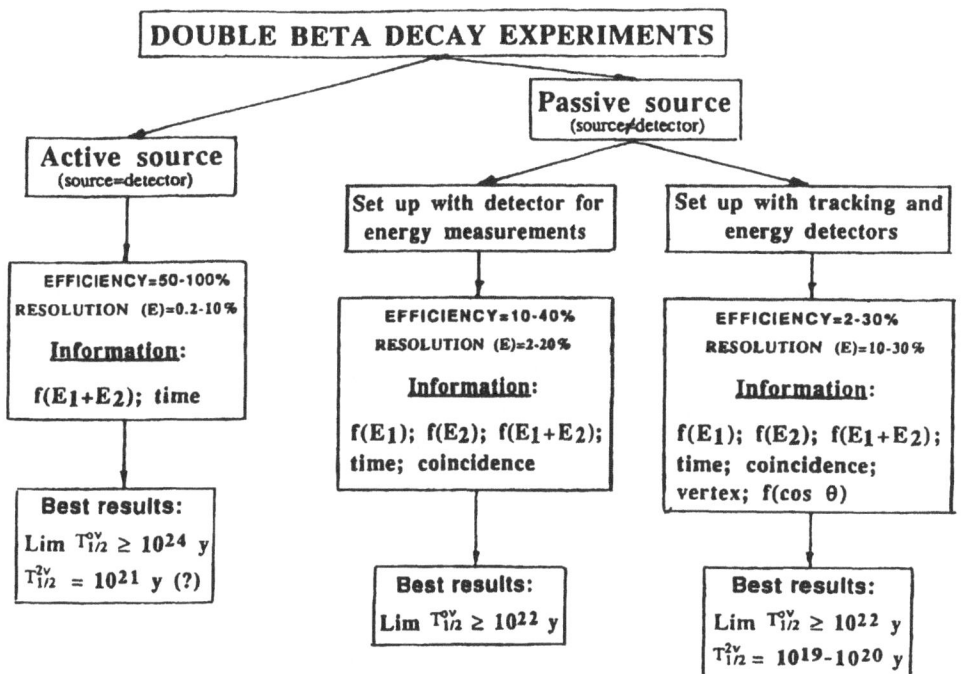

Figure 4. Classification of $\beta\beta$ decay experiments

$$C_{mm}\frac{<m_\nu>^2}{m_e^2} + C_{\eta\eta}<\eta>^2 + C_{\lambda\lambda}<\lambda>^2 \qquad (7)$$
$$+ C_{m\eta}<\eta>\frac{<m_\nu>}{m_e} + C_{m\lambda}<\lambda>\frac{<m_\nu>}{m_e} + C_{\eta\lambda}<\eta><\lambda>$$

or, when neglecting the effect of right-handed weak currents, by

$$[T^{0\nu}_{1/2}(0^+_i \to 0^+_f)]^{-1} = C_{mm}\frac{<m_\nu>^2}{m_e} = (M^{0\nu}_{GT} - M^{0\nu}_F)^2 G_1 \frac{<m_\nu>^2}{m_e} \qquad (8)$$

where G_1 denotes the phase space integral.

The half-life for $0\nu\chi$ decay is given by

$$[T_{1/2}]^{-1} = (M^{0\nu}_{GT} - M^{0\nu}_F)^2 F^{0\nu\chi}(<g_{\nu\chi}>)^2. \qquad (9)$$

Here the neutrino-majoron coupling constant $<g_{\nu\chi}>$ is given by

$$<g_{\nu\chi}> = \sum_{i,j} g_{\nu\chi} U_{ei} U_{ej} \qquad (10)$$

and $F_{0\nu\chi}$ denotes the phase space. The half-life of the $0\nu\chi\chi$ decay is

$$[T_{1/2}]^{-1} = (f_{\chi\chi} - m_e)^2 (M^{0\nu}_{GT} - M^{0\nu}_F)^2 F^{0\nu\chi\chi}. \qquad (11)$$

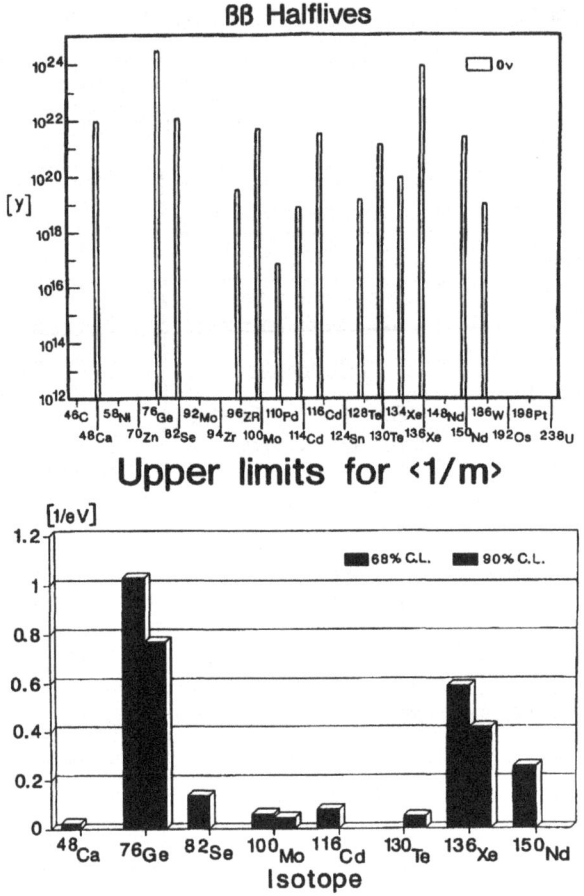

Figure 5. *Measured $0\nu\beta\beta$ half-life limits (a) and deduced mass limits (b)*

The coupling constant f_{xx} here is connected with the Zino mass according to

$$f_{xx} = \frac{g^2}{4M_Z \cos\Theta_W}. \tag{12}$$

After the major step of recognizing the importance of g.s. correlations for their calculation [16], in recent years the main groups used the QRPA model for calculation of $M^{0\nu}$. It was found that in most cases the uncertainties of the model were tolerable in the case of $0\nu\beta\beta$ decay (see [17] and Fig. 7). The problem of QRPA - extreme dependence of the calculated rate (for 2ν decay) on the choice of the renormalization of the particle-particle force, g_{pp}, seems recently to have been solved by applying the so-called Operator Expansion Method (OEM) [18, 19]

This method does not explicitly use the intermediate energy spectrum. This is the essential advantage over QRPA, since in this way the dependence on the pp force (which affects the distribution of β strength in the intermediate nucleus) is drastically reduced.

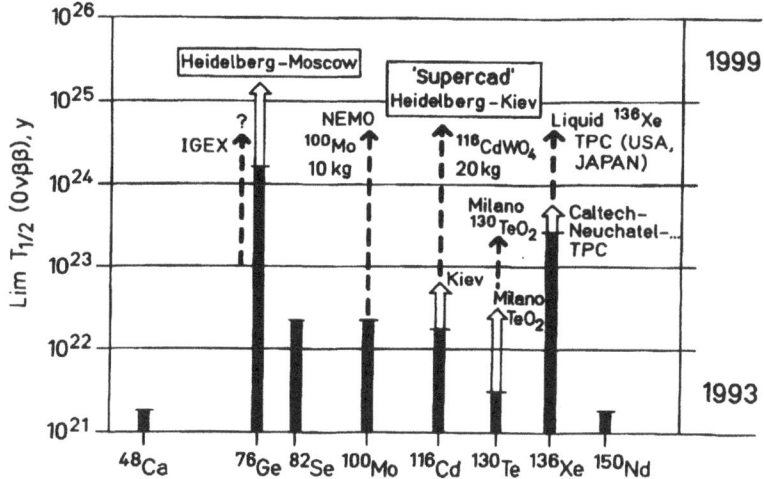

Figure 6. Present situation and perspectives of the most promising $\beta\beta$ experiments. Only for the isotopes shown $0\nu\beta\beta$ half-life limits $> 10^{21}$ y have been obtained. The thick solid lines correspond to the present, 1993, status, open bars and dashed lines to 'safe' and 'less safe' expectations for 1999, respectively (from [15]).

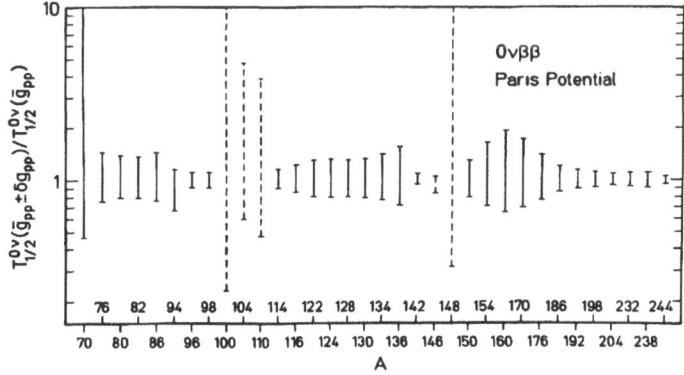

Figure 7. The uncertainty of QRPA-calculated $0\nu\beta\beta$ half-lives originating from the limited knowledge of the pp-force for the potential $\beta\beta$ emitters (from [17]).

The method allows (by ignoring many-particle scattering terms) to write the matrix element for 2ν decay in the form

$$M_{GT} = <0_F^+|\sum_{i\neq j} M_{ij}\tau_i^- \tau_j^-|0_I^+> = M_0 + M_1 \qquad (13)$$

$$M_{ij} = \frac{12(v_\sigma(r) - v_\tau(r))\Omega_0(ij)}{\Delta^2 - 16(v_\sigma(r) - v_\tau(r))^2} + \frac{4(2v_{\sigma\tau}(r)) - v_\sigma(r) - v_\tau(r))\Omega_1(ij)}{\Delta^2 - 16(2\nu_{\sigma\tau}(r) - \nu_\tau(r))^2}. \qquad (14)$$

Thus the matrix element involves the bare nucleon-nucleon interaction without any adjustable parameter. The wave functions of initial and final states we take from QRPA. Figure 8 shows the results for ^{100}Mo and ^{238}U. Calculations for all potential $\beta\beta$ emitters by this method are given in [19]. $0\nu\beta\beta$ matrix elements using OEM are very close to those of QRPA. That the dependence on the pp force is overestimated by QRPA has been shown recently also by comparing QRPA and shell model for ^{48}Ca [20]. It may be noted that the calculation of ^{238}U $\beta\beta$ decay by OEM yielding $T_{1/2}^{2\nu} = 0.9 \times 10^{21}\, y$ [19], was performed *before* the experimental result of [21] was known to us. The far-reaching conclusions drawn in that paper seem not to be on stable grounds.

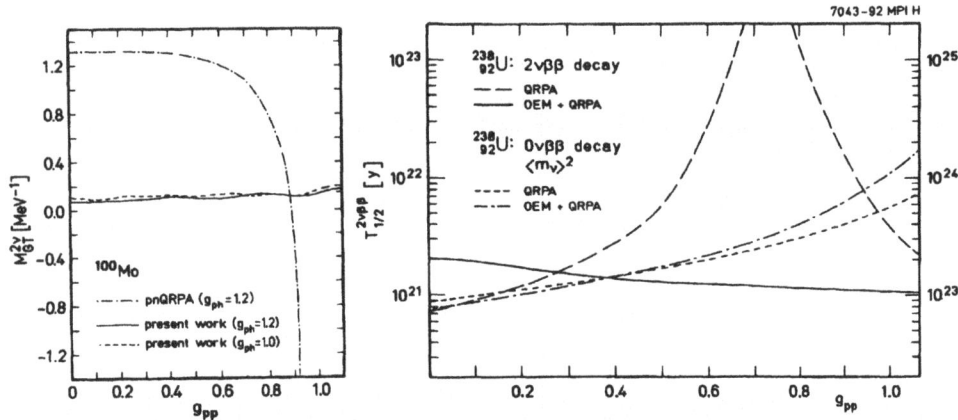

Figure 8. The calculated matrix elements $M_{GT}^{2\nu}$ of $2\nu\beta\beta$ decay of ^{100}Mo (a) and the half-lives of 0ν and 2ν decay of ^{238}U as function of the strength of the particle-particle force (g_{pp}) in QRPA and OEM (from [19]).

Status of the Heidelberg–Moscow experiment

Use of enriched ^{76}Ge (86%) instead of detectors from natural Ge (containing 7.8% of ^{76}Ge) could explore the half-life of $0\nu\beta\beta$ decay up to $\geq 10^{25}$ years and correspondingly the neutrino mass down to $\sim 10^{-1}$ eV, probing a class of left-right symmetric GUT models with a right-handed Majorana mass term of about 1 TeV, based on $SU(2)_L \otimes SU(2)_R \otimes U(1)$ (see [1], [11]). The HEIDELBERG-MOSCOW $\beta\beta$ experiment [11] makes use of 16.9 kg of ^{76}Ge metal enriched to 86%, corresponding to 14.5 kg of the isotope ^{76}Ge. Up to now one enriched detector of $\sim$ 1 kg, another of 2.9 kg (the largest ever produced Ge-detector at that time) and another three detectors (or crystals) of 2.5, 2.9 and 3.5 kg, respectively, have been produced. The first detector is running in the GRAN SASSO Underground Laboratory in Italy since end of July 1990, the second since September 1991, the third since August 1992 (Fig. 9).

Figure 9. *The $\beta\beta$ laboratory of the HEIDELBERG-MOSCOW experiment in the GRAN SASSO near Rome (a). Mounting of the enriched detectors under low-level conditions (b).*

0ν-decay: The results of the first 2649 days × kg of measurement are:
$T_{1/2}^{0\nu}(0^+ \rightarrow 0^+) > 1.9 \times 10^{24}$ (90%c.l.) (Fig. 10). The corresponding limit for the neutrino mass is $m_\nu < 1.1$ eV. The HEIDELBERG-MOSCOW experiment thus yields now the most stringent limit for the $0\nu\beta\beta$ half-life of ^{76}Ge and the sharpest limit for the neutrino mass

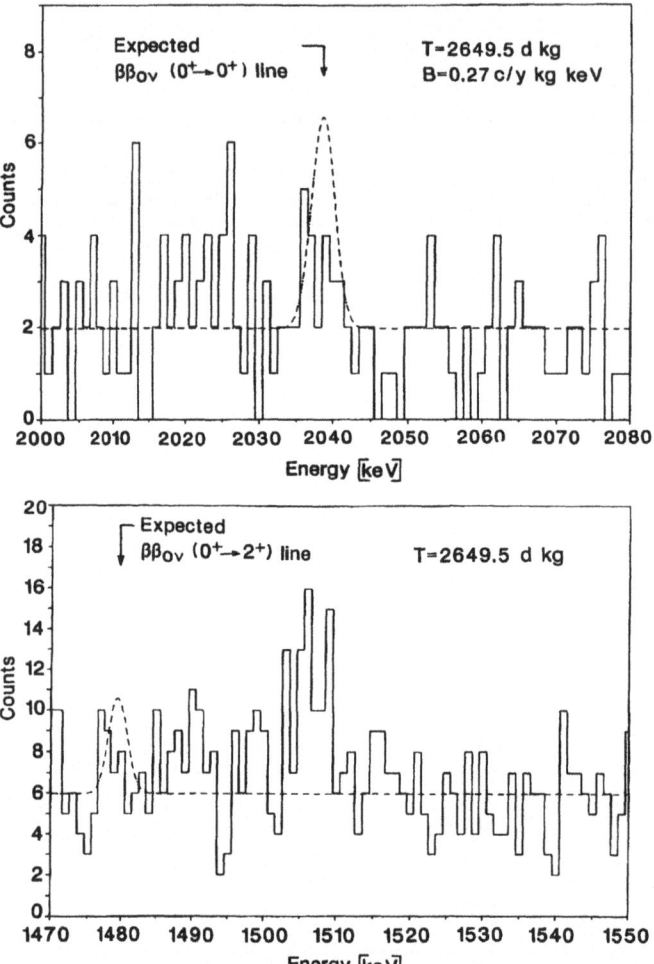

Figure 10. Details of the spectrum of the *HEIDELBERG-MOSCOW* collaboration in the region of neutrinoless double beta decay to the ground and first excited state of the daughter nucleus after a measuring time of 2649 kg d. Dashed-dotted peaks are excluded with 90% c.l. ,

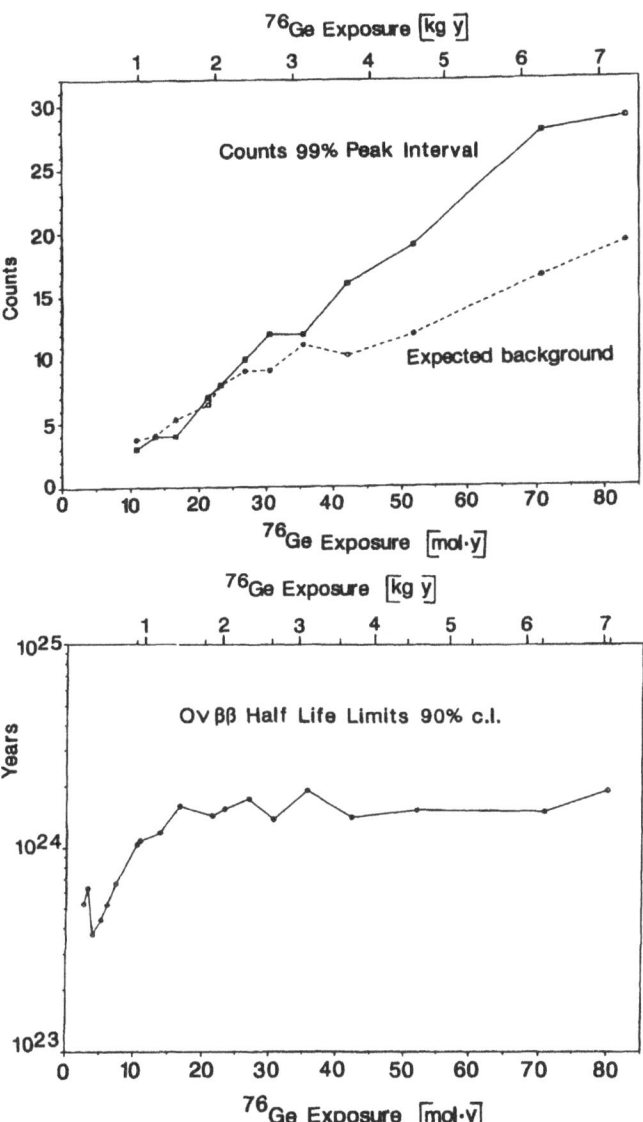

Figure 11. (a) Number of expected background counts (dashed line) and measured total number of counts (solid line) in the 3σ energy range of a $0\nu\beta\beta$ ($0^+ \to 0^+$) transition of ^{76}Ge, as function of measuring time in the HEIDELBERG-MOSCOW experiment. The expected background is extrapolated from $\sim$ 30 keV ranges left and right from the centroid energy (2038.6 keV). The deviation under a no-line hypothesis between observed and expected counts corresponds to $\sim$ 2.4 σ (from [28]). (b) Deduced $0\nu\beta\beta$ half-life limit for ^{76}Ge as function of measuring time in the HEIDELBERG-MOSCOW experiment (from [28]).

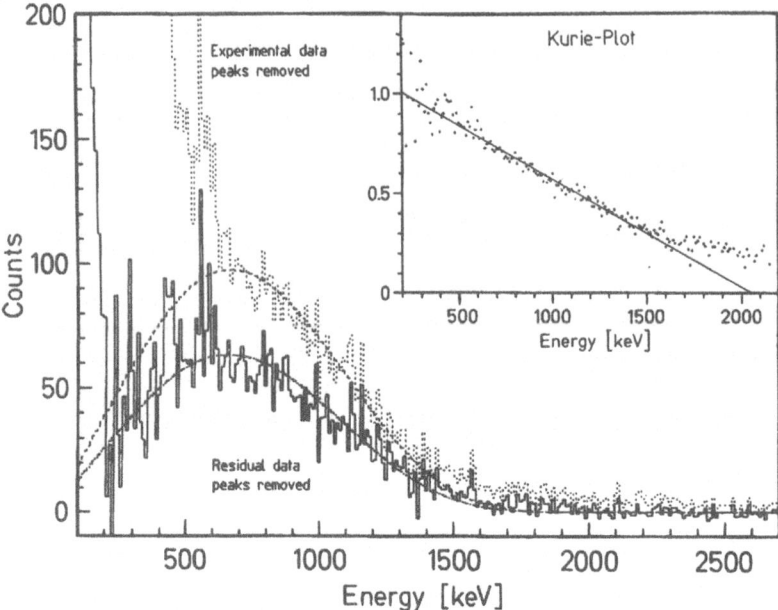

Figure 12. *Measured spectrum after 615 kg d (dotted curve, lines cut out) with the second detector, the resulting spectrum after subtraction (see [24]) of the background (solid histogram), and a fitted 2ν spectrum with $T^{2\nu}_{1/2} = 1.42 \times 10^{21}$ y (lower dashed-dotted curve). The dashed curve corresponds to a half-life of 0.9×10^{21} y claimed by [34].*

from detector experiments (for details see [11]). On the other hand we seem to observe (Fig. 11a) a systematic excess of events beyond the expected background precisely at the energy of a $0\nu\beta\beta$ $(0^+ \to 0^+)$ transition (present significance $\sim 2.4\,\sigma$). This coincides with the observation of a stagnating $0\nu\beta\beta$ half-life limit (Fig. 11b) as function of measuring time inspite of decreasing background. These effects require further attention. For their further investigation we have developed [22] a method of pulse shape analysis which will allow rejection of $\sim 80\%$ of multiple (non-$\beta\beta$) events.

The background level which characterizes the quality of the set-up is 0.29 events/(kg · year · keV) year keV (Fig. 10) in an 80 keV interval around the hypothetical 2038.5 keV $0\nu\beta\beta$-line for the total spectrum, and ~ 0.2 events/(kg · year · keV) at present. As an upper limit for a $\beta\beta$ transition to the first excited state of ^{82}Se which should occur at 1479.5 keV, we deduce a half-life limit $T^{0\nu}_{1/2}(0^+ \to 2^+) > 8.0 \times 10^{23}$ (90%c.l.), i.e. far beyond the half-life of 2.5×10^{22} years claimed by [23] (Fig. 10).

2ν-decay: For $2\nu\beta\beta$ decay subtraction of the background contributions from cosmogenically produced isotopes, natural decay chains, ^{210}Pb etc. from the spectrum measured with the second detector in 615 kgd yields [24] the spectrum shown in Fig. 12 and $T^{2\nu}_{1/2}(0^+ \to 0^+) = (1.42 \pm 0.03\,(\text{stat}) \pm 0.13\,(\text{syst})) \times 10^{21}$ y. This is probably the first undoubtable evidence for this up to now rarest nuclear decay mode. In the derivation of the experimental 2ν spectrum of Fig. 12 for the first time a parameter-free background model basing on quantitative identification and localisation of the different radioactive impurities

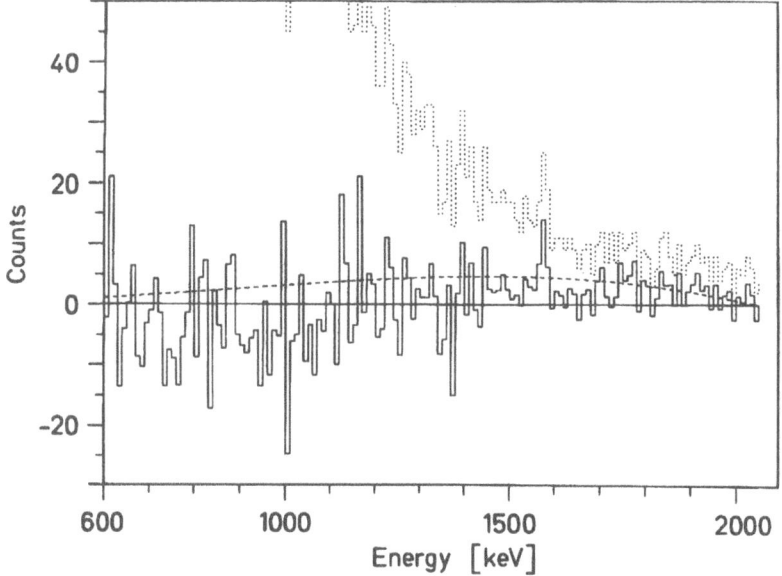

Figure 13. Spectrum remaining after subtraction of the 2ν spectrum from the solid histogram of Fig 12, together with a calculated $0\nu\chi$-spectrum with $T_{1/2}^{0\nu\chi} = 1.66 \times 10^{22}$ y (dashed curve) Also shown is the measured spectrum (dotted histogram)

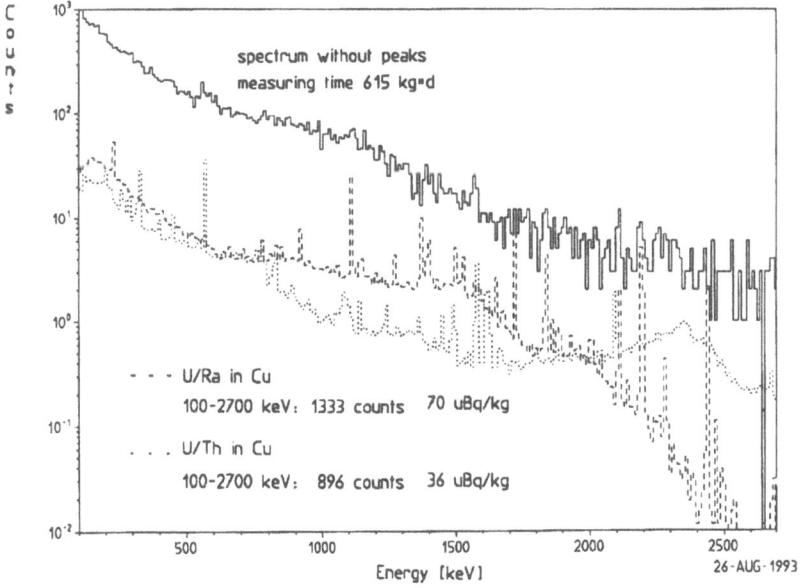

Figure 14. (a)

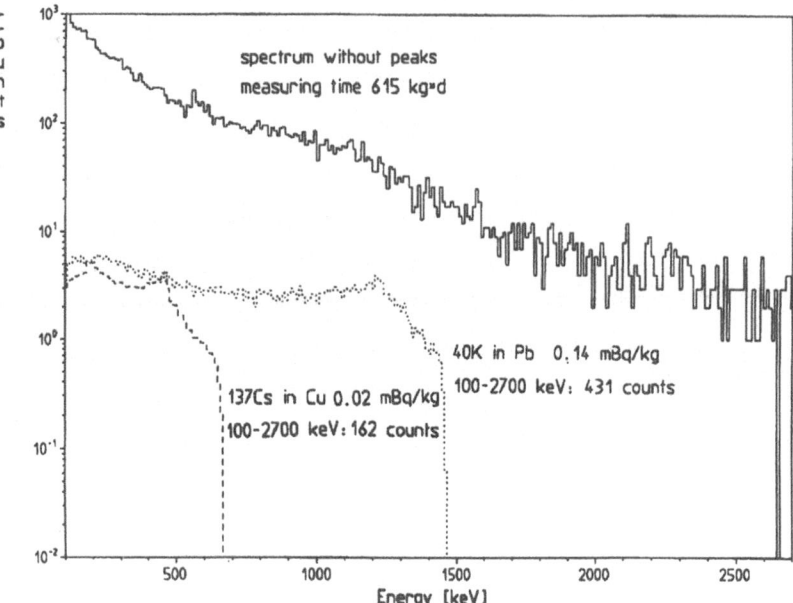

Figure 14. *(b)*

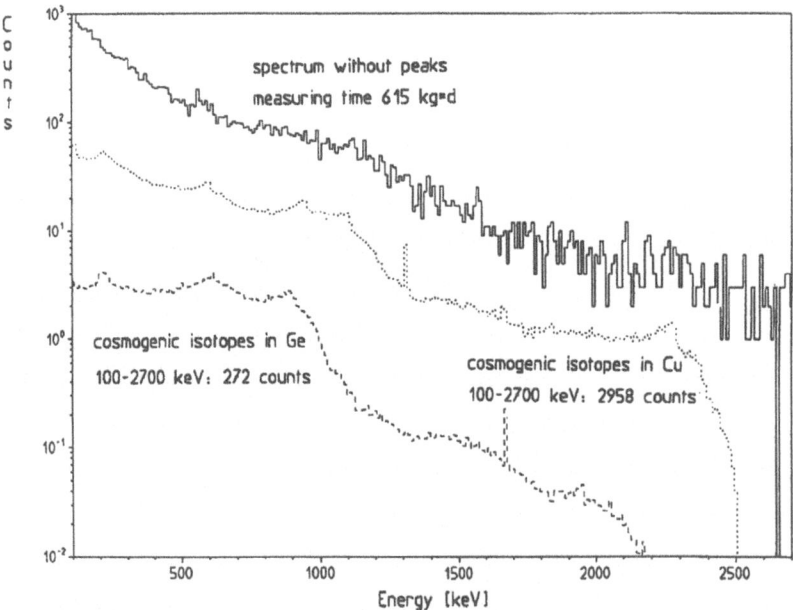

Figure 14. *(c)*

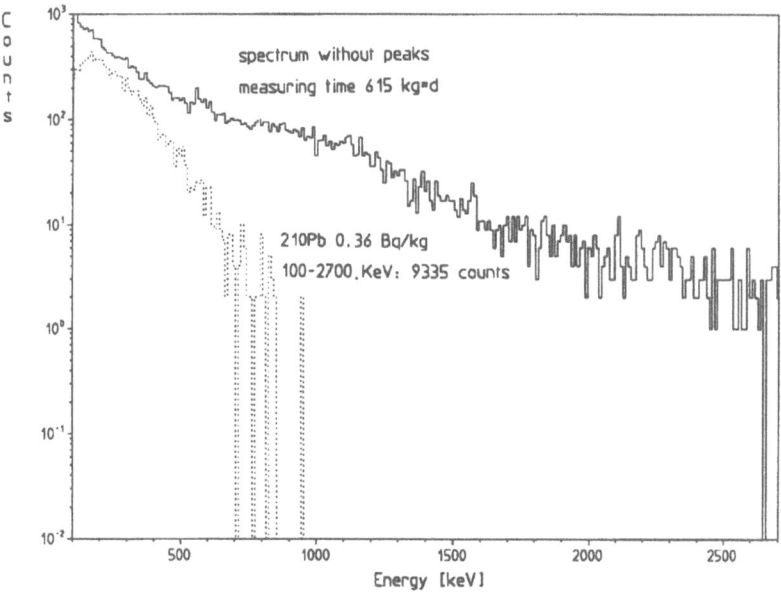

Figure 14. *(d) Fig. 14 (a) – (d): Some of the identified contributions to the observed spectrum from various radioactive impurities according to Monte Carlo calculations (from [29]).*

and estimate of their effect by the code GEANT has been applied. For details see [29] (see Figs. 13 and 14).

$0\nu\chi$ **-decay**: For Majoron-accompanied, $0\nu\chi$-$\beta\beta$ decay, subtraction of the 2ν-spectrum from Fig. 12 leads to the spectrum of Fig. 13 and an upper limit $T_{1/2}^{0\nu\chi} > 1.7 \times 10^{22}$ y (90% c.l.) and for the neutrino-majoron coupling constant to $<g_{\nu\chi}> < 1.8 \times 10^{-4}$ (90% c.l.) [25] thus excluding some recently discussed [26] indication of this decay mode.

e^-**-decay**: As a by-product of the $\beta\beta$-experiment we obtain [27] for the decay mode $e^- \to \gamma + \nu_e$ $T_{1/2} > 1.63 \times 10^{25}$ y (68% c.l.) which is the sharpest laboratory limit existing.

In total we hope to have 12 kg of enriched detectors in operation in the Gran Sasso by end of 1994. These would correspond to a non-enriched Ge experiment of at least 1.2 tons. In five years of measurement we could probe the neutrino mass down to ~ 0.1 eV.

Application of $\beta\beta$-technology to dark matter search and γ-line astrophysics

The results of the HEIDELBERG–MOSCOW experiment show for the first time that the technology of production of enriched HP Ge detectors can be handled. This has promising consequences for dark matter search and galactic γ-spectroscopy satellite missions (see [11, 30, 31, 32]). Enriched ^{76}Ge and ^{73}Ge detectors would allow to improve the LEP limits on Dirac- and Majorana WIMPs (weakly interacting massive particles). Use of enriched ^{70}Ge detectors in the ESA project INTEGRAL (International Gamma Ray Astrophysics Laboratory) would allow to reduce the β-background dominating in orbit, by an order of magnitude [31]. This has recently been checked in balloon flights [32] (see Fig. 15). The

INTEGRAL satellite mission aims at the study of a variety of astrophysical questions by high resolution γ-spectroscopy.

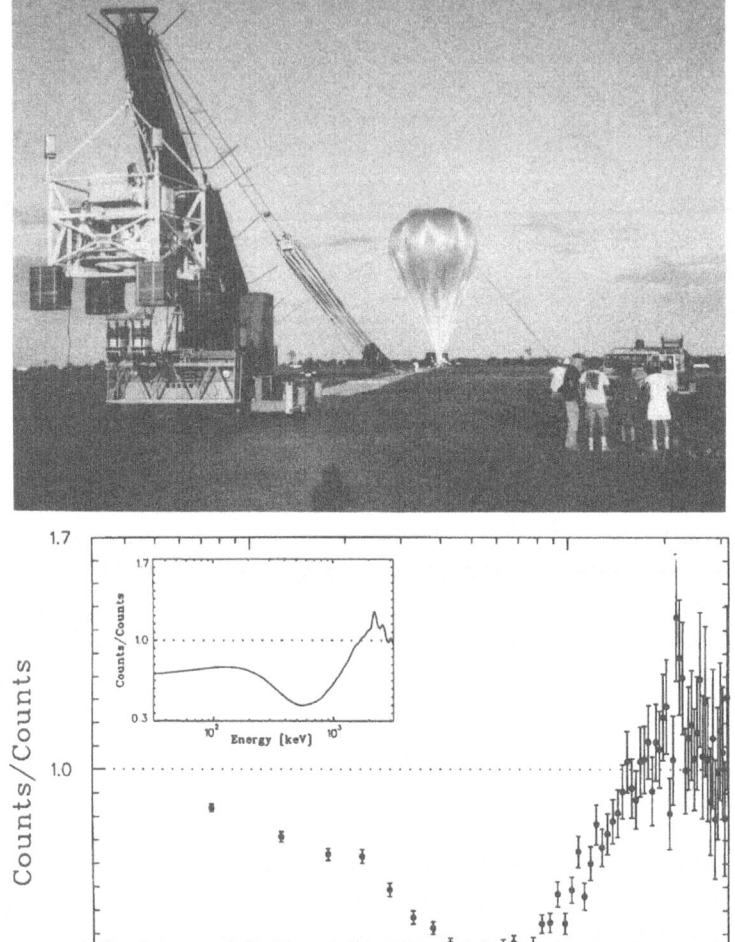

Figure 15. *Balloon campaign GRIS in Alice Springs (Australia), May 1992, with one enriched ^{70} Ge detector see [32] (upper half). Background ratio of enriched (in ^{70} Ge to 95%) to natural detector in the GRIS balloon campaign in Australia, 1992 (lower half).*

The HEIDELBERG-MOSCOW experiment allowed for the first time a search for dark matter with isotopically enriched material. Data taken with a 2.9 kg detector in 166 kgd were used to set limits on spin-independently interacting WIMPs [28]. A background level of 0.102 ± 0.005 events/(kg·d·keV) was achieved (average value between 11 and 30 keV), which is better by a factor of 5-7 than in other dedicated dark matter experiments with Ge

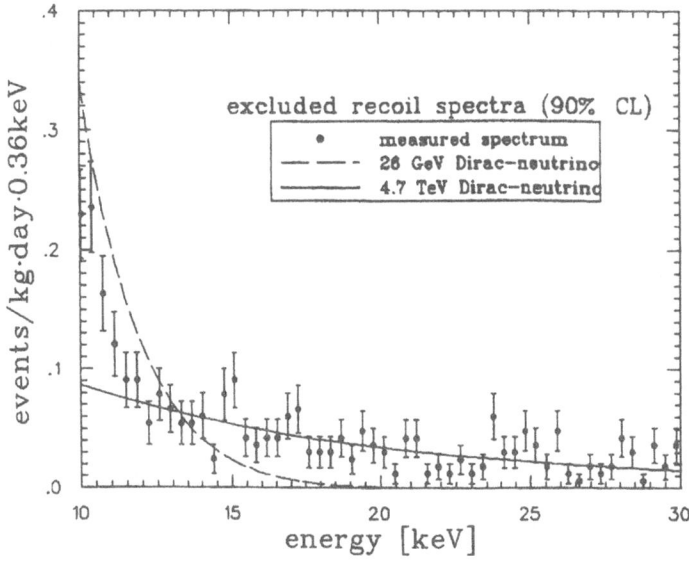

Figure 16. Low energy spectrum of the second enriched detector corresponding to a measuring time of 165 6 kgd The background counting rate in the energy range 11-30 keV is 0 102 ± 0 005 events/(kg d keV) Also shown are the expected recoil spectra of Dirac neutrinos with 26 GeV (dashed line) and 4 7 TeV (solid line) (from [28])

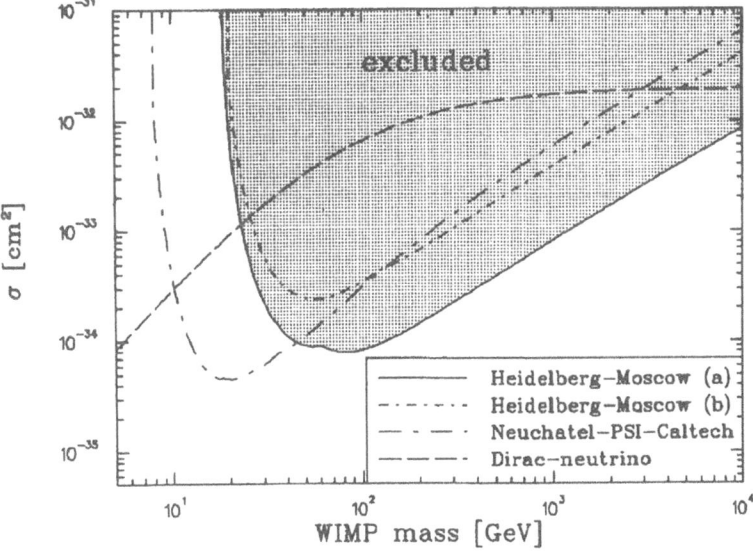

Figure 17. Exclusion limits for elastic scattering cross sections of WIMPs on ^{76}Ge from the HEIDELBERG-MOSCOW experiment (a) not corrected (b) corrected for loss of coherence, compared with the best result obtained with a natural Ge detector [33] (dashed-dotted line) WIMPs of a given mass with cross sections σ beyond the lines are excluded The expected cross section for Dirac neutrinos with standard weak coupling is shown as a dashed line (from [28])

detectors (Fig. 16). The existing cross section limits for WIMP masses above ~ 150 GeV were improved (Fig. 17) and Dirac neutrinos could be excluded as the dominant component of the dark halo in the mass range 26 GeV to 4.7 TeV.

Acknowledgement:

The author would like to thank Profs. Bellotti, Cabibbo, Maiani and Monacelli for their continuous generous support of the experiment. We also thank the German Ministry of Research and Technology (BMFT) and the Russian Atomic Energy Committee for their substantial support. The author is grateful to Prof. Yu. Zdesenko for fruitful discussions.

References

[1] P. Langacker in: *Neutrinos*, p.71, New York, 1988, edited by H.V. Klapdor, (Springer, Heidelberg, 1988)
[2] K. Grotz, H.V. Klapdor, *The Weak Interaction in Nuclear-, Particle and Astrophysics* (Adam Hilger, Bristol, New York, 1990)
[3] R. N. Mohapatra, P. B. Pal, *Massive Neutrinos in Physics and Astrophysics*, (World Scientific, Singapore, 1991)
[4] V. Kuzmin, V. Rubakov, M. Shaposhnikov, *Phys. Lett.* **185**, 36 (1985), M. Fukugita, T. Yanagida, *Phys. Rev. D* **42**, 1285 (1990), G. Gelmini, T. Yanagida, *Phys. Lett. B* **294**, 53 (1992), B. Campbell et al., *Phys. Lett. B* **256**, 457 (1991)
[5] R. N. Mohapatra, *Phys. Rev. D* **34**, 3457 (1986)
[6] J. Steinberger, *Phys. Rep.* **203**, 345 (1991)
[7] G. B. Gelmini, M. Roncadelli, *Phys. Lett. B* **99**, 411 (1981)
[8] Z. B. Berezhiani, A. Yu. Smirnov, J. W. F. Valle, *Phys. Lett. B*, **291**, 99 (1992)
[9] S. L. Glashow, *Phys. Lett. B* **256**, 255 (1991)
[10] K. S. Babu, R. N. Mohapatra, *Phys. Rev. Lett.* **67**, 1498 (1991), D. Choudhury, U. Sarkar, *Phys. Rev. Lett.* **68**, 2875 (1992)
[11] H. V. Klapdor-Kleingrothaus in: *Proc. Int. Symp. on Weak and Electromagnetic Interactions In Nuclei*, p. 201, Dubna, 1992 (WEIN '92) (World Scientific, Singapore, 1992) and in: *Proc. NEUTRINO '92, Nucl. Phys. B (Proc. Suppl.)* **31**, 72 (1993)
[12] H.T. Wong, *J. Phy. G.* **17** *Suppl.* (1991)
[13] M. K. Moe, *Preprint UCI-Neutrino* **93-1**
[14] H. Ejiri et al., *Phys. Lett. B* **258**, 17 (1991)
[15] Yu. Zdesenko in: *Proc. Neutrinos in Cosmology, Astro, Particle and Nuclear Physics*, Erice, (Plenum Press, 1993)
[16] K. Grotz, H. V. Klapdor, *Nucl. Phys. A* **460**, 395 (1986)
[17] A. Staudt, K. Muto, H. V. Klapdor-Kleingrothaus, *Europhys. Lett* **13**, 31 (1990)
[18] X. R. Wu, A. Staudt, H. V. Klapdor-Kleingrothaus, C. R. Ching, T. H. Ho, *Phys. Lett. B* **272**, 169 (1991)
[19] M. Hirsch, X. R. Wu, H. V. Klapdor-Kleingrothaus, C. R. Ching, T. H. Ho, *Phys. Rep.*, in press and *Z. Phys. A* **345**, 163 (1993)
[20] K. Muto, E. Bender, H. V. Klapdor-Kleingrothaus, *Z. Phys. A* **339**, 435 (1991)
[21] A. L. Turkevich et al., *Phys. Rev. Lett.* **67**, 3211 (1991)
[22] F. Petry et al., *Nucl. Instrum. Meth. AA* **332**, 107 (1993)
[23] J. Busto et al., *Nucl. Phys. A* **513**, 291 (1990)
[24] A. Balysh et al. (HEIDELBERG-MOSCOW collaboration), *Phys. Lett. B*, in press
[25] M. Beck et al. (HEIDELBERG-MOSCOW collaboration), *Phys. Rev. Lett.* **70**, 2853 (1993)
[26] M. K. Moe et al., *UCI-Neutrino 92-1, preprint* (1992) and *Proc. NEUTRINO '92, Granada, Nucl. Phys. B (Proc. Suppl.)* **31**, 68 (1993)

[27] A. Balysh et al. (HEIDELBERG-MOSCOW collaboration) *Phys. Lett. B* **298**, 278 (1993)
[28] M. Beck et al. (HEIDELBERG-MOSCOW collaboration), to be published
[29] B. Maier, *Dipl. Thesis*, Univ. of Heidelberg (1993)
[30] H.V. Klapdor-Kleingrothaus in: *Proc. Int. Symp. on Gamma Ray Line Astrophysics*, Saclay, 1990, *AIP Conf. Proc.* **232** 464 (1991)
[31] N. Gehrels, *Nucl. Instrum. Meth. A* **292**, 505 (1990)
[32] S. D. Barthelmy, L. M. Bartlett, N. Gehrels, M. Leventhal, B. J. Teegarden, J. Tueller, S. Belyaev, V. Lebedev, H. V. Klapdor-Kleingrothaus, *AIP Conf. Proc* **280** 1166 (1993)
[33] D. Reusser et al., *Phys. Lett. B* **255** 143 (1991)
[34] F. T. Avignone et al., *Phys. Lett. B* **256** 559 (1991)

ANHARMONIC AND DEFORMATION EFFECTS IN $2\nu\beta\beta$ DECAY

A. A. Raduta

Institute of Atomic Physics, Bucharest, POB MG6, Romania

Introduction

The beta decay [1-12] was always a central subject of nuclear physics since it requires an accurate description of both leptonic and baryonic components. Indeed, the single beta decay is one of the sensitive tests for the wave functions describing the single nucleon motion. On the other hand the process can be explained only by assuming a deep understanding of the properties of neutrinos as well as of its electroweak interaction.

Among many interesting topics of this field, the double beta decay provides a special attraction. Such a process may happen by two distinct channels. In one case the daughter nucleus is accompanied by two electrons and two antineutrinos while in the second case antineutrinos are missing from the final states. Conventionally the two processes are called $2\nu\beta\beta$ and $0\nu\beta\beta$ modes, respectively.

The experiments investigating the double beta decay are based either on direct counting or on geochemical techniques. The laboratory counter experiments allow one (in principle) to separate the two modes. However no evidence for the $0\nu\beta\beta$ decay was provided yet. The geochemical measurements determine the total abundance of the final nucleus giving therefore the total decay rate and a lower limit for the partial lifetimes of the two modes. Using these data and the nuclear matrix elements one may calculate the upper limits for the effective mass of ν as well as for the effective right handedness of its interaction. If they are reliable, these results can be used as arguments for the existence of the $0\nu\beta\beta$ mode. Indeed, the process may occur under one of two circumstances: a) neutrino has a nonvanishing mass and consequently the flip of helicity is allowed; b) neutrino is coupled to a right handed current.

In virtue of these facts the process is forbidden in the $SU(2)_L \times U(1)$ gauge model of the electroweak interaction in which the neutrinos are massless and there exists no right handed current. The SU(5) model also predicts a vanishing mass as well as the absence of the right handed interaction. However, the two properties of ν, which are sufficient conditions for double beta decay, are predicted by some Grand Unified Theories (GUT). To give an example, the SO(10) model contains a family of 16 fermions among which a neutral one is interpreted as a right handed neutrino. According to this model, the neutrino has a nonvanishing mass.

To conclude, the study of $0\nu\beta\beta$ decay could confirm or reject the predictions of the GUT and by this gives an answer to the question whether ν is a Dirac ($\nu \neq \tilde{\nu}$) or a Majorana ($\nu = \tilde{\nu}$) particle.

The drawback of the predictions for the upper limits of the neutrino mass and the effective right handedness of the electroweak interaction is that no reliable test for the nuclear matrix elements exists. However, the nuclear matrix elements responsible for the 0ν mode are related to those involved in the two neutrino double beta decay ($2\nu\beta\beta$) transitions. Therefore, an indirect test for the nuclear matrix elements would be to use them for computing the rates of the $2\nu\beta\beta$ decay and subsequently the predictions are compared to the existing data [10,11].

The general theory of both 0ν and 2ν decays has been reviewed by Primakoff and Rosen[1], Haxton and Stephens[2], and more recently by Tomoda[12]. Here I shall restrict my considerations to the case of two neutrino double beta decay.

The $2\nu\beta\beta$ mode has two electrons and two antineutrinos in the final state and is expected to appear in the standard model in second order (i.e. the process takes place via two consecutive β^- virtual transitions). According to this view point, a quantitative description of the process requires a careful study of the nuclear states of the initial, final as well as of the intermediate odd-odd nucleus. As a matter of fact, the model for these states discriminates between various theoretical formalisms.

The $2\nu\beta\beta$ is strongly hindered in the spherical shell model formalism in medium and heavy nuclei. This happens since the Fermi seas for protons and neutrons are different. Smearing out the occupation probabilities by switching on the pairing correlations, the process is allowed to proceed. However the BCS formalism predicts for the decay rate a value which is much larger (≈ 200) than the corresponding data. The proton neutron QRPA formalism has been applied to describe the beta and double beta decay by several authors using various microscopic Hamiltonians [13-28]. It was found that including the RPA correlations in the ground state an important suppression factor is obtained, but not enough to obtain a satisfactory agreement with the experimental data.

It is well known that the traditional RPA calculations ignore the particle-particle (pp) correlations. However, as it was shown in ref. 21 by Cha, the β^+ decay rate is very sensitive to the strength of the pp correlations. This effect was recently used by several autors in connection with the $2\nu\beta\beta$ decay, using either a schematic [15,16] or a realistic two body interaction [17,19,20]. The common conclusion was that, for a certain strength ($g_{pp}^{(1)}$) of the pp 1^+ interaction, the Gamow-Teller matrix element is vanishing. For the case of Bonn [17] and Paris [19,20] potentials this happens for $g_{pp}^{(1)} = 1$. Therefore the ph (particle-hole) and pp matrix elements of these realistic interactions have the right strengths to produce the needed cancellation. Another suppression mechanism was investigated in ref. 18 by considering the RPA quadrupole correlations in the ground states of the mother and daughter nuclei, otherwise the states describing the odd-odd nucleus being of a proton-neutron TDA type. The connection between the two mechanisms is commented in refs. 17 and 24.

To conclude, the realistic interactions produce too much suppression for $g_{pp}^{(1)} = 1$. Moreover $g_{pp}^{(1)} = 1$ lies close to the value (≈ 1.2) where the RPA breaks down. Consequently the stability of the RPA results against adding the next higher corrections is, at least in principle, questionable. A mechanism inducing only a moderate supression of the transition amplitude for $g_{pp}^{(1)} = 1$ is needed.

In the first part of my talk I will present the results showing the quantitative correction of the Gamow-Teller (GT) transition amplitude caused by the first higher RPA terms. To stress the importance of the higher RPA effects we also considered the transition leaving the daugther nucleus in the first excited state 2^+. Indeed such a transition is forbidden at the level of RPA approach.

Another feature which will be discussed in my lecture is that of nuclear deformation. Intuitively, nuclear deformation must influence the decay rate. Indeed, the pairing correlations for deformed and spherical orbits are different from each other. Moreover, the distortion of the mean field may produce proton-neutron quasiparticle pairs of low energy which increase the transition amplitude.

It is worth mentioning that most of the papers devoted to the $2\nu\beta\beta$ process, use a spherical single particle basis, although some of their applications refer to deformed nuclei. However there are few publications using either Nilsson [14] or Woods-Saxon [28] deformed states. Therein the RPA states describing the intermediate odd-odd nucleus are K-states, but do not have good angular momentum. Moreover, from the existent analysis, the effect coming exclusively from nuclear deformation can not be evaluated.

In the second part of my lecture I shall describe a possible way of estimating the effect of nuclear deformation on the GT matrix element. The signature of the formalism is that the RPA states have good angular momentum, although the deformation is explicitly accounted for by the mean field of the single particle motion [29-32]. The present approach allows for an unified description of the $2\nu\beta\beta$ process in the deformed and spherical nuclei. Therefore it is easy to estimate the contribution of the nuclear deformation to the M_{GT} matrix elements. The effect of the nuclear deformation is studied not only at the RPA level, but also by including the first higher order corrections through the boson expansion method. We study both cases, when the final state of the process is either 0_f^+ or 2_f^+. The major point of this formalism is the use of projected spherical single particle states. Therefore, the second part of my speech starts by describing the single particle basis.

My talk will be ended by summarising the main conclusions of our investigations of anharmonic and deformation effects on the rate of double beta decay.

Anharmonic Effects on the $2\nu\beta\beta$ Decay Rate

Here I shall sketch the results obtained for the $2\nu\beta\beta$ decay $^{82}_{34}Se \rightarrow ^{82}_{36}Kr$. As we have already mentioned, this transition is supposed to take place through two consecutive single β^- transitions involving the intermediate odd-odd nucleus $^{82}_{35}Br$. The mother nucleus is in the ground state (0_i^+) while the daughter nucleus may be left either in its ground state (0_f^+) or in the first excited state 2_f^+. Here we restrict our considerations to the Gamow-Teller transitions which, as a matter of fact dominate the Fermi ones in medium and heavy nuclei. The transition operator

$$\beta_\mu^- = <p|\sigma^\mu|n> b_p^+ a_n \qquad (1)$$

connects the ground state of the mother nucleus to the proton-neutron dipole states of the intermediate odd-odd nucleus. The states involved by the transition in two steps are eigenstates of a many body Hamiltonian consisting of three types of terms: i) A one body term describing the independent motion of the nucleons in a Woods-Saxon (WS) potential including corrections due to the Coulomb interaction. The parameters defining the WS potential are those of ref. 33. ii) The proton-proton and neutron-neutron 2^λ-pole interaction with $\lambda = 0, 2$. iii) The proton-neutron dipole interaction. The 2^λ pole (λ=0,1,2) interaction is taken as the Brueckner G matrix elements $\langle(ab)J|G|(cd)J\rangle$ calculated from the Bonn one-boson exchange potential by solving the Bethe-Goldstone equation. Concerning the other possible terms we would like to make the following comments. We did not include the proton-neutron pairing interaction. According to ref. 26, it has a negligible effect because protons and neutrons occupy different shells. Also we do not consider here the pn quadrupole interaction, which would lead to quadrupole states in the odd-odd nucleus. However these states can not be excited through the Gamow-Teller transition operator. The single particle space was restricted to the $3\hbar\omega$ and $4\hbar\omega$ shells. Such a truncation produces a renormalisation of the two body interaction which can be described by the following parametrisation. The pairing G-matrix for the proton system is multiplied by g_{pair}^p while that of the neutron system by g_{pair}^n. The strength for dipole G-matrix is multiplied either by $g_{ph}^{(1)}$, if particle-hole states are involved, or by $g_{pp}^{(1)}$, when the matrix corresponds to the space of particle-particle states. The quadrupole G-matrices are multiplied by $g_{ph}^{(2)}$ and $g_{pp}^{(2)}$ for both the proton and neutron systems. In our calculations the parameters $(g_{pair}^p, g_{pair}^n, g_{ph}^{(1)}, g_{ph}^{(2)}, g_{pp}^{(2)})$ have the values (1.01, 1.143, 1.1, 0.825, 0.825).

The $2\nu\beta\beta$ process is characterised by two quantities: i) The transition rate which is expressed as the square of the Gamow-Teller reduced transition amplitude $|M_{GT}^{(0\lambda)}|^2$ ($\lambda = 0, 2$). ii) The half-life $T_{\frac{1}{2}}^{0\lambda}$ of the mother nucleus with respect to this $\beta\beta$ decay channel. The expression for the half-life is very much simplified if in the energy denominator of the transition probability the lepton energy is replaced by the average value:

$$\Delta E = mc^2 + \frac{1}{2} Q_{\beta\beta}, \qquad (2)$$

where mc^2 is the rest energy of the electron and $Q_{\beta\beta}$ denotes the Q-value for the $\beta\beta$ decay. Indeed this approximation allows for a separation of the nuclear part from the kinematical factor and the inverse of the half-life can be written in a factorised form.

$$(T_{1/2}^{0\lambda})^{-1} = F |M_{GT}^{(0\lambda)}|^2, \qquad (3)$$

where F is a lepton phase integral and

$$M_{GT}^{(0\lambda)} = \sqrt{3} \sum_{k,m} \frac{_i\langle 0 \| \beta^{i+} \| km \rangle_i \; _i\langle km | km' \rangle_f \; _f\langle km' \| \beta^{f+} \| 0 \rangle_f}{(E_{km} + \Delta E_\lambda)^{\lambda+1}}$$

$$\equiv \sum_{k=1,2,3} {}_k M_{GT}^{(0\lambda)}, \lambda = 0, 2, \qquad (4)$$

where

$$(\beta^+)_\mu = (\beta^-)^+_{-\mu}(-)^\mu. \qquad (5)$$

The states involved in (4) were obtained as it follows. First the model Hamiltonian was treated through BCS approach and then the QRPA formalism defines the dipole and quadrupole bosons:

$$\Gamma^+_{1\mu}(l) = \sum_k [X_l(k) A^+_{1\mu}(k) + Y_l(k) A_{1-\mu}(k)(-)^{1+\mu}]$$

$$\Gamma^+_{2\mu}(l) = \sum_k [R_l(k) A^+_{2\mu}(k) + S_l(k) A_{2-\mu}(k)(-)^\mu], \qquad (6)$$

where $A^+_{2\mu}$ are operators for quadrupole pairs of quasiparticles of the same type (pp or nn) while $A^+_{1\mu}$ are those for dipole proton-neutron quasiparticles. Since the BCS states of the mother nucleus are different from those of the daughter nucleus, they are labelled by a distinguishing index ϵ taking the values i (initial) and f (final). Correspondingly, the RPA states carry the index i or f depending on whether the BCS ground state is that of initial or final nucleus. Also, the transition operator β^+ is accompanied by an upper index i if it acts on the states characterising the mother nucleus and by f when the transition to the daughter nucleus is involved.

If the RPA approach is adopted, the transition operators β^+ can be expressed as a linear combination of the RPA bosons $\Gamma_{1\mu}^+$ and $\Gamma_{1\mu}$. In this way only the single boson states $|1m\rangle_i, |1m\rangle_f$ contribute. Due to this fact the transition $0_i^+ \to 2_f^+$ is forbidden within the RPA approach. Using the boson expansion technique, the RPA expression of β^+ was modified by adding the first order corrections which are terms of second and third degrees in the bosons $\Gamma_1^+, \Gamma_2^+, \Gamma_1, \Gamma_2$. These operators activate the following dipole states in the intermediate nucleus:

$$|2, m_2\mu\rangle_\epsilon = (\Gamma_1^+(k_1)\Gamma_2^+(k_2))_{1\mu}|0\rangle_\epsilon,$$
$$|3, m_3\mu\rangle_\epsilon = N_3[\Gamma_1^+(k_1)(\Gamma_2^+(k_2)\Gamma_2^+(k_3))]_{1\mu}|0\rangle_\epsilon, \epsilon = i, f \quad (7)$$

with $m_2 = (k_1, k_2), m_3 = (k_1, k_2, k_3)$. It is obvious now that the transition $0_i^+ \to 2_f^+$ is no longer forbidden.

The energy denominator of the transition amplitude consists of two terms: a) E_{km}, the average of the energies corresponding to the states $|km\rangle_i, |km'\rangle_f$ normalised to the energy of the first RPA state 1^+, b) $\Delta E_0 = \Delta E + E_{1^+}$, $\Delta E_2 = \Delta E_0 - \frac{1}{2}E_{2^+}$, where E_{1^+} and E_{2^+} denote the experimental values for the energies of the states 1^+ and 2^+, respectively. In the last part of the relation (4) we denoted by $_kM_{GT}^{(0,\lambda)}$ the terms generated by the k-boson intermediate states (k=1,2,3).

We studied the dependence of $_kM_{GT}^{(0,\lambda)}$ on the strength of the dipole particle-particle interaction $g_{pp}^{(1)}$. As was already mentioned, $_1M_{GT}^{(00)}$ goes to zero when $g_{pp}^{(1)}$ approaches unity. Adding the corrections due to the two and three boson states, the amplitude $M_{GT}^{(00)}$ is significantly corrected. Indeed, for $g_{pp}^{(1)} = 1$, the corrected $M_{GT}^{(00)}$ is only moderately suppressed. It is only 2.2 times smaller than the RPA value corresponding to $g_{(pp)}^{(1)} = 0$. That increases the half-life by a factor of 4.6, improving the agreement with the experimental data. Thus for $g_{pp}^{(1)} = 1$ the discrepancy is reduced to a factor of 3.3. The zero of $_1M_{GT}^{(00)}$ which appears at $g_{pp}^{(1)} = 1$ is shifted to the area where the QRPA collapses.

Concerning the transition $0_i^+ \to 2_f^+$, the leading terms are those generated by one and two dipole intermediate states, respectively. The term $_2M_{GT}^{(02)}$ is vanishing for a value of $g_{pp}^{(1)}$ which is close to 1. The overall contribution $M_{GT}^{(02)}$ is a linearly increasing function of $g_{pp}^{(1)}$. Its magnitude is small comparing it to that of $M_{GT}^{(00)}$. For example, for $g_{pp}^{(1)} = 0$ one obtains $M_{GT}^{(02)} = 0.03 \, \text{MeV}^{-3}$.

The N-Z sum rule is obeyed within RPA and higher QRPA with two bosons. The higher QRPA with three boson terms exceeds N-Z by 17 per cent. This is a sign that the multi-boson states overcount the single particle degrees of freedom, i.e. violate the Pauli principle.

The Effect of Nuclear Deformation on the $2\nu\beta\beta$ Decay Rate

A Projected Single Particle Basis

It is well known that the microscopic description of deformed nuclei requires the use of a deformed single particle basis. But a wave function in a deformed intrinsic system has not a good angular momentum. Thus an angular momentum projection is needed.

The projection can be performed at the level either of the many body states or of the single particle states, but then coupling them with the core to good angular momentum. In a series of publications I have formulated a formalism which is based on a projected single particle basis[29-32]. The underlying ideas will be briefly sketched below. To this aim we consider the particle core Hamiltonian:

$$\widetilde{H} = H_{sm} + H_{core} - A_C \sum_\mu (b_{2\mu}^+ + (-)^\mu b_{2-\mu}) r^2 Y_{2-\mu}(-)^\mu \tag{8}$$

with H_{sm} the spherical shell model term describing the single particle motion and H_{core} a quadrupole boson Hamiltonian describing a phenomenological core. The two subsystems interact with each other through a qQ term.

It can be shown that the projected states:

$$\Phi_{nlj}^{IM}(d) = N_{nlj}^I P_{MI}^I [|nljI> \psi_g] \tag{9}$$

approximate quite well the eigenstates of H. Here $|nljm>$ are the spherical shell model states and ψ_g is a quadrupole boson $(b_{2\mu}^+, b_{2\mu})$ coherent state defined by

$$\psi_g = exp[d(b_{20}^+ - b_{20})]|0> \tag{10}$$

with $|0>$ standing for the vacuum state. d is a real parameter which simulates the nuclear deformation. The projection operator has the standard definition

$$P_{MK}^J = \frac{2J+1}{8\pi^2} \int D_{MK}^{J\,*}(\Omega) \hat{R}(\Omega) d\Omega \tag{11}$$

The normalisation factor is denoted by N_{nlj}^I.

For a suitable choice of the strength A_C the energies of Nilsson levels with $\Omega = I$ are reproduced by the expectation value of $H' = \widetilde{H} - H_{core}$ with the states $\Phi_{nlj}^{IM}(d)$:

$$\epsilon_{nlj}^I = \langle \Phi_{nlj}^{IM}(d) | H' | \Phi_{nlj}^{IM}(d) \rangle. \tag{12}$$

It is easy to show that the same result for $\epsilon_{nlj}^I(d)$ is obtained if the state $|nljI\rangle$ would be replaced by its time reversed state. The time reversed state for Φ_{nlj}^{IM} is $(-)^{I-M} \Phi_{nlj}^{I-M}$.

Consider now a one body operator $T_{k\mu}$ of rank k and projection μ. Its reduced matrix elements corresponding to the basis (9) can be written in a factorised form:

$$< \Phi_{nlj}^I \| T_k \| \Phi_{n'l'j'}^I > = f_{nlj}^{n'l'j'} < nlj \| T_k \| n'l'j' > . \tag{13}$$

The first factor carries the dependence on deformation. When one performs the matrix element (12) the core states play the role of spectators. However, there is a trace of the core and that is the factor carrying the informations about the nuclear deformation.

At this stage we may introduce the fermionic states $\phi_{\alpha IM}$ with α symbolizing the set of quantum numbers nlj, which also characterise the single particle energies $\epsilon_\alpha^I(d)$ defined by (12). By definition, the reduced matrix elements between the states $\phi_{\alpha IM}$ and $\phi_{\alpha' I'M'}$ are taken equal to those given by the right hand side of (13). In principle, there should exist a (scalar) Hamiltonian satisfying the eigenvalue equation:

$$H_{eff} \phi_{\alpha IM} = \epsilon_{\alpha I} \phi_{\alpha IM}. \tag{14}$$

It is worth noting that the dimension of the space spanned by $\Phi_{\alpha IM}$ for a fixed value of j and $d \neq 0$ is $(2j+1)(2j+3)/4$ and not $(2j+1)$ as it should be. It was shown in several publications that we could correct for the overdimension of the single particle space by quenching the matrix elements of the nucleon-nucleon interaction by a factor which is nothing else but the occupation probability of the states involved. Indeed, since the state Φ_{nlj}^{IM} can be obtained by projection from either of two intrinsic time reversed states one should occupy it only with two particles when a many nucleon system is treated. Therefore the occupation probability associated to the above mentioned state is $2/(2I+1)$.

To give an example, let us write the one body operator $T_{\lambda\mu}$ in the second quantisation with respect to the states $\phi_{\alpha IM}$ with the creation operator $c_{\alpha IM}^+$:

$$\hat{T} = \sum_{\alpha,I,M} \frac{\hat{\frac{1}{2}}}{\hat{I}} < \phi_{\alpha IM}|T_{\lambda\mu}|\phi_{\alpha'I'M'}> \frac{\hat{\frac{1}{2}}}{\hat{I'}} c_{\alpha IM}^+ c_{\alpha'I'M'} \quad (15)$$

where $\hat{I}$ stands for $\sqrt{2I+1}$.

To conclude, by quenching the nuclear one and two body interaction by a factor accounting for the occupation probabilities of the states involved, the projected single particle states can be used in a many body treatment. In this way the formalism for a deformed system is identical to that corresponding to a spherical single particle basis. However the energies of the many body states as well as the transition amplitudes will depend on deformation. Due to this feature we may say that the present formalism provides an unitary description of the two neutrino double beta decay in spherical and deformed nuclei.

It is worth mentioning that the formalism yielding the projected single particle basis is a microscopic counterpart of the phenomenological model known in literature under the name of coherent state model (CSM)[34-38]. The CSM was proposed for the simultaneous description of three interacting collective bands. Although the model has a small number of parameters, it is able to describe in a realistic fashion the energies and the B(E2) values of both transitional and well deformed nuclei. In the vibrational limit the members of the bands are the first three highest seniority states while in the rotational regime the predictions of liquid drop model are recovered. Analogously, in the vibrational limit ($d \to 0$) the set of states given by (9) are going to the spherical shell model states while for large values of the deformation parameter d the functions (9) describe the strong coupling regime.

Deformation Effects Within a Schematic Model

In what follows, the amplitudes for the Gamow-Teller transitions $0_i^+ \to \lambda_f^+$ ($\lambda = 0, 2$) will be calculated by using a schematic many body Hamiltonian which is written in the second quantisation language relative to the single particle basis $\phi_{\alpha IM}$:

$$H = \sum_{\tau,I,M} \frac{2}{2I+1}(\epsilon_{\tau I} - \lambda_\tau)c_{\tau IM}^+ c_{\tau IM} - \sum_{\tau,I} G_\tau P_{\tau I}^+ P_{\tau I'}$$

$$+2\chi \sum_\mu \beta_\mu^- \beta_{-\mu}^+ (-)^\mu - \frac{1}{2} \sum_{\tau,\tau',\mu} X_{\tau\tau'} Q_{2\mu}^{(\tau)} Q_{2\mu}^{(\tau')+} \quad (16)$$

with single particle energies $\epsilon_{\tau I}$, the Fermi surface λ_τ for protons ($\tau = p$) and neutrons ($\tau = n$), with a pairing force between nucleons with the same charge ($\tau \equiv \tau\alpha$), a proton-neutron spin-isospin force with operators

$$\beta_\mu^+ = \sum_{n,p} \frac{\hat{\frac{1}{2}}}{\hat{I}_n} < nI_nM_n|\sigma^\mu|pI_pM_p> \frac{\hat{\frac{1}{2}}}{\hat{I}_p} c_{nI_nM_n}^+ c_{pI_pM_p} \quad (17)$$

and the usual quadrupole-quadrupole force with three different strength parameters X_{pp}, X_{nn} and X_{pn}. In (16) and (17) we have already included the quenching factors $[2/(2I+1)]^{1/2} = \hat{\frac{1}{2}}/\hat{I}$ as discussed before.

We consider now the GT transition amplitude (4), where the multi-boson states are obtained this time by the QRPA treatment of the model Hamiltonian (16). Here, the same transition as in the previous section is considered, i.e. $^{82}_{34}Se \rightarrow ^{82}_{36}Kr$. The strength parameters involved in the model Hamiltonian were fixed in the standard way: The pairing strengths are fixed so that the mass difference of the neighbouring even-odd nuclei are reproduced. The strength of the spin-isospin interaction χ is fixed by fitting the experimental energy of the Gamow-Teller giant resonance[39]. The QQ interaction strengths are taken equal. Their common value is chosen so that the experimental energy for the first 2^+ is obtained[40,41]. All the parameters $g_{pp}^{(\lambda)}, g_{ph}^{(\lambda)}$ ($\lambda = 0, 1, 2$) are taken equal to unity excepting for $g_{pp}^{(1)}$ which is a free parameter. The deformation parameter and the strength of the qQ interaction have been selfconsistently determined. In order to avoid unnecessary complications we take for mother and daughter nuclei the same parameters (d, A_C) equal to the average values, i.e. $d=1.8$, $A_C=0.18$ (MeV fm^{-2}).

We calculated the transition amplitudes $M_{GT}^{(0\lambda)}$ and the strengths of the single beta transitions of the mother nucleus:

$$\beta_k^{(-)} = \frac{1}{3}(_i<0\|\beta^{i+}\|1k>_i)^2 \qquad \beta_k^{(+)} = \frac{1}{3}(_i<0\|\beta^{i-}\|1k>_i)^2 \qquad (18)$$

Having these, one can check whether the sum rule

$$S_{RPA} \equiv \sum_k (\beta_k^{(-)} - \beta_k^{(+)}) = N - Z \qquad (19)$$

is satisfied.

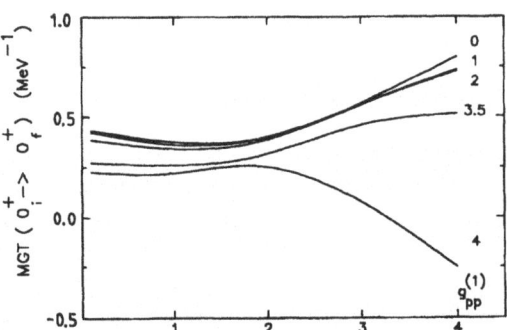

Figure 1. The $M_{GT}^{(00)}$ value is plotted as a function of d for several values of $g_{pp}^{(1)}$.

The dependence of $M_{GT}^{(00)}$ on deformation is shown in fig. 1 for different values of $g_{pp}^{(1)}$. There, only the RPA contribution is considered. The HRPA contribution is plotted in fig. 2 as a function of the strength of the particle-particle interaction. Again the contribution coming from the three boson components of the transition operator is quite large. The amplitude for the transition $0_i^+ \rightarrow 2_f^+$, is shown in fig. 3. In fig. 4 the sum rule S_{RPA} is plotted as a function of d. One sees that a good agreement to the N-Z value (=14) is obtained for $d=1.8$. We should remember that this value was obtained for $^{82}_{34}Se$ and $^{82}_{36}Kr$ by a selfconsistent procedure. We

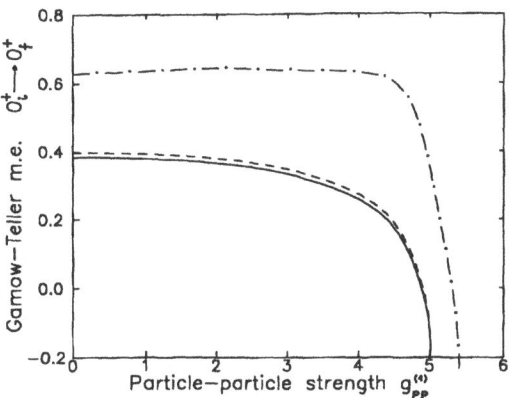

Figure 2. For $d = 1.8$, the amplitudes $_1M_{GT}^{(00)}$ (full curve), $_1M_{GT}^{(00)} +_2 M_{GT}^{(00)}$ (dashed curve) and $M_{GT}^{(00)}$ (dot-dashed curve) are plotted as function of $g_{pp}^{(1)}$.

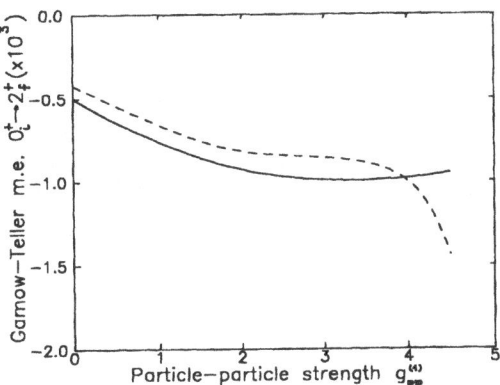

Figure 3. The amplitude $M_{GT}^{(02)}$ as function of $g_{pp}^{(1)}$.

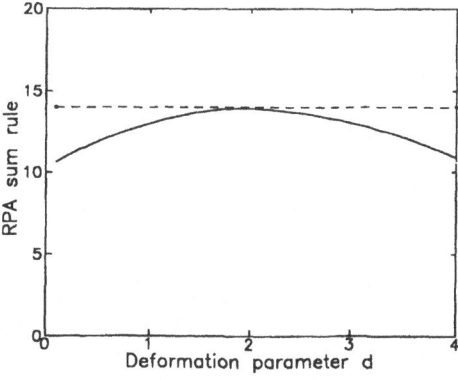

Figure 4. The $N - Z$ sum rule (19) as function of d.

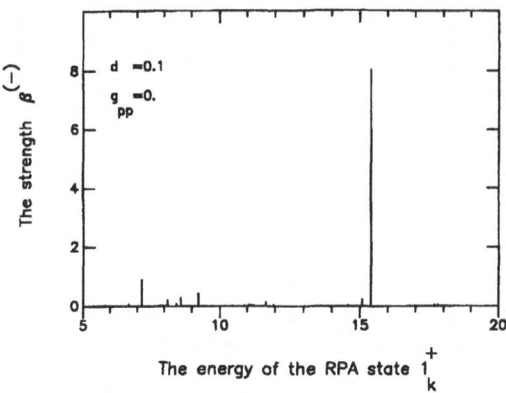

Figure 5. *The strength for the β^- transition of the mother nucleus versus the RPA energies. The predictions correspond to $d = 0.1$, $g_{pp}^{(1)} = 0$.*

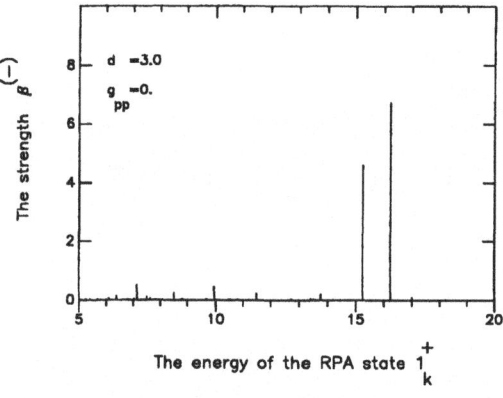

Figure 6. *The same as in Figure 5 but for $d = 3$.*

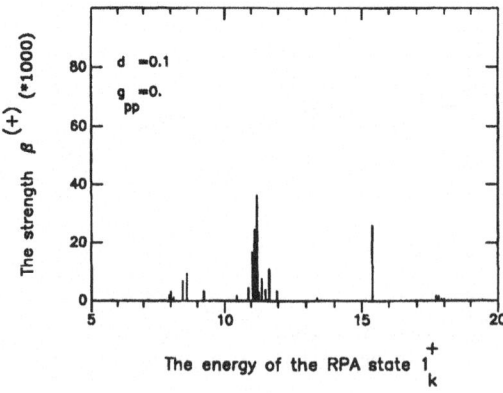

Figure 7. *The same as in Figure 5 but for β^+ transition.*

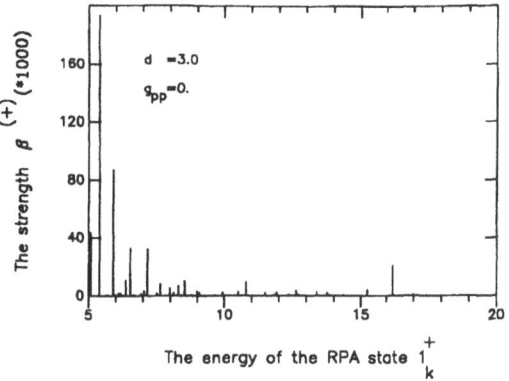

Figure 8. *The same as in Figure 7 but for d = 3.*

would like to stress on that this agreement pleads for the fact that the quenching factor which was considered for nucleon-nucleon interaction accounts for the overdimension of the single particle space in a realistic fashion.

It is interesting to see the influence of nuclear deformation on the distribution of the $\beta^{(-)}$ and $\beta^{(+)}$ strengths for the mother nucleus. These strengths are plotted in figs. 5-8 for two values of d (0.1 and 3.0). One remarks that the $\beta^{(-)}$ strength is fragmented when the deformation is increased. Although the $\beta^{(+)}$ strength has a small magnitude, it is very sensitive to the variation of the nuclear deformation. Indeed, increasing d the strength is moved to the region of lower energy states. Moreover the $\beta^{(+)}$ strength is increased by increasing d.

But also the particle-particle interaction produces fragmentation. Let us see what would be the result of considering simultaneously the two fragmentation effects. For $g_{pp}^{(1)} = 0$ there is a dominantly collective state (the case of d=0.1), then switching on $g_{pp}^{(1)}$, the state fragments and several collective states may appear. In case of $g_{pp}^{(1)} = 0$ there are two collective states (we meet this situation for d=3) of comparable strengths, the pp interaction destroys one collective state (the weaker one) and increases the collectivity of the other one.

Before closing this section, I advise those who are interested in learning details about the use of the projected single particle basis for the description of the $2\nu\beta\beta$ decay rate to consult my latest publications on this topic, i.e. refs. 42, 43.

Conclusions

The results presented here demonstrate that the ground state correlations induced by the higher RPA terms as well as by the nuclear deformation, play an important role when one aims at a realistic description of the $2\nu\beta\beta$ decay.

Acknowledgement

My interest for this field appeared after having many illuminating discussions with Prof. Amand Faessler. To finish the investigation on the nuclear deformation and beta decay, I received an important computational help from Dr. D. S. Delion. I also benefited of a careful and critical reading of the manuscript by Dr. I. I. Ursu. I express my gratitude to all of them.

References

[1] H. Primakoff and S. P. Rosen, *Rep. Prog. Phys.* **22**, 21 (1959)
[2] W. C. Haxton and G. J. Stephenson, Jr., *Prog. Part. Nucl. Phys.* **12**, 409 (1984)
[3] K. Ikeda, *Prog. Theor. Phys.* **31**, 434 (1964)
[4] I. Hamamoto, *Nucl. Phys.* **62**, 49 (1965)
[5] K. Ikeda, S. Fujita and J. I. Fujita, *Phys. Lett.* **3**, 271 (1963)
[6] J. I. Fujita and K. Ikeda, *Prog. Theor. Phys.* **36**, 288 (1966)
[7] J. A. Halbleib and R. A. Sorensen, *Nucl. Phys.* **98**, 542 (1967)
[8] F. T. Avignone III, R. L. Brodzinski, J. C. Evans Jr., K. Hensley, H. S. Miley and J. H. Reeves, *Phys. Rev. C* **34**, 666 (1986)
[9] D. O. Caldwell et al., *Phys. Rev. D* **33**, 2737 (1986)
[10] T. Kirsten in: *Proc. Int. Symp. of Nuclear Beta Decay and Neutrinos*, p.81, Osaka, Japan, 1986, edited by T. Kotani, H. Ejiri and E. Takasuki, (World Scientific, Singapore, 1986)
[11] S. R. Elliot, A. A. Hahn and M. K. Moe, *Phys. Rev. Lett.* **59**, 2020 (1987)
[12] T. Tomoda, *Rep. Progr. Phys.* **54**, 53 (1991)
[13] T. Tomoda, A. Faessler, K. W. Schmid and F. Grummer, *Nucl. Phys. A* **452**, 591 (1986)
[14] I. Krumlinde and P. Moller, *Nucl. Phys. A* **417**, 419 (1984)
[15] P. Vogel and P. Fischer, *Phys. Rev. C* **32**, 1362 (1985)
[16] P. Vogel and M. R. Zirnbauer, *Phys. Rev. Lett.* **57**, 3148 (1986)
[17] O. Civitarese, A. Faessler and T. Tomoda, *Phys. Lett. B* **194**, 11 (1987)
[18] O. Civitarese, A. Faessler, J. Suhonen, X. R. Wu, *Phys. Lett. B* **251**, 333 (1990)
[19] K. Muto and H. V. Klapdor, *Phys. Lett. B* **200**, 420 (1988)
[20] K. Muto, E. Bender and H. V. Klapdor, *Z.Phys. A* **334**, 177 (1989)
[21] D. Cha, *Phys. Rev. C* **27**, 2269 (1987)
[22] W. M. Alberico, M. B. Barbaro, A. Boltino and A. Molinari, *Ann. of Phys.* **187**, 79 (1988)
[23] J. Suhonen, T. Teigel, A. Faessler, *Nucl. Phys. A* **486**, 91 (1988)
[24] A. A. Raduta, A. Faessler and S. Stoica, *Nucl. Phys. A* **534**, 149 (1991)
[25] A. A. Raduta, A. Faessler, S. Stoica and W. Kaminsky, *Phys. Lett. B* **254**, 7 (1991)
[26] T. S. Sandhu and M. L. Rustgi, *Phys. Rev. C* **14**, 675 (1976)
[27] J. Suhonen, O. Civitarese and A. Faessler, *Nucl. Phys. A* **543**, 645 (1992)
[28] H. V. Klapdor, T. Oda, J. Metzinger, W. Hillebrandt, F. K. Thielemann, *Z. Phys. A* **299**, 213 (1981)
[29] A. A. Raduta and N. Sandulescu, *Nucl. Phys. A* **531**, 299 (1991)
[30] N. Sandulescu and A. A. Raduta, *Journal of Modern Phys. A* **7**, 125 (1992)
[31] A. A. Raduta in: *Building Blocks in Nuclear Structure*, Amalfi, 1992, edited by A. Covello, (World Scientific, Singapore)
[32] A. A. Raduta, D. S. Delion and N. Lo Iudice, *Nucl. Phys. A* **551**, 93 (1993)
[33] A. Bohr and B. R. Mottelson, *Nuclear structure*, vol.I, (Benjamin, New York, 1963)
[34] A. A. Raduta, V. Ceausescu, A. Gheorghe and R. M. Dreizler, *Phys. Lett. B* **99**, 444 (1981); *Nucl. Phys. A* **381**, 253 (1982)
[35] A. A. Raduta, C. Sabac, *Ann. Phys.* **148**, 1 (1983)
[36] A. A. Raduta, A. Faessler, Th. Koppel and C. Lima, *Z. Phys. A* **312**, 23 (1983)
[37] A. A. Raduta, *Rev. Roum. Phys.* **28**, 195 (1983)
[38] A. A. Raduta, S. Stoica and N. Sandulescu, *Rev. Roum. Phys.* **29**, 56 (1984)
[39] R. Madey et al., *Phys. Rev. C* **40**, 540 (1989)
[40] J. Barette et al., *Nucl. Phys. A* **235**, 154 (1974)
[41] J. Keinemen et al., *Nucl. Phys. A* **375**, 246 (1982)
[42] A. A. Raduta, D. S. Delion and Amand Faessler, *Phys. Lett. B* **312**, 13 (1993)
[43] A. A. Raduta, Amand Faessler and D. S. Delion, *Nucl. Phys.*, in press

NUCLEAR ASPECTS OF DOUBLE BETA DECAY

S. Stoica

Institute of Atomic Physics, Department of Theoretical Physics, Bucharest, POB MG 6, Romania

1. Introduction

The double beta ($\beta\beta$) decay is considered at present as one of the most sensitive processes to test physics beyond the Standard Model. The study of this process may provide us with informations on a number of important issues as: i) lepton number non-conservation; ii) nature of the neutrino; iii) neutrino mass (if neutrino is a Majorana particle); iv) the existence of right-handed currents in the weak interaction; v) the spontaneous breaking of a global B-L symmetry; vi) check of some hypothesis related to the left-right symmetric and supersymmetric theories. This is why there is a growing interest for the $\beta\beta$ decay study both theoretically and experimentally.

As it is well known the $\beta\beta$ decay process is an intricate mixture between particle and nuclear physics effects. Theoretically, concerning the particle physics side, the investigations are related to the possible mechanisms which might contribute to the $\beta\beta$ decay, while concerning the nuclear physics side the efforts are focused on improving the nuclear structure methods to perform accurate calculations of the nuclear matrix elements (n.m.e.) entering the $\beta\beta$ decay half-lives. Experimentally, a new generation of experiments are underway in Europe, US and Japan aiming at improving the lower limits for the neutrino mass ($<m_\nu>$), parameters connected to possible right-handed currents in the weak interaction ($<\lambda>$ and $<\eta>$), Majoron coupling constant ($<g_{\nu_\chi}>$), etc. The present status of the $\beta\beta$ decay searches has been described by several recent comprehensive reviews [1-8].

At present the following three double beta decay modes are mainly discussed:

$$X(A,Z) \to X(A,Z+2) + 2e^- + 2\bar{\nu}_e \tag{1}$$

$$X(A,Z) \to X(A,Z+2) + 2e^- \tag{2}$$

$$X(A,Z) \to X(A,Z+2) + 2e^- + \chi \tag{3}$$

In the frame of the Standard Model the first mode is allowed, irrespective of the nature of neutrino (Dirac or Majorana) or if it is massive or massless, while the second and third processes are forbidden. The second mode would imply the existence of at least one massive Majorana neutrino flavor if the forces are described by gauge theories [9] and may be also a signal for the existence of right-handed currents in the weak interaction. The third process would imply a spontaneous breaking of a global B-L symmetry. Even the measurements at LEP on Z^0-width [10] disagree with the existence of the doublet and triplet Majoron, however, the existence of a singlet or a mixture of a singlet and doublet Majoron is still possible.

Theoretically the $\beta\beta$ decay may be described as a second order process in the perturbation theory of the weak interaction induced by two succesive single-beta transitions via virtual intermediate states of the neighbouring nucleus. In the derivation of the half-life formulas for the processes (1) - (3), one can make some approximations to bring them in factorized forms. For the $2\nu\beta\beta$ decay mode one assumes with good approximation that the electrons and neutrinos are emitted in S-waves and neglects the nucleon recoil terms. Furthermore, assuming the two-nucleon mechanism, contributions only from left-handed currents and the impulse approximation for the hadronic currents, the expressions of the half-lives may be written as:

i) $2\nu\beta\beta$

$$\left[T_{1/2}^{2\nu}\right]^{-1} = F^{2\nu}|M_{GT}^{2\nu}|^2 \qquad (4)$$

One important quantity for $\beta\beta$ decay process is the momentum transfer at a weak interaction vertex, p, which is equal to the sum of the momenta of the electron and neutrino emitted at that vertex. In the case of $2\nu\beta\beta$ decay mode $p \leq Q_{\beta\beta} \sim 1 MeV$, ($Q_{\beta\beta}$ stands for the energy release of the decay) while for $0\nu\beta\beta$ decay mode the neutrino is a virtual particle and $p \sim \frac{1}{r_{nn}} \sim 100 MeV$ (r_{nn} is the mean internucleon distance). Consequently, in the case of $0\nu\beta\beta$ decay the recoil terms become important and have to be taken into account. Assuming again the nn mechanism, the half-life formulas may be also brought in a factorized form, but the expressions are now more complicated.

ii) $0\nu\beta\beta$

$$\left[T_{1/2}^{0\nu}\right]^{-1} = C_{mm}\left(\frac{<m_\nu>}{m_e}\right)^2 + C_{\lambda\lambda}<\lambda>^2 + C_{\eta\eta}<\eta>^2 + \qquad (5)$$
$$2C_{m\lambda}\left(\frac{<m_\nu>}{m_e}\right)<\lambda> + 2C_{m\eta}\left(\frac{<m_\nu>}{m_e}\right)<\eta> + 2C_{\lambda\eta}<\eta><\lambda>$$

For the half-life of the singlet Majoron mode we have [2]:

iii) $0\nu\chi$

$$\left[T_{1/2}^{0\nu\chi}\right]^{-1} = \left(M_{GT}^{0\nu} - M_F^{0\nu}\right) F^{0\nu\chi}(<g_{\nu\chi}>)^2 \qquad (6)$$

The M^{GT} and M^F occurring in the above formulas are the Gamow-Teller (GT) and Fermi (F) nuclear matrix elements (n.m.e.) for those processes and will be discussed in the next section in more detail. C coefficients are products of phase-space integrals and n.m.e. correponding to contributions originating from the existence of possible nonvanishing mass of the neutrino or small right-handed currents in the weak interaction. The quantities denoted by F are phase-space integrals related to the corresponding $\beta\beta$ decay modes. The detailed expressions of C and F quantities can be found, for example, in Refs. [1], [2], [8].

The above half-life formulas are for the $0^+ \to 0^+$ transitions. Similar expressions can be also written for the $0^+ \to 2^+$ transitions. It is worth mentioning that in the case of $0\nu\beta\beta$ decay mode, for the $0^+ \to 0^+$ transitions, the leading term is that derived from pure neutrino mass contributions while, for the $0^+ \to 2^+$ ones, the leading term is that proportional to $<\lambda>^2$. Thus, the experimental observation of these last transitions may be a signal of the existence of right-handed components in the weak interaction.

As one can see from the above formulas, to extract strigent limits from data for the neutrino mass and parameters related to the right-handed leptonic currents or for the Majoron coupling constant, we need reliable calculations of the n.m.e. involved. But, since there is at present no indisputable evidence for $\beta\beta$ decay modes beyond the Standard Model, we can check our nuclear structure calculations for $2\nu\beta\beta$ decay mode for which many experimental data are available [11-18].

In the present paper we will briefly review the nuclear structure methods used to compute the n.m.e. which enter the $\beta\beta$ decay half-lives, stressing on the QRPA-type methods. In section 3 a second-QRPA procedure based on a boson expansion (of Marumori type) of the single-beta transition operators is described. Further, this method is employed to calculate decay rates and half-lives for $2\nu\beta\beta$ decay transitions both on the ground states (g.s.) and first 2^+ excited states, for seven $\beta\beta$ emitters. The influence of the higher order QRPA corrections on the calculation of n.m.e. is also discussed. The last section is devoted to conclusions and some possible future applications of our formalism.

2. Nuclear Structure Methods

The nuclear structure part entering the double beta decay half-lives (eqs. (4) - (6)) depends essentially on the reduced matrix elements of the double F and GT operators:

$$M_F = < J_f^\pi || \frac{1}{2} \sum_{ij} \tau(i)\tau(j) || 0_i^+ > \tag{7}$$

$$M_{GT} = < J_f^\pi || \frac{1}{2} \sum_{ij} \sigma_i \sigma_j \tau(i)\tau(j) || 0_i^+ > \tag{8}$$

where σ and τ denote the Pauli and isospin operators, respectively. The double Fermi operator is proportional to the square of the total isospin raising/lowering operator and thus only connects states in the same isospin multiplet. This is why, for $2\nu\beta\beta$ decay mode, the role played by the reduced m.e. due to the successive Fermi transitions is minor and they may be wisely neglected. The only important transitions for this mode remain the GT ones involving only virtual 1^+ states of the intermediate odd-odd nucleus. Since in this channel the particle-hole (ph) and particle-particle (pp) nucleon correlations(which play an essential role in the case of charge-changing phenomena) have opposite signs, the n.m.e. are very sensitive to the strength of the last [4], [6].

In the case of $0\nu\beta\beta$ decay mode the nuclear tensor operators are of GT and F-type because the nucleon recoil terms require the p-wave for one of electrons. Since the exchange of the virtual neutrino gives rise to neutrino potentials, the n.m.e. contain now these potentials. Therefore, for this mode, besides the transitions via 1^+ intermediate states, other channels with $J^\pi \neq 1^+$ are also involved. In these last channels the ph and pp nucleon correlations have the same sign and consequently, the large sensitivity of the n.m.e. to pp correlations is now much suppressed [7], [8]. In the case of the Majoron mode it is worth mentioning that the n.m.e are the same as those for $0\nu\beta\beta$ mode if one neglects the exchange of heavy neutrinos.

The nuclear structure methods used to compute the above mentioned n.m.e. are basically of two types: i) shell model calculations and ii) QRPA-type calculations.

Shell model can provide us with microscopic wave functions (w.f.) which have quantum numbers (parity, angular momentum, isospin) corresponding to the symmetries of the

Hamiltonian. If one chooses a model space which is large enough, these w.f. will include all correlations of the interaction part of the Hamiltonian. The best example of a $\beta\beta$ emitter for which such calculations were performed successfully, in the frame of this model, is ^{48}Ca [1]. For other $\beta\beta$ emitters the valence space may become huge and a correct treatment of the residual interactions needs the introduction of too many shells. In these cases approximations to truncate the large basis are required, but the difficulty is that we do not know what are the "true" w.f. Usually one assumes a weak interaction between the systems of protons and neutrons [1], [22]). In this case each nucleus was described by a single product of proton and neutron w.f. and the proton-neutron (pn) correlations were taken into account by using linear combinations of products of these w.f. Other possible nuclei to be treated in the frame of shell model calculations might be ^{96}Zr and even ^{100}Mo particularly for $0\nu\beta\beta$ decay mode where the n.m.e are less sensitive to the pn correlations. The effects of pairing correlations as well as quadrupole deformation which are important in open-shell nuclei have been taken into account in a generalized version of the quasiparticle model (numerical codes MONSTER, VAMPIR) by Tuebingen group [23]. The s.p. spectrum is obtained in a consistent way and the number and angular momentum symmetries of the w.f. are restored by projection procedures. This method, which needs much computer time, was used to calculate $\beta\beta$ decay rates for several nuclei [24].

The pnQRPA was developed to describe the charge-changing processes [31]. In contrast to the usual particle-like QRPA [32], pairs of unlike-particles are assumed now to have boson properties. The pnQRPA phonons with angular momentum J and z-component M are defined by:

$$\Gamma^+_{JM}(l) = \sum_k \left(X_l(k) A^+_{JM}(k) + Y_l(k) A_{J-M}(k)(-)^{J-M} \right) \qquad (9)$$

where $k = (j_p j_n)_J$ and p and n distinguish between proton and neutron states. The energy eigenvalues ω_l and forward- and backward- going amplitudes X and Y are obtained by solving the QRPA equation:

$$\begin{bmatrix} A & B \\ -B & -A \end{bmatrix} \begin{bmatrix} X \\ Y \end{bmatrix} = \omega_l \begin{bmatrix} X \\ Y \end{bmatrix} \qquad (10)$$

with the normalization condition:

$$\sum_k = \left(\left(X^J_l(k) \right)^2 - \left(Y^J_l(k) \right)^2 \right) = 1. \qquad (11)$$

A and B submatrices contain products of combinations of BCS coefficients with pp or ph reduced m.e.

The pnQRPA is at present the most used method, since it is able to get agreement between the theoretical calculations and experimental results for a wide class of $\beta\beta$ emitters [26], [30], [33], [34]. Much progress has been made in the frame of this approach in understanding the importance of the pn correlations and of the way by which these correlations contribute to the suppression of the n.m.e. [4], [6], [26], The other avantages of this method are:

- The GT sum rule for the one body transition operators (which is model independent) is fulfilled with high accuracy.

- In the frame of this method it is easy to calculate $\beta^{\pm}$ strengths and compare them with the existing data. This comparison is a strong test of the reliability of the calculations of n.m.e.

However, the drawback of this procedure is the large sensitivity on the strength of pp correlations. This is why, in the recent past there were some attempts to overcome this lack. Civitarese et al. [35] developed a number projected QRPA but the influence of projection both at the level of BCS and QRPA of the w.f. do not change very much the pnQRPA results. Another method is the "Operator Expansion Method" (OEM) [7], [25]. This one takes advantage of the possibility of expanding the denominator of the $2\nu\beta\beta$ decay amplitude into an infinite geometrical series. In this way the summation over the intermediate states is replaced over an infinite series of commutators of the Hamiltonian with operators representing hadronic currents. The problem is to sum up exactly this infinite divergent series. To do this, one has to choose a very simple expression of the Hamiltonian of the process. In the same time, as a basic requirement, that Hamiltonian must admit simultaneously as eigenstates the states which can describe the parent, intermediate and daughter nuclei. The method has the nice feature that there is practically no dependence of the M_{GT} on the g_{pp} parameter (the strength of the pp interaction), but a further investigation of the approximations used, in order to clarify fully this dependence, is still needed.

Another attempt was to go beyond QRPA by taking into account higher order QRPA (HQRPA) corrections [28], [29]. Our goal was to see i) how the higher order terms influence the pnQRPA results, ii) whether a singular point for the M_{GT} m.e. remains if $g_{pp} = 0$, and at what value of this parameter the HQRPA procedure will collapse.

3. Second-QRPA Method

Our method is a second QRPA-type approach. This means that we used improved QRPA wavefunctions (by including HQRPA corrections) but we kept the BCS vacuum unchanged. The procedure is described in more detail in references [28], [29]. Its main ingredients will be reviewed here only shortly. To describe the properties of the parent, daughter and intermediate nuclei involved in the $\beta\beta$ decay process we used a many-body Hamiltonian with a one-body term describing the independent motion of the nucleons in a Woods-Saxon potential including Coulomb corrections and a nucleon-nucleon residual interaction taken as G matrices $< (ab)J|G|(cd)J >$, $(J = 0, 1, 2)$ calculated with Bonn OBEP. In our numerical calculations the single particle space is restricted to two full oscillator shells. This truncation requires a renormalisation of the two-body matrix elements which can be achieved by multiplying the G matrices $< (ab)J = 0|G_\tau|(cd)J = 0 >$ ($\tau = p, n$) by strength constants as follows: i) for pairing interaction (J=0) by g_{pp}^τ; ii) for dipole interaction (J=1) by $g_{ph}^{(1)}$ if b, d are hole states and by $g_{pp}^{(1)}$ when b, d are particle states; iii) for quadrupole interaction (J=2) by $g_{ph}^{(2)}$ and $g_{pp}^{(2)}$ both for the proton and neutron system. Accordingly to the previous experience for such type of calculations these constants are not far from unity and are fixed in the following way: $g_{pp}^{(\tau)}$ is chosen so that the experimental odd-even mass differences are reproduced. $g_{ph}^{(1)}$ is fitted so that the position of the GT resonance for the intermediate odd-odd nucleus is reproduced and $g_{ph}^{(2)}$ is determined by fitting the experimental energy of the first 2^+ state for the parent and daughter nuclei. $g_{pp}^{(2)}$ and $g_{ph}^{(2)}$ are set equal to unity while $g_{pp}^{(1)}$ is varied from 0. to 1.15. In the frame of the QRPA approach phonon states of a nucleus are described by applying the phonon operators defined by (9). $A_{JM}^+(k)$

are the pair operators and are defined as two quasiparticle tensors of rank J. As it is well known, QRPA can be viewed as the first term in a boson expansion of certain quasiparticle operator combinations. To introduce higher order QRPA corrections we have developed the quasiparticle pair operators $A^+_{JM}(k)$ in terms of QRPA boson operators $\Gamma^+_{JM}(k)$ and retained the next order beyond QRPA in the series expansion. In this way the properties of the initial, intermediate and final nuclei involved in the $\beta\beta$ decay are described by the following multiphonon states:

1) initial (i):

$$|0>_i; \qquad {}_i\Gamma^+_{2\mu}(k)|0>_i \qquad (12)$$

2) intermediate (int):

$$\begin{aligned}&{}_i\Gamma^+_{1\mu}(k)|0>_i \quad ; \quad \left({}_i\Gamma^+_{1\mu_1}(k_1){}_i\Gamma^+_{2\mu_2}(k_2)\right)_{1\mu}|0>_i \\ &{}_f\Gamma^+_{1\mu}(k)|0>_f \quad ; \quad \left({}_f\Gamma^+_{1\mu_1}(k_1){}_f\Gamma^+_{2\mu_2}(k_2)\right)_{1\mu}|0>_f\end{aligned} \qquad (13)$$

3) final (f):

$$|0>_f; \qquad {}_f\Gamma^+_{2\mu}(k)|0>_f . \qquad (14)$$

The two sets of multiphonon states which describe the intermediate nucleus come from two different QRPA procedures applied for the initial and final nuclei, respectively. The $2\nu\beta\beta$ decay rate for the transition $0^+_i \to 0^+_f$ has the expression:

$$M_{GT} = \sum_{im} \frac{{}_f<0\|\beta^+\|im>_f<im|im'>_i<im'\|\beta^+\|0>_i}{E_{int} - E_f + mc^2 + \frac{Q_{\beta\beta}}{2}} = \sum_{i=1,2} M^{(i)}_{GT} \qquad (15)$$

where i=1 and i=2 means one and two boson states, respectively. In order to calculate the reduced m.e. of the transition operators β^+ from (15), we have also to expand their QRPA expressions in terms of Γ boson operators. Thus, for the β^+ operator one obtains:

$$\begin{aligned}\beta^+_\mu =\ & \sum_k \left(B^{(10)}_k \Gamma^+_{1\mu}(k) + B^{(01)}_k \Gamma_{1-\mu}(k)(-)^{1-\mu}\right) \\ & + \sum_{k_1 k_2}\left(B^{(20)}_{k_1 k_2}\left(\Gamma^+_1(k_1)\Gamma^+_2(k_2)\right)_{1\mu} + B^{(02)}_{k_1 k_2}\left(\Gamma_1(k_1)\Gamma_2(k_2)\right)_{1\mu}\right) \\ & + \sum_{k_1 k_2}\left(B^{(12)}_{k_1 k_2}\left(\Gamma^+_1(k_1)\Gamma_2(k_2)\right)_{1\mu} + B^{(21)}_{k_1 k_2}\left(\Gamma^+_2(k_2)\Gamma_1(k_1)\right)_{1\mu}\right).\end{aligned} \qquad (16)$$

The coefficients B were obtained by requiring that the m.e. of the β^+_μ operators, in the boson basis, are identical to those corresponding to the rhs of (16) [36]. If in the expansion (16) one keeps only the linear terms, one finds the QRPA boson image of the fermion β^+_μ operators. We have neglected in (16) the terms of the type $\left(\Gamma^+_1\Gamma_2\right)_{1\mu}$ since they have a vanishing effect on the transition $0^+_i \to 0^+_f$. We have also neglected all three boson state contributions for the following reasons: i) If one takes into account only two boson state contributions to the HQRPA corrections, the sum rules, which single beta strengths must fulfil very accurate in the frame of QRPA ($\sum_l (<\beta^+>^2_l - <\beta^->^2_l = 3(N-Z)$), are violated only about 1%. ii) One can easily show [37] that the expression (16) of the β operators still commutes with the expressions of particle number and isospin operators taken in the same approximation as (6). If we add three boson state contributions, the commutation relations mentioned above hold no longer and we could have an additional source of spurious contributions during the computation. Accordingly with the previous results from boson expansion theory these three boson state contributions should be at most of the same order of magnitude as two boson state ones and hence do not qualitatively modify our conclusions about the influence that HQRPA corrections could have on GT m.e. Our numerical results refer to six $\beta\beta$ transitions: $^{76}Ge \to ^{76}Se$, $^{110}Cd \to ^{110}Pd$, $^{116}Cd \to ^{116}Sn$, $^{128}Te \to ^{128}Xe$, $^{130}Te \to ^{130}Xe$, $^{136}Ba \to ^{136}Xe$. The $M_{GT}^{2\nu}(g_{pp})$ functions are plotted in figs. 1-3: the upper curves (which drop steeply with respect to g_{pp}) represent the pnQRPA result of the computation of these m.e. while the lower curves represent HQRPA corrections as functions of g_{pp}. One can see that these corrections are functions more stable on this parameter. In spite of the small values of the HQRPA corrections, they become important by comparison with QRPA values just around $g_{pp} = 1.0$, where the desired suppression for M_{GT} is acquired, for all the transitions which are studied. By including them one gains stability in the computation of the GT m.e. since the zero point of the function $M_{GT}^{2\nu}(g_{pp})$ is shifted to the region where the QRPA procedure collapses. This happens because our formalism has included not only pn interactions (as in the standard pnQRPA) but also proton-proton and neutron-neutron interactions (through HQRPA corrections).

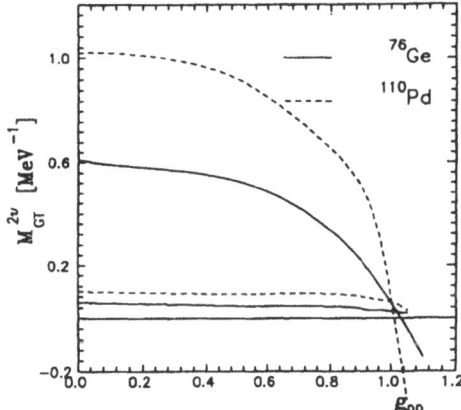

Figure 1. The GT m.e. M_{GT} are plotted as functions of $g_{pp}^{(1)}$ for ^{76}Ge (solid lines) and ^{110}Pd (dashed lines). The upper curves (which drop steeply with respect to g_{pp}) represent the pnQRPA result while the lower curves represent the HQRPA corrections. One can see that these corrections are functions more stable on this parameter and become important around $g_{pp} = 1$. Their inclusion shifts the zero point of the $M_{GT}^{2\nu}$ functions, computed in the the frame of QRPA, to the region where QRPA procedure collapses.

In the last time several experiments are searching for $\beta\beta$ decay transitions on excited states [19]-[21]. These experiments take advantage of the unambiguous signal due to the photons accompanying the deexcitation of the daughter nuclei on their ground states (g.s.). From the theoretical point of view the computation of the half-lives for the $2\nu\beta\beta$ transitions can be an additional test for the n.m.e. involved, when there exist experimental data for such transitions. On the other hand, the estimations of them for the cases not investigated up to

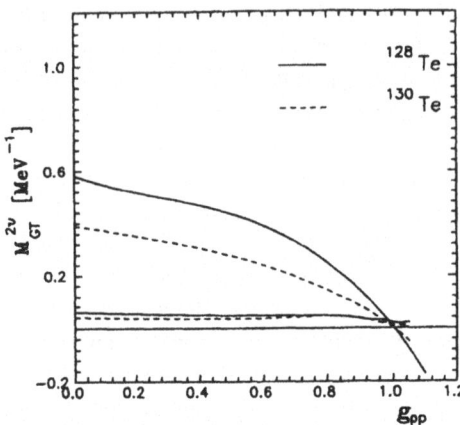

Figure 2. *The same as Fig. 1 but for ^{128}Te and ^{130}Te cases.*

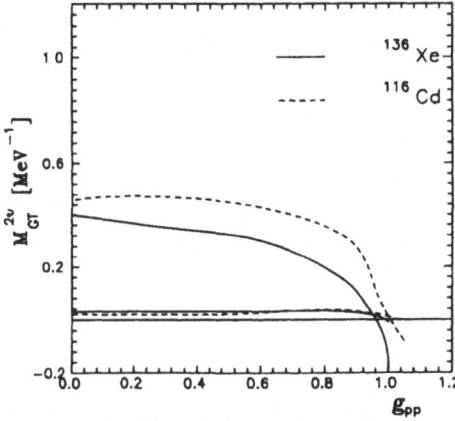

Figure 3. *The same as Fig. 1 but for ^{136}Xe and ^{116}Cd cases.*

now could be useful for experimentalists in planning future investigations on this line. Thus, we performed calculations of the $2\nu\beta\beta$ decay half-lives on the first 2^+ excited states with our procedure for the same $\beta\beta$ emitters. To fix the parameter g_{pp}, we have used the most recent data for the g.s. to g.s. $\beta\beta$ decay half-lives and our previous values for $M_{GT}^{2\nu}(0^+ \to 0^+)$ [29]. For the case of ^{110}Pd, where such experimental data have a large uncertainty, we fixed $g_{pp} = 1.0$. Further, we used this value to evaluate $M_{GT}^{2\nu}(0^+ \to 2^+)$ m.e. In our formalism, the n.m.e. for $0^+ \to 2_1^+$ transitions can be written as [29]:

$$\begin{aligned} M_{GT}^{2\nu}(0^+ \to 2_1^+) &= \sum_{k,m} \frac{<2_1^+\|\beta^+\|km>_f <km\|km'>_i <km'\|\beta^+\|0^+>_i}{(E_{int} - E_i + mc^2 + \frac{Q_{\beta\beta}^{02}}{2})^3} \\ &= \sum_{k=1,2} M_{GT}^{(k)}(0^+ \to 2_1^+) \end{aligned} \qquad (17)$$

Our results are presented in the table 1. Besides the half-lives for $2\nu\beta\beta$ decay on 2_1^+ excited states we present also the values for the GT n.m.e., computed with our code, for $\beta\beta$ transitions on g.s and 2_1^+ excited states of the daughter nuclei. One can observe that, except the case of ^{128}Te, the $\beta\beta$ decay half-lives on excited 2_1^+ states are of the order of $10^{23} - 10^{24}$ years and the most favorable cases are ^{82}Se, ^{116}Cd and ^{136}Xe for which these ones are of the order of 10^{23} years.

Table 1. *Two-neutrino double beta decay half-lives for transitions on 2_1^+ excited states. The values of the GT n.m.e., and g_{pp} are also specified. The numbers in parentheses denote the references from which the $2\nu\beta\beta$ decay half-lives for transitions on ground states are taken.*

Nucl	$Q_{\beta\beta}^{00}$ KeV	$Q_{\beta\beta}^{02}$ KeV	M_{GT}^{00} MeV^{-1}	M_{GT}^{02} MeV^{-1}	g_{pp}	$T_{1/2}^{2\nu}(0^+\to 0^+)$[y] exp	th	$T_{1/2}^{2\nu}(0^+\to 2_1^+)$[y] exp	th	$F_{2\nu}^0$ $y^{-1}MeV^2$	$F_{2\nu}^2$ $y^{-1}MeV^2$
^{76}Ge	2038	1478	0 163	0 097	0 95	1 42 10^{21} [13]		> 3 0 10^{21} (90%)[21]	2 36 10^{24}	3 8 10^{-20}	4 5 10^{-23}
^{82}Se	2995	2218	0 102	0 018	0 94	1 08 10^{20} [14]			1 97 10^{23}	1 26 10^{-18}	1 57 10^{-20}
^{110}Pd	2014	1356	0 09	0 006	1 0	> 6 10^{17} (68%)[15]	1 6 10^{20}		2 37 10^{24}	9 1 10^{-19}	1 17 10^{-21}
^{116}Cd	2802	1508	0 086	0 015	0 98	1 86 10^{19} [16]		> 1 7 10^{20} (68%)[21]	3 39 10^{23}	1 05 10^{-17}	1 31 10^{-19}
^{128}Te	868	425	0 011	0 002	1 02	7 7 10^{24} [18]			5 88 10^{30}	5 1 10^{-22}	4 25 10^{-26}
^{130}Te	2553	1997	0 032	0 008	0 99	2 7 10^{21}[15] [18]			2 56 10^{24}	1 1 10^{-18}	6 1 10^{-22}
^{136}Xe	2478	1660	0 075	0 017	0 94	> 1 6 10^{20} (95%)[19]			4 7 10^{23}	1 11 10^{-18}	7 35 10^{-21}

4. Conclusions

Concluding, one may say that within HQRPA procedure the suppression of the M_{GT} m.e. still occurs, but higher order corrections tend to stabilize the pnQRPA results mainly around the physical region. However, a further investigation of these corrections could be illuminating for reliable predictions of the GT m.e. For example additional meaningful correlations could be included in our formalism if an extended QRPA approach [38] would be employed where, beside the higher order corrections to the QRPA w.f., an improved BCS vacuum is also used. Our formalism could be also extended to be applied to calculate single-beta transitions for exotic nuclei or, in general, to treat other charge-changing processes.

References

[1] A. H. Haxton and G. J. Stephenson Jr., *Progr. Part. Nucl. Phys.* **12**, 409 (1984)
[2] M. Doi, T. Kotani and E. Takasugi, *Progr. Theor. Phys.* **83**, 1 (1985)
[3] J. D. Vergados, *Phys. Rep.* **133**, 1 (1986)
[4] A. Faessler, *Progr. Part. Nucl. Phys.* **21**, 183 (1988)
[5] F. T. Avignone III and Brodzinsky, *Progr. Part. Nucl. Phys.* **21**, 99 (1988)
[6] K. Muto and H. V. Klapdor, *Neutrinos*, edited by H. V. Klapdor, (Springer-Verlag, 1988)
[7] H. V. Klapdor-Kleingrothaus, *Proceedings of Zakopane School of Physics: Selected Topics in Nuclear Physics*, Zakopane, 31 August - 9 September 1992 (World Scientific, Singapore)
[8] T. Tomoda, *Rep. Progr. Phys.* **54**, 53 (1991)

[9] A. Bilenky and S. Petcov, *Rev. Mod. Phys.* **59**, 671 (1987)
[10] U. Amaldi, W. de Boer, H. Fuerstenau, *Phys. Lett.* **260B**, 447 (1991)
[11] T. Kirsten, *Proc. Int. Symp. on Nucl. Beta Decay and Neutrinos*, Osaka, Japan, p. 81, edited by T. Kotani, H. Fujiri and E. Takasugi (World Scientific, Singapore, 1986)
[12] S. R. Elliott, A. A. Hahn and M. K. Moe, *Phys. Rev. Lett.* **59**, 2020 (1987)
[13] A. Piepke, *Proc. XIII Moriod Workshop, Perspectives in Neutrinos, Atomic Physics and Gravitation*, Villars-sur-Alon, Jan. 30 - Feb. 7 1993
[14] S. R. Elliott et al., *J. Phys. G* **17**, 145 (1991)
[15] A.S. Barabash, *Preprint ITEP* **56**-1987
[16] H. Ejiri et al., *Proc. III Int. Symp. Weak and Electromagnetic Int. in Nuclei: WEIN '92*, Dubna, June 16-22 1992
[17] Yu. Zdesenko, *J. Phys. G* **17**, 243 (1991)
[18] T. Bernatowicz, J. Brannon, R. Brazzle, R. Cowsik, C. Hohenberg and F. Podosek, *Phys. Rev. Lett.* **69**, 2341 (1992)
[19] E. Belloti et al., *Phys. Lett.* **266B**, 193 (1991)
[20] M. Beck et al., *Z. Phys. A* **343**, 397 (1992)
[21] A. S. Barabash, *Phys. Lett.* **249B**, 186 (1990); A. S. Barabash et al., *Preprint LNPI* **1773** (1992).
[22] J. D. Vergados, *Phys. Rev. C* **13**, 865 (1976)
[23] K. W. Schmid, F. Grummer and A. Faessler, *Nucl. Phys. A* **431**, 205 (1984)
[24] T. Tomoda, A. Faessler, K. W. Schmid and F. Grummer, *Nucl. Phys. A* **452**, 591 (1986); T. Tomoda, *Nucl. Phys. A* **484**, 635 (1988), 591 (1986).
[25] C. R. Ching, T. H. Ho and X. R. Wu, *Phys. Rev. C* **40**, 304 (1989); X. R. Wu, A. Staudt, H. V. Klapdor-Kleingrothaus, C. R. Ching, T. H. Ho, *Phys. Lett. B* **276**, 274 (1992)
[26] P. Vogel and M. R. Zirnbauer, *Phys. Rev. Lett.* **57**, 3148 (1986)
[27] J. Suhonen, T. Taigel and A. Faessler, *Nucl. Phys. A* **486**, 91 (1988)
[28] A. A. Raduta, A. Faessler, S. Stoica and W. A. Kaminski, *Phys. Lett.* **254B**, 7 (1991); A. A. Raduta, A. Faessler and S. Stoica, *Nucl. Phys. A* **534**, 149 (1991)
[29] S. Stoica and W. A. Kaminski, *Phys. Rev. C* **47**, 897 (1993); S. Stoica and W. A. Kaminski, *Nuov. Cim.*, (in press)
[30] J. Suhonen, T. Taigel and A. Faessler, *Nucl. Phys. A* **486**, 91 (1988).
[31] J. A. Halbleib and R. A. Sorensen, *Nucl. Phys. A* **98**, 542 (1967)
[32] M. Baranger, *Phys. Rev.* **120**, 957 (1960); P. Ring and P. Schuck, *Nuclear Many Body Problem*, (Springer Verlag, New York, 1980)
[33] O. Civitarese, A. Faessler and Tomoda, *Phys. Lett. B* **194**, 11 (1987)
[34] K. Muto and H. V. Klapdor, *Phys. Lett.* **201B**, 420 (1988); K. Muto, E. Bender and H. V. Klapdor, *Z. Phys. A* **334**, 177 (1989)
[35] O. Civitarese, A. Faessler, J. Suhonen and X. R. Wu, *Nucl. Phys. A* **524**, 404 (1991); and *J. Phys. G* **17**, 943 (1991)
[36] T. Marumori, M. Yamamura and A. Tokunaga, *Progr. Theor. Phys.* **31**, 1009 (1964)
[37] A. Huffman, *Phys. Rev. C* **2**, 742 (1970)
[38] L. Zhao and A. Sustich, Preprint MSU (1991)

PROGRESS REPORT ON AN EXPERIMENT TO MEASURE PARITY MIXING OF THE $J^\pi T = 0^+1; 0^-1$ DOUBLET IN ^{14}N

M. Preiß and G. Clausnitzer

Strahlenzentrum der Justus-Liebig-Universität Gießen, D-35392 Giessen, Germany

The parity mixing of the $J^\pi T = 0^+1; 0^-1$ doublet in ^{14}N is investigated via $^{13}C(\vec{p}, p)^{13}C$ ($E_p \approx 1.15$ MeV) in order to study the isoscalar part of the weak nucleon-nucleon interaction, similar to an earlier investigation[1].

The strangeness conserving hadronic weak interaction is described in terms of a meson exchange model with one strong and one weak vertex. Since the strong coupling constants are empirically known, one can describe the weak nucleon-nucleon interaction using a parity non conserving (PNC) nucleon-nucleon potential in terms of seven weak coupling constants $h_\pi^1, h_\rho^0, h_\rho^1, h_{\rho'}^1, h_\omega^0, h_\omega^1$. The physics of W and Z exchange between the quarks is hidden inside of these coupling constants.

The PNC weak interaction between hadrons can be investigated in nuclear reactions by measuring observables which deviate from zero in case of parity violation. These PNC effects can be enhanced by nuclear structures. If isolated doublets of same spin and opposite parity are populated in light nuclei, the selection of different quantum numbers (in different reactions) offers a sensitivity to the different isospin components of the PNC potential and should allow a consistent determination of the weak coupling constants.

The discrepancy between the experimental result ($A_l = 8.6 \pm 5.9 \pm 2.4 \cdot 10^{-6}$)[1] and the theoretical expectation ($A_l \approx -2.8 \cdot 10^{-5}$)[2] in the A=14 case is motivation for a second experiment. The following preparations have been made:

1. regular analyzing power measurements in narrow energy steps nearby the 0^+1 resonance;
2. measurements of the polarization profile of the Lambshift Source beam for apparative asymmetry corrections;
3. reconstruction of the Giessen Lambshift Source into an Atomic Beam Source.

By using T-matrix formalism the regular observables were calculated and compared with the measurements. The longitudinal analyzing power was calculated within the OXBASH-Code. The most important systematic errors in this type of helicity experiments arise from residual transverse polarization gradients inherent in a nominally longitudinal polarized beam. So it is very important to know the regular analyzing power for corrections.

The minimum statistical error of the observed asymmetry appears at a scattering energy and angle where the product of the differential cross section and the square of the longitudinal analyzing power is at maximum. Figure 1 shows a contour plot of this figure of merit as a function of the proton energy and theta in the center of mass system. The dotted line marks the zero crossing of A_y. The largest longitudinal analyzing power, the best figure of merit and the smallest A_y are found at nearby $E_p = 1.158$ MeV, $\theta_{lab} = 90°$.

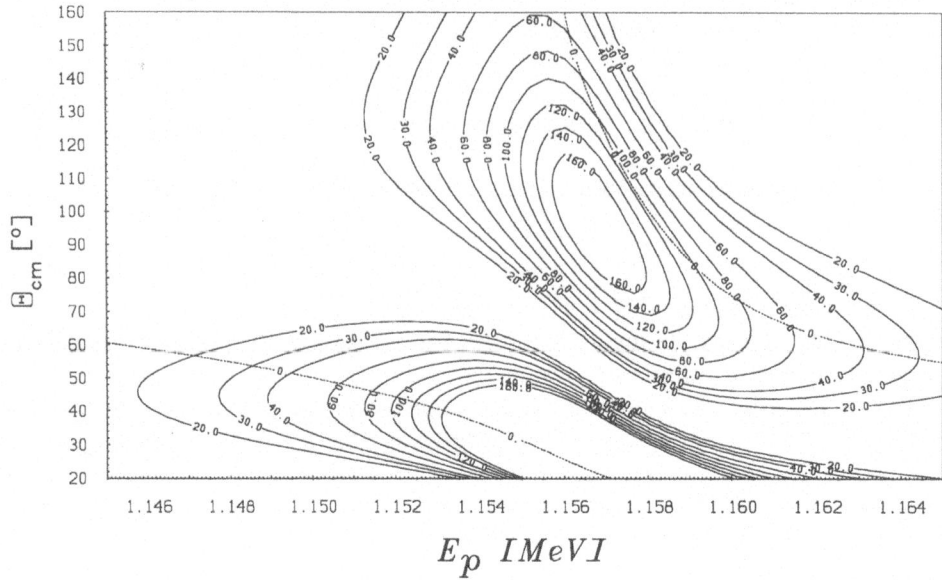

Figure 1. *Contour plot of the figure of merit as a function of the proton energy and θ_{cm}; the dotted line marks the zero crossing of A_y.*

All apparative preparations, selection of the optimum angle and energy and the preliminary determination of systematic asymmetries take into consideration that the predicted observable is in the range of $1 \cdot 10^{-5}$. Therefore the main problem will be the determination of apparative asymmetries during the experiment. We plan to measure the parity mixing of the $J^\pi T = 0^+1; 0^-1$ doublet immediately after the new Giessen Atomic Beam Source is operational (fall '93).

References

[1] Zeps et al., *A.I.P. Conf. Proc.* **176**, 1098 (1989)
[2] Adelberger et al., *Phys. Rev. C* **33**, 1840 (1986)

A NEW LOOK TO THE NUCLEAR STRUCTURE CALCULATIONS OF THE PNC CASES IN A=18-21 NUCLEI

M. Horoi[1,4], G. Clausnitzer[2], B. A. Brown[1] and E. K. Warburton[3]

[1] National Superconducting Cyclotron Laboratory, East Lansing, MI 48824
[2] Strahlenzentrum der Justus-Liebig-Universität, D-35392 Giessen, Germany
[3] Brookhaven National Laboratory, Upton, New York, 11973
[4] Institute of Atomic Physics, Bucharest, Romania

The experimental PNC results in ^{18}F, ^{19}F, ^{21}Ne and the actual theoretical analysis[1] show a discrepancy (see also Fig. 11 of Ref. 2). If one interprets the small limit of the experimentally extracted PNC matrix element ($<$ 0.029 eV) for ^{21}Ne as a destructive interference between the isoscalar and isovector contributions, then it is difficult to understand why the isovector contribution in ^{18}F is so small ($<$ 0.09 eV) while the isoscalar + isovector contribution in ^{19}F is relatively large (0.40 $\pm$ 0.1 eV). In order to understand the origin of this discrepancy a comparison of the calculated PNC matrix elements was performed. Previous nuclear structure calculations of the PNC matrix elements have been carried out (see Table 6 from Ref. 1) using all $(0+1+2)\hbar\omega$ configurations for ^{18}F and $(0+1)\hbar\omega$ for ^{19}F and ^{21}Ne. Table 1 presents results for ^{18}F obtained in the smaller ZBM model space[3], but including up to $4\hbar\omega$ configurations. One can see that the contributions from 3 and 4 $\hbar\omega$ configurations are significant.

Table 1. Partial contributions to PNC matrix elements (eV units). Here and in the following calculations DDH weak couplings[1] are used.

Nucleus	Interaction[3]	ΔT	$0\hbar\omega - 1\hbar\omega$	$2\hbar\omega - 1\hbar\omega$	$2\hbar\omega - 3\hbar\omega$	$4\hbar\omega - 3\hbar\omega$
	F-psd	1	1.045	-0.815	0.549	-0.187
^{18}F	Z-psd	1	1.119	-0.778	0.462	-0.148
	ZBMO	1	1.297	-0.669	0.430	-0.118

We performed also a large scale shell model calculation for ^{18}F using the Warburton-Brown interaction[4]: the first 4 major shells describe the shell model basis and up to $3\hbar\omega$ configurations have been included for the first time. Results presented in Fig. 1 confirm the necessity to include higher $\hbar\omega$ configurations.

Guided by these results we tried to see if the actual ZBM $(0+1+2+3+4)\hbar\omega$ calculations could give a hint about the above mentioned discrepancy. Table 2 presents different contributions to the PNC matrix element for ^{21}Ne. One can see that the isovector (IV) part is rather stable while the is isoscalar (IS) part fluctuates. These results are in contradiction with the usual interpretation[1] of the smallness of the PNC matrix element for ^{21}Ne.

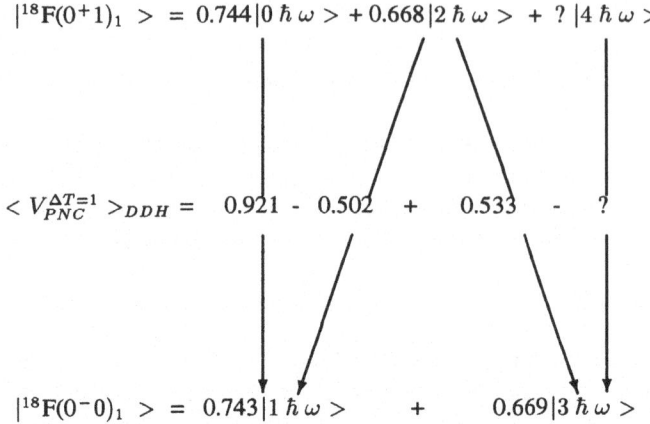

Figure 1. Wave function amplitudes and partial contributions to the PNC matrix element (eV) in a $(0+1+2+3)\hbar\omega$ calculation.

Table 2. Isoscalar and isovector contribution to the PNC matrix element (eV) in ^{21}Ne.

Nucleus	Interaction	IS	IV	Total
	F-psd	-0.113	0.556	0.442
^{21}Ne	Z-psd	0.071	0.359	0.430
	ZBMO	-0.010	0.339	0.329

To reproduce the experimental results one can introduce weighting factors

$$< V_{PNC} > = \alpha_{IS} < V_{PNC}^{DDH}(IS) > + \alpha_{IV} < V_{PNC}^{DDH}(IV) > . \tag{1}$$

These factors normalize the DDH weak coupling constants and the nuclear structure results also to the experimental values. An $(\alpha_{IS}, \alpha_{IV})$ plot, similar to that in Fig. 11 from Ref. 2, shows an overlapping region for the ^{18}F, ^{19}F and ^{21}Ne data. The $(\alpha_{IS}, \alpha_{IV})$ values in the overlapping region are in the range (0.45-0.75, < 0.1). In conclusion, we have proved the necessity to include up to $4\hbar\omega$ configurations in the calculations of the PNC matrix elements of the light nuclei and shown that the normalized ZBM calculations are not in contradiction with the experimental data.

MH and BAB thank Alexander von Humboldt Foundation for financial support.

References

[1] E. G. Adelberger and W. C. Haxton, *Ann. Rev. Nucl. Part. Sci.* **35**, 501 (1985)
[2] S. A. Page et al., *Phys. Rev. C* **35**, 1119 (1987)
[3] B. A. Brown et al., *MSUNSCL Report* **524** (1988)
[4] E. K. Warburton and B. A. Brown, *Phys. Rev. C* **46**, 923 (1992)

VI.

Nuclear Astrophysics

VARIOUS PROBLEMS OF NUCLEAR ASTROPHYSICS APPROACHED BY COULOMB DISSOCIATION EXPERIMENTS

H. Rebel

Kernforschungszentrum Karlsruhe, Institut für Kernphysik, P.O.B. 3640, D-76021 Karlsruhe, Germany

1. Introduction

The understanding of the nucleosynthesis in various sites of primordial and stellar burning scenarios requires the study of an ever increasing number of nuclear reaction rates. Their knowledge is a prerequisite of testing the consistency of models of the evolution of the early universe and of stars, at comparatively low temperatures in quiescent phases, in hot stellar processes like nova explosions, proton ingestion in He layers, more and more with the quest for studies of reaction rates of radioactive species.

The measurements of the relevant cross sections is, in general a rather difficult task since the expected values are among the smallest to be investigated in the laboratory. The standard laboratory approach involves the bombardment of rather thin targets with extremely-low-energy projectiles (Rolfs and Rodney 1988). Radiative capture processes are among the most important reactions for the formation of light elements. The Coulomb barrier strongly suppresses the cross section at low energies. As a matter of fact, the cross section values are almost always needed at energies far below which measurements can be performed in the laboratory. Usually they are obtained by extrapolations from the laboratory energy region over several decades downwards. The extrapolation procedures, using the concept of the astrophysical S-factor, are not free from theoretical bias and prejudice.

Table 1 illustrates the general situation by showing the limited knowledge about the rates for some selected examples of current interest in various astrophysical sites.

The $^3\text{He}(\alpha,\gamma)^7\text{Be}$ and $^7\text{Be}(p,\gamma)^8\text{B}$ capture rates affect substantially the solar neutrino flux at higher neutrino energies and bear strongly to the so-called solar neutrino puzzle.

This puzzle arises from twenty years of solar neutrino observations in the chlorine capture experiment indicating a substantial solar neutrino deficit: the ^{37}Cl capture experiment yields 2.1 ± 0.3 SNU (Rowley et al 1989), compared to 7.9 SNU of the Standard Solar Model of Bahcall and Ulrich (1988) or 6.4 SNU for the Turck-Chièze model (Turck-Chièze et al 1988). A deficit is confirmed by the Kamiokande II electron scattering experiment (Hirata et al. 1990) and the two Ga capture experiments, SAGE and GALLEX. SAGE (see Gavrin 1992) reported a flux of $58 \pm 17\ (24)\ (stat) \pm 14\ (syst.)$ SNU, GALLEX I (Anselmann et al 1992) $83 \pm 19\ (stat) \pm 8\ (syst)$ SNU and $97 \pm 23\ (stat) \pm 7\ (syst)$ SNU from GALLEX II (Anselmann et al 1993) with respect to 132 SNU of Bahcall's Standard Solar Model. New neutrino physics has been suggested to solve the puzzle. In addition of neutrino-flavour oscillations (see Mikheyev and Smirnov 1985) a non-zero value of the magnetic moment of

Table 1. *Some examples of radiative capture reactions of astrophysical interest.*

EXAMPLE	$E_{measured}$	ASTROPHYSICAL INTEREST
Hydrogen Burning $\alpha + {}^3He \rightarrow {}^7Be + \gamma$ $p + {}^7Be \rightarrow {}^8B + \gamma$ $E_0 \approx 20 - 30$ keV	≥ 107 keV ≥ 117 keV	Solar Neutrino Problem
Helium Burning $\alpha + {}^{12}C \rightarrow {}^{16}O + \gamma$ $E_0 \approx 300$ keV	≥ 0.94 MeV	Ashes of Red Giant (C/O Ratio)
Big Bang Nucleosynthesis $\alpha + t \rightarrow {}^7Li + \gamma$ $\alpha + d \rightarrow {}^6Li + \gamma$ $E_0 \approx 100$ keV	≥ 79 keV ≥ 1 MeV	Li Be B Production Test of the Standard Big Bang Model

the neutrino (associated with a non-zero rest mass) would enable a precession (a change of the helicity) of the neutrino in the magnetic field of the solar convective zone (Ohun et al 1986). However, serious conclusions about these speculations have also to consider the uncertainties of the nuclear cross sections entering in the solar model, especially of the ${}^7Be\,(p, \gamma)\,{}^8B$ rate since the deficit is due to high-energy neutrinos. Barker and Spear (1986) have presented evidence that the accepted value of the cross section at stellar energies is probably too large, in a way reducing the observed discrepancy by a significant amount.

Several years ago, Coulomb dissociation of fast projectiles was proposed as indirect method of several advantages to study radiative capture processes (Rebel 1985). During the last years, this method has proven to be a new access to a specific class of capture processes which are relevant for nuclear astrophysics. It is the purpose of this lecture to discuss the implications and the theoretical analysis of this kind of experiments and to point out the favourable conditions for experimental investigations. The progress in the field is discussed in a recent topical review (Baur and Rebel 1994). The method and possible background effects are scrutinized with general aspects, and some illustrative examples are discussed.

2. The Coulomb dissociation approach

The idea is to measure electromagnetic matrix-elements between bound and continuum nuclear states by means of electromagnetic excitation. These matrix-elements are directly related to those of photodisintegration, and, by time reversal, to those of radiative capture. In order to study the process

$$b + c \to a + \gamma \tag{1}$$

one uses the reaction

$$a + Z \to b + c + Z \tag{2}$$

where the Coulomb field of a (heavy) nucleus with proton number Z provides a copious source of equivalent photons. Since one can only start with a projectile nucleus a in the ground state, the present method is restricted to radiative capture which leads to the ground state of that nucleus a. The possibility to say something about the process $a^* + \gamma \to b + c$, where a^* denotes an excited state of nucleus a, by means of multiple electromagnetic excitation in reaction (2) seems to be remote. On the other hand, radiative capture processes induced by nuclei in excited state b^* (or c^*) are in principle accessible by the Coulomb dissociation method. Such processes can be of interest for nuclear astrophysics. Another point of interest arises from the fact that bare nuclei a, b and c appear in reaction (2). The effects of screening in laboratory and astrophysical environments have been proven to be an important issue (Blüge et al 1989); the results from Coulomb dissociation can represent the limit of no-screening.

In order to reach the nuclear continuum by means of electromagnetic exitation, the equivalent photon spectrum should contain the corresponding high frequency component. The "safe bombarding condition" of (sub)-Coulomb excitation implies a severe limitation of the equivalent photon spectrum to rather low energies. Therefore it is advantageous to use projectile velocities well above the Coulomb barrier, in order to overcome the adiabaticity condition

$$\hbar \omega \leq \frac{\hbar c}{b} \left(\gamma \cdot \frac{v}{c} \right)$$

where b is the impact parameter and $\gamma = 1/\sqrt{1 - (v/c)^2}$. Electromagnetic excitation processes occur dominantly for sufficiently large values of $b > R_1 + R_2$, the sum of the nuclear radii, i.e. at very forward angles. Nuclear excitation effects orginate from grazing collisions. These grazing collisions can in principle also lead to forward scattering angles and interfere with the electromagnetic excitation characterized by l-values much beyond grazing. Thus, the "footprints" of nuclear and electromagnetic exciation show up clearly in the angular distribution. These effects can be quantitatively studied in a DWBA approach, using well established computer codes. Simpler methods, of the eikonal type, can be very helpful to explore the physics of the interference of Coulomb and nuclear excitation (Glendenning 1984, Bertulani and Baur 1988).

A severe and challenging problem arises especially for direct transitions into the continuum: *"Post-acceleration"* or *"Coulomb final-state-interaction"* change the relative energy between the outgoing fragments which is determined by the asymptotic kinematics of coincident experimental observation. Since the strength of the Coulomb interaction scales with

$1/v$, these effects are expected to be expecially strong for low beam energies. As sufficiently high energies, which are of actual interest, a tremendous simplification occurs due to $\tau_{coll} \ll \tau_{nucl} = h/(E_n - E_0)$. Consequently we can apply the sudden approximation or Glauber theory ("frozen nucleus") (Baur et al 1992). These approaches have the merit that the response of the nucleus a, which is excited, is treated in a fully quantal way. The semiclassical picture of a classical relative motion of projectile and target can be retained. Interesting problems related to the time-dependent response of the projectile system to the transient Coulomb field of the target nucleus arise. It is a hope to use the time-dependent Coulomb field as a "clock" with investigating the time-dependence of nuclear excited states. The investigation of the time-dependent tunnelling problem is of special interest. It is well known (Rolfs 1973) that the radial matrix-elements of low-energy radiative capture peak at radii well beyond the range of nuclear interactions. A careful discussion of the time-dependence of the projectile wave function will be necessary to judge the relevance of the overlap of the wave-function of the break-up fragments with the target nucleus. A simple model has been developed (Shotter 1989). However, taking into account the time-dependence of the projectile wave- function, these strong interaction effects are considerably reduced (see Baur and Rebel 1994).

3. Specific features of astrophysically relevant cases

Table 2 presents a survey of various interesting cases and indicates the field, including dissociation reactions with neutrons in the breakup channel.

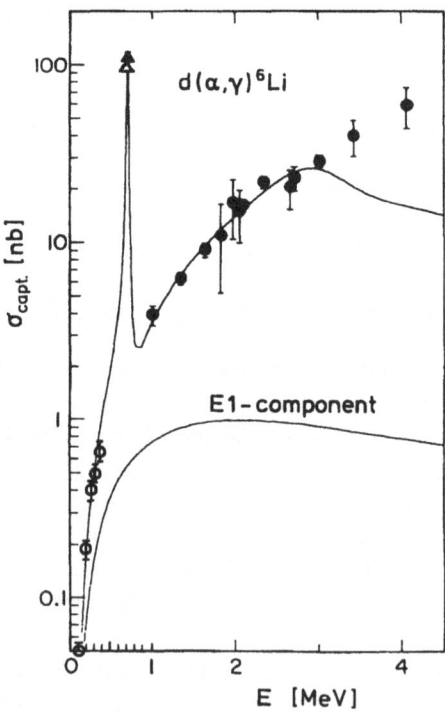

Figure 1. *Cross section for the $d(\alpha, \gamma)^6Li$ capture reaction. The low-energy data (open circles) are deduced from Coulomb dissociation of 156 MeV ^6Li projectiles (Kiener et al 1991 a).*

Table 2. *Radiative capture reactions of interest for light element synthesis accessible by Coulomb dissociation of fast projectiles.*

Reaction	$T_{1/2}$ (projectile)	Astrophysical Site	Reference
^{3}He$(\alpha,\gamma)^7$Be	53.3 d	Solar neutrino problem	Motobayashi and Gai 1992
^{7}Be$(p,\gamma)^8$B	770 ms	^{3}He abundancy	
^{7}Be$(\alpha,\gamma)^{11}$C	20.4 m		
^{4}He$(d,\gamma)^6$Li	stab.	Primordial nucleosynthesis of Li Be B-isotopes	Kiener et al 1989, 1991ab
^{6}Li$(p,\gamma)^7$Be	53.3 d		
^{5}Li$(\alpha,\gamma)^{10}$B	stab.		
^{4}He$(t,\gamma)^7$Li	stab.		Utsunomiya et al 1988, 1990ab
^{7}Li$(\alpha,\gamma)^{11}$B	stab.		
^{11}B$(p,\gamma)^{12}$C	stab.		
^{9}Be$(p,\gamma)^{10}$B	stab.		
^{10}B$(p,\gamma)^{11}$C	20.4 m		
^{7}Li$(n,\gamma)^8$Li	842 ms	Primordial nucleosynthesis in Inhomogeneous Big Bang	Sackett et al 1992
^{8}Li$(n,\gamma)^9$Li	178 ms		
^{12}C$(n,\gamma)^{13}$C	stab.		
^{14}C$(n,\gamma)^{15}$C	2.45 s		
^{14}C$(\alpha,\gamma)^{16}$O	stab.		
^{12}C$(p,\gamma)^{13}$N	10 m	CNO-cycles	
^{16}O$(p,\gamma)^{17}$F	65 s		Kiener et al 1993, Motobayashi et al 1991
^{13}N$(p,\gamma)^{14}$O	70.6 s		
^{20}Ne$(p,\gamma)^{21}$Na	22.5 s		
^{11}C$(p,\gamma)^{12}$N	11 ms	Hot p-p chain	Aguer and Kiener 1992
^{19}O$(\alpha,\gamma)^{19}$Ne	17 2 s	rp-process	Wiescher et al 1992
^{31}S$(p,\gamma)^{32}$Cl	291 ms		
^{12}C$(\alpha,\gamma)^{16}$O	stab.	Helium-burning	
^{16}O$(\alpha,\gamma)^{20}$Ne	stab.		
^{14}N$(\alpha,\gamma)^{18}$F	109 7 m		

We illustrate the general theoretical implications by particular astrophysically relevant examples.

a) *A test case with minimal post-acceleration effects:* $^6\text{Li} \to \alpha + d$

A full account of the successful Karlsruhe experiment at $E_{^6\text{Li}} = 156$ MeV has been published by Kiener et al (1991 a, b), demonstrating the useful conditions of the approach.

Though an experiment at lower energies ($E_{^6\text{Li}} = 60$ MeV; Hesselbarth and Knöpfle 1991) has found integrated cross sections which are in agreement with first order Coulomb excitation theory, the angular correlations of the nonresonant break-up exhibit considerable discrepancies. A more thorough theoretical analysis appears to be necessary as the experimental conditions do not guarantee the exclusion of substantial nuclear contributions for this case and higher-order effects should be included.

b) *A case exhibiting post-acceleration effects:* $^7\text{Li} \to \alpha + t$

This reaction has been investigated by Utsunomiya et al (1990 a b). A further experiment at again rather low beam energies, $E_{^7\text{Li}} = 54$ MeV, was performed by Gazes et al (1992) From simple classical considerations (see e.g. Baur and Weber 1989) strong post-acceleration effects have to be expected. Qualitative agreement of the experimental results is achieved with a rather classical model, which takes into account the distortions due to the target Coulomb field. However a quantal approach with convincing results is missing. At low beam energies, the sudden approximation is not applicable.

c) *A case with tractable post-accelerating effects:* $^8\text{B} \to ^7\text{Be} + p$

A current experiment at RIKEN attempts to measure the $^8\text{B} \to ^7\text{Be} + p$ dissociation at $E_{^7\text{Li}} = 50$ MeV/A (Motobayashi and Gai 1992). Since rather low equivalent photon energies are of astrophysical interest, the sudden approximation is well applicable at that beam energy. It is may be expected that a clean theoretical analysis of the potential experimental results is feasible. In view of the solar neutrino puzzle and recent GALLEX results (Anselmann et al 1993), the astrophysical S-factor for the $^7\text{Be}(p,\gamma)^8\text{B}$ reaction is of vital interest.

d) *A well established case with an exotic beam:* $^{14}\text{O} \to ^{13}\text{N} + p$

The results of two Coulomb dissociation experiments, at $E_{^{14}\text{O}} = 87.5$ MeV/A and $E_{^{14}\text{O}} = 70$ MeV/A (Kiener et al 1993), compare well with each other and with a recent capture experiment (Decrock et al 1991). The thermonuclear reaction rate is essentially dominated by the Γ_γ-width of the 5.17 MeV 1^- level in ^{14}O. The experiment yields $\Gamma_\gamma = 3.1 \pm 0.6$ eV (Motobayashi et al 1991), $\Gamma_\gamma = 2.4 \pm 0.9$ eV (Kiener et al 1993) and $\Gamma_\gamma = 3.8 \pm 1.2$ eV (Decrock et al 1991). The lower error bars in the Coulomb dissociation experiments are essentially due to the better statistical accuracy thanks to the fact that the Coulomb field of a nucleus is a prolific source of equivalent photons.

e) *Different equivalent photon spectra for different multipolarities:* $^{12}\text{N} \to ^{11}\text{C} + p$

The radiative capture process $^{11}\text{C}(p,\gamma)^{12}\text{N}$ is part of the reaction network in the hot p-p chain (Wiescher et al 1989). A microscopic cluster-model study has been recently performed (Descouvemont and Baraffe 1990). The ground state spin and parity of ^{12}N ($\tau = 11$ ms) are 1^+. Above the rather low $p + ^{11}\text{C}$ threshold at $E_{thr} = 0.6$ MeV there is a 2^+ level at 0.96 MeV, with a known γ-lifetime of 260 ± 40 fs. This is an M1 transition and the corresponding resonance strength is therefore known for astrophysical purposes. Since the E2 equivalent photon numbers are larger than the M1 photon numbers at realistic experimental conditions: $(v/c)^2 \ll 1$, the Coulomb excitation ($1^+ \to 2^+$) could proceed via E2 excitation. Angular distributions and the bombarding energy dependence may help to disentangle the two different contributions. For further potentially interesting levels, a 2^- ($E = 1.19$ MeV, $\Gamma = 118 \pm 14$ keV) and 1^- ($E = 1.8$ MeV, $\Gamma = 750 \pm 250$ keV), the excitation is purely by E1, and the above problem of multipole mixture is absent. The width of these two states is governed by the proton width, while the photon width is not directly known.

An experiment at GANIL is under way to determine the *gamma*-widths of these states (Aguer and Kiener 1992). A ^{12}N radioactive secondary beam of about 100 MeV/A interacts with a ^{208}Pb ("Primakoff"-) target. The ^{11}C and p fragments are observed in the forward direction. Due to the unequal charge to mass ratio of these fragments, noticeable post-acceleration effects might be expected. However, the decay length of the above mentioned resonances is long enough, so that the decay ^{12}N $\to p + ^{11}$C occurs essentially outside the Coulomb field.

f) *A really challenging case with a continuum transition of mixed multipolarities:* ^{16}O $\to \alpha + ^{12}$C

The importance of the ^{12}C (α, γ) ^{16}O radiative capture process for nuclear astrophysics is extensively discussed (Rolfs and Rodney 1988) with controversial results of the astrophysical S-factor at low energies. The Coulomb dissociation approach was theoretically outlined by Baur and Weber (1989). Due to the relatively high threshold ($Q = -7.162$ MeV) large incident ^{16}O energies are necessary to overcome the adiabaticity condition. The problem that E2 equivalent photon numbers are strongly dominating over E1 has been also emphasized (recently rediscovered by Shoppa and Koonin (1992)), and the sensitivity of angular correlations to E1-E2 interference has been pointed out (Baur and Weber 1989).

4. Outlook

The Coulomb dissociation method has been increasingly appreciated as a useful tool of photonuclear physics, using (equivalent) photons of energies below, above and in the giant resonance region. Applications to nuclear astrophysics are expecially rewarding. They need a precise knowledge of the relative energies between the outgoing fragments. In this respect the "magnifying glass effect" arising from the kinematical situation with three particle in the final state is very helpful, especially at the low relative energies characteristic for nuclear astrophysics. This contribution has put emphasis on the various theoretical implications of the analysis of the data. The problems require straightforward (but sometimes time consuming) procedures. There are also more subtle aspects of great intrinsic theoretical interest, e. g. the time-dependence of phenomena in a transient Coulomb field.

References

Aguer, P. and Kiener, J. (1992), private communication

Anselmann, P. et al. (1992): GALLEX collaboration, *Phys. Lett.* B **285**, 376 (1992); ibidem **285**, 390

Anselmann, P. et al (1993): GALLEX collaboration, *Phys. Lett.* B **314**, 445

Bahcall, J. N. and Ulrich, R. K. (1988), *Rev. Mod. Phys.* **60**, 297

Barker, F. C. and Spear, R. H. (1986), *Astrophys. J.* **307**, 847

Baur, G., Bertulani, C. A. and Rebel, H. (1986), *Nucl. Phys.* A **459**, 188

Baur, G. and Weber, M. (1989), *Nucl. Phys.* A **504**, 352

Baur, G., Bertulani, C. A. and Kalassa, D. M. (1992), *Nucl. Phys.* A **550**, 527

Baur, G. and Rebel, H. (1994), Topical Review: *Coulomb Dissociation as a Tool of Nuclear Astrophysics, J. Phys.* G **20**, 1

Blüge, G., Langanke, K., Reusch, H.-G. and Rolfs, C. (1989), *Z. Phys.* A **333**, 219

Decrock, P., Delbar, T., Duhamel, P., Galster, W., Huyse, M., Leleux, P., Licot, I., Liénard, E., Lipnik, P., Loiset, M., Michotte, C., Ryckewaert, G., Van Duppen, P., Vanhorenbeeck, J. and Vervier, J. (1991), *Phys. Rev. Lett.* **67**, 808

Descouvement, P. and Baraffe, I. (1990), *Nucl. Phys.* A **514**, 66

Gavrin, V. N. (1992), *26th Int. Conf. High Energy Physics*, Dallas

Gazes, S. B., Mason, J. E., Roberts, R. B. and Teichmann, S. G. (1992), *Phys. Rev. Lett.* **68**, 150

Glendenning, N.K. (1984), *Classical and Quantum Mechanical Description of Heavy-Ion Interaction, Proc. of the International Conference on Reactions between Complex Nuclei*, Nashville, Tennessee, June 10-14, edited by R. L Robinson et al. (North-Holland, 1984)

Hesselbarth, J. and Knöpfle, K. T. (1991), *Phys. Rev. Lett.* **67**, 2773

Hirata, K. S. et al. (1990), *Phys. Rev. Lett.* **65**, 1297

Kiener, J., Gils, H. J., Rebel, H. and Baur G. (1989), *Z. Phys.* A **332**, 359

Kiener, J., Gils, H. J., Rebel, H., Zagromski, S., Gsottschneider, G., Heide, N., Jelitto, H., Wentz, J. and Baur, G. (1991a), *Phys. Rev.* C **44**, 2195

Kiener, J., Gsottschneider, G., Gils, H. J., Rebel, H., Corcalciuc, V., Basu, S. K., Baur, G. and Raynal, J. (1991b), *Z. Phys.* A **339**, 489

Kiener, J., Lefebvre, A., Aguer, P., Bacri, C. O., Bimbot, R., Bogaert, G., Borderie, B., Clapier, F., Coc, A., Disdier, D., Fortier. S., Grunsberg, C., Kraus, L., Linck, I., Pasquier, G., Rivet, M. F., St Laurent, F., Stephyn, C., Tassan-Got, L. and Thibaud, J.P. (1993), in: *Radioactive Nuclear Beams, Sect. 4, 2nd Conf. on Radioactive Nuclear Beams*, Louvain-la-Neuve (1991), p. 311 - *Nucl. Phys.* A **552**, 63

Mikheyev, S. P. and Smirnov, A. Y. (1985), *Sov. J. Nucl. Phys.* **42**, 913

Murakami, H., Ando, Y., Iwasa, N., Kurokawa, M., Shirato, S., Ruan, J., Ichihara, T., Kubo, T., Inabe, N., Goto, A., Kubono, S., Shimoura, S. and Ishihara, M. (1991), *Phys. Lett.* B **264**, 259

Motobayashi, T. and Gai, M. (1992), RIKEN proposal, *private communication - RIKEN Accel. Prog. Report* **26**, 19

Okun, L.B., Voloshin, M.B. and Vysotsky, M.I. (1986), *Sov. J. Nucl. Phys.* **44**, 440

Rebel, H. (1985), *Workshop on Nuclear Reaction Cross Sections of Astrophysical Interest*, unpublished report, Kernforschungszentrum Karlsruhe (Febr. 1985)

Rolfs, C. (1973), *Nucl. Phys.* A **217**, 29

Rolfs, C. and Rodney, W.S. (1988), *Cauldrons in the Cosmos*, (University Press, Chicago, London, 1988)

Rowley, J. K., Cleveland, B. T. and Davis R. Jr. (1989) in: *Proc. XIIIth Int. Conf. Neutrinos Phys. and Astrophysics (NEUTRINO '88)*, p. 518, edited by Schneps, J. et al., (World Scientific, Singapore, 1989)

Sackett, D., Ieki, K., Galonsky, A., Kiss, A., Deak, T., Horvath, A., Seres, Z., Kolata, J., Kasagi, J., Bertsch, G., Bertulani, C. A., Sherril, B., Morrissey, D., Winger, J. and Orr, N. (1992), NSCL proposal

Shoppa, T. D. and Koonin, S. E. (1992), *Phys. Rev. C* **46**, 382

Shotter, A. C. (1989), *J. Phys. G: Nucl. Phys.* **15**, L41

Utsunomiya, H., Schmitt, R. P., Lui, Y. W., Haenni, D. R., Dejbakksh, H., Cook, L., Heimberg, P., Ray, A., Tamura, T. and Udagawa, T. (1988), *Phys. Lett.* **B 211**, 24

Utsunomiya, H., Lui, Y. W., Haenni, D. R., Dejbakksh, H., Cook, L., Srivastava, B. K., Turmel, W., O'Kelly, D., Schmitt, R. P., Shapira, D., Gomez del Campo, J., Ray, A. and Udagawa, T. (1990a), *Phys. Rev. Lett.* **65**, 847

Wiescher, M., Görres, J., Graff, S., Buchmann, L. and Thielemann, F. K. (1989), *Astrophys. J.* **343**, 352

Wiescher, M., Görres, J., Vouzukas, S., Aguer, P. and Kiener, J. (1992), GANIL proposal, *private communication*

WEAKLY INTERACTING MASSIVE PARTICLES AS THE DARK MATTER OF THE UNIVERSE

P. B. Price, D. P. Snowden-Ifft and E. S. Freeman

Physics Department, University of California, Berkeley, CA 94720, USA

Introduction

There are several reasons to think that we may be on the verge of a new Copernican revolution: not only are we not located at the center of the Universe; we may not even be made of the stuff of which > 90% of the Universe is made! Within the last decade the evidence has become overwhelming that dark matter exists within galaxies and in the Universe at large[1]. The mass density of the Universe Ω is measured in units of the critical density, $\rho_{crit} = 3H_0^2/8\pi G \approx 10^{-29}$ g cm^{-3}. Visible matter comprises only $\Omega_{vis} < 0.01$, whereas nucleosynthesis of light elements in the Big Bang is quantitatively accounted for if the present density of baryonic matter is $\Omega_b \approx 0.05$. Thus, some baryonic dark matter is believed to exist. Astronomical observations (galactic dynamics; x-ray and infrared satellite data) suggest that $\Omega > 0.5$. The inflationary model of the early universe predicts that $\Omega = 1$ exactly. Thus, as much as 95% of the Universe may be in the form of nonbaryonic particles.

Particle physicists need nonbaryonic particles in order to account for inconsistencies in the Standard Model. An attractive example is the supersymmetric grand unified theories, which contain weakly interacting massive neutral particles (WIMPs) such as neutralinos. These arise naturally as relics of the Big Bang. Suppose nonbaryonic particles denoted as δ are formed thermally in the early universe and are initially in equilibrium through annihilation and pair production. As the Universe expands and cools the annihilation rate decreases and eventually the δ and $\bar{\delta}$ freeze out of equilibrium. If particles δ make up the majority ($\Omega_\delta \approx 1$) of the dark matter, their annihilation cross section satisfies $<\sigma\frac{v}{c}>_{annih} \approx 10^{-37}$ cm^2, which is a typical weak cross section. Both the large-scale mass structure of the Universe and the fluctuations in the microwave radiation discovered by the COBE satellite are consistent with $\Omega = 1$, largely in the form of cold dark matter such as WIMPs.

WIMPs are believed to be gravitationally bound to the halo of our own Galaxy, with a mass density estimated to be ~ 0.3 GeV cm^{-3} on the basis of astronomical evidence. They constitute a dissipationless gas with a Maxwellian velocity distribution, $v_{rms} \approx 270$ km s^{-1}, truncated at an escape velocity of ~ 600 km s^{-1}. The earth moves through the WIMP gas at a velocity of ~ 230 km s^{-1}. How can one test this conjecture?

Searching for WIMPs in the Galaxy

A number of groups are developing detectors capable of responding to the weak signal that would result from elastic scattering of a WIMP off a nucleus[2]. The cross section for such a process is estimated to be in the range 10^{-32} to 10^{-38} cm^2 (or even lower) and the WIMP mass probably lies in the range from a few GeV to a few TeV. WIMPs with these properties would penetrate most of the earth and would produce a signal between $\sim 10^4$ and $\sim 10^{-4}$ recoils per kg per day in a detector. The energy of the recoil atom is typically only a few keV, and at least three times as much of the recoil energy is communicated to phonons as to ionization of atoms. Three groups, using natural Ge detectors at 77 K without phonon detection, have completed initial searches with enough sensitivity to rule out heavy Dirac neutrinos as the main constituent of dark matter[3]. To do better will require heroic efforts in developing larger, more sensitive detectors with very low noise and capable of discriminating against various types of background signals from radioactivity and cosmic rays. New detectors that will be operated at cryogenic temperatures appear capable of observing both the prompt ($\sim 10^{-6}$ s) ionization signal and the slow ($\sim 10^{-3}$ s) phonon signal.

We have been exploring the fascinating possibility of detecting the microscopic tracks left by recoil atoms struck by WIMPs passing through natural mica crystals deep underground[4]. For mica that has been storing such recoil tracks for a time of $\sim 10^9$ years, the collecting power far exceeds that attainable with electronic detectors. The main steps required are:

- to measure the storage time of tracks of recoil atoms in the sample being studied;
- to establish, by means of calibrations with neutrons and with ion beams, which of the several atom species in the mica produce recordable recoil tracks and with what efficiency;
- to devise a way to detect and distinguish tracks of WIMP-recoils from other kinds of tracks;
- to scan a large enough volume of mica to improve on existing limits on WIMP flux.

Known Types of Fossil Tracks in Mica

Muscovite mica, the most common variety, has the composition $KAl_3Si_3O_{10}(OH,F)_2$ and contains, as trace impurities, Fe and other heavy elements. Ancient tracks in mica are most conveniently studied by etching a freshly cleaved surface in concentrated hydrofluoric acid. (For a discussion of track-etching techniques and applications, see ref. 5.) The rate of surface removal in the absence of a track is very anisotropic: the rate parallel to the cleavage plane is about 50 times faster than the rate normal to the cleavage plane. Along a track the rate is still faster, and the resulting etchpit can be measured in an optical microscope if it is sufficiently long. As of 1986, three types of natural tracks had been discovered in mica:

- Spontaneous fission of ^{238}U atoms present in mica at atom concentrations $\sim 10^{-11}$ to 10^{-9} leads to tracks $\sim 20\mu$m long. This forms the basis of the fission track dating technique [5].
- Alpha decay of U and Th impurities in mica leads to recoiling daughter atoms with range a few tens of nm [6]. (The alpha particle does not produce enough radiation damage to leave a detectable track.) These etchpits are too shallow to be quantitatively measured with an optical microscope.
- Some of the highest-energy alpha particles in the Th decay chain undergo nuclear reactions with nuclei in mica, leading to compound nuclei that recoil while emitting a light particle [7]. The dominant events are recoils due to interactions of alphas with Al and Si nuclei, which have typical range 0.3 to 1.5 μm.

Let us designate the relative numbers of tracks of each type per unit surface as n_{SF}, n_α, and n_{int}. Some years ago it was reported[7] that, in the absence of any thermal fading, the relative proportions of fission tracks, alpha-recoil tracks, and alpha-interaction tracks are $n_{SF} : n_\alpha : n_{int} \approx 1 : 10^4 : 0.2$. In our current study we have found that in the absence of thermal fading n_{int}/n_{SF} ranges from ~ 0.1 to as high as ~ 2, depending on the proportion of Th to U in the sample.

WIMP-Recoil Track Collection Time of Mica

The fission-track dating method[5] gives the time interval over which spontaneous fission tracks from ^{238}U have been retained in a mica sample. One counts the surface density of natural fission tracks in the sample, irradiates it with a known dose of thermal neutrons, and counts the additional tracks due to induced fission of ^{235}U. From two equations one can calculate both the fission-track-retention age and the concentration of U. For our samples the track-retention age ranges from (0.3 to 2) $\times 10^9$ years, and the U concentration ranges from (0.7 to 7) $\times 10^{-12}$ atom fraction.

Tracks of recoil atoms, which consist of less heavily radiation-damaged material (lower dE/dx), are erased at lower ambient temperature than are fission tracks. In about 1/3 of the 50 mica samples we studied, we found that the recoil-track-retention age was as great as the fission-track retention age. Fortuitously, the radiation-damage rate (dE/dx) of alpha-interaction tracks is about the same as would be that of WIMP recoil tracks, so that the alpha-interaction track-retention age can be equated with the WIMP-recoil track-retention age. We assume that samples with $n_{int}/n_{SF} \geq 0.1$ have undergone no thermal fading, and for these samples we equate the WIMP-recoil track-retention age to the fission-track-retention age. The samples we are studying have recoil-track-retention ages $\sim$ (0.3 to 2) $\times 10^9$ years.

Response of Mica to Recoil Atoms

For all WIMP detectors the signal to be measured is the struck atom's recoil energy,

$$\Delta E = \frac{M \cdot m_\delta^2}{(M + m_\delta)^2} v^2 (1 - \cos\theta), \tag{1}$$

where M is the mass of the recoil atom, m_δ and v are the mass and velocity of the WIMP, and θ is the scattering angle in the center of mass. For $v_{rms} = 270$ km s^{-1}, M = 39 amu, and $m_\delta \gg $ M, the mean recoil energy is $\sim$ 30 keV. At this energy, more of the recoil energy is lost through hard-sphere collisions with atoms than through electronic stopping. Earlier experiments[8] showed that mica records tracks of ions at such low energies with an efficiency that increases with Z from a threshold at Z $\approx$ 8. Our recent experiments with beams of various ions have shown that ^{39}K, which comprises 5% of the mica structure, would be the major contributor to tracks of WIMP recoils. Fe and heavier elements are too rare to contribute, and Si, Al, O, and H are too lightly damaging to contribute.

To simulate the response of mica to WIMPs, we found that an exposure of mica to fast neutrons in a reactor is ideal. For WIMP masses greater than $\sim$ 50 GeV, the energy spectrum of recoil K atoms is similar when mica is exposed to unmoderated fast neutrons as when it is exposed to a Maxwellian distribution of WIMPs.

Recoil Atom Tracks with Atomic Force Microscopy

Recognizing that neither conventional optical nor electron microscopy is capable of quantitative measurements of ranges of recoil atoms in mica, we explored the idea of using the powerful new technique of Atomic Force Microscopy (AFM), which has proved to be an ideal instrument for our WIMP search[9, 10]. This instrument drags a nanostylus in a raster-scan over a surface by means of piezocrystal controls and accurately measures the up and down motions. For shallow etchpits such as result from etching recoil tracks in mica, one can achieve vertical resolution better than 0.1 nm, which permits individual atomic planes to be resolved. Figure 1 compares computer-generated images of two types of etched recoil tracks on the surfaces of mica crystals. At the top are alpha-recoil etchpits in a 300-million-year-old muscovite mica; at the bottom are etchpits in a mica that was first annealed to erase pre-existing tracks and then irradiated with 5.5×10^{13} fast neutrons in a reactor. The numbers next to the etchpit mouths give the depths in Angstroms. Because of the 50-fold greater rate of etching parallel to the cleavage plane than normal to it, an etchpit of a recoil track with a typical depth of $\sim$ 10 nm will have a mouth diameter of $\sim 1 \mu$m. This amplification is crucial for detection using the AFM. To convert the initial gray-scale image in the AFM to the simple image in Fig. 1 we first "flattened" the actual image in order to correct for curvature introduced by the piezocrystal device that made the scan. Pixels representing depths greater than a preset value (2 nm) were automatically grouped into contiguous clusters and the maximum depth of each cluster was determined. From such images one is able to determine the threshold for detection.

Imperfect correction for nonlinearity in the piezocrystal control system sets a present limit of $\sim 40\mu$m $\times$ 40 μm on the area of a single image, so that one must take hundreds or thousands of images in order to utilize fully the huge collecting power of mica. As AFM systems improve, the image size and scanning speed will increase. Another new imaging method — laser-feedback microscopy — also shows promise of providing accurate measurements of etchpit dimensions over large areas.

Rejection of Background due to Alpha-Recoil Tracks

Both ^{238}U and ^{232}Th contribute a troublesome background that must be recognized and discriminated against. (The contribution from ^{235}U is negligible.) A thermal neutron exposure followed by a count of induced fission tracks enables the U concentration to be determined. The ^{238}U decay chain to ^{206}Pb gives rise to eight connected recoil tracks in a three-dimensional random walk, with a mean extension normal to a mica cleavage plane of $\sim$ 40 nm. Th is about 4 times as abundant as U in mica, but because of its 3-fold longer half-life, the rate of production of alpha-recoil tracks by Th is about the same as by U. The ^{232}Th decay chain to ^{208}Pb gives rise to 6 connected recoil tracks, also with a mean extension of $\sim$ 50 nm normal to a mica plane.

As determined both from calculation and from our fast neutron calibrations, the typical extension of an etched WIMP recoil track normal to a mica plane should be only a few nm, much smaller than that for an alpha-recoil. We have developed a method that shows promise of being able to discriminate WIMP-recoils from alpha-recoils. The method involves determining the component of track-length normal to a cleavage plane. The approach is to scan both adjoining halves of a cleaved and etched mica crystal and to match the images so that the full extension of each recoil track crossing the cleavage plane can be determined. Monte

Figure 1. Binary AFM images of 40 µm × 40 µm areas showing alpha-recoil etchpits in unannealed mica (upper part) and neutron-recoil etchpits in mica annealed to erase pre-existing tracks (lower part).

Carlo simulations of random walks of chains of recoils show that it is extremely rare for all 6 (for ^{232}Th) or 8 (for ^{238}U) recoils to lie nearly on top of each other and parallel to a cleavage plane in such a way that they would fake the shorter range of a WIMP recoil. Consequently, a cut on etched depth should separate the characteristically shallow WIMP-recoil etchpits from the characteristically deep alpha-recoil etchpits.

Background of Recoils due to Fast Neutrons Underground

At shallow depths, neutrons due to cosmic ray interactions in the atmosphere and in rock could simulate a WIMP flux. Even at great depths where the cosmic ray neutron flux is negligible, unmoderated fast neutrons due to fission of ^{238}U in surrounding rock provide a background of recoil tracks that ultimately limits the sensitivity of mica to WIMPs. Since mica can be formed only at depths of several km of rock and resides for much of its lifetime at such depths, only the fission neutrons are important. Belli et al.[11] measured the energy distribution of neutrons in the Gran Sasso underground laboratory. They found a flux of 4×10^{-7} cm^{-2} s^{-1} ($\sim 10^{10}$ cm^{-2} over the lifetime of a typical mica crystal) with energy greater than 1 MeV. Based on our fast neutron calibrations, such a neutron flux would fake a WIMP-recoil density of ~ 1 etchpit per thousand fields of view at 40 μm $\times$ 40 μm. Because the mica may not have maintained any special orientation in space during its lifetime, there seems to be no way to distinguish the neutron-recoils from WIMP recoils on the basis of their angular distribution. Fortunately, the neutron-recoil background just estimated is so low that it does not represent a practical limitation on near-term mica experiments until the problem of scanning large areas of mica is solved.

The background of fast neutrons may be lower in mica-rich mineral deposits than in Gran Sasso. Large mica crystals grow in pegmatites with dimensions tens of meters in length and several meters in thickness. Since about 1 out of 10 atoms in muscovite is hydrogen, fission neutrons may be strongly moderated within such huge mica crystals, and neutron-recoils may not be a problem.

Background of Recoils Due to Underground Muons

Muons produced in cosmic ray interactions in the atmosphere and in rock penetrate much more deeply than do cosmic ray neutrons. The background due to elastic collisions of muons with K atoms in mica depends on the depth as a function of time, which may vary from mica to mica. To estimate the contribution of muon-recoils relative to that of WIMP recoils, we make the plausible assumption of a uniform erosion rate in which the burial depth of a mica sample decreases from 3 km to a final depth of D_o at a constant rate for a time of 10^9 years. One of our samples comes from a mine in Bihar Province, India, with $D_o = 150$ m; some samples come from shallow mines with $D_o \approx 10$ m; the provenances of some samples are unknown. Using as input the muon energy spectrum as a function of depth, the differential scattering cross section by muons as a function of energy, and the results of our calibration experiments with ion beams and fast neutron exposures, we calculated the surface density of etchpits at a cleavage plane due to muon-recoils. We conclude that, for WIMP masses $50 \leq m_\delta \leq 10^4$ GeV, for a WIMP cross section $\sigma = 10^{-34}$ cm^2, and for micas taken from depths $D_o > 150$ m, muon-recoils make a negligible contribution to background. For micas with $D_o = 10$ m, muon-recoils restrict the sensitivity to WIMPs: for a mass of $\sim 10^2$ to 10^3 GeV a WIMP cross section below $\sim 10^{-33}$ cm^2 would be inaccessible; for a mass of 10^4

GeV a WIMP cross section below $\sim 10^{-32}$ cm^2 would be inaccessible. For masses below $\sim 10^2$ GeV the WIMP-recoil detection efficiency in mica decreases with decreasing WIMP mass, and it becomes difficult for the mica method to compete with electronic methods.

To assess more definitively the possible muon-recoil background, we are investigating micas from mines deeper than 500 m.

Elastic Scattering Cross Sections of WIMPs

For dark matter consisting of heavy neutrinos, the cross section for scattering off of nuclei has terms proportional to the squares of the vector (g_v) and axial vector (g_A) coupling constants,

$$g_V = [N - (1 - 4\sin^2\theta_W)Z] \approx N \text{ and } g_A = \lambda[J(J+1)]^{1/2} \Sigma T_q^3 \Delta q \qquad (2)$$

where N and Z are the numbers of neutrons and protons in the target nucleus, θ_W is the Weinberg angle, λ is the effective spin parameter of the nucleus, T_q^3 is the third component of weak isospin, Δq is the fraction of the spin carried by the quark of type q, and the sum is over u, d, and s quarks. Since 4 $\sin^2\theta_W \approx 1$, we have $g_V \approx N$. Also, $g_A << g_V$ for ^{39}K and most other target nuclei. For the Dirac neutrino, scattering is mainly by Z_o-exchange:

$$\sigma_{\nu_D} \approx \frac{M^2 \cdot m_\delta^2}{2\pi(M + m_\delta)^2} G_F^2 N^2 |F_{mass}(q)|^2, \qquad (3)$$

where G_F is the Fermi coupling constant and F_{mass} is the form factor for the distribution of mass in the nucleus. For large momentum transfer ($q \cdot R > 1$), F_{mass} becomes < 1 and full coherency in scattering is lost. Using the Gaussian fit of $|F_{mass}|^2 = exp(-q^2/q_o^2)$ recommended in ref. 12, with $q_o = 3/R^2$, and R = 0.3 + 0.89$A^{1/3}$ fm, we find that for a ^{39}K nucleus, $|F_{mass}|^2 > 0.5$ for recoil energies less than $\Delta E = 100$ keV. The assumption of full coherence is approximately valid for all momentum transfers of interest in the mica experiment.

Figure 2, which will be referred to several times, compares calculated elastic scattering cross sections for several types of WIMPs with experimental limits based on the assumptions that the density of WIMPs in the Galactic halo is 0.3 GeV cm^{-3} and their velocity distribution is Maxwellian with $v_{rms} = 270$ km s^{-1}. As a comparison of the curves labeled "heavy ν_D" and "Ge limit (coherent)" shows, experiments with Ge detectors[3] have obtained null results that rule out heavy Dirac neutrinos with mass between ~ 10 GeV and ~ 2000 GeV and standard coupling.

Because it could scatter only by means of a spin-dependent interaction (g_A), a heavy Majorana neutrino would be very difficult to rule out, since $g_A << g_V$. The cross section is

$$\sigma_{\nu_M} \approx \frac{M^2 \cdot m_\delta^2}{8\pi(M + m_\delta)^2} \lambda^2 J(J+1) G_F^2 |F_{spin}(q)|^2 \qquad (4)$$

where F_{spin} is the form factor for distribution of spin in the nucleus. We assume, with Ellis and Flores[12], that $F_{spin}(q) \approx F_{mass}(q)$. For various target nuclei, they have tabulated values of $\lambda^2 J(J+1)$ based on the "odd-group" model[13], which is more accurate than the single-particle model for nuclei far from a closed shell.

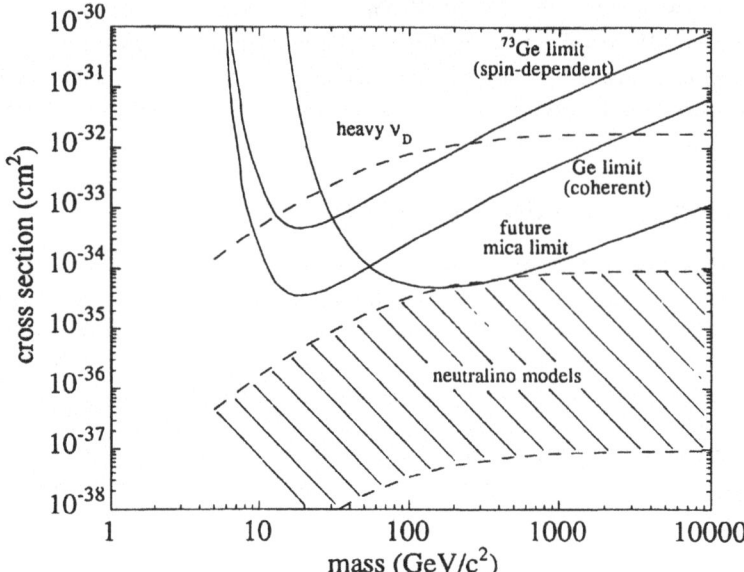

Figure 2. *Exclusion plot for scattering cross section vs WIMP mass. Dashed curves show expected cross sections for heavy Dirac neutrinos and neutralinos; solid curves show experimental upper limits on cross sections obtained with Ge detectors and limits that can be set in the present experiment if no ^{39}K-recoils are seen in mica down to a level set by neutron background.*

In searches for WIMPs that scatter only through a spin-dependent interaction, mica with ^{39}K as the target nucleus has advantages over Ge and most other detectors being developed. ^{39}K comprises 93% of all K (compared with ^{73}Ge, which has a natural abundance of only 7.8% of all Ge), and the factor $\lambda^2 J(J+1)$ in the odd-group model is 3 to 6 times larger for ^{39}K than for ^{73}Ge. The ratio $\sigma_{\nu_M}/\sigma_{\nu_D}$ is only $\sim 3 \times 10^{-4}$ for ^{39}K and much lower for most other detectors, which makes the heavy Majorana neutrino inaccessible to any searches in the near future.

The curve in Fig. 2 labeled "^{73}Ge limit (spin-dependent)" takes into account the 7.8% abundance of ^{73}Ge in the natural Ge detectors used to date.

In recent papers[12, 14] elastic scattering cross sections were estimated as a function of the many parameters that characterize supersymmetric particles such as neutralinos. The results depend crucially on the values of Δu, Δd, and Δs, which are quite controversial. The values based on the EMC study of muon scattering from protons lead to very different cross sections than when values obtained with the naive static quark model are used. Recent results at SLAC for electron scattering from polarized ^{3}He (which approximates a neutron target) are closer to the naive quark model. On this model, odd-Z target nuclei would scatter from neutralinos with much higher cross sections than would odd-N nuclei. Thus, ^{39}K is strongly favored over ^{73}Ge.

The broad shaded band in Fig. 2 gives an idea of the range of values of cross section for various models.

Limits on WIMPs Attainable with Mica

Six of our mica samples satisfy the criteria for use as WIMP detectors and are being subjected to intense study. They have track-retention ages of (0.3 to 2) $\times 10^9$ years, uranium concentrations of (0.7 to 7) $\times 10^{-11}$ atom fraction, and ratios of alpha-interaction track densities to fission track densities greater than 0.1. One is known to come from a depth larger than 100 m; three come from shallower depths; and two are from unknown locations. By comparing data on etchpits shorter than alpha-recoil etchpits in the different samples, we hope to be able to reject some background events. Densities of WIMP-recoils, muon-recoils, and neutron-recoils, which are produced dominantly by ^{39}K atoms, should correlate with track-retention age; in addition, densities of muon-recoils should anticorrelate with depth; and densities of alpha-recoils should correlate with concentration of U + Th and with age. After subtracting contributions due to background events, the sample with the lowest density per unit retention-age of remaining recoil tracks will be used to set an upper limit on the production rate of WIMP-recoils. The result would give only an upper limit, because neutron-recoils cannot be distinguished from WIMP-recoils.

The curve in Fig. 2 labeled "future mica limit" indicates the sensitivity we hope to achieve. Assuming the background due to muons and fast neutrons in at least one of the micas does not interfere, it may be possible to reach a cross section $\sim 10^{-34}$ cm^2 for WIMP masses ~ 50 to $\sim 10^3$ GeV. In the end, the lowest attainable limit may hinge on our ability to locate micas from very deep mines.

A Real-Time WIMP Search with Mica

The search with ancient mica crystals can only set an upper limit on WIMPs; it can never prove that K-recoils are due to WIMPs rather than to fast neutrons or muons. The same is true of the electronic detectors, all of which suffer from radioactive background and from electronic noise at the lowest deposited energy. Since 9 February 1993 we have been operating a real-time WIMP "telescope" in an underground location near Berkeley. A telescope mount continuously orients our detector normal to the direction of motion of the earth through the Galaxy to within an angle of 1°. The detector consists of 1 m^2 of a mica-silver-mica sandwich in an evacuated container. After a two-year exposure we will look for etchpits due to recoiled silver atoms in both the forward and backward mica collectors.

The collecting power (target volume × time × efficiency) of the mica array will be comparable to that of typical searches with electronic detectors, but it has the advantage of a distinctive WIMP signal: because the velocity of the earth through the galactocentric WIMP gas is of the same order as the rms WIMP velocity, struck silver atoms will be far more likely (our estimate is $\geq 10\times$) to recoil into the backward hemisphere than into the forward hemisphere. Thus, a WIMP signal would consist of an excess of recoil tracks in the trailing mica over those in the leading mica.

We chose silver for three reasons: it is available in very high purity, with a concentration of U + Th less than 1 ppb; its two stable isotopes, ^{107}Ag and ^{109}Ag, have a favorable spin factor for spin-dependent scattering by WIMPs; and it produces much deeper etchpits in mica than do ^{39}K and other atoms naturally occurring in the mica. Before the exposure began the mica crystals were annealed at a temperature of 650° C to erase alpha-recoil tracks. After the two-year exposure the mica will be etched and searched for recoil etchpits with the following rapid-scan method developed by Snowden-Ifft: the etched mica will be shadowed at an angle

$< 1°$ to the surface with a thin silver film. A silver etchpit (but not a potassium etchpit) is easily seen at relatively low magnification in an ordinary optical microscope by virtue of the absence of the otherwise continuous gold film on the surface.

If we find an excess of recoil tracks in one of the ancient micas, the collecting power of the mica-silver-mica detector can be scaled up to > 10 m^2 and > 5 years in order to search for a galactocentric WIMP signal.

Acknowledgements

We are indebted to Yudong He, Michael Solarz, Bernie Tower and Andrew Westphal for assistance and to the National Science Foundation for partial support (grant no. PHY-9307420).

References

[1] For reviews of theory and experiments, see J. R. Primack, D. Seckel and B. Sadoulet, *Ann. Rev. Nucl. Part. Sci.* **38**, 751 (1988); P. F. Smith and J. D. Lewin, *Phys. Repts.* **187**, 203 (1990)

[2] W. Seidel in: *Relativistic Astrophysics and Particle Cosmology*, Vol. 688, p. 632, edited by C. W. Akerlof and M. A. Srednicki, Annals New York Academy of Sciences, (New York Academy of Sciences, New York, New York, 1993)

[3] S. P. Ahlen et al., *Phys. Lett. B* **195**, 603 (1987); D. O. Caldwell et al., *Phys. Rev. Lett.* **61**, 510 (1988); D. Reusser et al., *Phys. Lett. B* **255**, 143 (1991)

[4] D. Snowden-Ifft and P. B. Price in: *Proc. 22nd International Cosmic Ray Conference*, Chap. 4, p. 746, Dublin Institute of Advanced Studies, Dublin, Ireland (1991)

[5] R. L. Fleischer, P. B. Price, and R. M. Walker, *Nuclear Tracks in Solids* (University of California Press, Berkeley, CA, 1975)

[6] W. H. Huang and R. M. Walker, *Science* **155**, 1103 (1967)

[7] P. B. Price and M. H. Salamon, *Nature* **320**, 425 (1986)

[8] J. Borg, J. C. Dran, Y. Langevin, M. Maurette and J. C. Petit, *Radiat. Eff.* **65**, 133 (1982)

[9] D. Snowden-Ifft, P. B. Price, L. A. Nagahara and A. Fujishima, *Phys. Rev. Lett.* **70**, 2348 (1993)

[10] R. V. Coleman, Q. Xue, Y. Gong, and P. B. Price, *Surf. Sci.* **297**, 359 (1993)

[11] P. Belli et al., *Nuovo Cim.* **101A**, 959 (1989)

[12] J. Ellis and R. A. Flores, *Phys. Lett. B* **263**, 259 (1991)

[13] J. Engel and P. Vogel, *Phys. Rev. D* **40**, 3132 (1989)

[14] See, e.g., K. Griest, *Phys. Rev. D* **38**, 2357 (1988)

RECENT TOPICS FROM NUCLEAR REACTIONS IN THE ENERGIES RANGING FROM keV TO GeV

Ken-ichi Kubo

Department of Physics,
Tokyo Metropolitan University,
1-1 Minami-osawa, Hachioji,
Tokyo 192-03, Japan

Introduction

The purpose of this lecture is to remark the recent developments of nuclear physics by focusing on the nuclear reaction studies in the three typical energy regions. At a few 100 keV energy region, the nuclear reaction cross sections in both experimental and theoretical works for the light-ion systems have been accumulated with a high accuracy. Those works have been motivated by positive participation to the astrophysical research field from the nuclear physics, namely as providing the basic data for the primordial nucleosynthesis network calculations. The nuclear data have contributed indicating a clear discrepancy among the existing Big Bang models. I will talk how this has been made for the recent five years by presenting our ^{9}Be nuclear abundance caculations in the first topic.

In the ordinary energy region of a few 100 MeV, the study of unstable nuclei has been developed. Furthermore the new techniques to inject the unstable reaction-products as the secondary reaction projectiles has been developed. This will be extended to some exotic new experiments which have never been observed as far as the stable nuclei are involved. If there exist such cases that the produced unstable nuclei are largely spin-polarized and those are ably used as the injectiles, a fancy enough exotic nuclear reaction can be proposed. I will indicate a possible enormous spin-polarization of the reaction-products in the second topic.

Spin physics at a few 100 GeV energy region is undoubtedly due to the quantum effects. This will be noted in the last topic. Can be the nucleus-nucleus collision well described in basis of the phenomenological nucleon-nucleon scattering amplitude? The answer is no. The reason seems related not to the nuclear structure effect, but rather to the nucleon structure effect. This statement could be the same for the π-nucleus scattering once starting from the π-nucleon phenomenological amplitude. I will show some systematics for these situations in the third topic.

This lecture will be then organized by presenting the following three independent topics in the wide energy range nuclear reactions from keV to GeV.

Contents
I Primordial Nucleosynthesis and Big Bang Models
   – keV; ^{9}Be abundance –
II Peripheral Scattering and Dynamical Spin-Polarization
   – MeV; Giant spin-polarization –
III High Energy Hadron Reactions
   – GeV; NN, NA - hadron productions –

I. Primordial Nucleosynthesis and Big Bang Models
– keV; ^{9}Be abundance –

I.1 Energy region we are approaching and why ^{9}Be

Let me begin my lecture by overviewing the thermal history of our universe. According to the Big Bang (BB) model, the quark-hadron QCD phase transition is considered to be occurred at about 10 μs after the BB, of which temperature was 10^{12} K, corresponding to about 150 MeV. This was birth of the neutrons and protons. The collisions of protons and neutrons produced D, He, Li, Be, and so forth, and the universe temperature went down to 10^9 K (100keV). This happened at a typical time of 3 minutes after the BB. Atoms were formed in 10^5 years after the BB, where the temperature came down to 10^4 K (~ 10 eV). After the time, the universe became a matter dominated world. The present age of the universe is considered to be about 1.5×10^{10} years.

From this BB scenario, we can see that a typical energy of the primordial nucleosynthesis is around 100 keV, therefore this is the typical energy region where we study the primordial nucleosynthesis.

In the flow of astrophysical nucleosynthesis starting from n and p, ^{7}Li is one of the key elements. Its theoretical nuclear abundances predicted by the two different BB models, the uniform density distribution (standard, UD) model[1] and the non-uniform density distribution (NUD) model[2], have been rigorously discussed[3] in comparison with the observed values. Further studies are, however, necessary for the ^{7}Li abundance not only by improving the theoretical models with confirming the parameters involved in the calculations, but also by improving the observations.

The other nucleus is ^{9}Be which sits in between lighter and heavier nuclear-mass regions in the flow of network. The two different BB models predict ^{9}Be abundance with difference by factor of 10^6. We may be then able to specify a preference of one model from the two by comparing the prediction of ^{9}Be abundance with the observation. Furthermore, ^{9}Be element in the old stars is considered keeping its original birth-situation, therefore we may be able to study more directly the basic problems of the early universe by accurate study of ^{9}Be abundance.

I.2 How complicated the reactions involved in the primordial ^{9}Be nucleosynthesis

The main flow producing ^{9}Be is the two-nucleon transfer reaction, ^{7}Li(^{3}H, n)^{9}Be. Experimental measurement of this reaction cross section is rather difficult, therefore any actual measurement at the astrophysical energy region did not exist until recent time, but very recently two experimental groups have reported the data. Before appeerence of the measurements, we have planned to work to estimate the theoretical cross section as precise as possible using the charge conjugate reaction ^{7}Li(^{3}He, p)^{9}Be, of which experimental data existed at the incident energy region we are approaching[4].

We will study these two reactions in detail hereafter for an especial interest that we can learn from these two reactions some general features existing in the study of reaction cross sections appearing in the primordial nucleosynthesis calculations in the light mass elements. In the following, we first concern the ^{7}Li (^{3}He, p) ^{9}Be reaction and then the ^{7}Li(^{3}H, n) ^{9}Be reaction in the energy range of 100 keV – 2 MeV. We will encounter mainly the following three complications.

Existence of the resonant states

In our cases, we have to consider each lowest three of ^{10}B (=^{7}Li + ^{3}He) and ^{10}Be (^{7}Li + ^{3}H) resonant states. The width of lowest 0.17 MeV 2^- state of ^{10}Be is not well known. We estimate its lower and upper limits as 0.06 and 0.60 MeV, respectively, and use these for the cross section calculations.

The transition amplitude is then given by sum of the direct and the compound processes.

$$\frac{d\sigma}{d\Omega}(D+R) = \left| \sum_{L_b L_a} \left(\sum G_D I_{Dj}^{L_b L_a} + \sum G_R I_{Rj}^{L_b L_a} \right) P_{L_b}^M(\theta) \right|^2$$

$G_{D,R}$: The geometrical factors,
$I_{D,R}^L$: The overlap integrals,

and D(R) for the direct 1st-order Born term (the resonance contribution) with the distorted waves, respectively. The $I_{Rj}^{L_b L_a}$ contains the nuclear resonance structure factor,

$$S_J^{L_b L_a}(J_{Li}, J_{He}; J) = 2ie^{-i\phi_\lambda}\sqrt{\Gamma_{\lambda L_b}^J \Gamma_{\lambda L_a}^J} / \left(E_\lambda - E - \frac{i}{2}\Gamma_\lambda \right).$$

The square-root of product of the resonance widths is assumed proportional to the penetration factor $P_{\ell_\lambda}(E)$ and the coefficient (g_λ) is regarded as a parameter to be fixed in the following calculation. The other parameter is the phase ϕ_λ(rad). These two parameters have been determined in this work by χ^2-fit of the angular differential cross sections of ^{7}Li(^{3}He, p)^{9}Be reaction to the existing experimental data for energy 1.0 – 2.5 MeV for $\lambda = 1 - 3$ corresponding to the resonance states from lower to upper levels. We coherently sum the direct and compound amplitudes, and the results will be shown later.

The reaction mechanisms, stripping and knock-on processes

For the ^{7}Li(^{3}He, p)^{9}Be reaction, two different reaction mechanisms can be considered; one is stripping and the other is knock-on mechanisms.
Stripping: The np-pair in the projectile ^{3}He is stripped out and trapped into the ^{9}Be bound state. Corresponding direct transition amplitude takes the following form of T-matrix element,

$$T_{st} = \langle f|V_{pd} + V_{pLi} - U_{pBe}|i\rangle \tag{1}$$

where i(f) indicates the initial (final) scattering states of ^{3}He + ^{7}Li (p + ^{9}Be) including the two intrinsic state wave functions.
Knock-on: On the other hand the same reaction is considered as the projectile ^{3}He knocks

out a proton of the target ^{7}Li and is captured by ^{6}He to make ^{9}Be. The direct amplitude for this case becomes

$$T_{ko} = \langle f | V_{p^3He} + V_{p^6He} - U_{pBe} | i \rangle. \tag{2}$$

We have to make a coherent sum of these stripping and knock-on amplitudes.

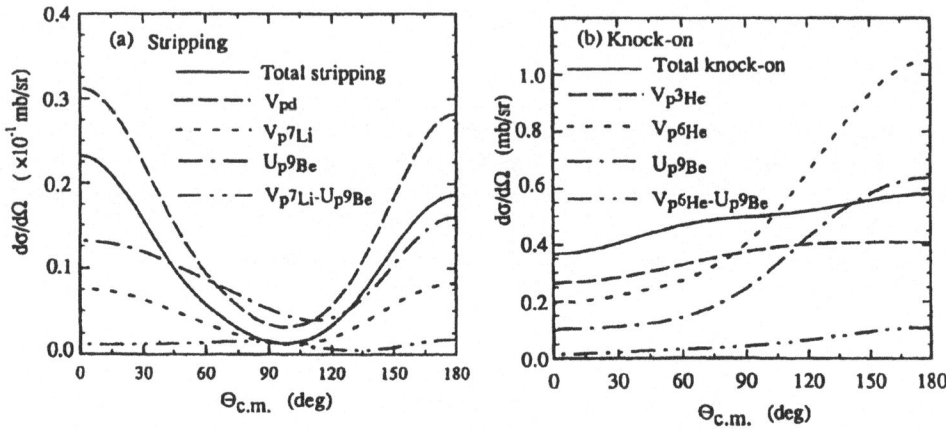

Figure 1. *The differential cross section of ^{7}Li(^{3}He,p)^{9}Be reaction at 2.5 MeV. (a) for the stripping and (b) for the knock-on mechanism calculations.*

Validity of the cancellation assumption

In the ordinary distorted-wave Born approximation, the 2nd and 3rd terms of eqs. (1) and (2) are assumed negligibly small by cancellation when masses of target and residual nuclei are large enough in comparison with those of projectile and ejectile. This assumption may lead to the actual picture of stripping or knock-on mechanism, since then the residual interaction V_{pd} or V_{p^3He} can be interpreted as interaction serving the stripping or knock-on transfer reaction, respectively.

However, for the reactions involved in the nucleosynthesis of lighter nuclei interested here, the masses of projectile and target are rather comparable, therefore the cancellation assumption may not be well valid and we have to check this point in the calculations.

Figs. 1a and b show the results of cross section calculation as a function of angles for the stripping model and the knock-on model, respectively. We can see that contributions from each V_{pLi} (the short dashed line) and U_{pBe} (the dot-dashed curve) are not so small, but the coherent sum of the two becomes very small and finally the sole V_{pd} contribution (the long-dashed curve) has a very similar shape as the angular distribution including all the interactions (the solid curve). Only the magnitude is different between the two cross sections obtained by exact calculation and by assuming the cancellation. Situation is the same for the knock-on mechanism as we can see in Fig.1b.

From these results, we may conclude the cancellation-assumption accepted in the usual 1st-order Born approximation with distortion can be approximately used unless the cross section magnitude is not concerned even for such lighter-nucleus reaction processes that play important roles in the primordial nucleosynthesis. This finding is quite useful, otherwise the calculation becomes much complicated enough. In the following calculations, however, we include all the interaction terms in the T-matrix, since we are interested in the cross section magnitude.

One noticeable result is the small cross section of the stripping mechanism. This may reduce further the problem that a certain double counting problem cannot be avoided in the simple sum of the stripping and knock-on amplitudes. However if one mechanism is smaller than the other, this problem may become small. We will include both mechanisms in the following calculations.

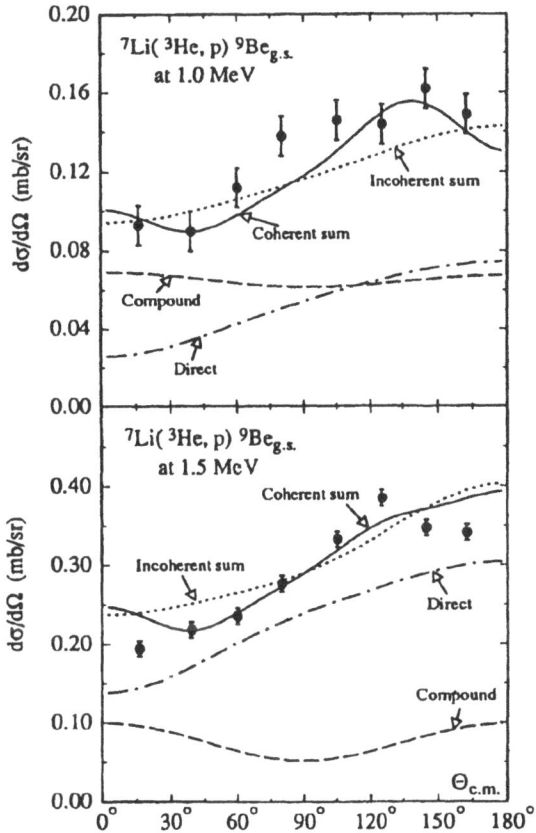

Figure 2. *The differential cross section of $^7Li(^3He,p)^9Be$ reaction at 1.0 and 1.5 MeV. Cross sections for the direct (stripping + knock-on) and compound processes and the coherent and incoherent sums are shown. The experimental data are from ref. 5.*

In Fig. 2, the differential cross sections of the $^7Li(^3He, p)^9Be$ reaction at 1.0 and 1.5 MeV are shown. The direct (stripping + knock-on) and the compound (resonance) cross sections are shown, and the two magnitudes are found to be rather comparable. Coherent sum of the direct and the compound processes is shown by the solid curve. We found that the incoherent

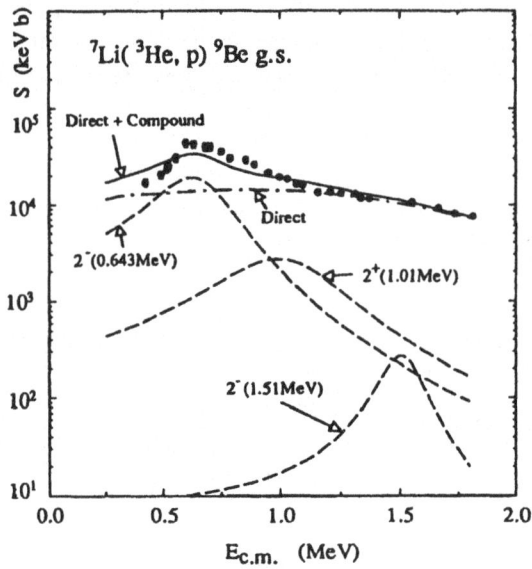

Figure 3. *The astrophysical S-factor for the $^7Li(^3He,p)^9Be$ reaction. Each direct and compound process contribution and the incoherent sum are shown.*

sum (the dotted curve) of the two does not show so much difference from that of the coherent sum, although the coherent sum gives better fits to the experimental data.[5]

I.3 The S-factor, ^{9}Be abundance and the Big Bang models

The astrophysical S-factor is defined by

$$S(E) = E\sigma(E)\exp\left(\frac{2\pi Z_1 Z_2 e^2}{\hbar v}\right),$$

where $\sigma(E)$ is the angular integrated cross section of the $^7\text{Li}(^3\text{He, p})^9\text{Be}_{g.s.}$ transfer reaction obtained from the above calculations. The obtained S-factor is shown and compared with the observed data[5] in Fig. 3.

We made the similar calculations for the $^7\text{Li}(^3\text{H, n})^9\text{Be}$ reaction by assuming the charge conjugation for the nuclear spectroscopic factors, the nuclear part of the optical model potentials, the resonance state parameters, and the interaction potentials. We have made coherent sum of the all possible transition amplitudes and obtained the angular integrated cross sections. From these we have calculated the astrophysical S-factor as a function of incident energies. The results are shown in Fig. 4 by the solid curves. The shadow area indicates ambiguity arising from the unknown width of the lowest state of ^{10}Be (= $^7\text{Li} + {}^3\text{H}$) compound state. We used the assumed widths 0.06 MeV as a lower limit and 0.6 MeV as an upper limit. From Fig. 4, we can summarize that (1) in the most important energy region (Gamow-window, 0.2 – 0.7 MeV), the direct process gives rather small contribution, (2) the lowest two resonances give a predominant contribution.

Recently, the experimental measurements of $^7\text{Li}(^3\text{H, n})^9\text{Be}$ reaction were reported by two groups[6]. The data are shown in Fig. 4. By comparision of the present calculations

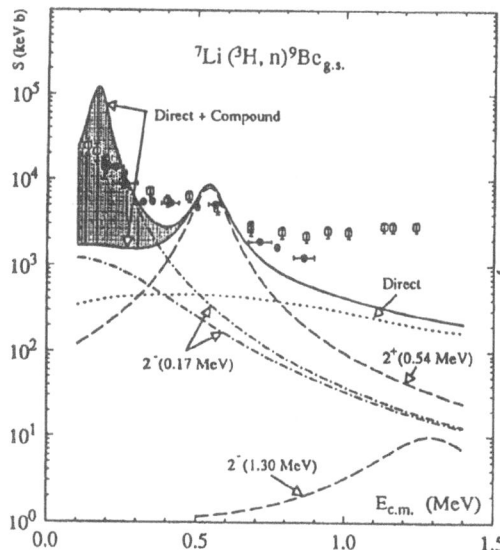

Figure 4. The astrophysical S-factor for the $^7Li(^3H,p)^9Be$ reaction. Direct and compound process contributions and the incoherent sum are shown. The hatched area corresponds to prediction due to upper and lower estimations of the width of 2^- (0.17 MeV) resonance state of ^{10}Be nucleus. The experimental data are from ref. 6.

with the data, we need more enhanced cross section from the direct reaction contribution with increasing energy. However, in the most important energy region (0.2 – 0.7 MeV), the present calculation is quite consistent with the data. The 2^- (0.54 MeV) resonance dominates the final result of S-factor.

Employing the S-factor thus obtained, we calculated the ^9Be abundance by solving the rate equation based on two different BB models[7]. The obtained number abundance $N(^9Be)/N(^1H)$ is shown in Fig. 5 as a function of $\eta = n_B/n_\gamma$, where n_B is the number of baryons and n_γ the number of photons. The NUD model (that is non uniform density distribution model) gives an enhanced abundance by factor of 10^6 compared with the UD model (the standard model with homogeneous density distribution in the range of 0.04 – 1 for Ω_B, the baryon mass density; $\Omega_B = 1.0$ means universe being closed by baryons, while $\Omega_B < 1.0$ means we need the dark matter). The shadowed area corresponds to ambiguity arising from the unknown width of the 2^- (0.17 MeV) state. The observed upper limit of ^9Be abundance has been reported by the two groups[8]. The present calculations are consistent with those.

I.4 Summary

In summary, we have studied the $^7Li(^3H,n)^9Be$ reaction for the energy $E_{cm} < 2$ MeV, by fully using the information established in the analysis of charge conjugate reaction $^7Li(^3He,p)^9Be$. We have pointed out the three complexities of this type of reaction treatment generally appearing in the primordial nucleosynthesis calculations. We have confirmed a large cancellation, even for such a light target and projectile reactions, of the two interaction terms

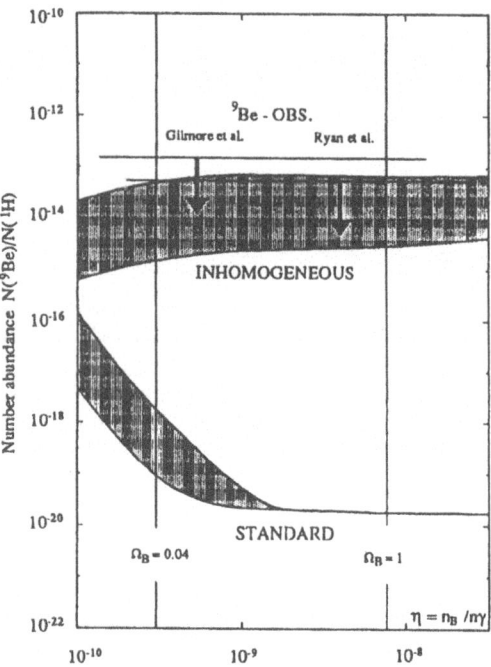

Figure 5. The ^{9}Be nuclear abundance. The hatched areas correspond to predictions due to upper and lower limits of the width of 2^- level of ^{10}Be. The observed upper limits of abundance are from ref. 8.

appearing in the T-matrix and assumed small in the usual distorted wave Born approximation. We have performed the cross section calculation as accurately as possible.

The obtained primordial abundance of ^{9}Be based on the two different models overlaps with the recent observation of upper limit. The inhomogeneous (NUD) model is found to predict a larger ^{9}Be abundance than that predicted by the standard (UD) model. The observation of the lower limit is strongly required for establishing priority or apparent differences between the two models.

References

[1] R. V. Wagoner et al., *Astrophys. J.* **148**, 3 (1967)
[2] J. H. Applegate et al., *Phys. Rev. D* **31**, 3037 (1985); G. M. Fuller et al., *Phys. Rev. D* **37**, 1380 (1988)
[3] T. Kajino and R. N. Boyd, *Astrophys. J.* **359**, 267 (1990); *Phys. Rev. Lett.* **14**, 125 (1991)
[4] Y. Yamamoto et al., *Phys. Rev. C* **47**, 846 (1993)
[5] D. P. Rath et al., *Nucl. Phys. A* **515**, 338 (1990)
[6] C. R. Brune et al., *Phys. Rev. C* **43**, 875 (1991); S. Bar houmi et al., *Nucl. Phys. A* **535**, 107 (1991)
[7] J. Yang et al., *Astrophys. J.* **281**, 493 (1984); C. R. Alcock et al., *Astrophys. J.* **320**, 439 (1987)
[8] G. Gilmore et al., *Astrophys. J.* **378**, 17 (1991); S. G. Ryan et al., ibid. **388**, 184 (1992)

II. Peripheral Scattering and Dynamical Spin-Polarization
– MeV; Giant spin-polarization –

II.1 Giant Spin-polarization in HI-1NT Reactions

Unstable nuclei and exotic reaction products are recently attractive subjects in the nuclear physics study with motivation aiming to use such products as the secondary projectiles for new reaction experiments. If there exist certain cases of spin of the reaction products being largely polarized, they are much interesting and we also want to know the dynamics providing the spin polarization. In this section, you will see our recent theoretical finding for the large spin-polarization cases[1].

The $^{12}C(^{13}C, ^{12}C)^{13}C_{g.s. 1/2^-}$ reaction at 100 MeV

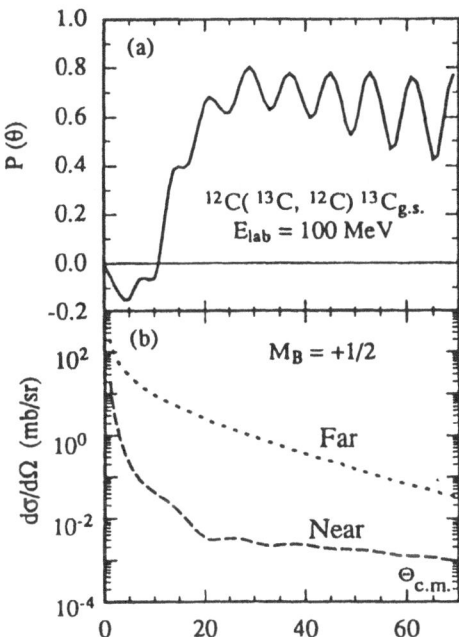

Figure 6. *The calculated spin-polarization of $^{13}C_{g.s.}$ $1/2^-$ in (a) and the far-near decompositions of the predominant $M_B = +1/2$ substate cross section in (b).*

We first look at the case of $^{12}C(^{13}C, ^{12}C)^{13}C_{g.s. 1/2^-}$ reaction at the incident energy of 100 MeV. The polarization of ^{13}C residual nucleus becomes large with positive constant sign as it is shown in Fig. 6a. The near-far decompositions of the dominant ($M_B = +1/2$) substate population cross section are shown in Fig. 6b. The present incident energy is sufficiently above the Coulomb barrier. Then the far-side trajectory dominates the transition, and we find predominance of the $M_B = +1/2$ state excitation for this case. The z axis is defined normal to the scattering plane.

The $^{48}\text{Ca}(^{13}\text{C}, ^{12}\text{C})^{49}\text{Ca}^*$ (2.02 MeV 1/2$^-$) at 47 MeV

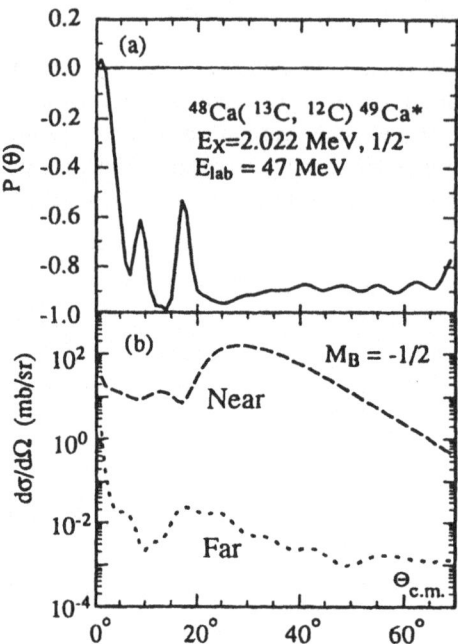

Figure 7. *The calculated spin-polarization of $^{49}\text{Ca}^*$ (2.22 MeV, 1/2$^-$) in (a) and the far-near decompositions of the predominant $M_B = -1/2$ substate cross section in (b).*

The other example is shown in Fig. 7a for the $^{48}\text{Ca}(^{13}\text{C}, ^{12}\text{C})^{49}\text{Ca}^*$ (2.02 MeV 1/2$^-$) reaction at 47 MeV. Spin-polarization of the residual nucleus is again enormously large but now with the negative sign. This change of sign coincides with the change of preferred trajectory from the far-side to the near-side as shown in Fig. 7b. The present incident energy is below the Coulomb barrier.

The origin producing spin-polarization for the present one-nucleon transfer reactions is spin-transfer itself carried by the transferred nucleon. Such a large magnitude with almost the constant sign can be realized by the predominance of one particular trajectory, far-side or near-side, depending on the incident energy. This phenomenon, therefore, may be referred to as the dynamical spin-polarization associated with the nucleon-transfer reaction.

The other cases

The two reactions mentioned above correspond to the nucleon transfer $p_{1/2} \rightarrow p_{1/2}$. Similar large spin-polarization occurs for the other cases. One example is polarization of $^{13}\text{C}_{\text{g.s.}}$ in the $^{12}\text{C}(^{12}\text{C}, ^{13}\text{C})^{11}\text{C}_{\text{g.s. 3/2}^-}$ reaction at 25 MeV/N, where nucleon transfer is $p_{3/2} \rightarrow p_{1/2}$.

For experimental measurement of the spin-polarization of reaction products, the particle-γ angular correlation technique may be used. The γ-ray measurement for decaying of the 1/2 spin-state requires a special techniques because we need the circular polarization measurement. The γ-ray angular distribution asymmetry associated with the decay of ^{49}Ca($1/2^-$) is predicted in our calculation by 50% at the maximum, which may be sufficiently enough for the spin-polarization measurement.

II.2 Dynamics producing the large spin-polarization and its sign-change

As we saw above for the $p_{1/2}$ to $p_{1/2}$ transitions, the sign of polarization is positive for the energy above the Coulomb barrier, whereas it is negative below the Coulomb barrier. Why is it so sensitive to the energy and the trajectory? This can be easily understood by noting the characteristic structure of the transition amplitude. Its essential form shows,

$$T_{fi}^{M_B} = \sum_{L_\alpha M_\alpha} \cdots \left\{ I_{L_\alpha}^{\ell=0} + M_B M_\alpha I_{L_\alpha}^{\ell=1} \right\},$$

where M_B and M_α are the magnetic substate q.n. of the spin-polarized nucleus and the scattering partial orbital angular momentum L_α, respectively. The transition amplitude is described by a coherent sum of the two possible ℓ-transfers. The z-axis is taken normal to the scattering plane. Then M_α is positive for the far-side trajectory and negative for the near-side trajectory when the scattering is observed in the left wing.

Consider one example of a constructive interference of the two ℓ-transfer amplitudes for a certain set of the M_B and M_α. Then change the trajectory, say from far-side to near-side, by changing the energy. If you want to have a similar constructive interference after this change, you have to change sign of M_B, hence the substate cross section becomes predominant for the other M_B, accordingly the spin-polarization changes sign. This is the reason why the sign of the polarization is sensitive to the trajectory, so that to the incident energy. It is possible to predict a definite sign for a particular set of reaction and energy[1] by using Brink's M_z matching condition argument[2].

II.3 Summary

We have predicted a possible large spin-polarization of the reaction products. We have clarified the origin providing such a large spin-polarization and the reason for sensitivity of its sign to the energy by noting predominance of the particular trajectory. The process is referred to as the dynamical spin-polarization. The experimental confirmation of the present prediction is strongly desired.

References

[1] Y. Yamamoto and K.-I. Kubo, *Phys. Rev. Lett.* **68**, 2588 (1992); and *Phys. Rev. C* in print
[2] D. M. Brink, *Phys. Lett. B* **40**, 37 (1972)

III. High Energy Hadron Reactions
– GeV; NN, NA - hadron productions –

III.1 Still enormous spin-effect in the π-production

Semi-classical estimation of the proton spin-effect

As an introduction to the high energy hadron reactions, let us estimate the spin effects in the high energy collisions whether it is expected negligibly small or still large. I will use two ways. First, let us regard the spinning of proton as a rotating motion around its intrinsic axis. Then the rotating energy is estimated as,

$$E_R = L^2/2F,$$

where $L^2 = S(S+1)\hbar^2$ ($S = 1/2$) and $F = 3m_p R^2/5$ with $R = 0.8$ fm. m_p is the proton mass. From these, E_R is estimated as about 40 MeV, which is a much lower energy in comparison with the collision energy (we are around GeV or more). Therefore, we may see very small spin effects in the NN collision at GeV energy region.

Second, we estimate the time period (T_{sp}) of the proton spinning, of which the rotating energy is about $E_R = 40$ MeV estimated above. Using the angular frequency ω obtained from E_R, we get,

$$T_{sp} = \frac{2\pi}{\omega} = 13/c,$$

where c is the velocity of light. In the NN collision at 1 GeV energy, the interaction time T_{int} which corresponds to the time of a proton passing through a distance of nucleon size $2R$ with $R = 0.8$ fm is estimated as

$$T_{int} = 1.6/c.$$

Therefore the spinning period T_{sp} is much longer, about ten times more, than the interaction time T_{int}. This means that, during the NN collision process, proton's intrinsic rotating motion looks very slow or almost stopped.

From these estimations, almost no spin-effects are expected in the high energy NN collision and the proton could be regarded as spinless particle. Is that so?

Still enormous spin-effects

In Fig. 8, the pion production experimental data are shown[1]; $\vec{p}p \to \pi X$ with $\pi = \pi^\pm$ and π^0 productions at 200 GeV/c momentum. The left-right asymmetry is shown as a function of the x_F, where,

$$x_F = k_L(\pi)/k_R(p),$$

the fraction of the incident proton's longitudinal momentum carried by the produced pions. First, the asymmetries become rapidly large as a function of x_F. Contrary to the above semi-classical estimation, the spin-effect is very large in such high energy collisions. Second, the asymmetries of π^+ and π^- show almost the same value but different signs and π^0 comes in between the two.

Parton model interpretation based on the SU(6) proton wave function

Let me take one case say the $\vec{p}p \to \pi^+ X$ reaction to see the x_F dependence of the asymmetry. We follow the parton model proposed by DeGrand el al[1,2]. We can express the reaction in terms of quarks,

$$\vec{p}(uud) + p(uud) \to \pi^+(u\bar{d}) + X.$$

The u-quark in the incident proton picks up $\bar{d}$ sea-quark from the vaccum. Through the pick-up, the $\bar{d}$ quark getting an external force F from the u-quark is accelerated into the same direction along the incident proton getting a velocity V up to the same velocity as the u quark. This means that the $\bar{d}$ quark begins the so-called Thomas precession. The precession

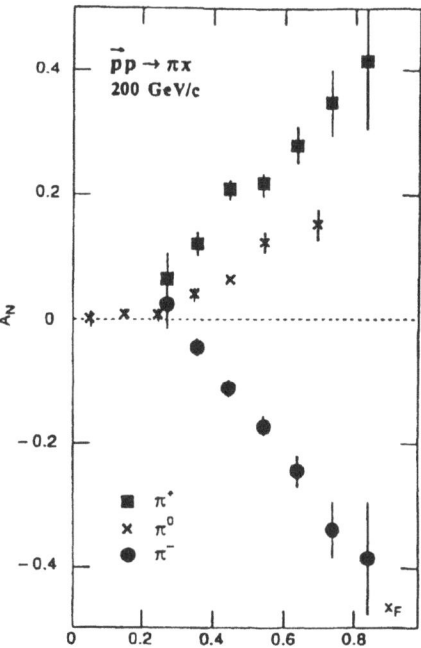

Figure 8. *The observed pion asymmetries in the spin-polarized $\vec{p}p \to \pi x$ collision at 200 GeV/c incident momentum.*[1]

motion couples with the intrinsic spin of $\bar{d}$ quark and provides a spin-orbit coupling type potential, which yields the asymmetry of the reaction products π. This is the basic idea of DeGrand et al.

However for the practical interpretation of the phenomena, they used the phenomenological rule which is; 'Fast-spin-up and slow-spin-down'. This means that the cross section becomes preferential for the parton-pair produced by the faster velocity parton (u) with spin-up and the slower velocity parton ($\bar{d}$) with spin-down. Together with the property of proton SU(6) wave function, this empirical rule can predict a positive or negative sign for the π^+ and π^- production, respectively.

Also other work exists. Boros et al.[3] used another form of the empirical rule which can be shortened as 'Spin-up-left and spin-down-right'.

However I would say that, for a complete interpretation of such a phenomenon, we need a more fundamental study on the qq-interaction basis, and then we can approach the quark dynamics through the high energy hadron production reactions.

III.2 A strange phase

Difficulty in the AA scattering based on the NN scattering

The forward NN scattering amplitude can be well parametrized as

$$f_{ij}(q) = \frac{k_N \sigma_T}{4\pi}(i+\rho)e^{-aq^2/2}$$

where q is the momentum transfer, σ_T the NN total cross section, ρ the real-imaginary ratio and a is a real constant. Even if the a is extended into a complex number, its imaginary part does not affect the NN cross section since the cross section is given by the square of the absolute value of $f_{ij}(q)$.

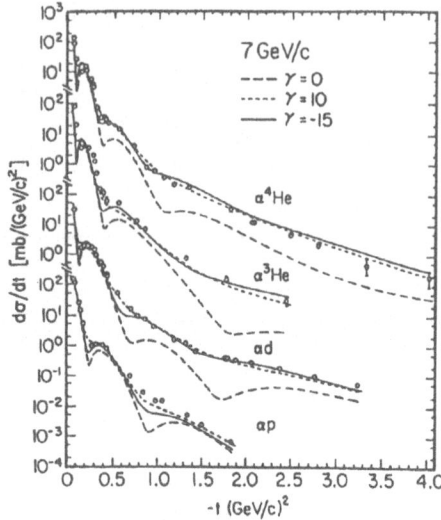

Figure 9. *The experimental and calculated differential cross sections for α elastic scattering on p, d, ^{3}He and ^{4}He targets at 7 GeV/c incident momentum (1 GeV/N incident energy).*[4]

The nucleon-nucleus (NA) and the nucleus-nucleus (AA) scattering at high energy have been studied in the basis of NN scattering amplitudes. One example[4] is shown in Fig. 9 for the A + ^{4}He elastic scattering with A = p, d, ^{3}He and ^{4}He at the incident energy of 1 GeV/N. The long dashed curves show the results obtained by Glauber model calculation using the NN scattering amplitude with parameters which were fixed from the NN data analysis. The calculations cannot fit the data for all four cases.

Then the constant a has been extended into the complex number, $a = \beta + i\gamma$, where β is the parameter to be fixed from the NN scattering cross section analysis and γ an arbitrary real value introduced here. The NN scattering amplitude with γ thus introduced does not change the NN cross section at all, but changes the AA scattering since the multiple NN scattering terms appear in the AA collisions. Therefore, the phase factor $e^{-i\gamma q^2/2}$ is not the overall phase but does affect the cross section.

In fact, for the A + ^{4}He elastic scattering, the finite value of γ changes the cross sections and $\gamma = 10$ (fm^2) improves the fit of the calculation to the data excellently[4]. Although we have only the limited cases, we may conclude that the γ thus introduced becomes constant if the collision energy per nucleon is the same, even if the collision system is different.

The similar situation has been found[5] in the π + ^{12}C scattering at $p_\pi = 0.8$ GeV/c. The Glauber model calculation using the πN scattering amplitude is shown in Figs. 10(a) and (b) for the elastic and the inelastic scattering. Much improved fits to the data are obtained for both the elastic and inelastic scattering, if γ is -25. Again we see the constant value of γ for the fixed collision energy, even for different scatterings. In Fig. 11, the π + ^{4}He scattering at the same momentum is shown. There exists a clear difference between $\gamma = 0$ and $\gamma \neq 0$

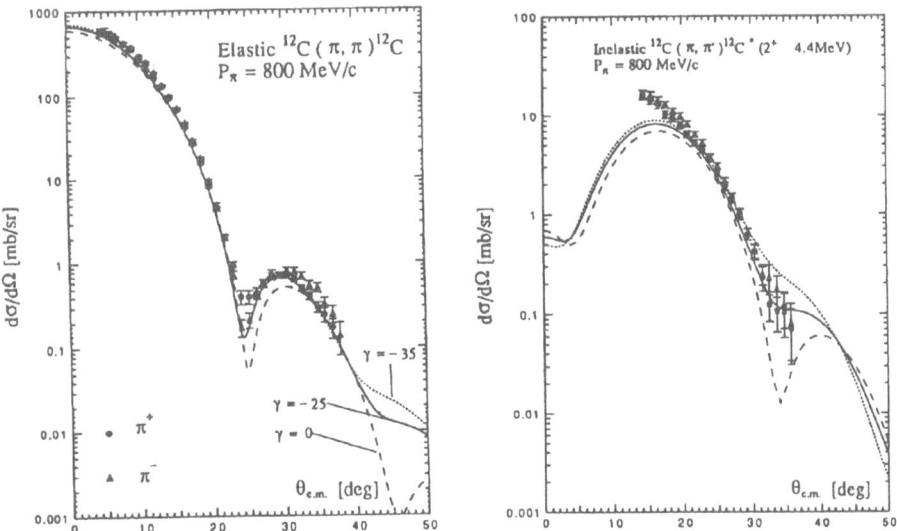

Figure 10. *The experimental and calculated pion scattering on ^{12}C at 800 MeV/c incident momentum The elastic and the 4.43 MeV 2^+ inelastic scattering are shown in left and right, respectively.*

cases. We have no measurement. Actual measurements and comparison with our results are strongly desired.

From where the phase

The origin of the γ parameter has not been understood yet. One speculation may be as follows. During the AA scattering, the NN interaction distorts the intrinsic structure of the nucleon. This means there exist two motions; one is the distortion of the intrinsic structure of the nucleon in the nucleus and the other is the NN relative motion in nucleus. The total Hamiltonian is then expressed by

$$H = H_{intrs} + H_{rel} + V_{coupl}$$

If one of the two Hamiltonians corresponds to a very slow motion compared with the other motion, we may introduce an adiabatic approximation treatment to solve the motion of the total Hamiltonian. If it is the case, we may find a phase arising from the adiabatic motion and it has been well known that the phase thus appears is purely real like the γ found above. Such a phase is referred to as the Berry phase and has been noticed in the wide range of physics and chemical phenomena. So this indicates a possible chance to study nucleon structure and then the quark dynamics through the nucleus-nucleus collision at high energy.

We are now studying origin and property of the γ-parameter in detail in this direction.

III.3 Summary

The asymmetries of π^+, π^- and π^0 associated with the spin-polarized NN collision have been discussed in terms of the parton model interpretation. It is extremely interesting that the phenomenological rule with the SU(6) proton wave function can interpret the qualitative property observed. Fundamental study of such phenomenological rules and further quantative

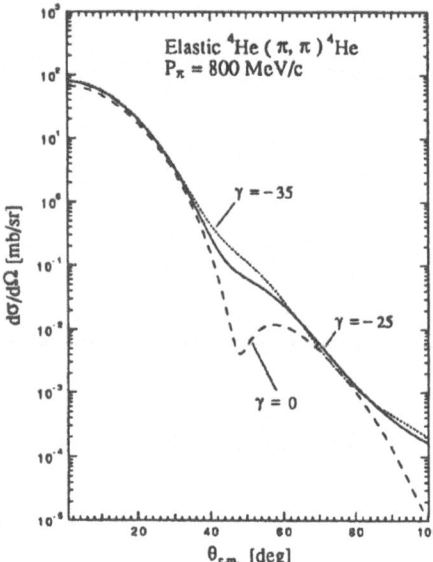

Figure 11. *The calculated pion elastic scattering cross section on ^{4}He at 800 MeV/c incident momentum for various γ values.*

study as a function of the x_F parameter in a wide range would be desired to get more accurate information about quark structure of the hadrons and its quark dynamics.

The purely real phase γ introduced here improves extremely well the nucleus-nucleus and π-nucleus scattering calculations, which cannot fit the observed data if the calculations are made accepting only the phenomenological NN and pN scattering amplitudes. The parameter γ has been found the same value for the different collision systems if the E/N is the same. Structure property of nucleon, not of the nucleus, may be considered as its origin, therefore the quark property of nucleon.

References

[1] D. L. Adams et al., *Phys. Lett.* B **246**, 462 (1991)
[2] T. A. DeGrand and H. I. Miettinen, *Phys. Rev.* D **24**, 2419 (1981)
[3] C. Boros et al., *Phys. Rev. Lett.* **70**, 1751 (1993)
[4] V. Franco and Y. Yin, *Phys. Rev. Lett.* **55**, 1059 (1985)
[5] M. Mizoguchi et al., to be published

AN ELECTROMAGNETIC CALORIMETER FOR SPECTROSCOPY OF TEV COSMIC RAYS MUONS

I.M. Brâncuş[1], H.J. Mathes[2], J. Wentz[2], H. Bozdog[1], M. Duma[1], M. Kretschmer[2], G. Pascovici[1], M. Petcu[1], H. Rebel[2] and K. W. Zimmer[1]

[1] *Institute of Atomic Physics Bucharest, Romania*
[2] *Kernforschungszentrum Karlsruhe, Germany*

Introduction

In principle, the energy of muons from cosmic rays can be deduced from the frequency and the energy of secondary showers, produced by the muons in thick absorber layers. The main interaction processes of high-energy muons leading to observable energy losses are:

- ionization
- pair generation
- bremsstrahlung
- inelastic interactions of muons with nuclei.

The frequent ionization collisions with very small energy transfers produce a localized trail of ionization ("visible" as muon track in charged particle detectors). The Landau distribution of energy losses by ionization has a high-energy tail due to stochastic collisions with large energy transfers. The high-energy knock-on-electrons produce electromagnetic shower cascades.

The theoretical cross sections for muon interaction with matter indicate that the amount of energy lost by the muon passing through matter is increasing with the muon energy, in particular due to the increase of the radiative processes.

The determination of the muon energy by means of the frequency of interactions with different energy transfers in absorber layers was theoretically investigated by Kokoulin and Petrukhin[1]. The application of the Maximum Likelihood Method (MLM) to infer the muon energy from the observed number of radiative interactions or from the distribution of their energy losses has been considered under the condition that the variation of muon energy may be neglected, and that only events with relative energy transfers less than 0.01 are taken into account in the dominant regime of pair production processes.

In context of the KASCADE project of cosmic ray studies[2], the interest for the muon spectrometry in TeV range prompted the idea to set up a prototype of a burst detector in IFA Bucharest to observe secondary electromagnetic showers. The detector arrangement is planned to consist of successive layers of absorbers Pb (1 cm) and scintillators (3 cm). The latter will allow measurements of the frequency of the secondary showers and the energy loss in the active layers.

Monte-Carlo simulations with the GEANT code[3] have been performed to study a response of a detector consisting of 30 layers of Pb (1 cm) and 30 NE114 scintillator layers (3 cm) for incident muon energies in the range 300 GeV – 30 TeV and for electromagnetic showers (e^+ e^-, γ) with 200 MeV to 100 GeV [4]. The energy determination of muons with 5 TeV and 30 TeV was studied by analyzing the correlated distribution of the number and of the total energy of bursts with a 1 GeV threshold. In addition[5] the possibility to discriminate between electrons or very (e^+ e^-) asymmetric pairs, with a confidence better than 80%, has been demonstrated.

In this contribution we give an account of the basic design, the present status and methods of data analysis with the Bucharest electromagnetic calorimeter used as "burst counter".

Experimental setup

The detector system proposed for the sampling calorimeter consists of a stack of lead absorber plates alternating with scintillator plates as sensitive layers. This detection system has a modular structure where each calorimeter module covers an area of 100×100 cm^2.

Figure 1 describes the layout of one scintillator module. The scintillator plate is read out by wave length shifters. Two of them are fed via a light guide into a photomultiplier type XP2081B-Valvo with green extended sensitivity. Both photomultipliers of one scintillator module are arranged at opposite corners of the module. The chosen read-out principle allows an nearly uniform measurement of energy deposits (pulse heights) whereas the time resolution decreases a little bit.

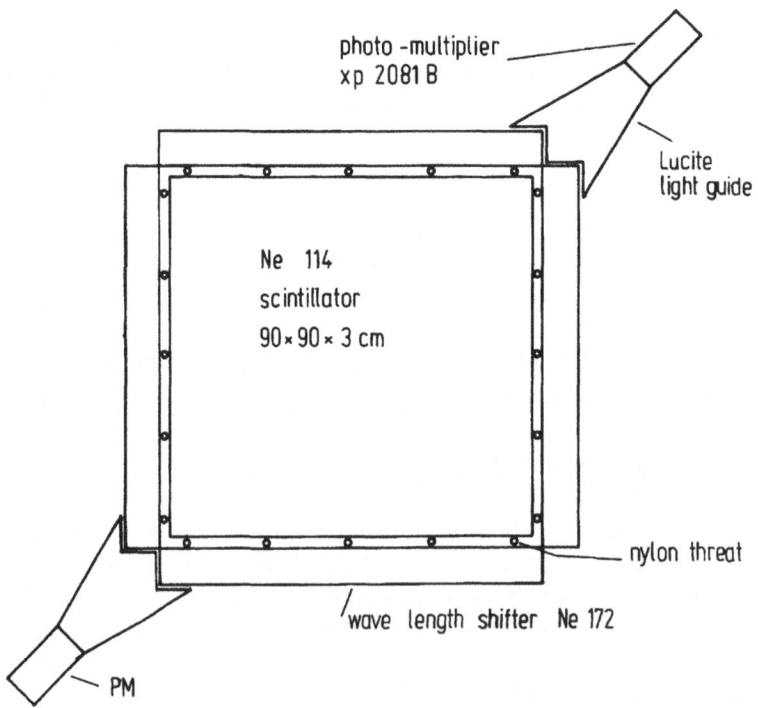

Figure 1. *Layout of a scintillator module.*

To achieve a large dynamic range in photomultiplier pulse read-out two signals are taken from the divider circuitry, namely an anode and a dynode signal. After preamplification both signals are fed into further amplifiers and shaping modules with a sample and hold circuit at the end of them. The signals from the sample and hold circuits are then converted by a VME based ADC board fabricated by Eltec. The read-out and data acquisition is done by a Force 68K CPU housed also in the VME system. Another additional signal from the photomultiplier preamplifier is used as input for the discriminator circuits which are generating the trigger for the read-out system. As future option the discriminator signals can be used for a timing system. This will allow us to measure the position of single particle passage through a scintillator plate by means of evaluating the time difference of 'left' and 'right' photomultiplier signals.

Methods of data analysis

Information on the muon energy given by the energy deposit

An important aspect is the calibration of the detector used as an electromagnetic sampling calorimeter. The energy deposit measured in scintillator layers has to be related to the energy of electromagnetic particles which have initiated the electromagnetic showers.

From the distribution of the energy deposit E_d in the scintillator as a function of the incident energy E_0 results a linear relation of the form:

$$E_d = 0.186 E_0 . \tag{1}$$

From the distributions of total energy deposit in scintillator, the resolution of the sampling calorimeter is deduced to be:

$$\sigma/E_d = 0.137/\sqrt{E_0} . \tag{2}$$

The response of the detector to the traversing muons has been simulated for muons with energies of 100 GeV, 300 GeV, 1 TeV, 5 TeV, 10 TeV and 30 TeV, with statistical accuracy of 620, 475, 231, 121, 30, 64 and 36 events.

Figure 2 presents the fluctuation of the total energy deposit of muons having incident energies in the range 300 GeV – 10 TeV in 30 cm lead absorber. One can recognize that it may be possible to separate with 2σ confidence level of muon energies differing by a factor of about 20 for the energy range of interest.

Determination of muon energy by measuring the correlations between the number of secondaries, their energy spectra and their total energy for various energy cut-offs

For 500 GeV, 1 TeV, 5 TeV, 10 TeV and 30 TeV muons, with the statistical accuracy of 246, 20, 66 and 6 events, the complete development of the induced electromagnetic cascades has been recorded on magnetic tape.

The detector response for a selected case corresponding to the muons of 10 TeV is displayed in Figure 3. The correlated distributions of the number of secondaries and of the total energy loss of muons in detector may lead to a better muon resolution.

In Figure 4 the distribution of the number of secondaries and their total energies is displayed for energy cut-offs of 200 MeV and 1 GeV. It results the possibility to separate muons with 5 TeV and 30 TeV.

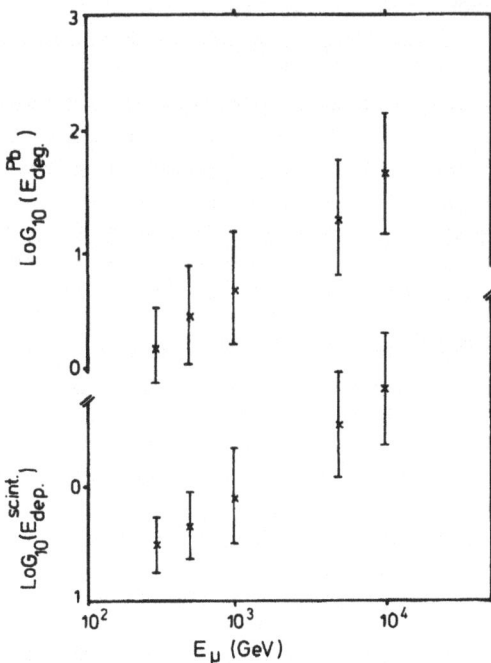

Figure 2. *Fluctuation of the energy deposit from muons in the range 300 GeV – 10 TeV in 30 cm lead.*

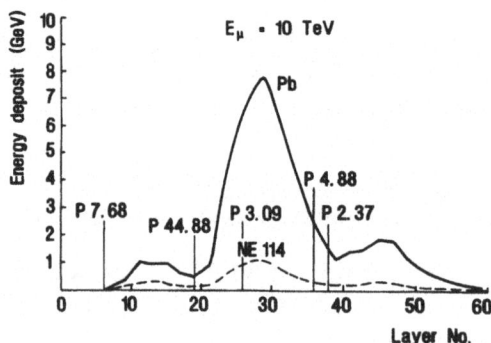

Figure 3. *Detector response to a 10 GeV muon. P denotes pair production and the energy released in a single interaction.*

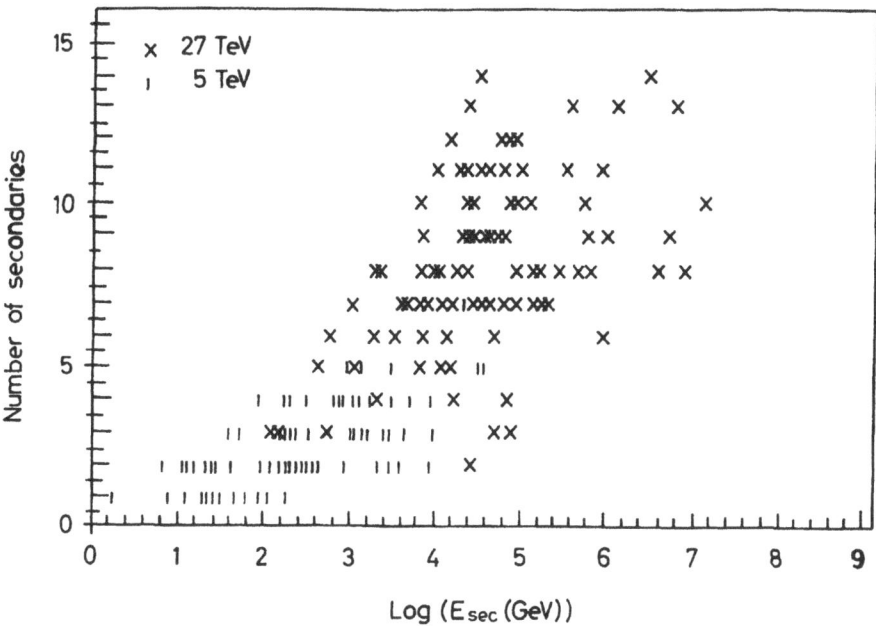

Figure 4. *Number of secondaries and their energies.*

Estimate of the energy of cosmic ray muons by using the maximum likelihood method for analysis

The application of maximum likelihood method (MLM) to determine the muon energy from the observed number of radiative interactions or from the distributions of their energy losses was investigated by Kokoulin and Petrukhin[1], under the condition that the variation of muon energy may be neglected and only events with relative energy transfers less than 0.01 are taken into account in the dominant region of pair production processes.

Figure 5 shows the energy dependence of various contributions to the total interaction probability, as a function of E/ϵ, displaying the E-dependence for a particular value of E/ϵ. The dominant role of pair production in the range $\epsilon/E = 0.001 - 0.1$ is obvious.

To determine the muon energy from the observation of a sequence of M shower events with energy losses $\epsilon_1, \epsilon_2, \ldots, \epsilon_M$, a logarithmic likelihood function (LLF) could be expressed by:

$$L = \sum_{i=1}^{M} \left\{ \ln \left[X_a \sum_{j=1}^{4} g_j(\epsilon_i, E_0) \right] - \ln(i) \right\} - X_a \int_{\epsilon_{min}}^{\epsilon_{max}} g_j(\epsilon', E_0) \, d\epsilon', \qquad (3)$$

where $g_j(\epsilon_i, E_0)$ specifies the probability for an event of a particular energy transfer ϵ_i from a muon with a kinetic energy E_0 to produce a specific interaction. The index j means

$j = 1$: pair production
$j = 2$: bremsstrahlung
$j = 3$: ionization
$j = 4$: nuclear interaction.

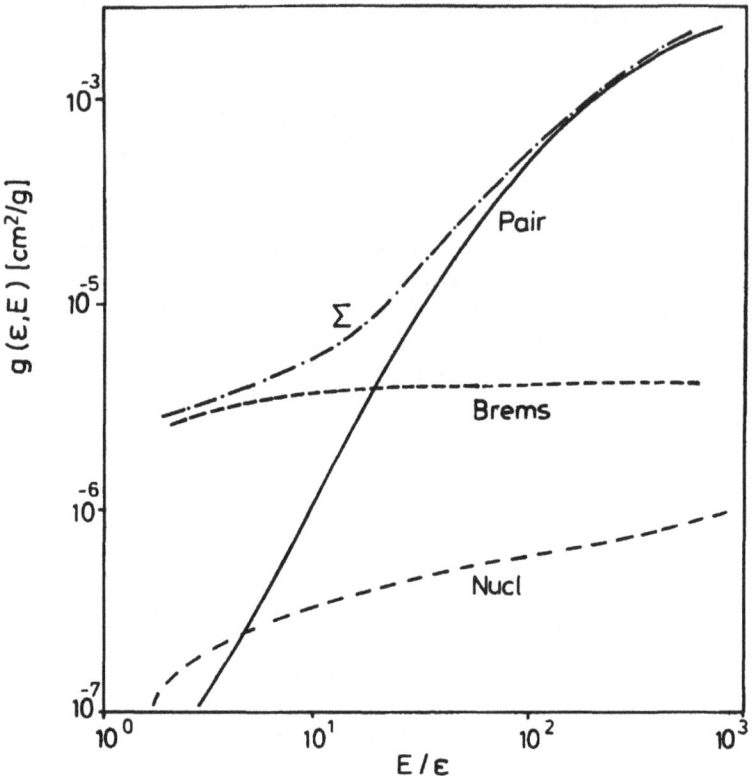

Figure 5. *Energy dependence of various contributions to the total interaction probability.*

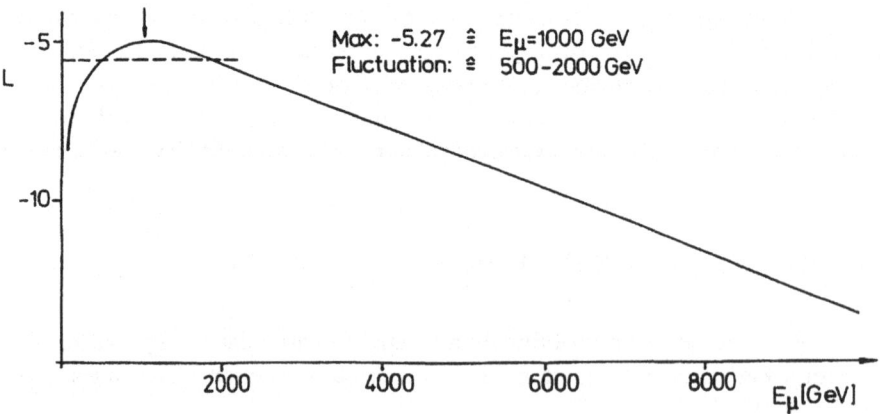

Figure 6. *Energy dependence of the calculated logarithmic likelihood function (LLF).*

X_a is the total thickness of the absorbing material, the ϵ_{min}, ϵ_{max} - limits of the integral depend on the incident muon energy and on the cut-off in the kinetic energy of secondaries.

Then the most probable value of E_0 for a given observed sample $\epsilon_1, \epsilon_2, \ldots, \epsilon_M$ is at the maximum of L. Figure 6 displays the energy dependence of LLF calculated for the case of events generating 2 secondaries with energies between 1 and 2 GeV. The dotted line shows the region where the LLF value is smaller than 0.5, representing a confidence level of 68% (1σ). The width of the corresponding energy interval indicates the accuracy in estimating the muon energy. The maximum is around 1 TeV, but the function is asymmetric.

In our first analysis of feasibility of MLM to estimate the muon energy, the simulations of the detector response are used to provide the number of secondaries and their energy transfers. The detector consisted of 30 layers of Pb, 1 cm thickness, alternating with 30 layers of scintillator, 3 cm thickness. Two cut-offs for the kinetic energy of secondaries have been used, 0.1 GeV and 1 GeV.

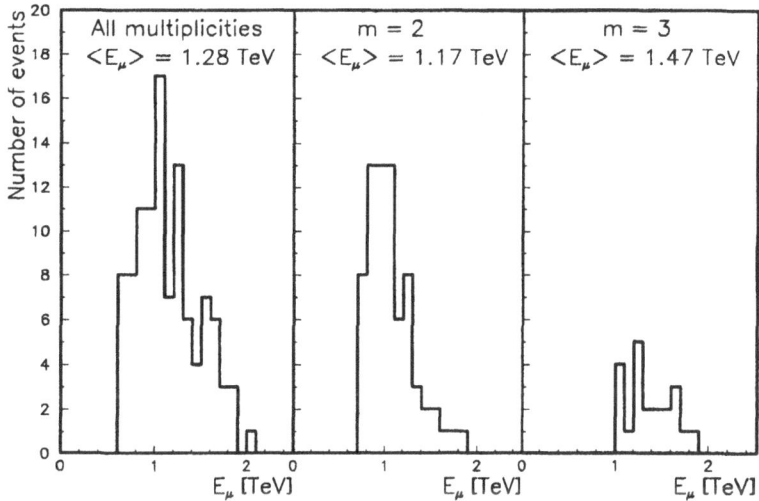

Figure 7. *The reconstructed energy of 1 TeV muons using the MLM.*

Figure 7 presents the results obtained by MLM using input values from the simulations of detector response to 1 TeV muons with a statistics of 100 events; the major contribution is given by events consisting in 2 secondaries with energies in the range 1 – 2 GeV. In the histograms the abscissa indicates values of the muon energy estimated from the application of MLM for various numbers of secondaries. From that a mean value for the muon energy in the vicinity of 1 TeV with an uncertainty of about 70% is deduced.

Comparing this result with the behavior of LLF shown in Figure 6, a good agreement is seen, indicating that for events producing 2 showers with energy transfers in the range of 1 – 2 GeV, the most probable mean energy of the incident muon is 1 TeV.

Unfortunately, for higher incident energies we found an underestimation of the reconstructed muon energy by this method, may be due to an underestimation of the number of showers as a consequence of a small number of layers.

Status and outlook

Presently three of the described detection layers are accomplished and connected with signal processing electronics and the data acquisition system. The high voltage system is currently under preparation and will be attached to the detector, soon. In future the production of further absorbing and detecting layers is necessary to achieve a calorimeter stack of at least 10 sensitive layers.

On the side of the theoretical analysis and the improvement of the energy determination by better statistical accuracy, a higher cut-off of secondaries and a new expression of LLF by taking into account the energy loss of muons are needed.

References

[1] R. P. Kokoulin and A. A. Petrukhin, *Nucl. Instr. Meth. A* **263** (1988) 46
[2] P. Doll et al., *KfK Report 4686* (1990), Kernforschungszentrum Karlsruhe ISSN 0303-4003
[3] R. Brun et al., *Cern Report* DD/EE(84-1) (1987)
[4] I. M. Brâncuş et al., *Internal Report* (1992), Kernforschungszentrum Karlsruhe
    G. Pascovici et al., *Internal Report* (1991), Kernforschungszentrum Karlsruhe
[5] K. W. Zimmer et al., *Contr. to DPG Meeting*, Mainz, March 22-26, 1993

THERMONUCLEAR REACTION RATE UNCERTAINTIES FROM NUCLEAR MODEL CALCULATIONS

V. Avrigeanu[1] and A. Harangoza[2]

[1] *Institute of Atomic Physics, PO Box MG-6, Bucharest, Romania*
[2] *Atomic and Nuclear Physics Department, University of Bucharest*

Studies of nucleosynthesis in evolving stars and supernovae rely heavily on reaction cross sections computed by means on nuclear models. A broad range of nuclear parameters and even model assumptions are required in this respect, and the most elaborated evolution models still suffer from major both astrophysical and nuclear physics uncertainties[1]. The statistical model of nuclear reactions is largely used to calculate the host of unmeasured reaction rates involved in the modeling of neon, oxygen, and silicon burning. Especially non-explosive as well as explosive Si burning comprises a very complex pattern of nuclear reactions, when the full nuclear statistical equilibrium is reached between all the nuclear species of the medium resulting in the production of the iron peak nuclei. The accuracy of the calculated cross sections for nucleon and alpha-particle induced reactions on $1f_{7/2}$ shell nuclei is discussed in this work.

An increased predictive capability of Hauser-Feshbach-Moldauer calculations[2,3] has been obtaind by (*a*) use of consistent sets of input parameters, determined or validated by means of various independent types of experimental data, and (*b*) unitary account of a whole body of related experimental reaction cross sections for isotope chains and neighboring elements as target nuclei.

Consequences of different classes of nuclear model assumptions and parameters on calculated cross sections are shown for example in Figure 1 for the (n,α) reactions on the target nuclei ^{51}V and ^{54}Fe in the incident energy range increased up to 20 MeV for completeness. The large experimental data basis for these reactions is well described by the calculated full curves, and the effects of various nuclear level densities (*a*), global optical model potentials for neutrons (*b*), and alpha-particles (*c*), as well as intranuclear transition rates related to the imaginary potential well (*d*) are compared. The two target nuclei correspond to quite different values of the asymmetry parameter (N-Z)/A, distinct behavior and increased pre-equilibrium emission weight[4] in the latter case being obvious. It could be underlined that using global or early parameter values (dotted curves in Figure 1*d*) the respective results can compensate (^{51}V) or add together (^{54}Fe) giving large differences with respect to the correct description of the experimental data. Nevertheless, these both circumstances lower the accuracy of calculated cross sections and thermonuclear reaction rates, and prove the usefulness of consistent analyses.

References

[1] M. Arnould, *Nucl. Phys. A* **538**, 493c (1992)

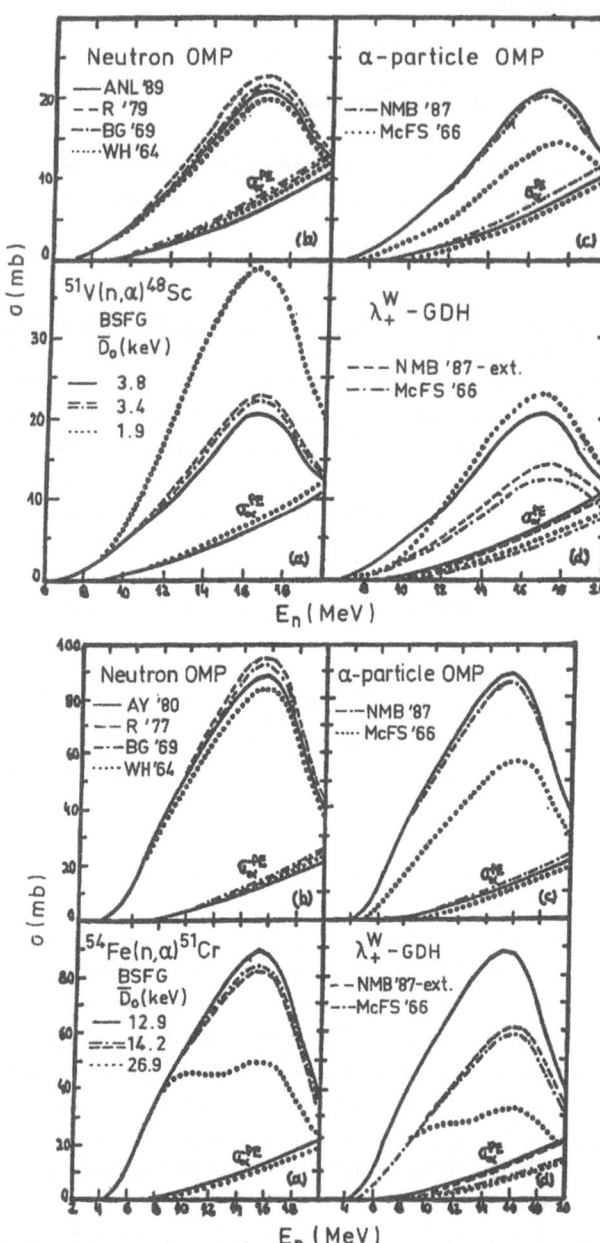

Figure 1. Calculated cross sections for the reactions $^{51}V(n,\alpha)^{48}Sc$ and $^{54}Fe(n,\alpha)^{51}Cr$.

[2] M. Avrigeanu, M. Ivascu and V. Avrigeanu, Z. Phys. A **329**, 177 (1988)
[3] M. Avrigeanu, M. Ivascu and V. Avrigeanu, Z. Phys. A **335**, 299 (1990)
[4] M. Avrigeanu and V. Avrigeanu in: *Nuclear Data for Science and Technology*, p.993, edited by S. M. Qaim, (Springer-Verlag, Berlin, Heidelberg, 1992)

VII.

Heavy Ion Collisions

NUCLEAR MOLECULAR PHENOMENA IN HEAVY ION COLLISIONS

Jae Young Park[1], Artur Thiel[2], Werner Scheid[3] and Walter Greiner[4]

[1] Department of Physics, North Carolina State University Raleigh, North Carolina 27695-8202, U.S.A.
[2] Bayer AG, D-5090 Leverkusen, Germany
[3] Institut für Theoretische Physik der Justus-Liebig-Universität, D-35392 Giessen, Germany
[4] Institut für Theoretische Physik der Johann-Wolfgang-Goethe-Universität, D-6000 Frankfurt am Main, Germany

Introduction

One of the interesting and outstanding aspects of heavy-ion nuclear physics is that nuclear molecules can be formed during collisions between heavy ions[1]. The narrow resonances observed in the energy dependence of the $^{12}C + ^{12}C$ reaction cross sections at energies near and above the Coulomb barrier have provided the first evidence for the existence of nuclear molecules[2]. Since this pioneering work 33 years ago, a large number of resonances was discovered[3], not only in the $^{12}C + ^{12}C$ system but also in many other heavy-ion systems, for example, $^{12}C + ^{16}O$, $^{16}O + ^{16}O$, $^{12}C + ^{24}Mg$ and $^{28}Si + ^{28}Si$. A nuclear molecule is a system of two (or more) nuclei which are bound together on their surfaces in the quasibound or bound states of a quasimolecular potential. In the microscopic picture a nuclear molecule is understood as a nuclear configuration in which the outermost bound (valence) nucleons orbit around both nuclei, and constitutes a bonding just like homopolar or covalent bonding in atomic molecules. The molecular configurations can be described in the framework of the two-center shell model (TCSM)[4,5]. The model describes the binding of the molecule in terms of a molecular single-particle states. Therefore, molecular single- particle configurations play an important role in understanding the structure and formation of nuclear molecular configurations in heavy-ion collisions.

The Two-Center Shell Model and Realistic Two-Center Single-Particle Energy Level Diagrams

A simple model which is appropriate for a microscopical description of heavy ion reactions and nuclear molecular configurations is the two-center shell model (TCSM)[4,5]. In the TCSM one assumes a single-particle potential (mean field) having a two-center shape, as shown in Fig. 1. The potential depends on the relative distance of the nuclear centers. The single-particle energies depend on the internuclear distance and on the frequency of the potential.

In an adiabatic collision the frequency has to be adjusted for all internuclear distances so that the volume enclosed in an equipotential surface remains constant (volume conservation).

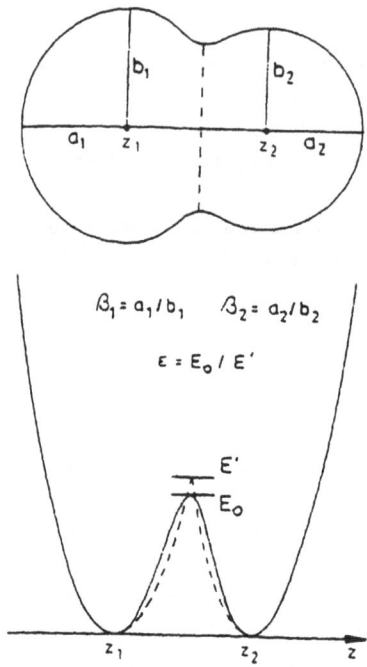

Figure 1. *Oscillator-like potential of the asymmetric TCSM and the associate nuclear shape.*

A prerequisite for the investigation of molecular single-particle effects and Landau-Zener transitions are realistic two-center level diagrams. Park et al.[6] calculated the level diagrams for the systems ^{12,13}C + ^{16,17}O. The parameters of the TCSM were adjusted to the experimental neutron single-particle energies for inter-nuclear distances $R \to 0$ and $R \to \infty$ and interpolated in-between. Fig. 2 shows the calculated adiabatic level diagram for the system ^{17}O + ^{12}C $\to$ ^{29}Si. At $R = 7.8$ fm one finds an avoided level crossing of two $\Omega = 1/2$-states approaching the single-particle states $1d_{5/2}$ and $2s_{1/2}$ at $R \to \infty$. Based on this fact Park et al.[6] predicted an enhanced transition of the loosely bound neutron of ^{17}O to the first excited $1/2^+$ state of ^{17}O in the ^{12,13}C + ^{17}O reactions.

The Landau-Zener Effect

The Landau-Zener effect is an enhanced transition between two adiabatic molecular levels, as it is well known from atomic collisions. As function of the relative distance between the nuclei (or atoms) the molecular levels do not cross in general if they have the same quantum number related to a symmetry. For spherical nuclei this symmetry is the rotational symmetry about the internuclear axis and in case of equal nuclei, in addition, the parity with respect to the center-of-mass of the system. The point of the nearest approach of two adiabatic levels with the same quantum number Ω of projection of angular momentum on the internuclear

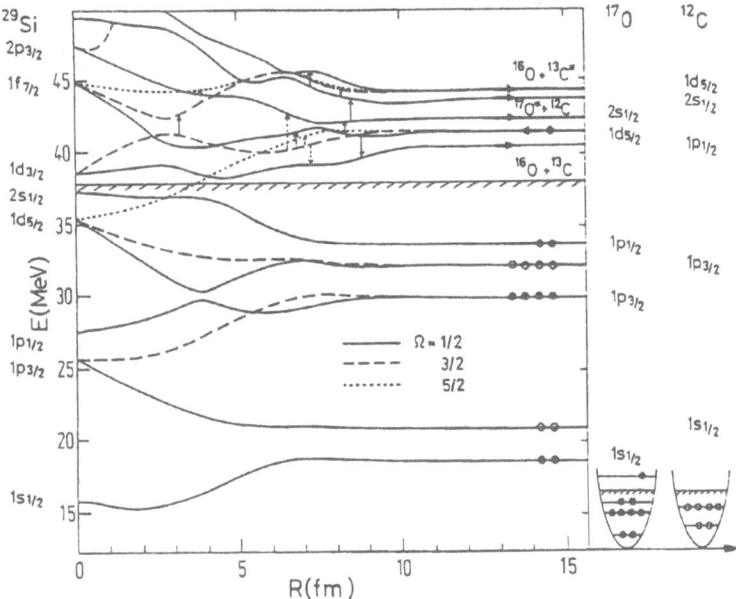

Figure 2. Adiabatic two-center neutron level diagram of the system $^{17}O + ^{12}C \to\ ^{29}Si$ as function of the internuclear distance R.

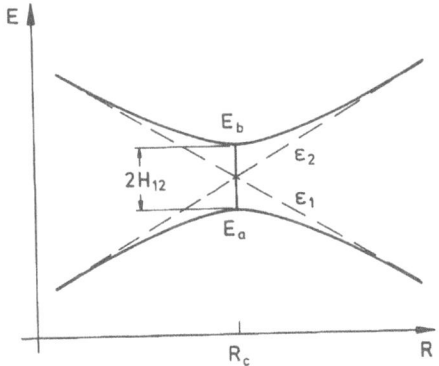

Figure 3. Avoided crossing of two adiabatic levels with energies E_a and E_b as function of the internuclear distance R. The dashed lines are the diabatic levels with energies ε_1 and ε_2. The interaction between the diabatic states at $R = R_c$ is denoted by H_{12}.

axis is denoted as avoided level crossing in literature. An avoided crossing of two adiabatic levels a and b is depicted in Fig. 3. In addition the corresponding diabatic levels ε_1 and ε_2 are shown by dashed lines. Here, H_{12} is the half of the energy splitting of the adiabatic levels at the avoided crossing at $R = R_c$.

As studied by Landau, Zener and Stückelberg[7] the transition probability from the lower adiabatic state to the higher one by a two-way passage at the crossing point is given by

$$P_{a \to b} = 2 P_{LZ}(1 - P_{LZ}) \tag{1}$$

where

$$P_{LZ} = e^{-2\pi G}, \quad G = \frac{|H_{12}|^2}{\hbar v |\frac{d}{dR}(\varepsilon_1 - \varepsilon_2)|}. \tag{2}$$

It is assumed in the derivation of the Landau-Zener transition probability that the point of avoided level crossing is far from the turning point of the relative motion of the nuclei and hence the relative velocity is constant in the region of avoided level crossing. The relative velocity is determined by the nucleus-nucleus potential and depends on the incident energy and the impact parameter. The enhancement of the transition between two states due to the Landau-Zener effect is a signature that molecular states are formed during a collision.

Molecular Reaction Theory with Particle-Core Model

For reactions between light nuclei we can use a molecular reaction theory. In the following we describe a theory of single-particle excitations of the nuclei which applies two-center shell model states for the single-particles. This reaction theory is denoted as the molecular dynamical particle-core model in literature[8]. It was first introduced by Park et al.[8], and Terlecki et al.[9] then formulated the model so that it could be applied to the elastic and inelastic scattering of ^{13}C on ^{13}C described by two ^{12}C cores with loosely bound neutrons. The extension of the model for neutron transfer reactions has been worked out by Park et al.[10]. Könnecke et al.[11] and Thiel et al.[12] have shown that the cross sections for inelastic scattering and the one neutron transfer in the ^{13}C + ^{13}C system measured by Balamuth et al.[13] and Korotky et al.[14] can be consistently explained within this molecular reaction theory.

The Hamiltonian and the Wave Function

The particle-core model divides the nucleus-nucleus system into two cores and N valence nucleons. Since the TCSM Hamiltonian, used for the description of the valence nucleons, is formulated in an intrinsic coordinate system with the z'-axis lying along the relative coordinate between the nuclei, we have to introduce a coordinate system rotating with the relative coordinate. The Hamiltonian has the form[12]

$$H = T + U(R) + iW(R) + H_{TCSM} + V_{res}, \tag{3}$$

$$T = -\frac{\hbar^2}{2\mu R}\left(\frac{\partial}{\partial R} + D\right)^2 R + \frac{1}{2\mu R^2}\left(\vec{I} - \vec{J}\right)^2, \tag{3a}$$

$$D = \frac{1}{A}\left(A_2 \sum_{i=1}^{N_1} \frac{\partial}{\partial z'_{icm}} - A_1 \sum_{i=N_1+1}^{N} \frac{\partial}{\partial z'_{icm}}\right), \tag{3b}$$

$$\vec{J} = \sum_{i=1}^{N_1}\left(\vec{r}'_{icm} - \frac{A_2}{A}\vec{R}\right) \times \vec{p}'_{icm} + \sum_{i=N_1+1}^{N}\left(\vec{r}'_{icm} + \frac{A_1}{A}\vec{R}\right) \times \vec{p}'_{icm}, \tag{3c}$$

$$H_{TCSM} = \sum_{i=1}^{N} h_{TCSM}(i) = \sum_{i=1}^{N} \frac{p'^2_{icm}}{2M} + V_{TCSM}(\vec{r}'_{icm}, R). \tag{3d}$$

Here, the coordinates $\vec{r}'_{icm}$ and momenta $\vec{p}'_{icm}$ of the valence nucleons are measured with respect to the total center-of-mass and are referred to the rotating coordinate system. The orientation of this system with respect to the laboratory system is given by the three Euler angles ϕ, θ (spherical polar angles of the relative coordinate $\vec{R}$), and ψ (superfluous and unphysical third Euler angle). The rotational kinetic energy in Eq. (3a) contains the total angular momentum operator $\vec{I}$ depending on the Euler angles and the angular momentum operator $\vec{J}$ of the valence nucleons measured with respect to the centers of the individual nuclei. H_{TCSM} is the Hamiltonian of the two-center shell model and V_{res} represents a residual interaction between the valence nucleons.

The Hamiltonian H can be used for solving the stationary scattering problem within the formalism of coupled channels. As wave function we take the following ansatz[12]

$$\Psi_M^I = \sum_\kappa R_\kappa^I(R) \psi_{\kappa IM},\qquad(4)$$

where the channel functions are given by

$$\psi_{\kappa IM} = \left(\frac{2l+1}{8\pi^2}\right)^{1/2} \sum_{M'} (l0JM'|IM') D_{MM'}^{I*}(\phi,\theta,\psi)\Phi_{\kappa M'}.\qquad(5)$$

The structure of the channel functions is the same as in strong coupling models (Nilsson model). The functions $\Phi_{\kappa M}$ are built up by the single-particle states of the TCSM and coupled to a total spin J for large separations of the nuclei. The antisymmetrization of the valence nucleons and the symmetry for core exchange if necessary is taken into account (for details, see e.g., Thiel et al.[12]).

The Coupled Channel Equations

For a given incident energy E and total angular momentum I the coupled channel equations for the radial wavefunctions are obtained by the projection

$$\left\langle [i^l Y_l \otimes \Phi_{\alpha J}]_M^{(I)} | H - E | \Psi \right\rangle = 0.\qquad(6)$$

Applications of the Molecular Particle-Core Model to the $^{17}O + ^{12}C$ Systems

In the coupled channel calculation we have used the TCSM with parameters as determined by Park et al.[6] except a different neck parameter $\varepsilon = 1$. Figure 4(a) shows the radial behaviour of the energies of the TCSM levels explicitly taken into account in the coupled channel calculations. We notice in figure 4(a) the pronounced avoided level crossing between the $2s_{1/2}$ and $1d_{5/2}$ ^{17}O levels with $|\Omega| = 1/2$. Only between these two levels does a radial coupling matrix element arise in our truncated set of molecular states. Its radial behaviour is shown in fig. 4(b). In order to demonstrate the nuclear Landau-Zener effect in the coupled channel calculations we also used a second radial coupling, shown by the dotted curve in fig. 4(b), where the peak is artificially suppressed.

$^{17}O + ^{12}C$ Scattering

We applied the molecular particle-core model to the $^{17}O + ^{12}C$ reaction[16-18]. The particle-core model decomposes the ^{17}O nucleus into a ^{16}O core and a valence neutron described by the eigenfunctions of the TCSM Hamiltonian for the $^{17}O + ^{12}C$ system. In the coupled channel calculations we included the elastic channel and the inelastic excitation

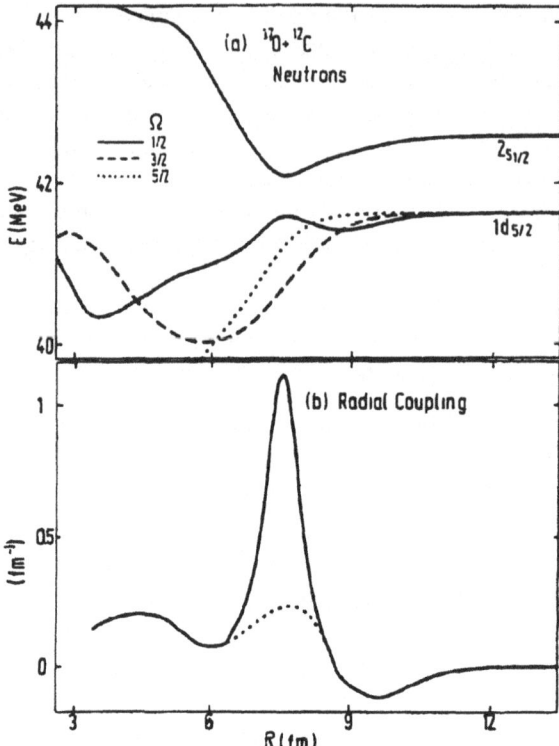

Figure 4. *(a) TCSM level diagram for neutron states of the $^{17}O + ^{12}C \to {}^{29}Si$ system. Full, dashed and dotted curves denote TCSM levels with $|\Omega| = 1/2$, $3/2$ and $5/2$, respectively. (b) Radial coupling matrix element between the elastic and inelastic channels.*

of the ^{17}O nucleus to the first excited $1/2^+$ state at 0.871 MeV as well as the one neutron transfer leading to the $^{16}O + {}^{13}C$ system.

In the elastic channel the valence neutron of ^{17}O occupies the molecular states approaching the $1d_{5/2}$ state of ^{17}O for $R \to \infty$, and in the inelastic channel the $2s_{1/2}$ state. Figures 5-7 display, respectively, angular distributions at the bombarding energy $E_{lab} = 50$ MeV for the elastic scattering of ^{17}O on ^{12}C, the inelastic scattering of ^{17}O to its first excited $1/2^+$ state and the one neutron transfer leading to the $^{16}O + {}^{13}C$ system. The solid lines represent the result of coupled channel calculations which include, in addition to the elastic and inelastic scattering of ^{17}O on ^{12}C, also the single neutron transfer to the $^{16}O + {}^{13}C$ system. The dotted curves in figs. 5 and 6 show the results of calculations[17] in which, only the elastic and inelastic scattering of ^{17}O on ^{12}C were treated explicitly. We note that the additional inclusion of the one- neutron transfer leads to an overall decrease in the calculated elastic angular distribution and it slightly damps the more forward oscillations in the calculated inelastic angular distributions. The characteristic structures which we identify as signatures of the nuclear Landau-Zener effect, namely the elastic backward angle rise and the oscillatory structure of the inelastic angular distributions, are present in our calculations.

The transfer angular distribution resulting from the present coupled channel calculation is shown by the solid line in fig. 7. It reproduces the shape of the experimental transfer angular distribution, but lies slightly above the data. Since the relevant contributions of this

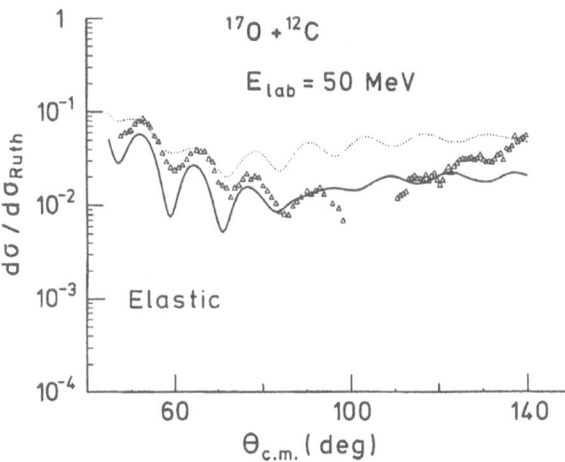

Figure 5. Angular distributions for the elastic scattering of ^{17}O on ^{12}C at the bombarding energy of $E_{lab} = 50$ MeV. The triangles denote the data of Freeman et al.[15]. Full and dotted curves are obtained as described in the text.

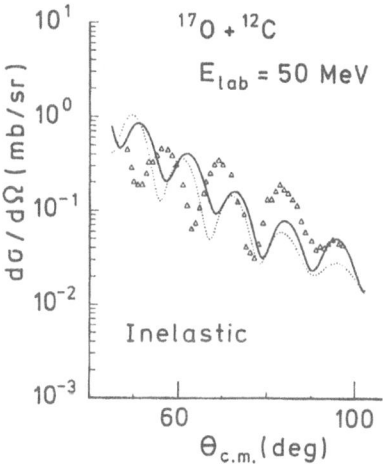

Figure 6. Inelastic angular distribution $^{12}C(^{17}O, ^{17}O^*(1/2^+))^{12}C$ at $E_{lab} = 50$ MeV. Full and dotted curves are obtained as described in the text.

radial coupling are coming from the enhancement due to the pseudocrossing, a variation of the peak height can be simulated by multiplying this radial coupling potential (solid line in fig. 4(b)) by a factor[17]. A reduction of the height of the Landau-Zener peak which enhances the transition from the ground state to the first excited $1/2^+$ state of ^{17}O, i.e. a reduction of the corresponding radial coupling strength by a factor 0.75 and 0.5 gives the dashed-dotted and dotted curves, respectively, in fig. 7.

It is found that when the Landau-Zener effect is suppressed by smoothing out the peak in the corresponding radial coupling potential, the calculations yield a flat inelastic angular distribution and the elastic and transfer angular distributions decrease monotonously as a

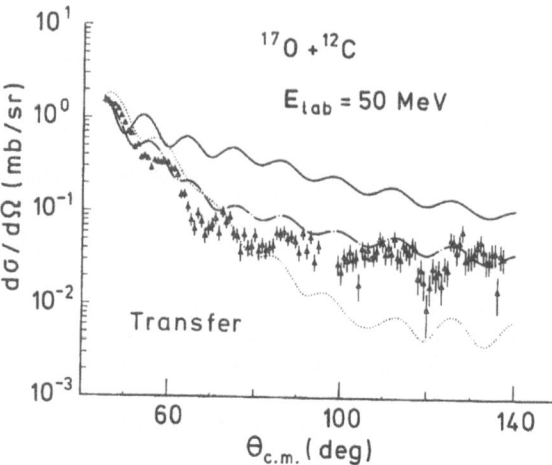

Figure 7. *Angular distribution for the single neutron transfer to the $^{16}O + ^{13}C$ system at $E_{lab} = 50$ MeV. The curves are explained in the text.*

function of the scattering angle, and that the oscillatory structure of the experimental inelastic angular distribution and the backward angle yields in the elastic and transfer data can only be reproduced when the nuclear Landau-Zener effect is taken into account in our calculations.

Fusion of ^{17}O with ^{12}C

The fusion process for bombarding energies near the Coulomb barrier certainly involves the formation of nuclear molecular-like configurations in the entrance channels. Eyal et al.[19] observed that complete fusion cross sections for the $^{18,17,16}O + ^{12}C$ systems, increase with increasing mass of the oxygen isotopes. One possible explanation of such observations is that the outermost bound nucleons lower the fusion barrier by forming a neck[19]. This neck formation is a molecular process. Fusion enhancement due to molecular orbital formation by the valence nucleons can be studied by fusion cross section calculations within the molecular particle-core model. The absorption cross section can be written as an integral of the product of the density with the imaginary part W of the optical potential

$$\sigma_{Abs} = \frac{2}{\hbar} \int |\Psi|^2 W d\tau . \qquad (7)$$

As employed by Udagawa et al.[20] the fusion cross section is defined as that part of the absorption cross section which is obtained by cutting off this integral at a finite radius R_F. Our calculations show that the inclusion of the inelastic excitation, as well as of the transfer of the valence neutron of ^{17}O in the coupled channel calculations, leads to an enhancement in the calculated fusion yields which becomes less pronounced if the nuclear Landau-Zener effect is suppressed.

Figure 8 shows the fusion cross sections for a fusion radius of 7 fm. The solid line results from coupled channel calculations with the nuclear Landau-Zener effect adjusted to the Strasbourg data[15] and by including the elastic scattering and inelastic excitation of ^{17}O on ^{12}C and the one neutron transfer to the $^{16}O + ^{13}C$ system. If the nuclear Landau-Zener effect is suppressed in the calculation, the dashed curve is obtained. The dotted curve refers

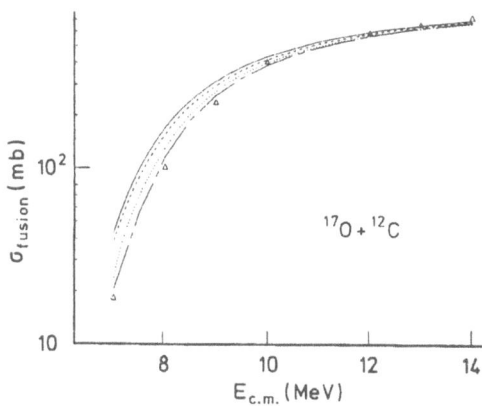

Figure 8. *Fusion cross sections calculated with a fusion radius $R_F = 7$ fm.*

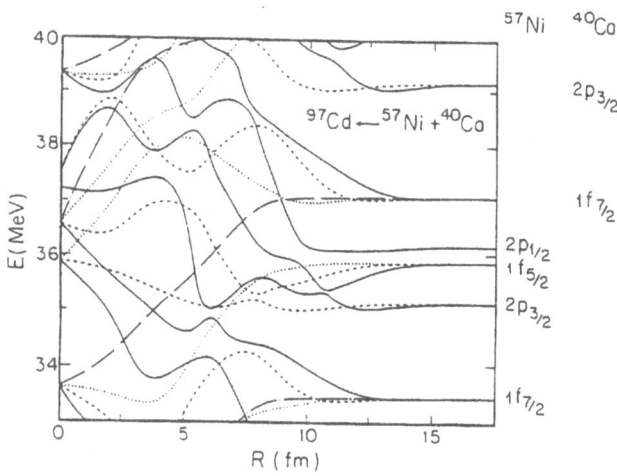

Figure 9. *TCSM neutron levels for $^{57}Ni + {}^{40}Ca \to {}^{97}Cd$ as functions of the internuclear distance. Levels with $|\Omega| = 1/2, 3/2, 5/2$ are denoted by the full, dashed and dotted lines, respectively.*

to a coupled channel calculation including only the elastic and inelastic scattering of ^{17}O on ^{12}C.

The short-long dashed curve is obtained by an optical model calculation without any couplings. The triangles represent the fusion data of Eyal et al.[19]. All fusion cross section calculations shown in fig. 8 overestimate the experimental fusion yields for energies below 10 MeV. Here, one should keep in mind, however, that we have used in the coupled channel calculations for the fusion cross section an optical potential which is adjusted to the Strasbourg data[15] in the energy range $E_{c.m.} = 16.5 - 29$ MeV, which lies well above the energy range considered in the fusion calculations.

Outlook

Up to now, the ^{28}Si + ^{28}Si system is the heaviest system in which resonant molecular-like structures have been experimentally observed in the scattering cross sections. The experimental search for nuclear molecular-like structures in reaction amplitudes of still heavier systems as ^{40}Ca + ^{40}Ca has not yet been successful.

In order to investigate if observable avoided level crossings near the Fermi level occur in scattering systems heavier than the ^{17}O + ^{12}C system we have calculated realistic two-center level diagrams for the following target projectile combinations: ^{17}O + ^{28}Si, ^{17}O + ^{40}Ca and ^{57}Ni + ^{40}Ca. As an example figure 9 displays the neutron level structure of the TCSM for the reactions ^{57}Ni + ^{40}Ca $\to$ ^{97}Cd. We recognize that for $R \to \infty$ the TCSM calculations reproduce the experimental energies of the valence orbitals of the separated nuclei. The TCSM level diagram for the ^{57}Ni + ^{40}Ca system reveals a very pronounced avoided level crossing of two $|\Omega| = 1/2$ levels of the $2p_{3/2}$ and $1f_{5/2}$ state of ^{57}Ni at a relative distance $R_c \approx 10.6$ fm. Signatures due to this pseudoscrossing should arise in the inelastic excitation of ^{57}Ni to its first excited $5/2^-$ state if it is assumed that ^{57}Ni can be decomposed in a ^{56}Ni core with a valence neutron occupying the $2p_{3/2}$ state in the ground state and the $1f_{5/2}$ state in the first excited state of ^{57}Ni.. The TCSM level diagram for ^{57}Ni + ^{40}Ca demonstrates that the nuclear Landau-Zener effect can also arise in other scattering systems heavier than ^{17}O + ^{12}C. The avoided crossing between the $|\Omega| = 1/2$ TCSM levels of the $2p_{3/2}$ and $1f_{5/2}$ states of ^{57}Ni shows that the occurrence of observable pseudocrossings is not restricted to collisions of s-d shell nuclei. The ^{57}Ni + ^{40}Ca scattering however, is difficult to measure because ^{57}Ni is unstable with a half-life of 36 hours. So this system could probably be investigated only by using a radioactive ^{57}Ni beam.

Conclusions

The formation of nucleon molecular orbitals in nucleus-nucleus collisions have been discussed by our group[10,16,18] and others[14,21]. In particular, the Landau-Zener transitions between these nuclear molecular orbitals have been studied in recent years by us[6,9,12,16-18,22-25] and by others[26]. The observation of traces of a nuclear Landau-Zener effect in cross sections of light heavy ion reactions is a signature that the nucleons form molecular orbitals during the collision.

We have studied scattering and fusion cross sections of the ^{17}O + ^{12}C system by coupled channel calculations within the molecular particle-core model. The dynamic reaction calculations presented here are in agreement with the statements that the structures of the elastic and inelastic scattering of ^{17}O on ^{12}C and for the one-neutron transfer have their origin in the Landau-Zener type transition between the considered molecular single-particle states. Its signatures, the backward angle yields in the elastic and transfer angular distributions and the oscillatory behavior of the inelastic angular distributions, produce the characteristics of the data from Strasbourg[15].

Our calculated fusion cross sections show enhancements due to the explicit treatment of inelastic excitation and transfer of the valence neutron of ^{17}O. The dominant fusion enhancement for energies near the Coulomb barrier comes from the neutron transfer, since it involves a positive Q-value of 0.8 MeV.

The future of the theory of nuclear molecules can be seen in the extension of the ideas developed for light systems to heavier ones. The systems formed by sd-shell nuclei (e.g.,

^{28}Si + ^{28}Si) show structures in these cross sections, which are due to the excitation of barrier resonances strongly coupled to the collective states of the nuclei. The molecular states, excited in these reactions, resemble very much to the high spin states of fissioning shape isomers similar to the rotational-vibrational states studied in lighter systems and suggested already for the giant nuclear molecules[27]. Here we recognize an interesting bridge between the lighter and heavier molecular systems. The continuation of the theory of nuclear molecules to heavier systems will probably reveal qualitative new effects and complement the existing phenomenological reactions theories in the explanation of specific molecular signatures in the experimental data. More work, both experimental measurements and dynamical reaction calculations as outlined in this report, are necessary to identify the nuclear Landau-Zener effect and consequently confirm molecular single-particle effects and firmly establish the formations of nuclear molecules in heavy-ion collisions.

References

[1] R. Y. Cusson, R. K. Smith and J. A. Maruhn, *Phys. Rev. Lett.* **36**, 1166 (1976)

[2] D. A. Bromley, J. A. Kuehner and E. Almqvist, *Phys. Rev. Lett.* **4**, 365 (1960)

[3] N. Cindro, ed. *Nuclear Molecular Phenomena* (North Holland Pub. Co., New York, 1978), N. Cindro, R. A. Ricci and W. Greiner, eds. *Dynamics of Heavy Ion Collisions* (North Holland Pub. Co., New York, 1981), K. A. Eberhard, ed. *Resonances in Heavy Ion Collisions* (Lecture Notes in Phys. 156) (Springer-Verlag, Berlin 1982), N. Cindro, W. Greiner and R. Caplar, eds. *Fundamental Problems in Heavy Ion Collisions* (World Scientific Pub. Co., Singapore, 1984), N. Cindro, W. Greiner, R. Caplar, *Frontiers of Heavy-Ion Physics* (World Scientific Publ Co., Singapore, 1987), N. Cindro, *Ann. Phys. Fr.* **13**, 289 (1988), R. Caplar and W. Greiner, eds. *Heavy-Ion Physics Today and Tomorrow* (World Scientific Pub. Co., Singapore, 1991)

[4] P. Holzer, U. Mosel and W. Greiner, *Nucl. Phys. A* **138**, 241 (1969); D. Scharnweber, W. Greiner and U. Mosel, *Nucl. Phys. A* **164**, 257 (1971)

[5] J. A. Maruhn and W. Greiner, *Z. Phys.* **251**, 431 (1972)

[6] J. Y. Park, W. Greiner and W. Scheid, *Phys. Rev. C* **21**, 958 (1980)

[7] L. D. Landau, *Z. Phys. Sov.* **2**, 46 (1932); C. Zener, *Proc. Roy. Soc. A* **137**, 696 (1932); E. C. G. Stückelberg, *Helv. Phys. Acta* **5**, 320 (1932)

[8] J. Y. Park, W. Scheid and W. Greiner, *Phys. Rev. C* **6**, 1565 (1972)

[9] G. Terlecki, W. Scheid, H. J. Fink and W. Greiner, *Phys. Rev. C*, **18**, 265 (1978)

[10] J. Y. Park, W. Scheid and W. Greiner, *Phys. Rev. C* **20**, 188 (1979)

[11] R. Könnecke, W. Greiner and W. Scheid, *Phys. Rev. Lett.* **51**, 366 (1983)

[12] A. Thiel, W. Greiner, J. Y. Park and W. Scheid, *Phys. Rev. C* **36**, 647 (1987)

[13] D. P. Balamuth et al., *Phys. Lett.* **140B**, 295 (1984)

[14] S. K. Korotky, K. A. Erb, R. L. Phillips, S. J. Willett and D. A. Bromley, *Phys. Rev. C* **28**, 168 (1983)

[15] R. M. Freeman, C. Beck, F. Haas, A. Morsad and N. Cindro, *Phys. Rev. C* **33**, 1275 (1986)

[16] A. Thiel, W. Greiner and W. Scheid, *J. Phys. G: Nucl. Phys.* **14**, L85 (1988); A. Thiel, W. Greiner, J. Y. Park and W. Scheid, *J. Phys. G: Nucl. Part. Phys.* **15**, 1833 (1989)

[17] A. Thiel, *J. Phys. G: Nucl. Part. Phys.* **16**, 867 (1990)

[18] A. Thiel, J. Y. Park and W. Scheid, *Phys. Rev. C* **43**, 1480 (1991); *J. Phys. G: Nucl. Part. Phys.* **17**, 1237 (1991)

[19] Y. Eyal et al., *Phys. Rev. C* **13**, 1527 (1976)

[20] T. Udagawa, B. T. Kim and T. Tamura, *Phys. Rev. C* **32**, 124 (1985)

[21] W. von Oertzen and H. G. Bohlen, *Phys. Rep.* **19**, 1 (1975); B. Imanishi and W. von Oertzen, *Phys. Rep.* **155**, 29 (1987)

[22] J. Y. Park, W. Scheid and W. Greiner, *Phys. Rev. C* **25**, 1902 (1982)

[23] Y. Abe and J. Y. Park, *Phys. Rev. C* **28**, 2316 (1983)
[24] J. Y. Park, K. Gramlich, W. Scheid and W. Greiner, *Phys. Rev. C* **33**, 1674 (1986)
[25] M. H. Cha, J. Y. Park and W. Scheid, *Phys. Rev. C* **36**, 2341 (1987)
[26] B. Imanishi, W. von Oertzen and H. Voit, *Phys. Rev. C* **35**, 359 (1987); N. Cindro, R. M. Freeman and F. Haas, *Phys. Rev. C* **33**, 1280 (1986); B. Milek and R. Reif, *Phys. Lett.* **157B**, 134 (1985); H. Voit, N. Bischof, W. Tiereth, I. Weizenfelder, W. von Oertzen and B. Imanishi, *Nucl. Phys. A* **476**, 491 (1988); B. Imanishi, S. Misono and W. von Oertzen, *Phys. Lett. B* **210**, 35 (1988) and **241**, 13 (1990); T. Tazawa and Y. Abe, *Phys. Rev. C* **41**, R17 (1990)
[27] P. O. Hess, W. Greiner, W. T. Pinkston, *Phys. Rev. Lett.* **53**, 1535 (1984)

ENERGY DEPENDENCE OF THE INVERTED SCATTERING POTENTIALS OF THE $^{12}C + {}^{12}C$ SYSTEM IN THE RANGE $E_{cm} = 8 - 12$ MeV

Barnabás Apagyi[1,2] and Werner Scheid[1]

[1] Institut für Theoretische Physik der Justus-Liebig-Universität, Giessen, Germany
[2] Quantum Theory Group, Technical University of Budapest, Hungary

Introduction: The quantum inverse scattering problem at fixed energy[1] is solved in two steps. First the elastic differential scattering cross section is analysed by a minimization procedure[2] to obtain a set of phase shifts. Then these phase shifts are used as input quantities by the inversion procedure to provide the optical potential. In this contribution we show that within the framework of the modified Newton-Sabatier method[3,4] it is possibile to unify these two separate steps.

Theory: The conventional method of model independent phase shift analysis minimizes the error square function

$$\chi^2 = \frac{1}{N} \sum_{i=1}^{N} \left(\frac{d\sigma(\theta_i)/d\Omega|_{exp} - d\sigma(\theta_i)/d\Omega|_{cal}}{\Delta d\sigma(\theta_i)/d\Omega|_{exp}} \right)^2, \quad (1)$$

with respect to the sets $\{S_l\}$ contained in the expression of the differential cross section $d\sigma(\theta)/d\Omega|_{cal} = |f(\theta) + f(\pi - \theta)|^2$ written for the case of elastic scattering of identical particles. Using standard notation, the scattering amplitude $f(\theta) = f_C(\theta) + f_n(\theta)$ consists of the Coulomb part f_C and the nuclear part $f_n(\theta) = (2ik)^{-1} \sum_l (2l+1) \exp(2i\sigma_l)(S_l - 1)P_l(\cos\theta)$. Evidently, only even partial waves ($l = 0, 2, ..., l_{max}$) contribute to Eq. (1).

One can minimize Eq. (1) also with respect to the spectral coefficients $\{c_l\}$ involved in the modified Newton method[3,4] defined by the following equations

$$U^B(\tilde{\rho}) = -\frac{2}{\tilde{\rho}} \frac{d}{d\tilde{\rho}} \sum_{l=0}^{l_{max}} c_l^B \varphi_l^0(\tilde{\rho}) \varphi_l^B(\tilde{\rho})/\tilde{\rho}, \quad \varphi_l^B(\tilde{\rho}) = \varphi_l^0(\tilde{\rho}) - \sum_{l'=0}^{l_{max}} c_{l'}^B L_{ll'}^B(\tilde{\rho}) \varphi_{l'}^B(\tilde{\rho}), \quad (2)$$

with $\tilde{\rho} = k_B r$ and the known matrix $L_{ll'}^B(\tilde{\rho}) = \int_0^{\tilde{\rho}} \varphi_l^0(\rho) \varphi_{l'}^0(\rho) d\rho/\rho^2$ where $\varphi_l^0(\tilde{\rho}) = \tilde{\rho} j_l(\tilde{\rho})$ is the regular solutions belonging to the reference potential $U_0^B \equiv 0$, and the analytical form of the solution functions is known beyond a radius r_0: $\varphi_l^B(r \geq r_0) = A_l^B \tilde{\rho} [\cos \delta_l^B j_l(\tilde{\rho}) - \sin \delta_l^B n_l(\tilde{\rho})]$ with $k_B = (2\mu E^B)^{1/2}/\hbar$, $A_l^B = const$, and $E^B = E_{cm} - Z_1 Z_2 e^2/r_0$. The key quantity in equations (2) are the (complex) spectral coefficients $\{c_l\}$ ($l = 0, 1, 2, .., l_{max}$) which are set by the minimization of Eq. (1). They uniquely determine the optical potential $V(r) = E_{cm}(U^B(\tilde{\rho}) + Z_1 Z_2 e^2/r_0)$ for $r < r_0$ and $V(r) = Z_1 Z_2 e^2/r$ for $r > r_0$, and thereby the cross section $d\sigma(\theta_i)/d\Omega|_{cal}$, too.

Application: We have chosen differential cross section data of the elastic scattering of ^{12}C nucleus by a ^{12}C target measured[5] at eight scattering energies in the range between $E_{cm} = 8$ to 12 MeV. The result of the inversion calculations is depicted in Fig. 1. In general, the shapes of the potentials differ from Woods-Saxon shapes. The real parts are characterized

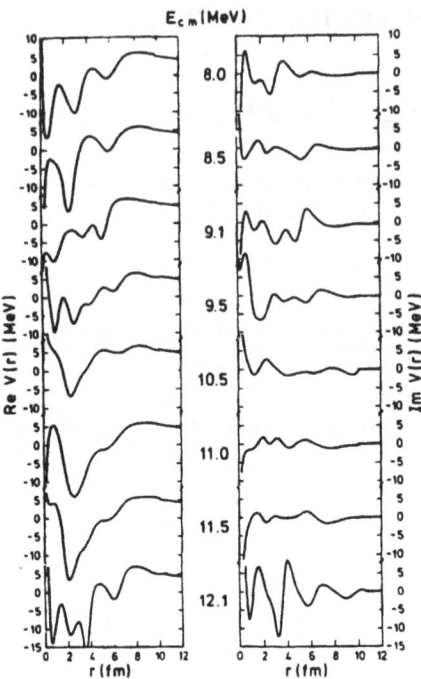

Figure 1. *Real part and imaginary part of the inverted $C^{12} + C^{12}$ potentials obtained by inverse methods for the eight energies indicated in the middle part of the figure. The real part is shown on the left side and the imaginary part on the right side. Matching radius has been kept fixed at $r_0 = 10$ fm. The differential cross sections re-calculated from the inverted potentials agree nicely with the experimental input of Ref. 5.*

by a pronounced minimum at relative distances of $r \approx 2.4 - 3$ fm which can be interpreted as originating from the configurations of the prolately deformed ^{24}Mg nucleus. At nearly all energies a shallow minimum in the real potential can be observed at internuclear distances of $5 - 6$ fm, which is possibly caused by quasimolecular ^{12}C cluster configurations. The real potentials show also the expected Coulomb barrier in the outer region around $r \approx 8 - 9$ fm. The imaginary parts exhibit positive maxima in those regions of radial distances where the real parts have minimum values which indicate a feed-back effect of flux to the elastic channel. In summary, the obtained elastic potentials for the $^{12}C + {}^{12}C$ system show a gradual and systematic energy dependence in the interval of energy considered.

Acknowledgment: This work has been supported in part by DFG/MTA (45), OTKA (517, 518, T7283), and GSI Darmstadt.

References

[1] R. G. Newton, *J. Math. Phys.* **3**, 75 (1962); P. C. Sabatier, *J. Math. Phys.* **7**, 1515, 2079 (1966)
[2] C. Marty, *Lecture Notes in Physics* **156**, 216 (1982)
[3] M. Münchow and W. Scheid, *Phys. Rev. Letters* **44**, 1299 (1980)
[4] K.-E. May, M. Münchow and W. Scheid, *Phys. Lett. B* **141**, 1 (1984)
[5] W. Treu, H. Fröhlich, W. Galster, P. Dück, and H. Voit, *Phys. Rev. C* **22**, 2462 (1980); W. Treu, H. Fröhlich, P. Dück and H. Voit, *Phys. Rev. C* **28**, 237 (1983)

THE Ni+Ni PUZZLE: RESONANCES IN THE SCATTERING OF MEDIUM HEAVY IONS?

N. Cindro[1], U. Abbondanno[2], Z. Basrak[1], M. Bettiolo[3], M. Bruno[4], M. d'Agostino[4], P. M. Milazzo[4], R. A. Ricci[3,5], W. Scheid[6], J. Schmidt[6], G. Vannini[2] and L. Vannucci[3]

[1] Rudjer Bošković Institute, 41001 Zagreb, Croatia
[2] Dipartimento di Fisica dell'Università di Trieste, 34127 Trieste, Italy, and Istituto Nazionale di Fisica Nucleare, Sezione di Trieste, 34127 Trieste, Italy
[3] Istituto Nazionale di Fisica Nucleare, Laboratori Nazionali di Legnaro, 35020 Legnaro, Italy
[4] Dipartimento di Fisica dell'Università di Bologna, 40126 Bologna, Italy, and Istituto Nazionale di Fisica Nucleare, Sezione di Bologna, 40126 Bologna, Italy
[5] Dipartimento di Fisica dell'Università di Padova, 35131 Padova, Italy
[6] Institut für Theoretische Physik der Justus-Liebig-Universität, 35392 Giessen, Germany

Abstract: Excitation functions and angular distributions of ^{58}Ni+^{58}Ni and ^{58}Ni+^{62}Ni scattering have been measured at energies just above the Coulomb barrier. Evidence for structure in the angle-integrated excitation functions has been found. Attempts are presented to interpret these results, including the presence of resonant states.

1. Introduction: Prediction of Resonances

Resonances in heavy-ion collisions are a phenomenon first observed more than 30 years ago in ^{12}C+^{12}C scattering and reactions[1]. Since that time, the domain of observation has been extended to systems as heavy as ^{28}Si+^{28}Si[2].

The specificity of heavy-ion resonances is that they imply the existence of discrete states high up in the continuum, states that do not mix with the surrounding high density statistical "grass". Such discrete states, hence, should be of a different nature. The first conjecture was that they represent simple configurations so different from the statistical background, that no mixing occurs and the states conserve their stability, hence, longevity.

A next step, taken very early in the study of heavy-ion resonances, was to associate these simple states with the so-called quasi-molecular configurations, pictured as a complex of two osculating spheres. The basis of this picture is shown in Fig. 1, exhibiting a collection of resonances observed in ^{12}C+^{12}C collisions in an E_x vs. L(L+1) plot. The rotational nature of resonances is apparent from the linear dependence

$$E_x = E_0 + K \cdot L(L+1) \tag{1}$$

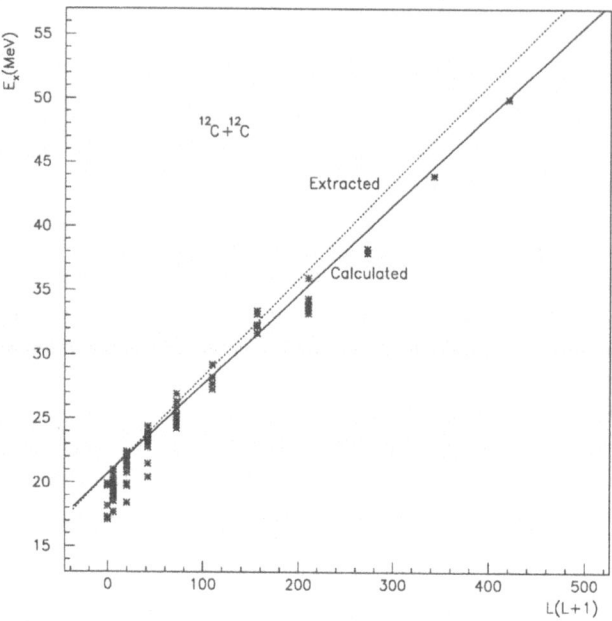

Figure 1. *Resonances observed in $^{12}C+^{12}C$ in a E_x of ^{24}Mg vs. $L(L+1)$ plot. The "extracted" line was obtained by fitting the slope of the experimental data; the "calculated" one was obtained by means of expression (1), using the model value of K (see ref. 2).*

typical for the spectrum of a quantum rotor. Interpreting the slope constant K in terms of

$$K = \frac{\hbar^2}{2\mathcal{J}_{eff}} \qquad (2)$$

one extracted a value of $\mathcal{J}_{eff}$ very close to that of the moment of inertia of two osculating ^{12}C spheres. In fact, ref. 2 lists a value of $\mathcal{J}_{eff} = (3.0 \pm 0.2)$ for the extracted moment of inertia, while this value for a rotating ^{12}C dumbbell ($r_0=1.3$ fm) is $\mathcal{J}_{calc} = 3.1$, all in 10^{-41} MeV s^2. The typical width of the resonances shown in Fig. 1 is 10-100 keV, corresponding to lifetimes around 10^{-20} s.

The quasimolecular picture is a simple model with some predictive power. A key step forward has been undertaken by the Frankfurt group[3], who introduced the concept of molecular windows. The underlying idea is that, if resonances are to be observed, their width Γ should be smaller than their spacing ΔE. The so called spreading width

$$\Gamma \downarrow \sim \imath W = 2\pi \ |<cn\ |\ V\ |\ el>|^2 \ \rho_{cn}(E_x,L) \qquad (3)$$

(W is the optical imaginary potential) was introduced with an empirical expression for the matrix element $<cn|V|el>$. The region of observation of the resonances in the E_x vs $L(L+1)$ plane corresponds to the loci of low level density ρ_{cn} in the composite system. These loci form a window through which we can peek at the resonances; hence the name molecular window.

Relating this idea to the osculating nuclei picture, the orbiting-cluster model (OCM)[4] was developed since 1979. In spite of its simplicity, this model showed considerable predictive power for lighter systems (up to ^{28}Si) and was recently used to predict resonances in heavier systems[5]. The most striking result of this calculation was the prediction of possible resonant behaviour of heavier colliding nuclei leading to composite systems with a closed neutron or proton $g_{\frac{7}{2}}$ shell (e.g. ^{28}Si+^{66}Zn leading to ^{94}Ru or ^{58}Ni+^{46}Ti leading to neutron deficient ^{104}Sn). Later on, this calculation was extended to still heavier systems[6]. ^{58}Ni+^{58}Ni turned out to be a favorable case, less so ^{58}Ni+^{62}Ni.

In the next section we report on the experimental investigation of the latter two systems.

2. Results and Discussion

The experiments described in this Section for both ^{58}Ni and ^{60}Ni targets have been performed at the Laboratori Nazionali di Legnaro XTU Tandem accelerator. The ^{58}Ni+^{58}Ni and ^{58}Ni+^{62}Ni elastic and inelastic scattering angular distributions around θ_{cm}=90° were measured in the laboratory energy range from 220 to 230 MeV in steps of 0.5 MeV. 10 μg.cm^{-2} ^{58}Ni and ^{62}Ni layers evaporated onto a 10 μg.cm^{-2} Carbon backing were used as targets. The combined thicknesses corresponded to an energy loss of the beam of 130 keV over the energy range of the experiment. The mass identification was performed using the kinematic-coincidence technique, by which the angles of emission and the kinetic energies of binary events were measured in coincidence. The experimental set-up and the data analysis have already been described in detail elsewhere [7]. We outline them here only briefly.

Two position-sensitive silicon detectors (100 μm thick, 47 mm long and 8 mm wide) were used in the experiment. The detectors for scattered and recoil nuclei were placed each at angles of ±45° with respect to the beam direction, at a distance of 155.5 mm from the target. The angular range covered in the laboratory system was 17.2°. The solid angle subtended by each detector was 15.5 msr and the overall angular resolution was 0.18°. Combining these data with the energy resolution (about 1%), the mass resolution turned out to be $\simeq$ 1 u. The good angular resolution allowed to discriminate elastic from inelastic events using the difference in the relative scattering angles.

2.1 ^{58}Ni+^{58}Ni

The excitation functions of the angle-summed (76° $\leq \theta_{cm} \leq$ 104°) elastic and inelastic scattering differential cross sections of ^{58}Ni+^{58}Ni are shown in Fig.2. In the elastic scattering, structures are visible at incident energies of 220.5, 222.5, 224.5, 226 and 229.0 MeV. The peaks at 220.5, 222.5 and 229 MeV are visibly correlated with peaks observed in the inelastic scattering; the correlation for the two middle peaks is less obvious. We keep in mind that the correlation in various exit channels is a necessary requirement for the presence of a resonance in the intermediate composite system.

The next figure (Fig. 3) shows angular distributions of elastically scattered ^{58}Ni on ^{58}Ni taken at the peaks in the excitation functions (notice that the same figure in ref. 8 (Fig. 4) is shown with a different vertical scale). Each measured angular distribution is almost perfectly fitted by squared single polynomial forms $[P_L(\cos \theta)]^2$ with L around 60. A statistical analysis (moving average correlation functions, ref. 9) was performed over the inelastic scattering distributions: the so treated angular distributions also show a periodicity of $\Delta \theta_{cm} \sim 3$, reflecting the same periodicity in the original angular distributions (where the periodicity was more difficult to establish due to large experimental uncertainties).

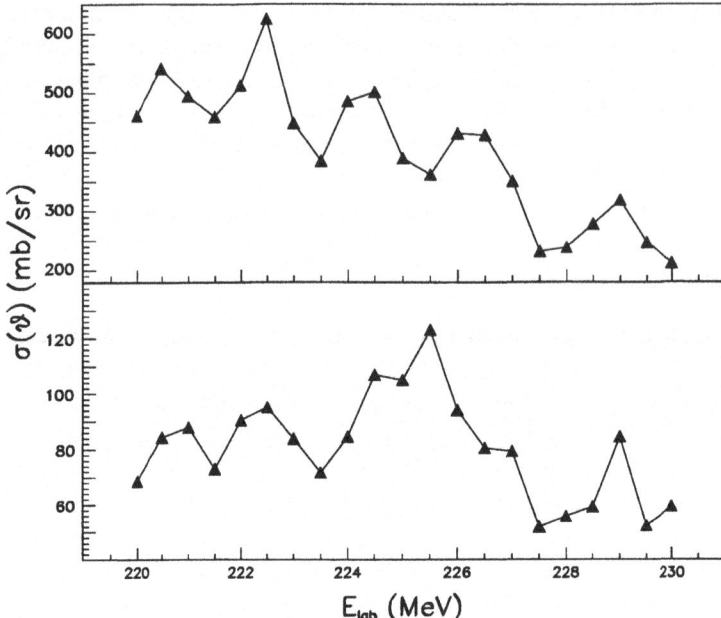

Figure 2. Excitation functions of the angle-summed ($76° \leq \theta_{cm} \leq 104°$) elastic (up) and inelastic (down) differential cross sections of ^{58}Ni+^{58}Ni.

Table 1. Analysis of the structure observed in ^{58}Ni+^{58}Ni elastic and inelastic data.

E_{lab} (MeV)	E_{cm} (MeV)	L (a)	L (b)	$L_{gr}(\hbar)$	$L_{OCM}(\hbar)$(c)
220.50	110.25	60± 2	58± 2	47	64
222.50	111.25	58± 2	58± 2	48	66
224.50	112.25	58± 2	58± 2	50	66
226.00	113.00	56± 2	56± 2	51	68
229.00	114.50	58± 2	58± 2	54	70

(a) from Legendre polynomial fits of the elastic data
(b) from statistical analysis of the inelastic data following ref. 9
(c) from OCM calculations following ref. 5

Taken at their face value, the above results could represent the first evidence of narrow (∼1 MeV) resonant states observed in nucleus-nucleus collisions leading to an intermediate system of A> 100 (^{116}Ba). We remind that the heaviest systems where heavy-ion resonances have been so far observed are ^{24}Mg+^{24}Mg and ^{28}Si+^{28}Si, leading to composite systems ^{48}Cr and ^{56}Ni, respectively. In favour of such an interpretation are the following facts:

- squared single Legendre polynomial forms $[P_L(\cos \theta)]^2$ with L∼60 render well the elastic scattering data. The results of this analysis are summarized in Table 1. We stress that the L-values obtained from the analysis of the data should be regarded solely as the order of the fitting polynomial; it would be premature to assign spin values on the basis of the above analysis;

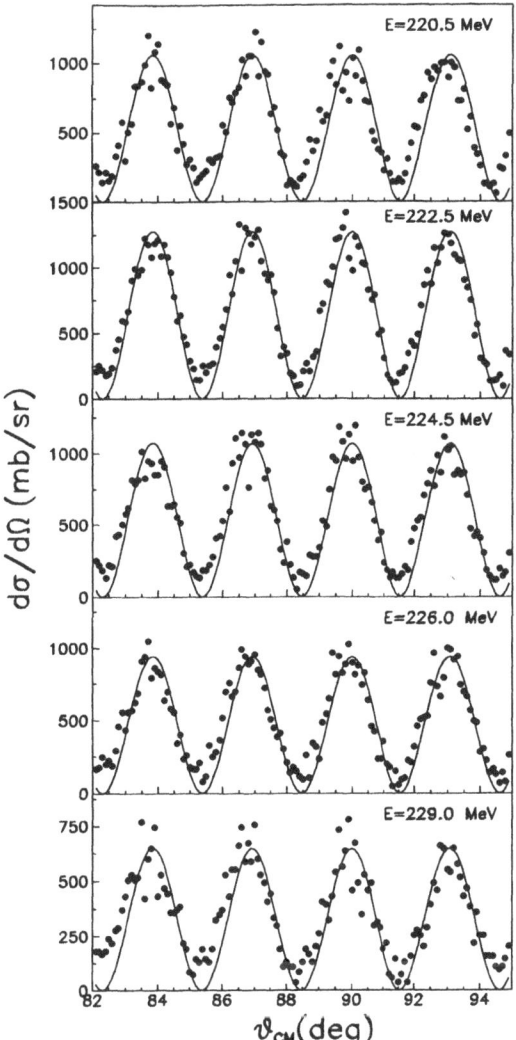

Figure 3. ^{58}Ni+^{58}Ni *elastic scattering angular distributions measured at the energies of the maxima in the excitation functions of Fig. 2. Lines are fits with a squared single Legendre polynomial.*

- good agreement in the periodicity between the elastic and inelastic angular distributions;
- marked disagreement (in the trend!) with the calculated values of the grazing angular momentum (see Table 1);
- finally, good agreement of the obtained L values with angular momentum predictions of the orbiting-cluster model (average predicted angular momentum L $\sim$ 64 $\hbar$).

There are, however, arguments against the resonance interpretation of the ^{58}Ni+^{58}Ni data:
- $[P_L(\cos\theta)]^2$ forms render the elastic scattering angular distributions also off the peaks in the elastic excitation function;
- other mechanisms could produce a similar behaviour.

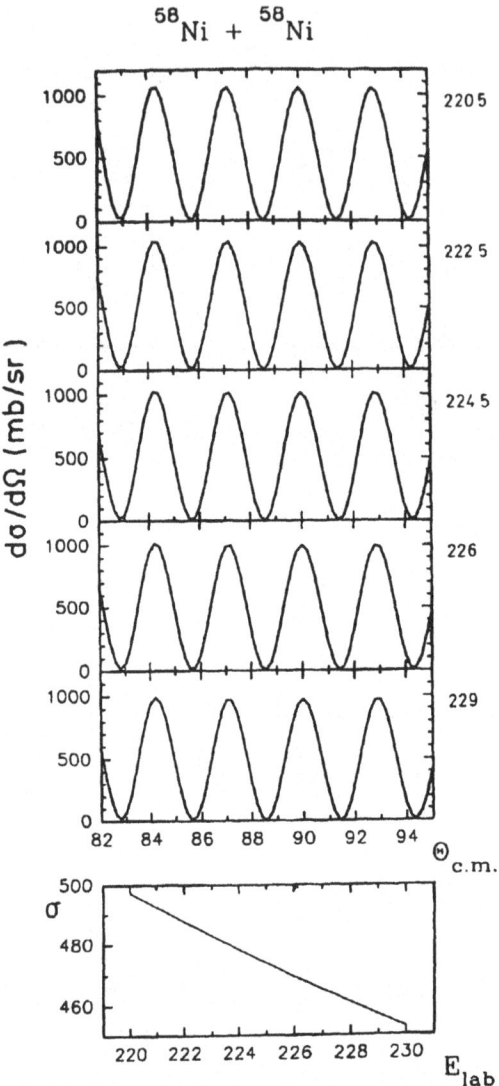

Figure 4. *Elastic (Coulomb) scattering of identical particles $^{58}Ni+^{58}Ni$. Upper part: calculated angular distributions at the energies of the peaks in Fig. 2; lower part: excitation function of the angle-integrated cross section for the above angular distributions.*

This latter point refers in particular to the fact that we deal with the scattering of identical spin-zero particles. The differential cross section for such a process, given by the general formula

$$\frac{d\sigma}{d\Omega} = \mid f(\theta) + f(\pi - \theta) \mid^2 , \qquad (4)$$

could exhibit interference phenomena similar to those observed, as shown in Fig. 4. As the curves in Fig. 4 were calculated using the Coulomb amplitude $f(\theta)$ only, it is not surprising that the angle-integrated cross section (bottom of Fig. 4) does not show any structure in the incident energy dependence.

2.2 ^{58}Ni+^{62}Ni

An obvious test of the identical particle scattering vs. resonance phenomena is the measurement of ^{58}Ni+^{62}Ni, a system less favoured for resonance observation by the OCM than ^{58}Ni+^{58}Ni, but composed of mutually different particles. So, the elastic scattering of ^{58}Ni on ^{62}Ni was also measured in the laboratory energy range from 220 to 230 MeV in steps of 0.5 MeV. A report of this measurement was presented at the Int. Conference on Atomic and Nuclear Clusters, Santorini, earlier this year[10].

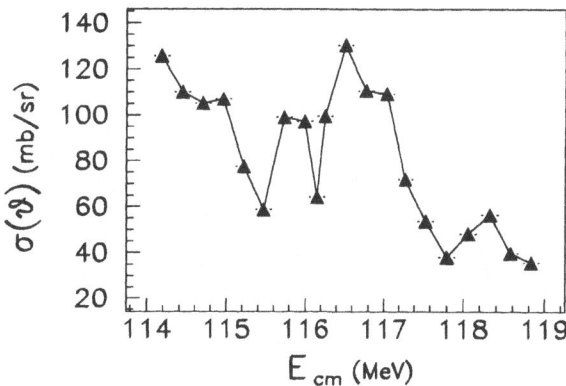

Figure 5. *Excitation function of the angle-summed ^{58}Ni+^{62}Ni elastic scattering differential cross sections for $72° \leq \theta_{cm} \leq 94°$.*

Fig. 5 shows the excitation function of the angle-summed ^{58}Ni+^{62}Ni elastic scattering differential cross section for $72° \leq \theta_{cm} \leq 94°$. Two structures are visible at incident energies of 224.0 and 225.5 MeV, respectively.

Unfortunately, due to the strong Coulomb contributions, elastic scattering angular distributions showed only weak oscillations, unsuitable for a Legendre polynomial analysis, so that the already mentioned statistical analysis, using the autocorrelation and correlation functions, had to be applied[9]. The oscillatory behaviour of the correlation function was indeed observed for energies corresponding to the structures at E=224.0 and 225.5 MeV; such behaviour is absent at energies far from the region of the observed structures (e.g. at E=229.5 MeV). These results are shown in Fig. 6. The periodicity of the measured angular distributions at the energies of the structures, $\Delta \theta_{per}$, deduced from the correlation analysis, is about 3°; we could, then, temptatively assign values of $L=\pi/\Delta \theta_{per}$ to the predominant partial wave(s) generating the angular distributions corresponding to the peaks in the excitation function.

The above results are summarized in Table 2. It is significant that the deduced values of L for the ^{58}Ni+^{62}Ni scattering are essentially identical to the values of L extracted from the ^{58}Ni+^{58}Ni scattering.

We stress that the values of L reported in columns 3 and 4 of Table 1 and column 4 of Table 2 should not be automatically understood as angular momentum quantum numbers. This would be the case only if the structures in the excitation functions in Figs. 2 and 5 would be definitely associated with virtual resonances in the potential or to resonant states in

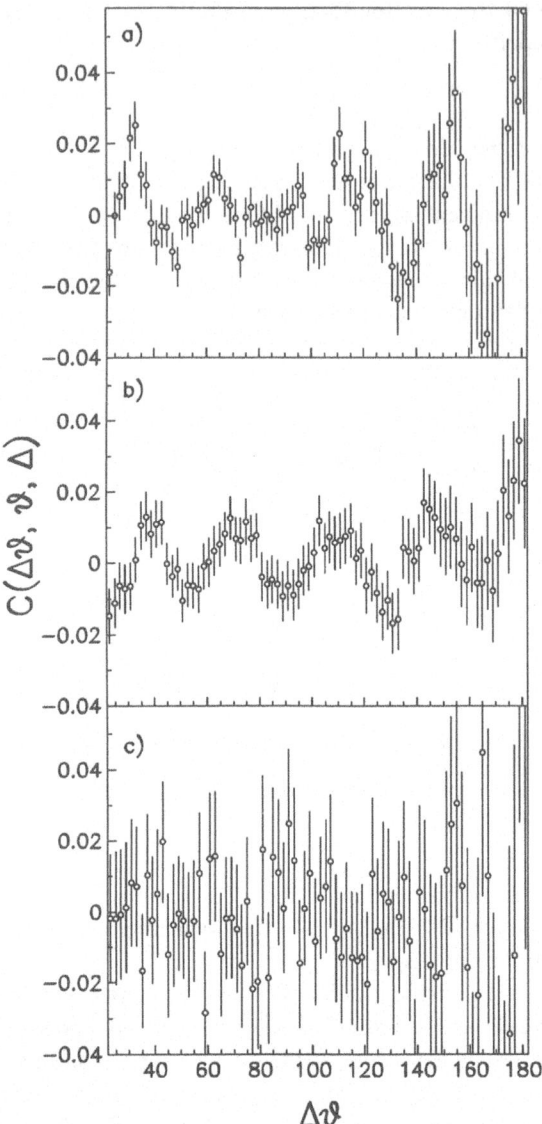

Figure 6. Correlation functions $C(\theta, \Delta\theta, \Delta)$ vs. the angular interval $\Delta\theta$ for (a) $E=224.0$, (b) $E=225.5$ and (c) $E=229.5$ MeV elastic scattering angular distributions of $^{58}Ni+^{62}Ni$.

the composite system. In ref. 5 we have discussed the latter possibility in the frame of the orbiting cluster model (OCM). We shall discuss the former in the next subsection.

2.3 Virtual states in a potential well

Trying to understand the phenomena described in the two preceding subsections, we now discuss the approach based on the search for pockets in the interaction potential of the two colliding nuclei. Such pockets would generate quasi bound states and virtual resonances which may be responsible for the structure observed in the excitation functions of the angle-

Table 2. Analysis of the structure observed in ^{58}Ni+^{62}Ni elastic scattering data.

E_{lab} (MeV)	E_{cm} (MeV)	$\Delta\theta_{per}$(deg)	L (a)	L_{OCM} ($\hbar$)(b)
224.0	115.7	2.95	61± 1	57
225.5	116.5	3.42	53± 1	59

(a) from statistical analysis of the elastic data following ref. 9
(b) from OCM calculations following ref. 5

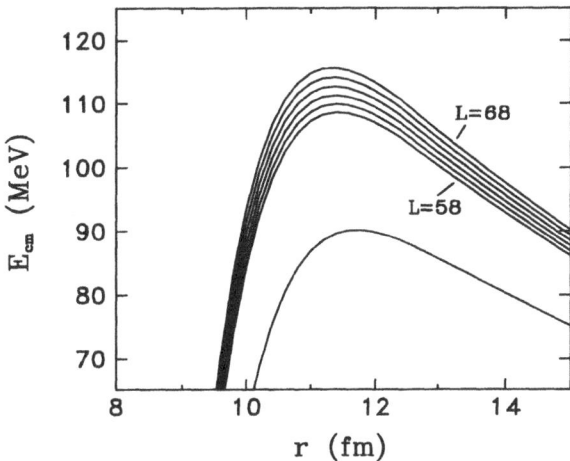

Figure 7. ^{58}Ni+^{58}Ni *interaction potentials for values of the angular momentum from L=58$\hbar$ to 64$\hbar$. The lowest curve is for L=0.*

summed data. Virtual states near the barrier were predicted by Könnecke et al.[11] to generate structure around E_{cm} =110 MeV in the ^{60}Ni+^{60}Ni elastic scattering. In the present work we have calculated the folding potential for ^{58}Ni+^{58}Ni (shown in Fig. 7) for values of angular momentum from L=58$\hbar$ to 68$\hbar$. The barriers of the calculated potential have heights around 110 MeV.

The potential for L=0, also shown in Fig. 7, was calculated by a M3Y plus knock-on-exchange, double-folding model with parameters μ_1 =0.40 fm, V_1 =-784.4 Mev fm, μ_2 =0.25 fm, V_2 =1578.75 MeV fm for M3Y and with a pseudopotential of δ-type of strength V_δ =-81 fm^3 [12]. In this calculation the density of the ^{58}Ni nuclei was assumed to be of the Fermi type, $\rho_0/(1+\exp[(r-R_0)/a])$, with ρ_0 =0.17 fm^{-3}, R_0 =4.06 fm and a=0.60 fm. Using the real potential of Fig. 7, we have carried out an optical model calculation of the ^{58}Ni+^{58}Ni elastic scattering cross section. In this calculation, the imaginary part of the optical potential had a Woods-Saxon form with a strength of W=-(8.0+0.023E_{cm}) MeV and geometrical parameters R=11 fm and a=4.5 fm. Partial waves up to L=300 were summed.

The resulting excitation function of the elastic scattering at θ_{cm} =90° shows structure (Fig. 8, upper part), although the calculated oscillations are somewhat wider than those observed in the experiment. These structures are not caused by a single partial wave as it was found for the gross structures in the corresponding 90° excitation functions for lighter systems. On the other hand, the angular distributions calculated by using the same optical potential show periodical behaviour; an example (E_{cm} =110 MeV) is given in Fig.8 (lower

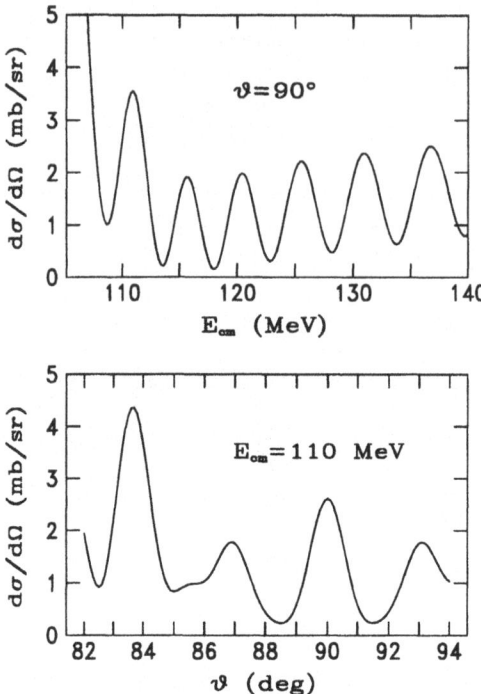

Figure 8. ^{58}Ni+^{58}Ni calculations with the potential of Fig. 7. Upper part: the elastic scattering excitation function at $\theta_{cm}=90°$; lower part: angular distribution at $E_{cm}=110$ MeV.

part). The calculated shape of the angular distribution is not as regular as the shapes observed in the experiment, but the average periodicity (3°- 4° in the c.m.) is similar.

3. Summary

Using the kinematic coincidence method we have measured the elastic and inelastic scattering of ^{58}Ni+^{58}Ni and the elastic scattering of ^{58}Ni+^{62}Ni around $\theta_{cm}=90°$ in steps of 0.5 MeV (lab.) from $E_{lab}=220$ MeV to $E_{lab}=230$ MeV. The measured excitation functions of the angle-summed differential cross sections show structure ~ 1.5 MeV (lab) wide and correlated for elastic and inelastic scattering data for the ^{58}Ni+^{58}Ni system. The angular distributions for the ^{58}Ni+^{58}Ni collision are well fitted by squared single Legendre polynomial forms $[P_L(\cos\theta)]^2$ with $L\sim 60$. The angular correlation analysis of the ^{58}Ni+^{62}Ni elastic scattering yields periodicities leading also to values of $L\sim 60$.

Speculations advanced to understand the origin of this behaviour include molecular-window type models (overall predictions of resonances in the investigated energy and angular momentum range), folding potential calculations (qualitative reproduction of the structure in the excitation functions and of the periodicity of angular distributions) and, for ^{58}Ni+^{58}Ni, identical particle Coulomb scattering (reproduction of the periodicity of the angular distributions). The extent of the data does not yet allow to firmly decide in favour of any of the advanced interpretations. Further investigations, in particular additional measurements of the ^{58}Ni+^{60}Ni and ^{60}Ni+^{60}Ni systems, as well as coupled channel calculations of the cross sections including the excited states of the Ni nuclei, have to be carried out for a deeper understanding of the measured data.

References

[1] E. Almqvist, D. A. Bromley and J.A. Kuehner, *Phys. Rev. Lett.* **4**, 515 (1960); D. A. Bromley, J. A. Kuehner and E. Almqvist, *Phys. Rev. Lett.* **4**, 365 (1960)
[2] For a review of the subject see, e.g., U. Abbondanno and N. Cindro, *Int. Jour. Mod. Phys. E* **2**, 1 (1993)
[3] W. Scheid, W. Greiner and R. Lemmer, *Phys. Rev. Lett.* **25**, 176 (1970)
[4] N. Cindro, *J. Phys. G* **4**, L23 (1978); D. Počanić and N. Cindro, *J. Phys. G* **5**, L25 (1979); N. Cindro and D. Počanić, *J. Phys. G* **5**, 359 (1980)
[5] N. Cindro and M. Božin, *Ann. of Phys.* **192**, 307 (1989)
[6] K. Šparavec, *Private communication*, Rudjer Bošković Institute (1992)
[7] U. Abbondanno et al., *J. Phys. G* **16**, 1517 (1990)
[8] N. Cindro et al., in: *Nuclear Physics of Our Times, Proc. Intl. Conf. on Nucl. Phys.*, Sanibel Island, FLA, USA, Nov. 17 - 21, 1992, edited by A. Ramayya, p. 423 (World Scientific, Singapore, 1993)
[9] G. Pappalardo, *Phys. Lett.* **13**, 320 (1964); E. Gadioli, I. Iori, M. Mangialaio and G. Pappalardo, *N. Cim.* **38**, 1105 (1965)
[10] L. Vannucci et al., *Proc. Intl. Conf. on Atomic and Nuclear Clusters*, Santorini, Greece, Jun 28 - Jul. 2, 1993, to be published in *Z. Phys. A*
[11] R. Könnecke, W. Greiner and W. Scheid, in: *Nuclear Molecular Phenomena*, p. 109, edited by N. Cindro, (North Holland, Amsterdam, 1978)
[12] G. R. Satchler and W. G. Love, *Phys. Rep.* **55**, 183 (1979)

QUASIMOLECULAR STATES - A PARTICULAR CASE OF THE NEW-CLASS RESONANT STATES

Cornelia Grama, N. Grama and I. Zamfirescu

Institute of Atomic Physics, PO Box MG-6, RO-76900 Bucharest, Romania

In [1], [2] new-class S-matrix resonant state poles have been identified by the global analysis of the pole function $k = k_l(g)$ in the case of a short range potential with Coulomb and centrifugal barrier. The general properties of these poles and of their corresponding wave functions have been studied. These new-class resonant state poles have properties that differ from the properties of the usual resonant state poles. Being stable with respect to the change of the potential strength the new-class resonant state poles in the neighborhood of the so called "stable points" and the corresponding resonant states are of exceptional interest. Among the stable points the closest to the real k-axis is the most important because it has the largest influence on the cross section. An approximate expression for this stable point $k_z^{(1)}(c,l)$ was obtained that is valid for a large range of values of the Coulomb parameter $c = Z_1 Z_2 e^2 M/\hbar^2$ and of the orbital angular momentum l, provided that c is large. This approximation can be used in order to obtain the energy and width of a new-class resonant state corresponding to a pole in the neighborhood of the stable point $k_z^{(1)}(c,l)$ (for the sake of simplicity the notation c and k are used for cR and kR, respectively, where R is the radius of the potential):

$$E \approx \frac{\hbar^2}{2MR^2}\left(k_B^2 - \alpha_1[k_B^2 + l(l+1)]^{2/3} - \frac{\alpha_1^2}{2}\frac{[k_B^2 + l(l+1)]^{4/3}}{k_B^2} - \frac{\alpha_1^2}{2}\frac{k_B^2}{[k_B^2 + l(l+1)]^{2/3}}\right) \quad (1)$$

$$\Gamma \approx \frac{\hbar^2}{2MR^2}\left(2\alpha_2[k_B^2 + l(l+1)]^{2/3} - \alpha_1\alpha_2\frac{[k_B^2 + l(l+1)]^{4/3}}{k_B^2} - \alpha_1\alpha_2\frac{k_B^2}{[k_B^2 + l(l+1)]^{2/3}}\right) \quad (2)$$

where $k_B = [2c + l(l+1)]^{1/2}$, $\alpha = \left(\frac{9\pi}{8}\right)^{2/3} exp(i\pi/3)$, $\alpha_1 = Re(\alpha)$ and $\alpha_2 = Im(\alpha)$. From Eq.(1) one can see that the new-class resonant states corresponding to a pole in the neighborhood of $k_z^{(1)}(c,l)$ with given c form a rotational band with energies below the barrier top represented by the first term in (1). The other terms give a correction to the linear dependence of the energy on $l(l+1)$ that is important for light nuclear systems (small c) at low values of the angular orbital momentum l. It results from (2) that the widths of the resonant states corresponding to a pole in the neighbourhood of the stable point $k_z^{(1)}(c,l)$ increase with increasing c and l. For a given system the width Γ has a slower variation with respect to l as l increases.

A comparison of the experimental excitation energies of a large number of quasimolecular states reported in literature and the calculated excitation energies (1) with the same prescription for the radius $R = r_0(A_1^{1/3} + A_2^{1/3})$ with $r_0 = 1.3$ fm shows a good agreement for all the l values both below and above the Coulomb barrier. Our approach of the quasimolecular states as new-class resonant states provides thus an unitary treatment of the quasimolecular states both below and above the Coulomb barrier. A comparison of the calculated and experimental widths of the quasimolecular states is difficult to be done due to the process of spreading of the single particle excitation on a large energy region. It would be necessary to determine and analyze all the fragments of the single particle resonance spread on an interval of several MeV. There are few reported data concerning the elastic reduced widths of the fragments of the quasimolecular states. For the available data there is a good agreement of the calculated widths (2) and the "experimental" single particle widths recomposed from the components spread over a large energy region.

Eq.(2) shows that for the light heavy ion systems there is a narrow l window of observability at small values of the orbital angular momentum, due to the rapid increase of the width as l increases. For heavier systems the widths of the resonant states increase slowly as l increases and consequently the states can be observed even at very large values of the orbital momentum.

The wave function localization outside the well region for the new-class resonant states corresponding to the poles in the neighborhood of the stable points confers them stability against the dissolution into the complicated states of the continuum. Thus the doorway character of the quasimolecular states results naturally.

The best fit parameters of the optical model potential used to explain the heavy ion scattering [3] satisfy the condition for the corresponding potential strength to be inside the absorption window for which the new-class resonant states could exist.

References

[1] C. Grama, N. Grama and I. Zamfirescu, *Global method for all S-matrix poles identification. New classes of poles and resonant states*, Ann. Phys. (N.Y.) **218**, 346 (1992)

[2] C. Grama, N. Grama and I. Zamfirescu, *New class of resonant states for potentials with Coulomb barrier. Quasimolecular states*, Ann. Phys.(N.Y.), in press.

[3] R. H. Siemssen, *Surface transparency and resonance phenomena and the heavy ion optical model* in: *Nuclear Molecular Phenomena. Proceedings of the International Conference on Resonances in Heavy Ion Reactions, Hvar, Yugoslavia 1977*, edited by N. Cindro (North-Holland, Amsterdam, New York, Oxford, 1978)

CLASSICAL CLUSTER FORMATION, APPLICATION TO NUCLEAR MULTIFRAGMENTATION

J. B. Garcia, C. Cerruti, S. Gazaix

Centre de Recherche Nucleaire, 67000 Strasbourg, FRANCE

Introduction

Our aim consists in reproducing some nuclear multifragmentation features with a classical, statistical and *static* model. In heavy ions reactions, above 60 MeV/u, a large production of Intermediate Mass Fragments (IMF) has been experimentally observed. It is believed that after the collision the compound nucleus undergoes a compression phase and then a thermalized expansion phase in which clusters are thought to be formed. After shortly introducing the Classical Cluster Formation procedure, some general behaviour and a comparison with experimental results will be shown.

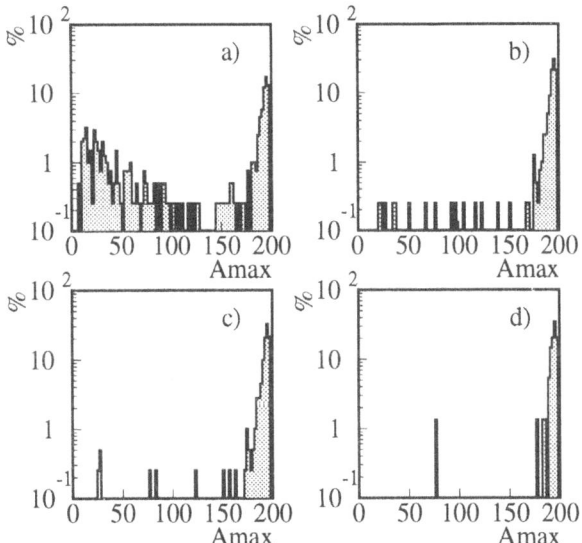

Figure 1. *Simulations of ^{197}Au for 400 events in the ground state, percentage of the mass of the biggest cluster. 1a) only 1-1 bonds, 1b) 1-1+1-2 bonds, 1c) 1-1...1-3 bonds, 1d) 1-1...1-6 bonds.*

General Process

Our approach is based on a classical binding condition (including phase-space dependence and nucleon-nucleon interaction through Yukawa potential) that enables one to decide whether

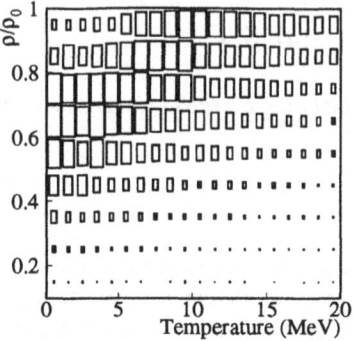

Figure 2. *Average IMF multiplicity as a function of T and ρ/ρ_0.*

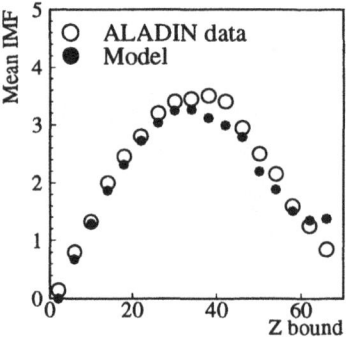

Figure 3. *Average IMF multiplicity.*

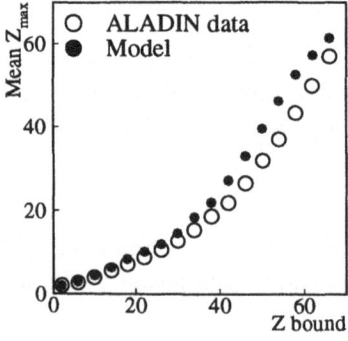

Figure 4. *Average charge of the biggest cluster.*

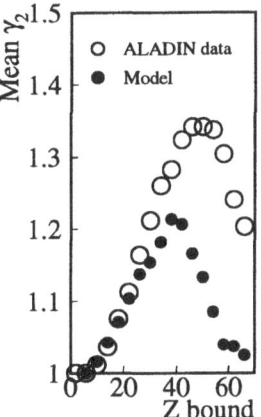

Figure 5. Average γ_2

two nucleons are bound or not (1-1 bond). This criterion can be easily extended to treat in-medium effects (1-2,1-3,...,1-n bonds) and even cluster-cluster bonds. Applying such a process to a set of nucleons (drawn in a Fermi distribution), one is left with a set of clusters. In our calculations a nucleon evaporation-like process has also been included.

General Trends

Since our three internal parameters depend on the bulk properties, one has to check whether they are recovered or not. The binding energy is correctly reproduced. The distribution of the mass of the biggest clusters exhibits a sharp peak around the mass of the system, when one includes *in-medium* effects up to 1-3 bonds (fig. 1c). Now, if one varies the temperature as well as the nuclear density (fig. 2), one can distinguish three domains which correspond to evaporation, multifragmentation and vaporization.

Comparison with Experiments

In order to perform an experimental comparison, we have chosen the ALADIN data [2]. The free parameters needed (temperature, size of the projectile ...) have been provided by a BUU calculation. We have also included an experimental filter. Most of observables are function of Z_{bound}, which is the sum of detected charges whithin an event. The overall agreement is rather good (figs. 3 and 4), except for the biggest values of Z_{bound}. This slight disagreement in fig. 3 and fig. 4 causes a bigger discrepancy in fig. 5. Note that γ_2 is a measure of the fluctuations of the charge distribution.

References

[1] J.B. Garcia, C. Cerruti, *to be published*
[2] P. Kreutz et al., *GSI-Preprint* **92-33**, March 1992

AZIMUTHAL ANISOTROPIES OF PIONS IN HEAVY-ION COLLISIONS: A NEW CHANCE OF PROBING THE HOT AND DENSE REACTION PHASE?

St. A. Bass[1,2], C. Hartnack[2,3], H. Stöcker[1] and W. Greiner[1]

[1] *Institut für Theoretische Physik der Universität, PO Box 11 19 32, 60054 Frankfurt, Germany*
[2] *GSI Darmstadt, PO Box 11 05 52, 64220 Darmstadt, Germany*
[3] *Laboratoire de Physique Nucléaire, Nantes, France*

New experimental facilities at GSI and Berkeley (LBL) enable for the first time the experimental investigation of correlations of secondary particles like pions and kaons with the outgoing baryon resonance matter.

The azimuthal angular emission pattern of pions in Au+Au heavy ion collisions at 1 GeV/nucleon is analysed in the framework of the IQMD model, an upgrade to the Quantum Molecular Dynamics model [1, 2], incorporating explicit isospin treatment and pions. In the IQMD model pion production and scattering are treated via the Δ-resonance. The fate of the pion is governed by two distinct processes:

1. pion absorption: $\pi N N \rightarrow \Delta N \rightarrow N N$
2. pion scattering: $\pi N \rightarrow \Delta \rightarrow \pi N$.

The delta forming the intermediate state in the pion scattering is called a *soft* delta in order to distinguish it from deltas produced in inelastic nucleon–nucleon collisions which are called *hard* deltas. The masses of the deltas produced in inelastic NN collisions are choosen according to a Breit-Wigner distribution.

At projectile- and target-rapidities we observe an anticorrelation in the in-plane transverse momentum between pions and protons (figure 1a)), a phenomenon known and investigated since the first measurements of asymmetric systems by the DIOGENE collaboration [3, 4, 5]. This anticorrelation only takes place in semiperipheral and peripheral collisions. Our investigation shows that it is due to multiple π N scattering in large chunks of spectator matter [6]. This anticorrelation can be seen as a *reflection* of pions by the spectator matter. The freeze out of these pions occurs in the intermediate reaction phase – they are useful for investigating the behaviour of pions in nuclear matter but are no probe for the hot and dense early reaction phase.

At CM-rapidity, however, we find that high p_t pions are being preferentially emitted perpendicularly to the event-plane (figure 1b)). Again, this effect is only observed in semiperipheral and peripheral collisions [7]. This prediction has been confirmed by the TAPS and KaoS collaborations at GSI [8, 9]. Our analysis shows that this anisotropy is dominated by the reaction $\Delta N \rightarrow N N$, the final step in the pion absorption process. Pions emitted in the event-plane have a higher probability of being absorbed than those being emitted perpendicularly to the event-plane. The excess of pions emitted perpendicularly to the event-plane stems from pions which either have been emitted from a *hard* Δ or which have scattered less than three times (figure 1c)). We conclude that these pions might therefore be more sensitive to processes taking place in the early hot and dense reaction phase than isotropically emitted low p_t pions which freeze out in the late reaction stages.

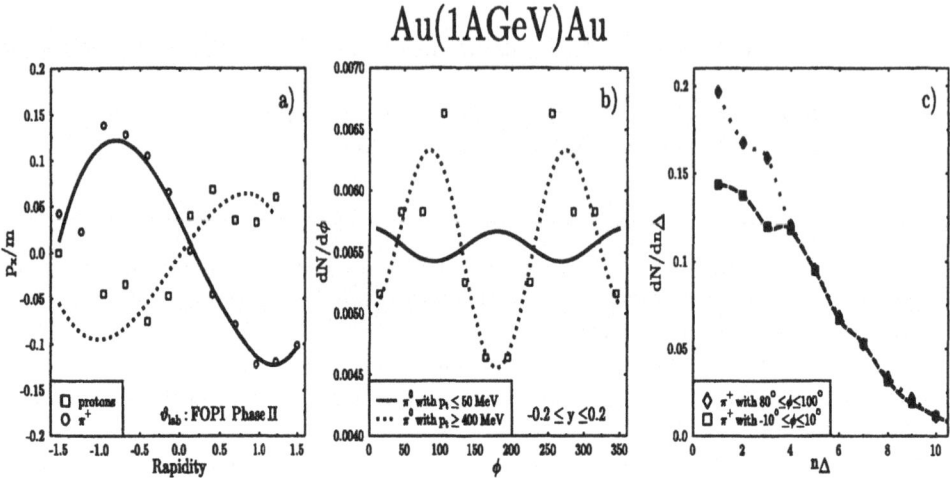

Figure 1. Frame a) shows the anticorrelation of $\langle p_x \rangle / m$ for π^+ and p with an angular cut according to the FOPI Phase II Spectrometer; frame b) shows the azimuthal angular distribution of π^0 at midrapidity for low and high p_t and frame c) depicts the number of Δ-generations, a π goes through before it freezes out, for π emitted in the reaction plane and perpendicularly to it.

References

[1] C. Hartnack, L. Zhuxia, L. Neise, G. Peilert, A. Rosenhauer, H. Sorge, J. Aichelin, H. Stöcker, W. Greiner, *Nucl. Phys. A* **495**, 303 (1989)
[2] J. Aichelin, *Phys. Reports* **202**, 233 (1991)
[3] J. Gosset and the DIOGENE collaboration, *Phys. Rev. Lett.* **62**, 1251 (1989)
[4] Ch. Hartnack, H. Stöcker, W. Greiner in: *Proceedings of the International Workshop on Gross Properties of Nuclei and Nuclear Excitations XVI*, edited by H. Feldmeier, (Hirschegg (Austria), 1988)
[5] B. A. Li, W. Bauer, G. F. Bertsch, *Phys. Rev. C* **44**, 2095 (1991)
[6] S. A. Bass, C. Hartnack, R. Mattiello, H. Stöcker, W. Greiner, *Phys. Lett. B* **302**, 381 (1993)
[7] S. A. Bass, C. Hartnack, H. Stöcker, W. Greiner, *Phys. Rev. Lett.* **71**, 1144 (1993)
[8] D. Brill and the KaoS collaboration, *Phys. Rev. Lett.* **71**, 336 (1993)
[9] L. Venema and the TAPS collaboration, *Phys. Rev. Lett.* **71**, 835 (1993)

THE RATIO OF THE Φ AND J/Ψ MESON YIELDS AS A POSSIBLE SIGNATURE OF THE QUARK-GLUON PLASMA FORMATION

I. Lovas

Department for Theoretical Physics, Lajos Kossuth University, H-4010 Debrecen, Hungary

It is very probable that in hadronic matter at high enough density and/or temperature a phase transition takes place which leads to the formation of a plasma consisting of deconfined quarks, antiquarks and gluons. We believe that the QGP phase is present in the central region of some neutron stars and it existed in the early Universe. We do hope that it may occur in head-on collisions of high energy heavy ions, too. If the hadronic matter at high energies is transparent then the QGP will contain almost as many antiquarks as quarks, in other words the baryon number density will be low. If the hadronic matter stops during the collision a baryon rich plasma will be formed. According to the analysis of the CERN experiments with ^{32}S ions of 200 GeV/n energy at least the half of the hadronic matter was stopped, consequently one may assume that the QGP produced in heavy ion collision will be baryon rich at least in this energy region. In future experiments to be performed in CERN and in BNL by the help of colliding beams one hopes to reach the critical values of physical parameters of the QGP formation. These estimated critical values are the following: energy density: $\mathcal{E} \approx 1\text{--}3$ GeV/fm^3, temperature: $T \approx 150\text{--}200$ MeV, baryon density: $\rho \approx (5\text{--}10)\,\rho_{nucl}$. During the last decade a great number of suggestions has been formulated for the observable signature of the QGP formation. The most extensively studied possibilities are the emission of lepton pairs, the strangeness enhancement, the production of strange antibaryons or exotic particles, the Φ enhancement, the J/Ψ suppression,etc. Here we will concentrate on the combination of the last two possibilities.

The vector mesons Φ and J/Ψ are especially interesting since they are bound states of $s\bar{s}$ and $c\bar{c}$ pairs, respectively, while the nuclei of the colliding heavy ions carry such quark pairs only as components of the "sea". A few years ago we have proposed the measurement of the ratio of the Φ and J/Ψ meson production as a signature of the quark-gluon plasma formation in high energy heavy-ion collisions [1]. In the last years the production and the absorbtion of the Φ and J/Ψ mesons was analysed in a number of papers [2, 3] and a great deal of experimental information has been accumulated [5, 4] which seems to corroborate the relevance of our suggestion. On the other hand by the advent and the fast development of the RICH (Ring Image Cherenkov) [8] detector an extremely powerful experimental tool has been found to measure with high accuracy the Φ and J/Ψ meson yields by the same detector. This provides an excellent chance to measure the ratio of the Φ and J/Ψ production. In the present paper we analyse the effect of the assumed QGP formation and that of the hadronic absorption on the azimuthal asymmetry of the ratio of the Φ and J/Ψ meson yields. The idea to measure the ratio of the Φ and J/Ψ meson yields in order to detect the QGP formation was based upon two independent observations made on one hand by Shor [9] and on the other hand by Satz and Matsui [10].

(i) In the hot QGP $s\bar{s}$ quark pairs can be produced by quark-quark, quark-gluon, and gluon-gluon collisions. In the rehadronization of the plasma neutral Φ mesons with hidden flavor may be formed by the combination of strange quark and antiquark pairs. Mesons with hidden strangeness are also produced in the hadronic processes, however, the production probability is rather low because of the Okubo–Zweig–Iizuki rule. It was pointed out some years ago by Shor that the Φ meson formation is not inhibited by the OZI rule during the rehadronization process since the Φ mesons are formed from already existing "prefabricated" constituents [9]. By these considerations Shor has arrived at the conclusion that the QGP formation leads to the enhancement of the Φ meson production.

(ii) Analysing the J/Ψ meson production in relativistic heavy-ion collisions Matsui and Satz discovered that the observed number of the J/Ψ mesons must decrease if QGP formation takes place [10]. The J/Ψ mesons in the plasma will be destroyed, since the $c\bar{c}$ pairs cannot remain in stable bound states because of the Debye–screening of the gluonic interaction induced by the plasma. Contrary to the case of the $s\bar{s}$ pairs, the production of $c\bar{c}$ pairs by quark-quark, quark-gluon and gluon-gluon collisions at typical temperatures of the QGP ($T \sim 0.2$ GeV) is negligible because of the large mass of the charmed quarks ($M \sim 1.5$ GeV). Consequently, the formation of J/Ψ mesons in the course of the rehadronization is negligible.

At this point we must note that according to the results of the CERN experiments both predictions seem to be correct. A real Φ enhancement and a real J/Ψ suppression has been observed. These observations, however, may not be interpreted as a proof for the QGP formation since the results can be understood also by taking into account the effect of the hot and dense hadronic medium produced in the collision. Further experiments with heavier projectiles and/or colliding beams are needed to obtain unique signature of the QGP formation.

According to the previously outlined arguments the effect of the QGP formation on the yields of the Φ and J/Ψ mesons is just the opposite, consequently, the ratio of the two yields must be even more sensitive to the formation of the QGP. Considering the feasibility of this kind of experiment it must be pointed out that the RICH detectors developed recently allow for the observation of Φ and J/Ψ mesons in the same detector. The neutral vector mesons decay with a certain branching ratio into an electron-positron pair. By measuring the momentum of the electron-positron pair the rest mass of the vector meson can be determined. In the CERES spectrometer developed in the last few years the momentum of the electron and the positron can be measured rather accurately also in an intensive hadronic background [7]. This type of spectrometer consists of two RICH detectors and a strong magnetic field beween them. In the RICH detector the charged particles produce Cherenkov radiation. The intersection of the Cherenkov cone and the surface of the radiator define a ring. By measuring the Cherenkov photons distributed along this ring the centre of the ring can be determined. The momenta of the particles change in a well defined manner in the strong magnetic field between the two RICH detectors. The change of the momentum of the electron and the positron can be obtained with very high precision by measuring the relative shift of the centres of the Cherenkov rings. In this way the energy spectrum of the $e^+ - e^-$ pairs can be measured and the maxima of the spectrum can be identified with the vector mesons.

In this paper the collision of two identical nuclei is considered in their center of mass system. We focus our attention to a scenario with baryon rich QGP. For the sake of simplicity, we assume that the hadronic matter in the reaction zone is completely stopped forming a fireball and the fraction of the volume of the fireball with density exceeding some critical value ρ_c is converted into QGP. The fireball is contained in a volume defined by the overlap of two identical, Lorentz-contracted spheres with shifted centers. The distance of their centers is the impact parameter b. The volume of the fireball and also the number of its nucleons is determined by the impact parameter b and the nuclear radius R. The impact parameter b can be determined experimentally by measuring the multiplicity of the charged fragments. This means that the volume of the fireball can be determined in each individual collision event. Furthermore we assume that the plasma is confined in a region, defined by $\rho(\vec{r}) > \rho_c$ where $\rho(\vec{r})$ is the density of the fireball at position $\vec{r}$. The plasma fraction of the fireball is controlled by the critical density ρ_c. The number of the J/Ψ mesons observed in the reaction plane of the heavy-ion collision perpendicular to the relative momentum, i.e. in the direction x, is denoted by N_Ψ^x. Similarly, the number of J/Ψ mesons observed perpendicularly out of the reaction plane, i.e. in the direction y, is denoted by N_Ψ^y. The number of J/Ψ mesons observed in the x direction can be expressed as follows:

$$N_\Psi^x = \int d^3r \rho_\Psi(\vec{r}) \exp(-\mu_\Psi l^x(\vec{r})) \Theta^x(\vec{r}). \tag{1}$$

Here the number of the J/Ψ mesons produced in a volume element at the point $\vec{r}$ is denoted by $\rho_\Psi(\vec{r}) d^3r$, the absorption coefficient is given by μ_Ψ and the distance of the point $\vec{r}$ from the surface of the fireball measured in the x direction is denoted by $l^x(\vec{r})$. The total absorption of the J/Ψ mesons by the plasma is expressed by means of the step function $\Theta^x(\vec{r})$, which is equal to 1 in the points of the fireball except for the volume of the plasma and for the shadow of the plasma, where its value is zero. The number of the Φ mesons observed in the x direction can be given as follows:

$$\begin{aligned} N_\Phi^x &= \int d^3r \rho_\Phi(\vec{r}) \exp(-\mu_\Phi l^x(\vec{r})) \Theta^x(\vec{r}) \\ &+ \int d^3r \rho_\Phi(\vec{r}) \exp(-\mu_\Phi l^x(\vec{r})) \Theta_p(\vec{r}), \end{aligned} \tag{2}$$

where the step function $\Theta_p(\vec{r})$ is zero everywhere except for the volume of the plasma where its value is 1. Here the first term is the analogue of expression (1), while the second term accounts for the mesons produced in the rehadronization of the plasma. The ratio of the number of Φ mesons to that of the J/Ψ mesons can be expressed in the following form:

$$N_\Phi^x/N_\Psi^x = P(\text{hadron}) A^x(\text{hadron}) + P(\text{plasma}) A^x(\text{plasma}), \tag{3}$$

where the production ratios P and the absorption ratios A^x are defined by

$$\begin{aligned} P(\text{hadron}) &= \rho_\Phi(\text{hadron})/\rho_\Psi(\text{hadron}), \\ P(\text{plasma}) &= \rho_\Phi(\text{plasma})/\rho_\Psi(\text{hadron}), \end{aligned} \tag{4}$$

and

$$A^x(\text{hadron}) = \frac{\int d^3r \exp(-\mu_\Phi l^x(\vec{r})) \Theta^x(\vec{r})}{\int d^3r \exp(-\mu_\Psi l^x(\vec{r})) \Theta^x(\vec{r})},$$

$$A^x(\text{plasma}) = \frac{\int d^3r \exp\left(-\mu_\Phi l^x(\vec{r})\right)\Theta_p(\vec{r})}{\int d^3r \exp\left(-\mu_\Psi l^x(\vec{r})\right)\Theta^x(\vec{r})}. \tag{5}$$

Here the production density $\rho_\Psi(\vec{r})$ is approximated by its average value ρ_Ψ (hadron) taken over the volume of the hadronic fraction of the fireball. Similarly, the production density $\rho_\Phi(\vec{r})$ is approximated by its average values ρ_Φ(hadron) and ρ_Φ(plasma) taken over the volumes of the hadronic and the plasma fraction of the fireball, respectively. Similar expression is valid for the ratio of mesons observed in the direction y. By measuring the azimuthal asymmetry of the charged reaction products the reaction plane can be fixed in every collision event and thus the directions x and y can be specified. Let us assume that the colliding nuclei are Lorentz-contracted to the thickness $2r_0$ with the proton radius $r_0 = 0.9$ fm (high-energy collision) and that the initial kinetic energy of the participants is equally distributed among them in the fireball. Then QGP is present where the density $\rho(\vec{r})$ in the fireball exceeds the critical value ρ_c. The density is calculated in a conservative manner by summing the densities of the overlapping Lorentz-contracted nuclei. The critical density ρ_c is a free parameter of order 1 fm^{-3}, regulating the plasma fraction $F(b,\rho_c) = V_{\text{plasma}}(b,\rho_c)/V_{\text{fireball}}(b)$ of the volume of the fireball. Taking the total cross sections $\sigma_{\Psi N} = 6.5$ mb and $\sigma_{\Phi N} = 8$ mb [6, 12], we obtain the following absorption coefficients: $\mu_\Psi = 0.20$ fm^{-1} and $\mu_\Phi = 0.24$ fm^{-1}. Here the value of the total cross section $\sigma_{\Psi N}$ corresponds to the weighted average of the values obtained by a recent analysis of hadron-, photon-, and nucleus-nucleus data [6]. At present there is no reliable method to estimate the ratio $w = \rho_\Phi(\text{plasma})/\rho_\Phi(\text{hadron})$, therefore its value must be considered as a free parameter. Calculations for $w = 1.0$ and $w = 10.0$ have been performed. The first case corresponds to the situation when Φ enhancement due to the plasma rehadronization does not occur at all. The critical density has been varied to consider cases with $F(b=0,\rho_c) = 0.00, 0.15, 0.28, 0.35, 0.50$ for $R = 7$ fm.

In Figs. 1 and 2 the asymmetry in the ratio of Φ to J/Ψ production

$$A_{xy} = (N_\Phi^x/N_\Psi^x) / (N_\Phi^y/N_\Psi^y) \tag{6}$$

is shown. The direction x is perpendicular to the initial momenta of the colliding nuclei in the reaction plane, the direction y is perpendicular both to the initial momenta and the reaction plane. In the case of a pure hadronic scenario of the heavy-ion collision, the asymmetry A_{xy} increases slowly with increasing impact parameter (decreasing multiplicity). For the realistic value $\sigma_{\Psi N} = 6.5$ mb it remains small even for large impact parameters. (Calculations for a pure hadronic scenario were performed by setting $\rho_c \to \infty$). In the case of the presence of the QGP in a certain fraction of the fireball volume, the asymmetry depends strongly on the Φ enhancement w due to plasma rehadronization. For the pessimistic case of no Φ enhancement, i.e. $w = 1.0$, no additional asymmetry occurs as compared to a purely hadronic scenario for cases with plasma fraction $F(b=0,\rho_c) \leq 0.50$ Assuming an enhancement $w = 10.0$ in Φ production due to plasma rehadronization alters the picture significantly for the same plasma fraction. The asymmetry increases with increasing impact parameter more remarkably, but it drops suddenly at the impact parameter for which the QGP fraction of the fireball vanishes.

Summarizing, we conclude that the azimuthal asymmetry in the ratio of Φ to J/Ψ production can signal the presence of QGP even if it is present only in a small fraction of the reaction zone. It would be interesting to study this question in a realistic heavy-ion reaction

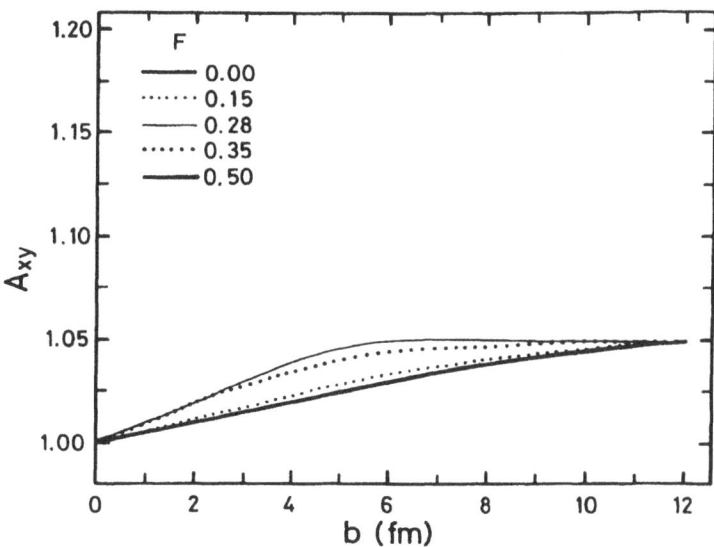

Figure 1. The asymmetry $\mathcal{A}_{xy} = (N_\Phi^x/N_\Psi^x)/(N_\Phi^y/N_\Psi^y)$ as the function of the impact parameter b. The enhancement factor w for the Φ meson production in QGP compared to hadronic matter has the value $w = 1.0$. The plasma fraction F is defined as the ratio of the volumes: $F = V_{plasma}/V_{fireball}$.

model without the assumption of full stopping and incorporating hydrodynamic expansion. According to our simple scenario, the QGP is contained in a connected region of the fireball. A similar analysis can be performed for the case with disjunct "bubbles" of QGP. This kind of picture would naturally arise if colour ropes are formed as it was discussed recently in [13]. Finally we note that both the LHC and the RHIC are planned now to be equipped with RICH detectors which hopefully will permit to perform the measurements outlined above.

Acknowledgement

The author is indebted to K. Sailer and to Z. Trócsányi for their cooperation.

References

[1] I. Lovas, K. Sailer and Z. Trócsányi, *J. Phys. G* **15**, 1709 (1989); I. Lovas, K. Sailer and Z. Trócsányi, *Phys. Letters B* **298**, 419 (1993)

[2] P. Koch, M. Heinz and J. Pisut, *Phys. Letters B* **243**, 149 (1990); P. Koch and M. Heinz, *Nucl. Phys. A* **525**, 293 (1991)

[3] J. Cleymans, *Nucl.Phys. A* **525**, 205 (1991); J. Cleymans, H. Satz, E. Suhonen and D. W. von Oertzen, *Phase Structure of Strongly Interacting Matter*, edited by J. Cleymans, (Springer, Heidelberg, 1990)

[4] C. Baglin et al., NA38 Collaboration., *Phys. Letters B* **272**, 49 (1991)

[5] C. Baghi et al., NA38 Collaboration, *Phys.Letters B* **272**, 449 (1991); C. Baglin et al., NA38 Collaboration, *Phys. Letters B* **268**, 453 (1991)

[6] C. Gerschel, J. Hüfner, Prep. IPNO-DRE 92-01 (1992)

[7] U. Faschingbauer et al., CERN/SPSC 88-25 (1988); J. Schukraft et al., CERN/SPSC 88-40 (1988); A. Drees et al., *Spring Meeting on Nuclear Physics*, Salzburg 1992, *Verhandl. DPG (VI)* **27**, 197 (1992)

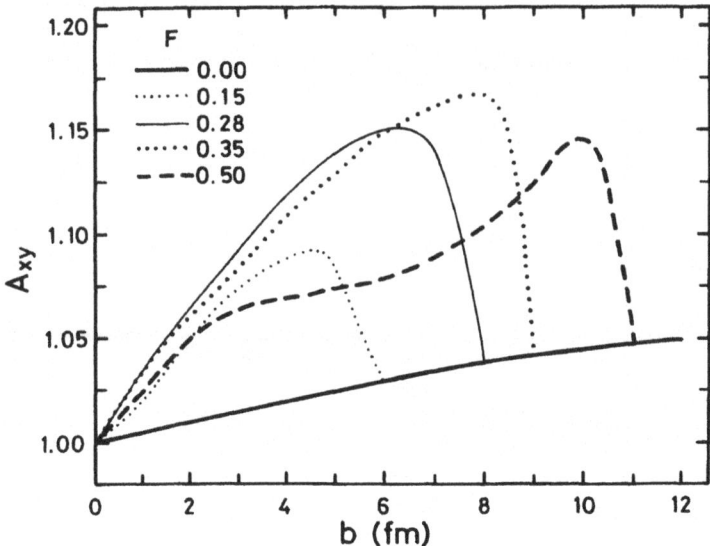

Figure 2. The asymmetry $\mathcal{A}_{xy} = (N_\Phi^x/N_\Psi^x)/(N_\Phi^y/N_\Psi^y)$ as the function of the impact parameter b. The enhancement factor w for the Φ meson production in QGP compared to hadronic matter has the value $w = 10.0$. The plasma fraction F is defined as the ratio of the volumes: $F = V_{plasma}/V_{fireball}$.

[8] A. Breskin et al., *Nucl.Instr. and Meth.* **263**, 237 (1988); R. Arnold et al., *Nucl.Instr. and Meth.* **273**, 466 (1988); A. Drees et al., *Nucl.Instr. and Meth.* **273**, 793 (1988); E. Chesi et al., *Nucl.Instr. and Meth. A* **283**, 602 (1989); J. Seguinot et al., *Nucl.Instr. and Meth. A* **297**, 133 (1990)

[9] A. Shor, *Phys. Rev. Lett.* **54**, 1122 (1985)

[10] T. Matsui and H. Satz, *Phys. Lett. B* **178**, 416 (1986); F. Karsch, M. T. Mehr and H. Satz, BNL-40122 preprint (1987); L. B. Kovacs, T. G. Kovacs and I. Lovas, *Few Body Systems* **9**, 67 (1990)

[11] H. Sorge, A. v. Keitz, R. Mattiello, H. Stöcker, W. Greiner, *Z. Phys. C* **47**, 629 (1990)

[12] H. Behrend et al., *Phys. Lett. B* **56**, 408 (1975)

[13] H. Sorge, M. Berenguer, H. Stöcker, W. Greiner, Prep. LA-UR-92-1078 (1992)

PROCESSES IN PERIPHERAL ULTRARELATIVISTIC HEAVY-ION COLLISIONS

Martin Greiner[1], Mario Vidović[2] and Gerhard Soff[2]

[1] Institut für Theoretische Physik der Justus-Liebig-Universität, Heinrich-Buff-Ring 16, D-35392 Gießen, Germany
[2] Gesellschaft für Schwerionenforschung, Planckstraße 1, Postfach 110552, D-64220 Darmstadt, Germany

1. Introduction

The proposed ultrarelativistic collider LHC at CERN, which is mainly destined for a pp collider mode, will also be used for a heavy-ion collider mode. The heavy ions could be accelerated up to 3.5 TeV/u. Such ultrarelativistic heavy-ion collisions could be used to investigate the formation and decay of a possible quark-gluon plasma; another application could be the production of new particles. In peripheral collisions new particles could be produced coherently via the strong transverse electromagnetic fields; cross sections scale with $Z^4 \approx 10^7$, where Z is the nuclear charge number, with respect to corresponding e^+e^- collider production cross sections. The maximum frequency contained in the electromagnetic fields goes as the inverse of the Lorentz contracted nuclear radius $\omega_{max} = \gamma/R$, which turns out to be $\omega_{max} \approx 100 GeV$ for LHC; as a consequence two photon fusion processes could produce particles, which have masses of the same order of magnitude as heavy nuclei. Such exotic particles are for example Higgs bosons, supersymmetric particles or glueballs.

In the following we will mainly concentrate on the electromagnetic particle production processes occurring in the ultrarelativistic heavy-ion collisions. A convenient way to estimate these cross sections is given by the equivalent photon method, sometimes also called Weizsäcker-Williams method[1]. In section 2 we will derive this method directly from QED and generalize it to include an impact parameter dependence. The electromagnetic production of certain exotic particles is discussed in section 3. An extended outlook on other processes in peripheral ultrarelativistic heavy-ion collisions will be given in section 4.

2. Foundations of the Equivalent Photon Method

A nucleus moving at nearly the speed of light has transverse electromagnetic fields; the electric and magnetic fields have the same absolute value and are perpendicular to each other. Therefore an observer can not distinguish between these tranverse electromagnetic fields and an equivalent swarm of photons. Equating the energy flux of the electromagnetic fields through a transverse plane with the energy content of the equivalent photon swarm yields the equivalent photon distribution

$$n(\omega) = \frac{4Z^2\alpha}{\omega} \int \frac{d^2k_\perp}{(2\pi)^2} \left(\frac{F\left(\omega^2/\gamma^2 + \vec{k}_\perp^2\right)}{\omega^2/\gamma^2 + \vec{k}_\perp^2} \right)^2 |\vec{k}_\perp|^2 \quad ; \qquad (1)$$

α is the finestructure constant and F represents the nuclear charge formfactor, which we will choose to be the one of a homogeneously charged sphere. Photon frequencies above $\omega_{max} = \gamma/R$ are highly suppressed. – Given now the equivalent photon distribution, it is straightforward to estimate the electromagnetic production cross section of particles in a heavy-ion collision in the following way:

$$\sigma^{WW}_{A_1A_2\to A_1A_2X} = \int d\omega_1 \int d\omega_2 \; n_1(\omega_1)n_2(\omega_2)\sigma_{\gamma\gamma\to X}(\omega_1,\omega_2) \quad ; \qquad (2)$$

two photons stemming from the two colliding swarms of equivalent photons fuse to produce the particles X. $\sigma_{\gamma\gamma\to X}$ represents the two-photon fusion cross section. The cross section (2) is called the equivalent photon cross section.

At this stage we could insert two-photon fusion cross sections of interest and calculate the total electromagnetic production cross sections for particles in a heavy-ion collision. However some caution has to be taken:

The masses of the exotic particles, we are interested in, have masses, which come close to the maximum frequencies contained in the electromagnetic fields; therefore they will be produced, where the electromagnetic interaction energy is largest. That occurs at small impact parameters in the vicinity of the nuclear surfaces. On the other hand, central collisions have to be excluded in order to have a rather clean experimental trigger on peripheral collisions. Naturally the question arises: How much do small impact parameters contribute to the total equivalent photon cross section (2)? – The equivalent photon cross section (2) consists of a quantum mechanical part ($\sigma_{\gamma\gamma\to X}$) and a classical part ($n_1 n_2$). It is not clear, how good this ansatz approximates the full quantum mechanical cross section $\sigma_{A_1A_2\to A_1A_2X}$. – Therefore, there is a need to derive the equivalent photon cross section directly from QED and to generalize this method to include an impact parameter dependence!

We will use the external field approximation, where the electromagnetic potentials are treated classically. The two nuclei are supposed to move with constant velocities on straight parallel trajectories, which have a distance $\vec{b}$. The electromagnetic potentials then follow from Poisson's equation $\Box A^\mu = j^\mu$. – The total cross section for the electromagnetic production of particles in a heavy-ion collision is given as the integral of the transition probability, which is the square of the S-matrix element, over the phase space of the outgoing produced particles and the impact parameter $\vec{b}$:

$$\sigma_{A_1A_2\to A_1A_2X} = \int d^2b \left(\prod_i \int d^3\tilde{p}_i\right) \left| \int \frac{d^4k_1}{(2\pi)^4} \int \frac{d^4k_2}{(2\pi)^4} [A_1^\mu(k_1)\Gamma_{\mu\nu}(k_1,k_2,p_i)A_2^\nu(k_2)] \cdot \right.$$

$$\left. \cdot (2\pi)^4 \delta^4\left(k_1 + k_2 - \sum_i p_i\right) \right|^2 \qquad (3)$$

with $d^3\tilde{p} = (1/2E)d^3p/(2\pi)^3$ for bosons and $d^3\tilde{p} = (m/E)d^3p/(2\pi)^3$ for fermions; the electromagnetic potentials do enter in the S-matrix element as does the transition current $\Gamma_{\mu\nu}(2\pi)^4\delta^4(k_1 + k_2 - \sum_i p_i)$, where the δ-function guarantees energy-momentum conservation and the vertex function $\Gamma_{\mu\nu}$ describes the interaction of the two virtual photons with the produced outgoing particles. Expressions for $\Gamma_{\mu\nu}$ are given elsewhere[2].

Within the external field approximation the expression (3) is still exact. It is now to introduce reasonable approximations, which will lead to the equivalent photon cross section (2). It is our goal to seperate the fusion process of two virtual photons into an emission process of equivalent photons and a successive fusion process of two real photons. Real photons are massless since the square of their four-momentum $k = (\omega, 0, 0, \omega)$ is zero; the first component of k represents the photon energy, whereas the last three components represent the photon momentum, which we have chosen along the z-axis. Actually the relation $k^2 = 0$ is the result of an exact cancellation of two big quantities. The four-momentum of the virtual photons $k = (\omega, \vec{k}_\perp, \omega/v)$, where v is the velocity of the moving heavy-ion, is determined by the classical electromagnetic potentials. The virtual photons are not massless, since $k^2 = -\omega^2/\gamma^2 - \vec{k}_\perp^2$ as $v \to c = 1$; again this relation goes back to a cancellation of two big quantities, but this time not exactly to zero. This motivates the following assumption:

$$\mathcal{O}\left(|\vec{k}_\perp|\right) = \mathcal{O}\left(\frac{\omega}{\gamma}\right) = \frac{1}{\gamma}\mathcal{O}(\omega) = \frac{1}{\gamma}\mathcal{O}(k_\parallel) \quad ; \tag{4}$$

the Lorentz contraction factor $\gamma = (1 - v^2)^{-1/2}$ takes on values of several thousands in our case of interest (LHC). Applying the relation (4), does not mean that $|\vec{k}_\perp|$ and ω/γ can simply be set equal to zero in the invariant matrix element $A_1^\mu \Gamma_{\mu\nu} A_2^\nu$; in fact, if we would do so, the invariant matrix element would vanish. This becomes clear, if we again have a closer look on the expression (1) for the equivalent photon distribution; the factor $|\vec{k}_\perp|^2$ has still to be extracted from the invariant matrix element! As a consequence very special care has to be taken, when applying the relation (4) to the invariant matrix element. This calculation is straightforward, but rather tedious; details are given elsewhere[2]. We end up with

$$\begin{aligned}\sigma_{A_1 A_2 \to A_1 A_2 X} &= \int d\omega_1 \int d\omega_2 \int d^2b \left[n_\parallel(\omega_1, \omega_2; \vec{b})\sigma^\parallel_{\gamma\gamma \to X}(\omega_1, \omega_2) \right. \\ &\left. \quad + n_\perp(\omega_1, \omega_2; \vec{b})\sigma^\perp_{\gamma\gamma \to X}(\omega_1, \omega_2)\right] \\ &= \int d\omega_1 \int d\omega_2 \, n_1(\omega_1) n_2(\omega_2) \sigma_{\gamma\gamma \to X}(\omega_1, \omega_2) \quad ,\end{aligned} \tag{5}$$

where $\sigma^\parallel_{\gamma\gamma \to X}$ and $\sigma^\perp_{\gamma\gamma \to X}$ are polarized two-photon fusion cross sections and add up to give the total two-photon fusion cross section $\sigma_{\gamma\gamma \to X} = (\sigma^\parallel_{\gamma\gamma \to X} + \sigma^\perp_{\gamma\gamma \to X})/2$. The two-photon distribution

$$n_\parallel(\omega_1, \omega_2; \vec{b}) = \frac{1}{\pi^2 \omega_1 \omega_2} \int d^2 x_\perp \left|\vec{E}_{1\perp}(\omega_1, \vec{x}_\perp - \vec{b}) \cdot \vec{E}_{2\perp}(\omega_2, \vec{x}_\perp)\right|^2 \tag{6}$$

depends on the transverse electromagnetic fields $\vec{E}_{1\perp}$ and $\vec{E}_{2\perp}$ of the two heavy ions; for $n_\perp(\omega_1, \omega_2; \vec{b})$ the dot product in the expression (6) simply has to be replaced by a cross product. – The second step in eq.(5) illustrates, that we have indeed derived the equivalent photon cross section (2) from QED directly. If we skip the integration over the impact parameter in eq.(5), we come up with an impact parameter dependent differential equivalent photon cross section!

3. Electromagnetic Production of Higgs Bosons, Supersymmetric Particles and Glueballs

The expressions for the relevant polarized two-photon fusion cross sections needed for the calculation of the electromagnetic production cross section (5) of various exotic particles will not be given here; they can be found in another reference [2].

With the derivation of an impact parameter dependent differential cross section, it is now possible to exclude central collisions. Experimentally this is necessary in order to trigger on nearly intact heavy ions. We introduce a simple cutoff, where all impact parameters smaller than twice the nuclear radius are discarded, and obtain the reduced cross section

$$\sigma^{red}_{A_1 A_2 \to A_1 A_2 X} = \int_{2R}^{\infty} \frac{d\sigma_{A_1 A_2 \to A_1 A_2 X}}{db} db = \int_{2R}^{\infty} \frac{d^2 \sigma_{A_1 A_2 \to A_1 A_2 X}}{d^2 b} 2\pi b \, db \quad . \quad (7)$$

For heavy exotic particles ($m \geq 50 GeV$) the reduced cross sections turn out to be 20-70% smaller than the total equivalent photon cross section (2) for Pb+Pb collisions at LHC energies. Except when considered for smaller heavy ion collider energies as for example at RHIC ($200 GeV/u$) the reduction of the equivalent photon cross sections for the production of lighter exotic particles is generally negligible.

Together with the top quark the Higgs boson is one of the missing links of the Standard Model of electroweak and strong unification. In the intermediate mass regime ($90 GeV \leq m_{H_{sm}} \leq 160 GeV$) its reduced electromagnetic production cross section at ultrarelativistic heavy-ion colliders amounts to several tens of pb for LHC energies. Thus for an expected LHC luminosity of $\mathcal{L} = 2 \cdot 10^{27} cm^{-2} sec^{-1}$ and an assumed running time of $10^6 sec/year$ only about 0.1 produced SM Higgs bosons with an assumed mass of $m_{H_{sm}} = 100 GeV$ are to be expected for LHC. – The situation might change completely for the Higgs bosons of the minimal supersymmetric extension of the Standard Model (MSSM) [3]. Here the reduced electromagnetic production cross sections depend very sensitively on the chosen free parameters of the MSSM. Within the current experimental limits on these parameters there exist parameter regions, where the reduced cross sections are enhanced even by a factor of 100 over the corresponding SM Higgs cross sections.

Interacting in lowest order with photons, supersymmetric particles behave like heavy bosons or fermions; it is their charge which couples to the photon and it is their mass which determines the cross section. The masses of the supersymmetric particles will lie above 20-50 GeV [4]. For fermion masses of $m_{f\pm} = 50$ and 100 GeV we find reduced cross sections of $\sigma^{red}_{Pb+Pb \to Pb+Pb+f\pm} = 3.4$ and 0.26 nb at LHC; corresponding boson cross sections are smaller, i.e. $\sigma^{red}_{Pb+Pb \to Pb+Pb+b\pm} = 0.74$ and 0.007 nb for $m_{b\pm} = 50$ and 100 GeV at LHC. With the expected collider performance mentioned before about 10 fermions with mass $m_{f\pm} = 50$ GeV and about 2 charged (spin 0) bosons with mass $m_{b\pm} = 50$ GeV could be generated at LHC. If such small mass charged supersymmetric particles do exist, these production rates might still be sufficient; once a supersymmetric particle is produced it can only decay into the lightest SUSY particle, which can hardly be detected. Therefore, in principle, one would always trigger on missing energy and momentum events together with a multiplicity of lepton and quark-jet events. A full acceptance detector would be needed for such an investigation.

One prediction of QCD is the existence of exotic bound states containing several gluons and quarks. Bound states, which predominantely contain gluons compared to quarks, are called glueballs. Glueballs should be produced in reactions involving J/Ψ decays. The electromagnetic production of glueballs should also be studied, because this could contribute a strong background to a possible signal for a quark-gluon plasma. – We consider here two glueball candidates, the $\eta(1440)$ meson with mass $m = 1440$ MeV and the $f_2(1720)$ meson with mass $m = 1713$ MeV. The electromagnetic production cross sections are expected to be in-between 0.01 mb and 100 mb depending on the heavy-ion collision energy and the unknown two-photon decay widths of these mesons; confer figure 1. At LHC about 10^6

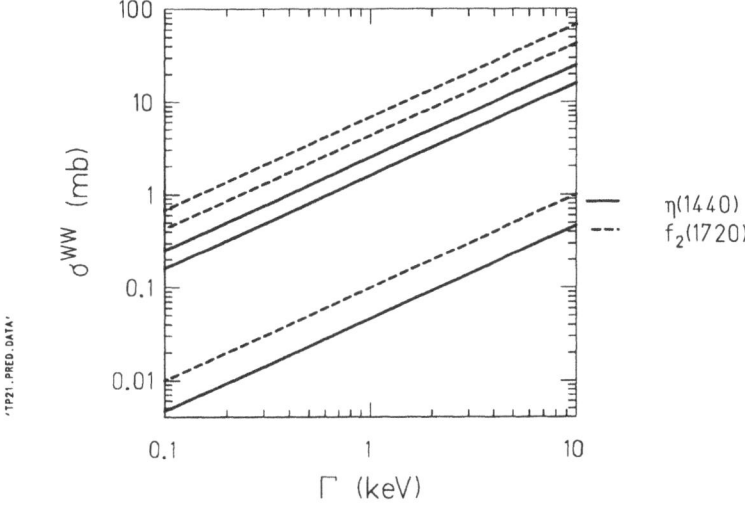

Figure 1. *Equivalent photon cross-section for the production of the two glueball candidates $\eta(1440)$ (solid lines) and $f_2(1720)$ (dashed lines) in a Pb+Pb collision in dependence of the two-photon decay width Γ for SSC, LHC and RHIC energies, from top to bottom.*

glueball mesons could be produced per year and even at RHIC with a designated luminosity of $\mathcal{L} = 10^{26} cm^{-2} sec^{-1}$ and an assumed running time of 10^7 sec/year about 10^5 glueball mesons are to be expected.

4. Conclusions and Outlook

We have derived the equivalent photon method directly from QED and, as a by-product, we have also obtained an impact parameter dependent generalization of this method. The electromagnetic production cross sections for exotic particles as for example Higgs bosons, supersymmetric particles and glueballs in ultrarelativistic heavy-ion collisions have been calculated; they are large compared to corresponding hadronic cross sections in pp collisions. So far only the glueball meson production rates are reasonable, so that their detection should be feasable. – In future one should further study the electromagnetic production of heavy mesons like b-mesons or the Υ. They should still be produced with high rates at LHC; on the other side at RHIC their production should be suppressed because the accessible photon frequencies are too low. In this respect also a careful background analysis for the detectability of these exotic particles has to be performed. Of interest could also be combined electromagnetic and hadronic processes in peripheral colllisions. For example photon-gluon fusion[5], where the photon from the one nucleus penetrates into the other nucleus and fuses with a gluon, could be used to study the gluon distribution of a nucleon inside a nucleus.

Acknowledgement

One of us (M.G.) wants to thank the organizers of this NASI summer school on "Frontier Topics in Nuclear Physics" for their invitation to experience the friendly and inspiring atmosphere of this meeting.

References

[1] E. Fermi, *Z. Phys.* **29**, 315 (1924); E. J. Williams, *Proc. Roy. Soc. A* **139**, 163 (1933); C. Weizsäcker, *Z. Phys.* **88**, 612 (1934)
[2] M. Vidović, M. Greiner, C. Best and G. Soff, *Phys. Rev. C* **47**, 2308 (1993)
[3] M. Greiner, M. Vidović and G. Soff, *Phys. Rev. C* **47**, 2288 (1993)
[4] Particle Properties Data Booklet, *Phys. Rev. D* **45**, Part 2 (June 1992)
[5] C. Hofmann, G. Soff, A. Schäfer and W. Greiner, *Phys. Lett. B* **262**, 210 (1991)

ELECTRON-POSITRON PAIR CREATION IN RELATIVISTIC ATOMIC HEAVY ION COLLISIONS

Joachim Thiel, Johannes Hoffstadt, Norbert Grün and Werner Scheid

Institut für Theoretische Physik der Justus-Liebig-Universität, Giessen, Germany

Introduction

The production of electron-positron pairs in relativistic atomic heavy ion collisions has been studied with increasing interest in the last years. With the notation "atomic collisions" we mean collisions with an impact parameter larger than twice the nuclear radius. Theoretical work on the electron-positron creation in collisions of fast charged particles started with the work of Landau and Lifschitz (for references of the former work see e.g. Heitler [1]). With the development of colliders for relativistic heavy ions new elaborate work on electron-positron creation came up [2, 3]. Pair creation with capture of the electron into a bound state of an ion changes the charge of the ion and is one of the main processes for the loss of ions in relativistic heavy ion colliders. The cross section is of the order of 100 barn for RHIC energies ($E_{cm} = 100$ GeV/nucleon). A nonperturbative calculation usually enlarges the cross section for pair creation with capture. Therefore, nonperturbative calculations are urgently needed for high incident energies. Reviews on this field have been given by Bertulani and Baur [4] and Eichler [5].

In the case of projectiles and targets with lower charge numbers and larger impact parameters one can use the perturbation theory of first order to calculate the probabilities for pair creation. However for collisions of ions with very high charge numbers and at small impact parameters it became evident in the last years that the electron-positron creation is a nonperturbative process. A 3-dimensional, fully nonperturbative treatment of lepton-pair production has been carried out by Strayer et al. [6] solving the time-dependent Dirac equation with a basis spline expansion. Their probabilities for μ-pair production with capture into the K-shell of the target ion are by several orders of magnitude larger than the prediction of the first order perturbation theory. In this contribution we present three nonperturbative methods, which yield probabilities for electron-positron pair creation enhanced by one or two orders over the results of the perturbation theory. In the second section we discuss the coupled channel method worked out by Momberger et al. [7], and Rumrich et al. [8]. This method is also used by Hoffstadt [9] and extended to higher incident energies by Baltz et al. [10]. In the third section we present the finite difference method for solving the Dirac equation on a two-dimensional grid and apply it to a collision of U^{92+} (10 GeV/nucleon) on U^{91+} with very small impact parameters [11]. The fourth section gives an explanation of the nonperturbative character of the electron-positron creation by showing that the energy of the target K-shell states dives into the negative continuum due to the relativistically enhanced field of the projectile. For CERN energies of $E_{lab} = 200$ GeV/nucleon we present coupled channel calculations for the creation of free pairs in the fifth section, where the electron-positron pairs are treated as bosons.

Coupled Channel Calculations for free Electron-Positron Pair Production and with Capture

We make use of the semiclassical approximation. This means that we treat the motion of the nuclei classically assuming the target nucleus as fixed at the origin of the coordinate system and the projectile nucleus moving with constant velocity v_P in the z-direction of the target rest system on a straight line with an impact parameter b. The electron field is quantum mechanically described by the time-dependent Dirac equation ($\hbar = m = c = 1$):

$$(H_T + V_P(t))\Psi(\vec{r}, t) = i\partial\Psi(\vec{r}, t)/\partial t \tag{1}$$

with
$$H_T = -i\vec{\alpha}\vec{\nabla} + \beta - \frac{Z_T e^2}{r}, \tag{2}$$

$$V_P = -Z_P \gamma e^2 (1 - v_P \alpha_z)/ \mid \vec{r}\,' - \vec{R}'(t) \mid, \tag{3}$$

$$\vec{r}\,' = (x, y, \gamma z), \quad \vec{R}'(t) = (b, 0, \gamma v_P t).$$

H_T is the target Hamiltonian, V_P the interaction between electron and projectile and γ the relativistic Lorentz factor: $\gamma = (1 - v_P^2)^{-1/2}$. The time-dependent Dirac equation (1) is solved by expanding Ψ in a complete set of eigenstates ϕ_j of the target Hamiltonian

$$\Psi_\iota(\vec{r}, t) = \sum_j a_{j\iota}(t)\phi_j exp(-iE_j t). \tag{4}$$

We choose the functions ϕ_j as Coulomb-Dirac wave functions which are analytically known. For the positive and negative continuum we discretize these functions by means of wave packets of the form [12]

$$\phi_{E_j}(\vec{r}, t) = \frac{1}{\sqrt{\Delta E_j}} \int_{E_j - \frac{\Delta E_j}{2}}^{E_j + \frac{\Delta E_j}{2}} dE \phi_E(\vec{r}) exp(i(E_j - E)t). \tag{5}$$

Insertion of the expansion (4) into the Dirac equation (1) and projection on a particular basis state j yields coupled equations for the expansion coefficients:

$$i\dot{a}_{j\iota} = \sum_k \langle \phi_j \mid V_P \mid \phi_k \rangle exp(i(E_j - E_k)t) a_{k\iota}. \tag{6}$$

These equations have to be solved with the initial conditions $a_{j\iota}(t \to -\infty) = \delta_{\iota j}$. In the case of electron-positron pair production we have to treat a system with an infinite number of electrons which initially occupy the negative continuum ($E_F = -mc^2$). Since the amplitudes $a_{j\iota}$ are assumed as single-particle amplitudes, the probability for finding an electron in an unperturbed state with $E_\iota > E_F$ can be expressed as [13]

$$P_\iota(t \to \infty) = \sum_{E_k < E_F} \mid a_{\iota k}(t \to \infty) \mid^2 \tag{7}$$

and the probability for a hole (=positron) in an unperturbed state with $E_i < E_F$ as [13]

$$P_i(t \to \infty) = \sum_{E_k > E_F} \mid a_{ik}(t \to \infty) \mid^2 . \tag{8}$$

Expressions (7) and (8) are the inclusive probabilities to find a single electron or positron, respectively, in a certain state i independent of the creation of further electron-positron pairs which can be created from the initial vacuum state with other energies. Also probabilities for the production of a single, two or more electron-positron pairs can be calculated from expressions constructed with the amplitudes $a_{ji}(t \to \infty)$.

The probability to observe a K-shell electron from pair production is given by

$$P_{1s} = \sum_{E_j < -mc^2} \mid a_{j,1s}(t \to \infty) \mid^2 . \tag{9}$$

In this formula use has been made of the time reversal symmetry so that only a single time integration of eq.(6) starting with the $1s_{1/2}$-state is necessary to evaluate the probability (9).

The calculation of the transition matrix elements in eq.(6) has been carried out by a method based on the procedure of Amundsen and Aashamar [14] and described by Momberger et al. [12]. The matrix elements $\langle \phi_i \mid V_P \mid \phi_j \rangle$ are transformed to momentum space and calculated in a multipole expansion. The method of the evaluation of matrix elements with continuum states involving the wave packets (5) is also given in Ref. [12].

First we consider the scattering of U^{92+} on U^{92+} at an incident energy of $E_{lab} = 10$ GeV/nucleon and an impact parameter of $b = 386$ fm (Compton wavelength $\hbar/(mc)$). This impact parameter approximately yields the maximum contribution to the total cross section in first order perturbation theory. The basis set contains 5 bound states ($1s_{1/2}$, $2s_{1/2}$, $2p_{1/2}$, $2p_{3/2}$, $3s_{1/2}$) and wave packets at 20 energies for the continua ($E_j = 1.3, ..., 7.3; -1.3, ..., -6.1; \Delta E = 0.6$; unit mc^2) for each of the three angular momenta $s_{1/2}, p_{1/2}, p_{3/2} (\kappa = -1, +1, -2)$. The number of coupled coefficients of this basis set is 172 including all magnetic substates. The restriction in angular momentum summation is presently the most important shortcoming of these calculations. From first order perturbation theory it is known that the summation should be extended up to about $\mid \kappa \mid = 10$, corresponding to 20 different angular momenta. However, coupled channel calculations with basis sets of that size are not feasible for us. Recently, large basis calculations were carried out by Baltz et al. [10] for the Pb on Pb system at ultrarelativistic energies.

In Fig. 1 we show results for the differential probability to observe a positron of energy E_{e^+} as a function of this energy. The histogram-like curves show the total probability and its decomposition into contributions from pair creation of free pairs and with simultaneous capture. The smooth curves are the results of a calculation in first order perturbation theory, where the same angular momentum quantum numbers are taken into account as in the coupled channel calculation. We find the astonishing result that couplings of higher orders increase the positron spectrum by more than one order of magnitude, especially the capture from vacuum is raised by roughly two orders of magnitude.

In Fig. 2 we show probabilities for pair production with K-shell capture and for the ionization of a K-shell electron in a collision of Pb on Pb at an incident energy of E_{lab}=1.2 GeV/nucleon in comparison with results of the perturbation theory in first order [8]. For the pair production we find an increase of a factor of 50 at zero impact parameter above the result of the first order perturbation theory. With increasing impact parameter the coupled channel

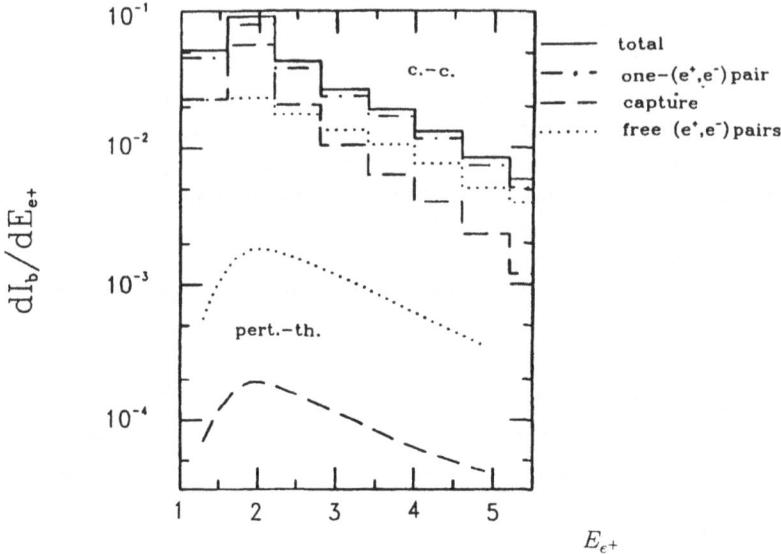

Figure 1. Differential probability for the observation of a positron of energy E_{e^+} (in units of $mc^2 = 511$ keV) in the collision $U^{92+} + U^{92+}$ at $E_{lab} = 10$ GeV/nucleon and $b = 386$ fm. The histogram-like curves show the coupled channel results, the continuous curves below are the results of the first order perturbation theory. The solid curve gives the total positron spectrum and the dotted-dashed curve the contribution from the production of a single electron-positron pair during the collision. The dotted curves show the probabilities for the production of free pairs and the dashed curves those for the production of pairs with simultaneous inner shell capture.

results approach the results of the perturbation theory as expected. The total cross section for pair production with K-shell capture is obtained as 0.15 barn and 1.0 barn with perturbation theory and coupled channel calculations, respectively. This increase of the cross section is remarkable. Very recent experiments of Belkacem et al. [15] with U on Au collisions at $E_{lab} = 0.96$ GeV/nucleon yielded a cross section for pair production with K-shell capture of 2.19 ± 0.25 barn, which is in the order of our theoretical nonperturbative result.

Finite Difference Method for Electron-Positron Pair Production with Capture

Now we present results of a numerical solution of eq.(1) by the finite difference method [11]. For reasons of simplicity we assume that the projectile nucleus moves with constant velocity v_P along the z-axis and the target nucleus is fixed at $z = 0$ which means that the impact parameter is set equal to zero. This is a good approximation for atomic collisions with small values of b in the order of two times the nuclear radius. Thus we solve a rotationally symmetric problem with the finite-difference method on a two-dimensional grid. The numerical method and tests of the accuracy are published by Becker et al. [16].

With the finite difference method we calculated the probability for pair creation with K-shell capture and ionization in nearly central collisions of U^{92+} on $U^{92,91+}$ at $E_{lab} = 10$ GeV/nucleon. The Dirac equation (1) was solved on a two-dimensional grid by starting

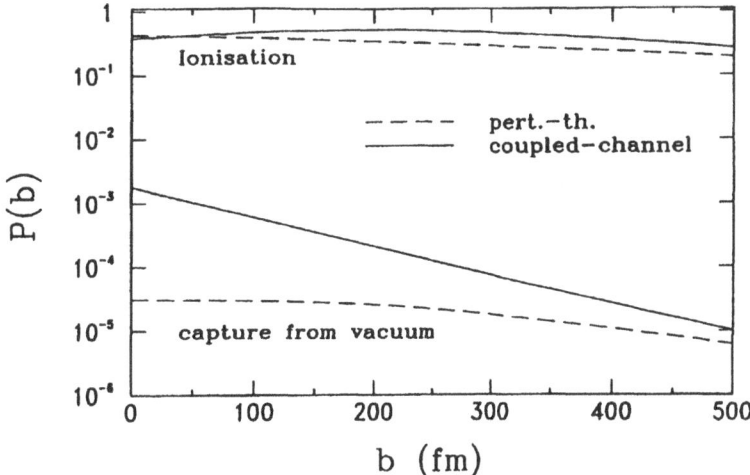

Figure 2. *Probabilities for pair creation with capture into the K-shell and for K-shell ionization in the collision of Pb on Pb at E_{lab} =1.2 GeV/nucleon.*

with the $1s_{1/2}$ wave function bound at the target ion. The grid had a measure of 600x300 meshes with a size of $\Delta z = \Delta \rho = 5 \cdot 10^{-4} a_o = 26.5$ fm. Fig.3 shows the time development of the absolute square of the wave function in the $z\rho$-plane (cylindrical coordinates z and ρ) for the times -8,0,6 and 11 (unit $\hbar/(mc^2) = 1.288 \cdot 10^{-21} s$). For further explanation see the figure caption. One can see that in part the wave function follows the projectile indicating large probabilities for ionization and pair production with high linear momenta of the ionized electron and created positron in the direction of the collision axis.

The wave function can be used to obtain the amplitudes for the analytical eigenstates of the target Hamiltonian. By projecting on excited bound states we get the probabilities for excitation and by projecting on the positive and negative continuum states the probabilities for ionization and pair production with capture in the $1s_{1/2}$ bound state of the target, respectively. The exact target eigenfunctions with quantum numbers κ (Coulomb-Dirac functions) are not the optimum basis states to describe ionization and pair production with high linear momenta in the forward direction. Therefore, we also projected with Sommerfeld-Maue functions which are approximate solutions of the target Hamiltonian. They contain an incoming spherical wave and a plane wave with linear momentum and helicity, but they are not orthogonal.

Fig. 4 shows the probability of pair production with capture for the nearly central collision of U^{92+} (10 GeV/nucleon) on U^{92+} as a function of time. The solid curve represents the values obtained by summing the probability obtained with Coulomb-Dirac wave functions with $\mid \kappa \mid \leq 5$. Values for the probability obtained with Sommerfeld-Maue functions were only calculated for t=0, 6 and 11 because of the limited computer time. They are shown by full dots and are interpolated by a dashed-dotted curve to guide the eyes. The probability is compared with those of a calculation using first order perturbation theory (dashed curve) and of a coupled channel calculation (dotted curve). Again we find the strongly nonperturbative character of the pair production in relativistic collisions of very heavy ions.

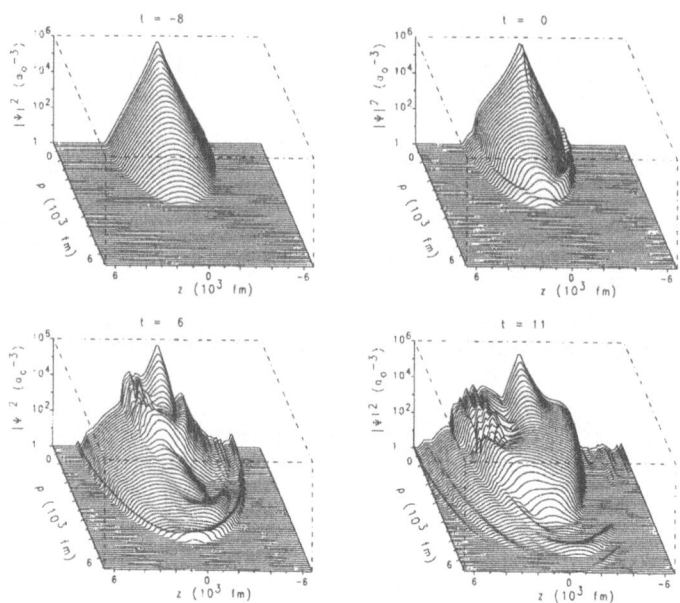

Figure 3. *Probability density of the wave function of the electron at times $t=-8,0,6$ and 11 (units $1.288 \cdot 10^{-21}$ s) for a nearly central collision of U^{92+} on U^{91+} with an incident energy of $E_{lab} = 10$ GeV/nucleon. The projectile moves from the right to the left hand side and is located at $z = -3.1, 0, 2.3$ and $4.2 \cdot 10^3$ fm, respectively. The density is given in units of a_o^{-3} ($a_o = 5.3 \cdot 10^4$ fm).*

The fair agreement of the finite difference method with the coupled channel method seems to be astonishing since only angular momenta with $\kappa = \pm 1, \pm 2$ were used in the coupled channel calculation. We found that continuum states with high angular momenta get occupied in the finite difference calculation at larger times ($t > 0$) via a transfer of the probabilities from smaller to higher angular momenta, whereas for t=0 the finite difference and coupled channel methods yield similar dominant contributions of the $p_{1/2}(\kappa = 1)$ continuum states to the total probability. This may be the reason for the agreement of both methods in the total probability for pair production with capture.

Nonperturbative Character of Pair Production

The reason for the failure of the perturbation theory in first order can clearly be seen in Fig.4 where the probability increases up to the order of unity near t=0 although the final probability is very small. Therefore, the perturbation theory becomes invalid around the point of closest approach for relativistic very heavy ions and small impact parameters.

Fig. 5 shows the expectation value of the energy of the initial $1s_{1/2}$ state, calculated with the time-dependent numerical wave function [11]. We note that this expectation value immerses into the negative continuum around the point of closest approach because of the relativistically enhanced electromagnetic field of the projectile. Therefore, one expects a large overlap of this state with the states of the negative continuum and, therefore, a large probability to create electron-positron pairs. This situation has strong connections with the effect of the spontaneous production of electron-positron pairs in heavy ion collisions near the

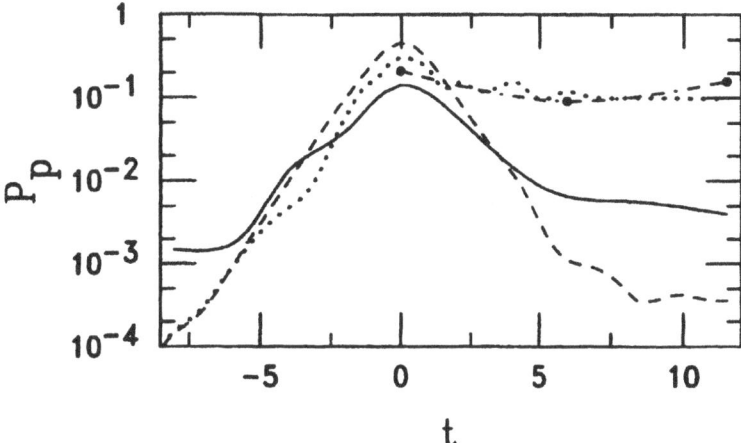

Figure 4. *The probability of pair production with capture for nearly central collisions of U^{92+} (10GeV/nucleon) on U^{92+} is shown as a function of time. The unit of time is $1.288 \cdot 10^{-21}$ s. The solid curve is calculated by projecting with Coulomb-Dirac eigenfunctions of the target Hamiltonian up to $|\kappa|=5$. The values calculated by projecting with Sommerfeld-Maue functions are shown by full dots and are interpolated by a dashed-dotted curve to guide the eyes. The result is compared with those obtained by first order perturbation theory (dashed curve) and a coupled channel calculation (dotted curve).*

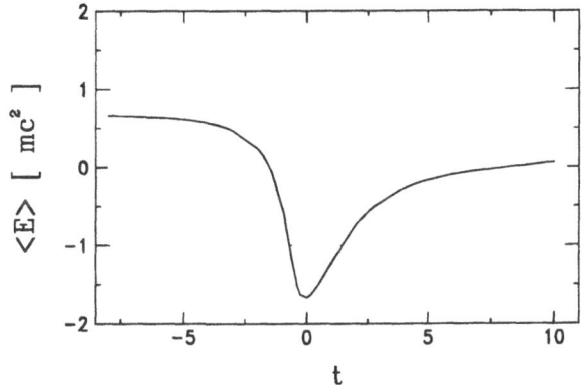

Figure 5. *Expectation value of energy calculated with the solution obtained with the finite difference method for U^{92+} (10GeV/nucleon)+U^{91+}.*

Coulomb barrier which has been studied by Greiner and his Frankfurt school [17]. Obviously, this diving of the expectation value of energy into the negative continuum is the reason for the strong nonperturbative behaviour of the pair production process.

Coupled Channel Formalism in a Bosonized Space

It was shown [4] that perturbative results at ultrarelativistic incident energies ($E_{lab} \geq 500$ GeV/nucleon) and impact parameters of $b = 386$ fm violate unitarity. Because of this behaviour, multiple-pair creation is expected to get important for high energies and small impact parameters of the order of the Compton wavelength. In the nonperturbative theories described above multiple-pair creation can only be treated with a very large numerical effort. For example, the number of channels in a coupled channel calculation has to be much larger than the expected number of pairs in each region of the electron and positron energies. The Pauli principle allows only one electron per channel, although each channel includes a large number of physical states. In order to avoid this artificial effect the fermionic theory can be bosonized. This means, each electron-positron pair can be treated as a boson and the Pauli principle neglected. If as a further approximation any rescattering of the electrons or positrons of the created pairs is ignored [18, 19, 20], a Poisson distribution is obtained for the multiplicity of the pairs, with the mean number of pairs given by the lowest order perturbation theory. This procedure restores unitarity by incorporating nonperturbative processes (loop graphs and multiple creation of pairs), but it is actually only consistent up to lowest order perturbation theory.

In this section we will present a consistent nonperturbative bosonic theory based on a coupled channel formalism, in order to study nonperturbative effects in the multiple-pair production. As usual we use the semiclassical approximation as described in the second section. The nuclei are assumed as point-like charged particles producing the classical electromagnetic field. The calculations are carried out in the coordinate frame where the nuclei have opposite constant velocities of the same amount moving on straight lines. To distinguish this frame from the target frame (or laboratory frame, lab), we will call it the collider frame, which is for identical ions the center-of-momentum frame. In the following the indices T and P stand for target and projectile, respectively, b is the impact parameter and v_0 the absolute value of the ion velocities.

The classical electromagnetic fields of the moving ions are given by

$$A_0^{(T,P)} = \gamma e Z_{T,P} / \sqrt{(x \pm b/2)^2 + y^2 + \gamma^2 (z \pm v_0 t)^2}$$
$$\vec{A}^{(T,P)} = \mp v_0 A_0^{(T,P)} \vec{e}_z, \tag{10}$$

where $\gamma = 1/\sqrt{1 - v_0^2}$ is the Lorentz factor and $Z_{T,P}$ the charge numbers of the nuclei. The upper signs are valid for the target field and the lower signs for the projectile field.

Standard boson mappings of fermionic theories map the fermionic space one-to-one into a bosonic one, where partly the Pauli principle is neglected [21, 22]. We use the boson mapping in order to reduce the number of channels in the coupled channel calculations. We choose a mapping which maps the physical states of a certain range in the electron and positron momenta to just one bosonic pair state.

Differently to the formalism of the last sections we introduce a field-theoretical treatment and use the interaction picture. The free leptonic field operator is expanded in plane wave solutions $\tilde{\varphi}_{p,s}$ for positive energy eigenvalues and $\tilde{\chi}_{p,s}$ for negative energy eigenvalues of the free Dirac equation $i\partial \Psi/\partial t = \{-i\vec{\alpha}\vec{\nabla} + \beta\}\Psi$, i.e.

$$\Psi(\vec{r}, t) = \sum_{s=-1/2}^{1/2} (2\pi)^{-3/2} \int d^3 p \left[b(p, s) \tilde{\varphi}_{p,s}(\vec{r}, t) + d^+(p, s) \tilde{\chi}_{p,s}(\vec{r}, t) \right]. \tag{11}$$

Here, p is the momentum eigenvalue and s the helicity. The creation and annihilation operators for the electrons are denoted by b^+ and b and those for the positrons by d^+ and d, respectively. They obey the canonical fermionic anticommutation relations. The interaction Hamiltonian, which describes the interaction between the lepton field and the classical electromagnetic field $A = A^{(T)} + A^{(P)}$, is given by $H_I = -e \int d^3x :\overline{\Psi}(x) A_\mu \gamma^\mu \Psi(x):$, where $:...:$ denotes normal ordering of the fermionic operators, and $\gamma^\mu = (\beta, \beta\vec{\alpha})$ the Dirac γ-matrices. Obviously the interaction operator consists of four different parts describing creation of pairs, annihilation of pairs, and rescattering of electrons or positrons.

We can neglect the Pauli principle under the condition that the number of created pairs in each bosonic pair state is much smaller than the number of underlying fermionic states. This condition must hold true for the reaction time. Then we are able to map the bi-fermionic pair creation operators $b^+ d^+$ to bosonic operators $A^+_{\alpha\beta}$,

$$\int_{\Delta_\alpha} d^3p \int_{\Delta_\beta} d^3p' \, b^+(p, s_\alpha) d^+(p', s_\beta) / \Delta^3 P \quad \longrightarrow \quad A^+_{\alpha\beta}. \tag{12}$$

Here we introduced pair channels $\alpha\beta$ with quantum numbers $\vec{p}_\alpha s_\alpha$ for the electron and $\vec{p}_\beta s_\beta$ for the positron, where α and β both represent momentum ranges Δ_α, Δ_β. The momentum spaces for electrons and positrons are divided into cubes of the same size $\Delta^3 P$. Each cube Δ_α is labeled by the mean momentum $\vec{p}_\alpha$ and the helicity s_α. The pair creation and annihilation operators $A^+_{\alpha\beta}$ and $A_{\alpha\beta}$ are assumed to fulfil bosonic commutation relations.

In a similar way we map the annihilation operators db and the rescattering operators b^+b, d^+d of the interaction Hamiltonian to the bosonic operators $A_{\alpha\beta}$, $R^{(+)}_{\alpha\beta}$, and $R^{(-)}_{\alpha\beta}$, respectively. It turns out that the rescattering operators can also be expressed in terms of A^+ and A: $R^{(+)}_{\alpha\beta} = \sum_{\gamma(E_\gamma<0)} A^+_{\alpha\gamma} A_{\beta\gamma}$ and $R^{(-)}_{\alpha\beta} = \sum_{\gamma(E_\gamma>0)} A^+_{\gamma\alpha} A_{\gamma\beta}$. The transition of the interaction Hamiltonian into the bosonic picture can be illustrated for the creation part:

$$\int_{\Delta_\alpha} d^3p \int_{\Delta_\beta} d^3p' \int d^3x \, \tilde{\varphi}^+_{p,s_\alpha} \gamma_\mu A^\mu \tilde{\chi}_{p',s_\beta} b^+_{p,s_\alpha} d^+_{p',s_\beta} = \int_{\Delta_\alpha} d^3p \int_{\Delta_\beta} d^3p' \mathbf{M}^{A^+}_{\alpha\beta} b^+_{p,s_\alpha} d^+_{p',s_\beta}$$

The basic idea is to average the matrix elements $\mathbf{M}$ over Δ_α and Δ_β, and after multiplying by $\Delta^3 P$, we get

$$\mathcal{M}^{(A^+)}_{\alpha\beta} \int_{\Delta_\alpha} d^3p \int_{\Delta_\beta} d^3p' b^+_{p,s_\alpha} d^+_{p',s_\beta} / \Delta^3 P = \mathcal{M}^{(A^+)}_{\alpha\beta} A^+_{\alpha\beta}.$$

The multiplicity distribution in each channel is a Poisson distribution, if the Pauli principle is neglected. Therefore, coherent states are a natural choice for the bosonic Fock space states:

$$|\Phi(t)\rangle = \exp\left[-\sum_{\alpha\beta} |v_{\alpha\beta}(t)|^2/2\right] \exp\left[\sum_{\alpha\beta} v_{\alpha\beta}(t) A^+_{\alpha\beta}\right] |0\rangle, \tag{13}$$

where $|0\rangle$ is the bosonic vacuum state. The mean number of created pairs in the channel $\alpha\beta$ is given by

$$N_{\alpha\beta} = \lim_{t\to\infty} \langle\Phi(t)| A^+_{\alpha\beta} A_{\alpha\beta} |\Phi(t)\rangle = \lim_{t\to\infty} |v_{\alpha\beta}(t)|^2. \tag{14}$$

Using the variational principle in the boson space,

$$\delta\left\{\int_{t_1}^{t_2} \langle\Phi(t)| H_I - i\partial/\partial t |\Phi(t)\rangle \, dt\right\} = 0, \tag{15}$$

and varying over the coefficients $v_{\alpha\beta}$, we derive coupled channel equations for the coefficients $v_{\alpha\beta}$:

$$i\dot{v}_{\alpha\beta} = \mathcal{M}_{\alpha\beta}^{(A^+)} + \sum_{\delta(E_\delta<0)} \mathcal{M}_{\beta\delta}^{(R^{(-)})} v_{\alpha\delta} + \sum_{\gamma(E_\gamma>0)} \mathcal{M}_{\alpha\gamma}^{(R^{(+)})} v_{\gamma\beta}. \quad (16)$$

We checked the accuracy of our matrix elements with perturbative calculations. The equations of the first order perturbation theory are obtained if all coefficients v on the right-hand side of equation (16) are set to zero. Second and higher order equations result by iteration. It can analytically be shown that the first order perturbation theory with free states gives no contributions to the pair production probability. The first non-vanishing order is the second one, which is equivalent to the Weizsäcker-Williams approximation in the collider frame in the high energy limit ($\gamma \to \infty$), used for example in Refs. [4, 20].

We applied the bosonic coupled channel formalism to Pb-Pb collisions at the CERN energy $E_{lab} = 200$ GeV/nucleon. The channels represent states out of cells of the momentum spaces for electrons and positrons with volumes of $(2 \text{ MeV/c})^3$. The momentum spaces are restricted to ellipsoids with radii of 6 MeV/c in transversal direction and 10 MeV/c in longitudinal direction, which yields 200 channels per helicity value. With this choice we have 400 electron and 400 positron channels. For the bosonic channels we took all combinations into account. This means that we solved 160000 coupled differential equations.

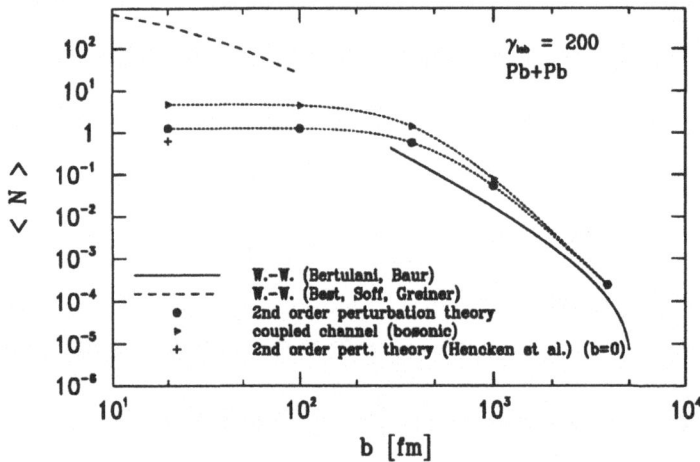

Figure 6. Mean number of created pairs in collsions of Pb^{82+} on Pb^{82+} at an incident energy of $E_{lab} = 200$ GeV/nucleon. The results of this work obtained in second order perturbation theory and by the bosonic coupled channel formalism are shown in comparison with other calculations (see legend).

Our results and other calculations are shown as functions of the impact parameter in Fig. 6. For impact parameters larger than the Compton wavelength the Weizsäcker-Williams result of Bertulani and Baur [4] agrees well with our results. However, the Weizsäcker-Williams approximation is not valid for smaller impact parameters. The results of Best et al. [20] using the Weizsäcker-Williams approach seem to be too large, since the Weizsäcker-Williams procedure is problematic at low impact parameters because of a singularity although it has been regularized by form factors. At low impact parameters our second order results are in agreement with a similar calculation of Hencken et al. [23] who applied second order perturbation theory with free plane waves at an impact parameter of zero. These authors stated that their result for $b = 0$ is useful as a good aproximation up to $b = 386$ fm as we have also found. The good agreement of our perturbative results with those of Hencken et al. [23] proves the accuracy of our matrix elements.

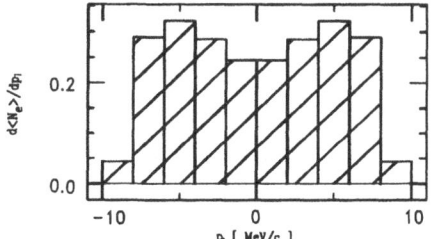
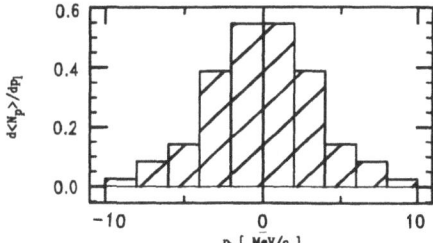

Figure 7. *Multiplicity distribution (unit c/MeV) as a function of the longitudinal momenta of the electrons (left part) and positrons (right part) obtained in coupled channel calculations for collisions of Pb on Pb at $E_{lab} = 200$ GeV/nucleon and an impact parameter of $b = 20$ fm.*

The nonperturbative enhancement at low impact parameters is not as strong as it was found for example by Momberger et al. [7] for pair production with capture. The nonperturbative coupled channel results are only enhanced by a factor of about three. The main difference between the nonperturbative and perturbative calculations are found in the momentum distribution of the electrons and positrons. Whereas these particles have mainly momenta lower than 2 MeV/c in the perturbative treatments, their distribution is strongly changed in the coupled channel calculations. As shown in Fig. 7, the electrons follow the nuclei assuming the same velocity, whereas the positrons tend to have different velocities because of the strong positive charges of the ions. This behaviour shows that the capture of electrons is important at high incident energies, too. Therefore, the incorporation of bound states into our basis should be the next step in our investigations.

Finally we calculated the cross section with the mean numbers of pairs shown in Fig. 6 via

$$\sigma = 2\pi \int_{2R}^{\infty} db\, b\, N(b), \qquad (17)$$

where R is the radius of the Pb-nuclei. The result is $1 \cdot 10^4$ barn in the case of second order perturbation theory and $2.4 \cdot 10^4$ barn in the case of coupled channel calculations.

Our result in perturbation theory can be scaled to the system S on Au yielding 223 barn, which can be compared to the experimental result of 82 ± 22 barn [24]. We want to point out that we agree with the experimental cross section if we restrict the momenta of the electrons and positrons to the measured ones. A large amount of the cross section arises from higher momenta up to about 8 MeV/c in the collider frame.

Acknowledgments

This work was supported by BMFT(06 GI 728), GSI (Darmstadt), HLRZ (Jülich), and the HRZs in Gießen, Darmstadt, and Kassel.

References

[1] W. Heitler, *The quantum theory of radiation* (Oxford University Press, London) 1957.
[2] G. Soff, *Electron-positron pair creation and K-shell ionization in relativistic heavy-ion collisions, Proceedings of the XVIII Winter School, Selected topics in nuclear structure*, eds. A. Balanda and Z. Stachura, (Bielsko-Biala, Poland 1980), p. 201.
[3] H. Gould, *Atomic Physics Aspects of a Relativistic Nuclear Collider*, Lawrence Berkeley Laboratory Technical Information, Rep. No. LBL 18593UC-28 (1984).
[4] C.A. Bertulani and G. Baur, *Phys. Rep.* **163**, 299 (1988).
[5] J. Eichler, *Phys. Rep.* **193**, 167 (1990).
[6] M.R. Strayer, C. Bottcher, V. E. Oberacker and A.S. Umar, *Phys. Rev.* **A41**, 1399 (1990); J.C. Wells, V.E. Oberacker, A.S. Umar, C. Bottcher, M.R. Strayer, J.S. Wu, G. Plunien, *Phys. Rev.* **A45**, 6296 (1992).
[7] K. Momberger, N. Grün and W. Scheid, *Z. Phys. D-Atoms, Molecules and Clusters* **18**, 133 (1991).
[8] K. Rumrich, K. Momberger, G. Soff, W. Greiner, N. Grün and W. Scheid, *Phys. Rev. Lett.* **66**, 2613 (1991).
[9] J. Hoffstadt, *diploma thesis, Universität Giessen* (1993).
[10] A.J. Baltz, M.J. Rhoades-Brown and J. Weneser, *Phys. Rev.* **A47**, 3444 (1993).
[11] J. Thiel, A. Bunker, K. Momberger, N. Grün and W. Scheid, *Phys. Rev.* **A46**, 2607 (1992).
[12] K. Momberger, N. Grün, W. Scheid and U. Becker, *J. Phys. B: At. Mol. Opt. Phys.* **23**, 2293S (1990).
[13] J. Reinhardt, B. Müller, W. Greiner and G. Soff, *Phys. Rev. Lett.* **43**, 1307 (1979).
[14] P.A. Amundsen and K. Aashamar, *J. Phys. B: At. Mol. Phys.* **14**, 4047 (1981).
[15] A. Belkacem, H. Gould, B. Feinberg, R. Bossingham, W.E. Meyerhof, *Phys. Rev. Lett.* **71**, 1514 (1993).
[16] U. Becker, N. Grün and W. Scheid, *J. Phys. B: At. Mol. Phys.* **16**, 1967 (1983).
[17] J. Reinhardt and W. Greiner, *Rep. Prog. Phys.* **40**, 219 (1977).
[18] G. Baur, *Phys. Rev.* **A42**, 5736 (1990).
[19] M.J. Rhoades-Brown, J. Weneser, *Phys. Rev.* **A44**, 330 (1991).
[20] C. Best, W. Greiner, G. Soff, *Phys. Rev.* **A46**, 261 (1992).
[21] D. Jansen, F. Dönau, S. Frauendorf, R.V. Jolos, *Nucl. Phys.* **172**, 145 (1971).
[22] P. Ring, P. Schuck, *The Nuclear Many Body Problem*, Springer Verlag, New York, Heidelberg, Berlin (1980).
[23] K. Hencken, D. Trautmann, G. Baur, submitted to *Phys. Rev.* **A** (1993).
[24] C.R. Vane, S. Datz, P.F. Dittner, H.F. Krause, C. Bottcher, M. Strayer, R. Schuch, H. Gao, R. Hutton, *Proceedings of the XIII ISIAC, Stockholm* (1993).

VIII.

Miscellaneous Topics

GAMMA AND MESON PRODUCTION BY CHERENKOV-LIKE EFFECTS IN NUCLEAR MEDIA

W. Stocker[2] and D. B. Ion[1]

[1] *Institute for Atomic Physics, Bucharest, Romania*
[2] *Sektion Physik, University of Munich, D-85748 Garching, Germany*

1. Introduction

Electromagnetic Cherenkov radiation from radio up to optical and even X-ray domains of frequencies is a well known effect in normal dielectrics with many applications [1]. Recently [2], we extended these considerations to nuclear media by introducing nuclear gamma Cherenkov radiation (NGCR) that should be emitted from charged projectiles moving through nuclei with a velocity larger than the phase velocity of photons in the medium. In addition, we developed the old idea of coherent meson production in hadronic reactions [3] and studied nuclear mesonic Cherenkov radiation (NMCR) [4].

Intuitively, real photons can be produced from the electromagnetic field around a charge by either bremsstrahlung, where the photons are shaken off by an inertial effect, by transition radiation with a sudden change of the photon phase velocity, or by the Cherenkov mechanism where the charge has to move with a velocity larger than the photon phase velocity leading to an emission of real photons from the charge. Such a simple picture can also be used in the mesonic case where - in the classical description - analogous concepts and field equations are valid.

More strictly speaking, three general conditions have to be fulfillled in order to get the Cherenkov mechanism. (i) The projectile must be coupled to a specific radiation field, that could be e. g. of (π, η, K etc.)-mesonic pseudoscalar type, (ρ, ω, K^* etc.)-vector mesonic type, (γ,weak boson etc.)-type; (ii) the radiation field must be modified inside the medium via multiple scattering from the constituents; and (iii) the particle source must move in the medium with a velocity higher than the phase velocity of the radiation field inside this medium.

The present article starts in Sect. 2 with classical approaches to the electromagnetic and scalar mesonic Cherenkov effect specialized to the nuclear medium. This section is followed by a treatment of the nuclear index of refraction (Sect. 3), both for photons and mesons, that is a basic quantity in the Cherenkov theory. The Sect. 4 is devoted to a quantum theoretical approach to the nuclear mesonic Cherenkov radiation. The Sect. 5 displays that NGCR might be an important contribution to high-energy cosmic gamma radiation.

2. Classical Theory of Cherenkov Radiation of Photons and Mesons in Nuclei

Classical electromagnetic field theory for the Cherenkov effect is based on Maxwell equations. In an analogous way, the semi-classical description of scalar and vector Cherenkov mesons is given by phenomenological Klein-Gordon and Proca equations, respectively. In this section we point out the parallelism between these descriptions of electromagnetic and mesonic Cherenkov mechanisms in nuclear media. For details of a consistent treatment we refer to the earlier investigations [2, 3] and references quoted therein.

Table 1. *Classical description of electromagnetic and mesonic Cherenkov effects (see also text).*

Electromagnetic case	Scalar mesonic case
Maxwell wave equations in nuclear medium	Klein Gordon equations in nuclear medium
$\left[\nabla^2 - \epsilon\mu \frac{\partial^2}{\partial t^2}\right] \epsilon A_0 = -4\pi\rho_e,$ $\left[\nabla^2 - \epsilon\mu \frac{\partial^2}{\partial t^2}\right] \mu^{-1}\vec{A} = -4\pi\vec{j},$ $\rho_e \equiv e\delta(x)\delta(y)\delta(z-vt),$ $\vec{j} = \rho_e\vec{v}, \quad \mu \equiv 1.$	$\left[\nabla^2 - n^2\left(\frac{\partial^2}{\partial t^2} + m^2\right)\right](\gamma\varphi) = -4\pi\rho_G\sqrt{1-v^2},$ $\rho_G \equiv G_{eff}\delta(x)\delta(y)\delta(z-vt),$ G_{eff} is the effective coupling constant, γ is an operator [4].
Energy loss per unit time into the region $b > a$	
$\left(\frac{dE}{dt}\right)_{b>a} = \frac{2}{\pi}\frac{(Ze)^2}{v} \cdot$ $\cdot Re \int_0^\infty i\omega s^* a K_1(s^*a) K_0(sa)\left(\frac{1}{\epsilon(\omega)} - v^2\right) d\omega,$ where $s^2 \equiv \frac{\omega^2}{v^2}(1-v^2\epsilon(\omega)), Res \geq 0.$	$\left(\frac{dE}{dt}\right)_{b>a} = \frac{2}{\pi} \cdot G_{eff}^2 \frac{(1-v^2)}{v} \cdot$ $\cdot Re \int_0^\infty |\gamma(\omega)|^{-2}(-i\omega)s^* a K_1(s^*a) K_0(sa) d\omega,$ where $s^2 \equiv \frac{\omega^2}{v^2}\left(1-v^2n^2(\omega)\left(1-\frac{m^2}{\omega^2}\right)\right), Res \geq 0.$
Spectral intensities	
$\frac{dN_\gamma}{d\omega} = \frac{Z^2\alpha}{v} Re\left\{-i\left(v^2 - \frac{1}{\epsilon(\omega)}\right)\left(\frac{s^*}{s}\right)^{1/2}\right\} \cdot$ $\cdot exp(-2aRes),$ where $\alpha = 1/137$.	$\frac{dN_\pi}{d\omega} \simeq G_{eff}^2 \frac{(1-v^2)}{v} |\gamma(\omega)|^{-2} Re\left(-i\left(\frac{s^*}{s}\right)^{1/2}\right) \cdot$ $\cdot exp\{-2aRes\}.$
Coherence condition	
$v \geq v_{ph}(\omega) \equiv \frac{\sqrt{2}}{(|n^2(\omega)|+Ren^2(\omega))^{1/2}} \cdot$	$v \geq v_{ph}(\omega) \equiv \frac{\omega}{q} \cdot \frac{\sqrt{2}}{(|n^2(\omega)|+Ren^2(\omega))^{1/2}}, q = (\omega^2-m_\pi^2)^{1/2}.$
Index of refraction	
$\epsilon(\omega) = n^2(\omega) = 1 + \frac{4\pi\rho}{\omega^2} C \bar{f}^{\gamma N}(\omega)$.	$n^2(\omega) = 1 + \frac{4\pi\rho}{q^2} C \bar{f}^{\pi N}(\omega)$.

For the electromagnetic case we describe the target nucleus by a dielectric function $\epsilon(\omega)$, and take the magnetic nuclear permeability $\mu(\omega) = 1$. The projectile with charge Ze moves with constant velocity v greater than the photon phase velocity $v_{ph}(\omega)$ through the target. From the Poynting vector of the moving point charge the energy loss per unit time $(dE/dt)_{b>a}$ due to electromagnetic collisions for impact parameter b greater than a is shown in Table 1. The spectral intensity $dN_\gamma/d\omega$ is obtained from the asymptotics of the Bessel functions $K_0(z)$ and $K_1(z)$. The Cherenkov coherence condition has to be fulfilled. It is equivalent to the condition

$$cos\theta = v_{ph}(\omega)/v \leq 1 \quad , \qquad (1)$$

where θ is the classical Cherenkov angle.

In a similar way the case of the scalar mesonic Cherenkov effect can be treated. The Poynting vector for the scalar meson field ϕ is given by $-\frac{1}{4\pi}\frac{\partial \phi}{\partial t}\nabla\phi$. In Table 1 the modifications of the resulting formulae - compared to the electromagnetic case - can be seen to concern in particular the effect of the non-zero rest mass and the scalar character of the emitted Cherenkov mesons.

For a weakly absorptive medium ($Imn^2 << Ren^2 - 1$) one obtains the expressions for the intensities,

$$\frac{dN_\gamma}{d\omega} = \frac{(Ze)^2}{v} \left(v^2 - \frac{1}{Ren^2(\omega)} \right) exp \left\{ \frac{-a\omega Imn^2(\omega)}{\left(Ren^2(\omega) - \frac{1}{v^2}\right)^{1/2}} \right\} \quad , v \geq v_{ph}(\omega) \quad , \qquad (2)$$

for the Cherenkov photons, and

$$\frac{dN_s}{d\omega} = G_{eff}^2 \cdot \frac{1-v^2}{v} |\gamma^{-2}| exp \left\{ \frac{-a\frac{q^2}{\omega}Imn^2(\omega)}{\left(\frac{1}{v_{ph}^2(\omega)} - \frac{1}{v^2}\right)^{1/2}} \right\} \quad , v \geq v_{ph}(\omega) \quad , \qquad (3)$$

for the Cherenkov scalar mesons.

3. Index of Refraction, Thresholds and Cross Sections for Cherenkov Photons and Mesons in Nuclear Media

As seen in Sect. 2 classical Maxwell theory predicts Cherenkov photons when the coherence condition (see Table 1) is fulfilled inside the nuclear medium, or, equivalently, when the kinetic energy T of the charged projectile is higher than the threshold energy $T_{thr}(\omega)$, given by

$$T_{thr}(\omega) = m_p \left[\left(1 - v_{ph}^2(\omega)\right)^{-1/2} - 1 \right] \quad , \qquad (4)$$

where m_p is the projectile rest mass.

For the calculation of the γ-nuclear refractive index n the multiple scattering approach [5] is used, i. e.

$$n^2(\omega) = \left(\frac{k}{q}\right)^2 = 1 + \frac{4\pi\rho}{\omega^2} \cdot C(\omega) \cdot \bar{f}^{\gamma N}(\omega) \quad , \qquad (5)$$

where ω is the photon total energy, q the free photon momentum, k the photon momentum inside the nuclear medium, $\rho = 0.17 fm^{-3}$ the nuclear density; $\bar{f}^{\gamma N}(\omega)$ is the spin-isospin averaged forward γN Compton scattering amplitude. The correlation factor $C(\omega)$ might be taken to be of Lorentz-Lorenz form (see e.g. Lax [5])

$$C(\omega) = \left[n^2(\omega) + 2\right]/3 \quad . \qquad (6)$$

Under these assumptions, using eq. (5) and the experimental data of Armstrong et al.[6] for the spin-isospin averaged Compton amplitudes $\bar{f}^{\gamma N}(\omega)$ as well as the dispersion relations data [7], the index of refraction of nuclear media and the threshold energies are obtained as displayed in Figs. 1 and 2.

It was found [8] that nuclear gamma Cherenkov radiation (NGCR) is possible in two photon energy bands. The first Cherenkov band, called CB-I, is just below the $\Delta(1236)$

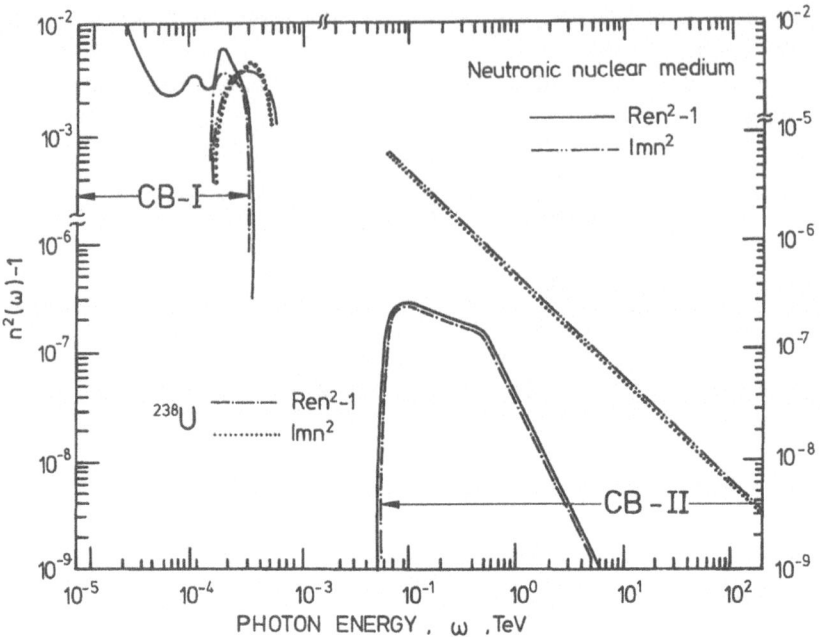

Figure 1. *Refractive index of photons inside ^{238}U and neutronic medium.*

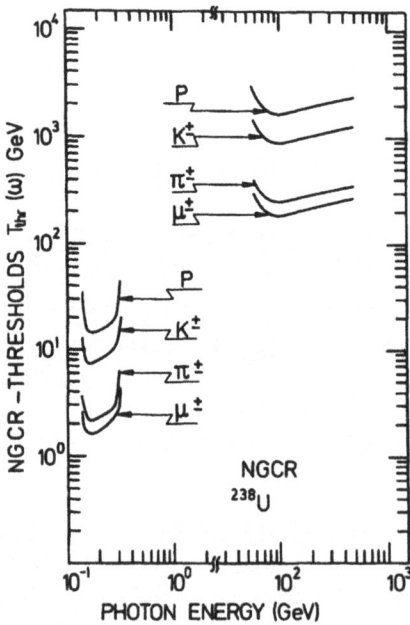

Figure 2. *NGCR threshold energies for various reactions with the target ^{238}U.*

resonance in the range

$$140\ MeV\ <\ \omega < 300\ MeV\ \text{for}\ ^{238}\text{U},\qquad(7)$$
$$20\ MeV\ <\ \omega < 300\ MeV\ \text{for a neutronic medium},$$

at projectile energies higher than the threshold energies given in Fig. 2, i. e. 15 GeV/nucleon for nuclear projectiles. The photons of the second band CB-II are predicted to be in the energy range

$$50\ GeV < \omega < 500\ GeV\ \text{or higher}\qquad(8)$$

for nuclear projectiles with energies higher than 1.5 TeV/nucleon.

For the refractive index beyond $\omega = 500\ GeV$, where the forward scattering amplitudes are not yet known experimentally, the following ansatz was used

$$Re\ n^2(\omega) = 1 + \frac{\omega_{or}^2}{\omega^2}\ ,\quad Im\ n^2(\omega) = \omega_{ot}/\omega\ .\qquad(9)$$

The parameters $\omega_{or} = 197.6\ MeV$ and $\omega_{ot} = 0.446\ MeV$ were determined from the values of the refractive index at $\omega = 500\ GeV$ that was obtained from the dispersion relation data [7]. The form (6) for n is used in literature since long as an approximate asymptotic form for the pionic refractive index in a hadronic medium. It can also be inferred from the results of ref.[9].

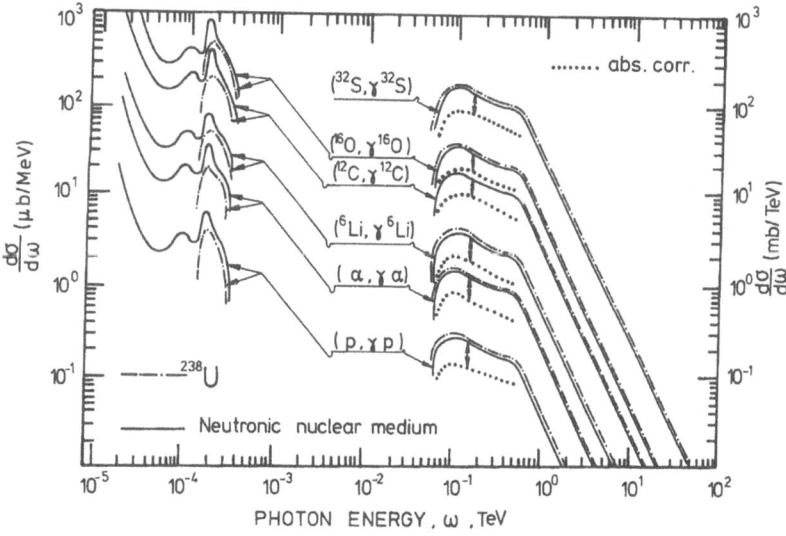

Figure 3. *NGCR differential cross sections $d\sigma/d\omega$.*

In Fig. 3 the numerical predictions for the NGCR cross sections for both bands are displayed for some nuclear reactions; targets are ^{238}U as well as a neutronic medium with

$A_T = 238$ neutrons. The cross sections are obtained starting from the spectral intensity (see Table 1) according to the formula

$$\frac{d\sigma}{d\omega} = \frac{V}{v} \cdot \frac{dN_\gamma}{d\omega} = \frac{4\pi Z^2 \alpha r_0^3}{3} \cdot \left(A_P^{1/3} + A_T^{1/3}\right)^3 \cdot \left\{1 - \left[\frac{v_{ph}(\omega)}{v}\right]^2\right\} F_A(\omega), \qquad (10)$$

where $e^2 = \alpha = 1/137$, and $V = \frac{4\pi r_0^3}{3}\left(A_P^{1/3} + A_T^{1/3}\right)^3$ is the collision volume with the nuclear radius parameter r_0; A_P and A_T are the numbers of nucleons in the projectile and target, respectively. $dN_\gamma/d\omega$ is the number of NGCR photons emitted per unit time into the energy interval $(\omega, \omega + d\omega)$. F_A is a general absorption factor that takes into account the absorption of NGCR before and after emission.

For the calculation of the refractive index of pions inside a nuclear medium we also start from the Foldy-Lax formula (5), where now the spin-isospin averaged πN scattering amplitudes in forward direction $\bar{f}^{\pi N}(\omega)$ have to be inserted. Two cases for the correlation factor $C(\omega)$ are assumed. (i) The scatterers are distributed randomly, which makes the factor C to be unity. (ii) The Lorentz-Lorenz correction is taken into account by the factor of form (6). The results obtained for the pionic index of refraction using the experimental data of Pedroni et al. [10] for the πN forward scattering amplitudes are presented in Fig. 4 of ref. [11].

Cherenkov pions were found to be possible in the energy band between 210 MeV and 320 MeV at incident projectile energies larger than 750 MeV per nucleon for both cases of C. The pion Cherenkov spectra $dN_\pi/d\omega$ in the case $\gamma(\omega) = 1$ are independent of pion energy when absorption is negligible; for $C = 1$ a narrow peak near the pion energy $\omega = 224 MeV$ is found that - for the case $C = C(LL)$ - is shifted to an energy below $\omega = 210\ MeV$, the lower limit of the Pedroni data [10].

It should be noted here (see also Sect. 4) that including higher energy data on πN scattering of ref. [12] a second and a third Cherenkov pion band show up.

4. Quantum Theoretical Approach to Cherenkov Pion Production in Nuclear Media

First, we establish the mesonic Cherenkov radiation condition using relativistic kinematics. For this in Fig. 4 the Cherenkov emission process is displayed schematically for a single

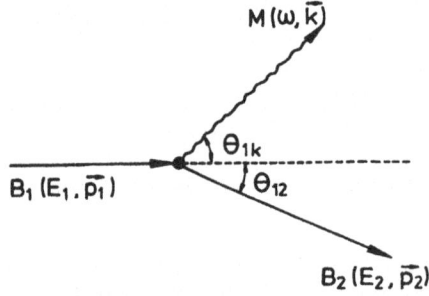

Figure 4. *The NMCR process.*

meson M (with energy ω and momentum $\vec{k} = Re\vec{k}$) that is radiated in the medium from an incident baryon B_1 (with energy E_1 and momentum $\vec{p}_1$) that itself goes over into a final baryon B_2 (with energy E_2 and momentum $\vec{p}_2$). The energy-momentum conservation in the nuclear medium requires

$$E_1 = \omega + E_2 \quad , \quad \vec{p}_1 = Re\vec{k} + \vec{p}_2 \quad . \tag{11}$$

For the initial and final baryons B_1 and B_2 we assume that the usual mass-shell relations $E_i^2 - |\vec{p}_i|^2 = M_i^2$, $i = 1, 2$, are also valid inside the nuclear medium, while for the meson M we take the energy-momentum relation

$$(Rek)^2 = (\omega^2 - m_M^2)(Ren)^2 \quad . \tag{12}$$

For the NMCR angle θ_{1k} (see Fig.4) we obtain ($|\vec{p}_1| = p_1, |Re\vec{k}| = Rek$)

$$cos\theta_{1k} = v_{ph}(\omega)/v_1 + Rek \cdot [1 - v_{ph}^2(\omega) + (M_2^2 - M_1^2)/(Rek)^2]/2p_1 \tag{13}$$

where $v_{ph}(\omega)$ is the meson phase velocity defined as

$$v_{ph}(\omega) = \frac{\omega}{Rek} = \frac{\omega}{qRen} \quad . \tag{14}$$

Now, from the condition that θ_{1k} must be a physical angle ($|cos\theta_{1k}| \leq 1$) we get the *quantum coherence condition*

$$\left| \frac{v_{ph}(\omega)}{v_1} + \frac{Rek}{2p_1}[1 - v_{ph}^2(\omega) + (M_2^2 - M_1^2)/(Rek)^2] \right| \leq 1 \quad . \tag{15}$$

At high projectile energy, when $E_1 \gg \omega$ or $p_1 \gg Rek$, the eqs.(13) and (15) go over into the expression for the classical Cherenkov angle, defined by $cos\theta_{1k} \cong v_{ph}(\omega)/v_1$, and into the classical Cherenkov coherence condition $v_{ph}(\omega) \leq v_1$, respectively. The thresholds for the kinetic projectile energy $T_1 = E_1 - M_1$, for the meson produced as NMCR in the nuclear medium, are given by

$$T_{thr}(\omega) = M_1[(1 - v_{1thr}^2)^{-1/2} - 1] \tag{16}$$

where $v_{1thr} \equiv v_{ph}(\omega)$ for the classical threshold $T_{thr}^C(\omega)$, and

$$v_{1thr} = v_1^0(\omega) \equiv \frac{v_{ph}(\omega)}{1 + F^2} + \frac{F}{(1 + F^2)^{1/2}}[1 - \frac{v_{ph}^2(\omega)}{1 + F^2}]^{1/2} \tag{17}$$

with

$$F = \frac{Rek}{2M_1}[1 - v_{ph}^2(\omega) + (M_2^2 - M_1^2)/(Rek)^2] \tag{18}$$

for the quantum threshold $T_{thr}^Q(\omega)$. We note that $v_1^0(\omega)$ in eq.(17) is the solution of the equality (15).

Now, using the dispersion relation predictions [12] as well as the Pedroni et al. data (P) [10], we made calculations on $v_{ph}(\omega), T^C_{thr}(\omega), T^Q_{thr}(\omega)$ for the π^+ emission as NMCR in the nuclear reaction $^{208}Pb(p, n\pi^+)^{208}Pb$. Our results show clearly the possibility of π^+ emission as NMCR in the following three energy bands: *CB1* in $\omega = 197 \div 313\ MeV$, *CB2* in $\omega = 910 \div 960\ MeV$, *CB3* in $\omega = 80 \div 1000\ GeV$. Similar results are obtained for π^0 emission and π^- emission as NMCR in the nuclear reactions $^{208}Pb(p, p\pi^0)^{208}Pb$ and $^{208}Pb(n, p\pi^-)^{208}Pb$, respectively, but only for the *CB1* and *CB3* pionic Cherenkov bands. For symmetric nuclei the π^+ *CB2* band is suppressed.

The transition probability (per unit time) for the emission of a pion in a NMCR process (see Fig.4) in the nuclear medium is given by the golden rule. The number of pions emitted per unit time by the Cherenkov process $N \to \pi N$ into the energy region $(\omega, \omega + d\omega)$ then follows as (see refs.[4] for details)

$$\frac{dN_\pi}{d\omega} = \frac{1}{2\pi v_1} \mid \tilde{H}_{fi} \mid^2 Rek \frac{dRek}{d\omega} \Theta(1 - v_{thr}(\omega)/v_1) \quad , \tag{19}$$

where $\tilde{H}_{fi}$ are matrix elements of the Hamiltonian describing the interaction between the effective quantized pionic field and initial and final nucleons. $\Theta(x)$ is the Heaviside step function. The matrix element $\mid \tilde{H}_{fi} \mid^2$ is calculated explicitly (see [4]),

$$\mid \tilde{H}_{fi} \mid^2 = \frac{G^2_{\pi NN}}{\mid n(\omega) \mid^2} \cdot \frac{1}{2\omega} S(E_1, \omega) \cdot F_I = \frac{G^2_{\pi NN}}{\mid n(\omega) \mid^2} \frac{k^2 - \omega^2 + (M_1 - M_2)^2}{4\omega E_1(E_1 - \omega)} \cdot F_I, \tag{20}$$

where $G^2_{\pi NN}/4\pi = 14.6$ is the usual pion-nucleon pseudo-scalar (ps) coupling constant and F_I is the isospin factor ($F_I = 1$ for $N \to \pi^0 N$, and $F_I = 2$ for $p \to n\pi^+$ or $n \to p\pi^-$ NMCR channels, respectively). Of course, the result (20) was obtained after summing and averaging over initial and final nucleon spin states, respectively. Therefore, combining (19) with (20) we obtain

$$\frac{dN_\pi}{d\omega} = \frac{G^2_{\pi NN}}{4\pi v_1} \cdot \frac{F_I}{\mid n(\omega) \mid^2} \cdot \frac{1}{v_{ph}(\omega)} \cdot \frac{dRek}{d\omega} \cdot \frac{(Rek)^2[1 - v^2_{ph}(\omega)] + (M_1 - M_2)^2}{4E_1(E_1 - \omega)} . \tag{21}$$

Now, the differential cross section $\frac{d\sigma}{d\omega}$ for the π^+ emission as NMCR in the nuclear reaction $^{208}Pb(p, n\pi^+)^{208}Pb$ can be obtained from eq.(21) by multiplying with the factor V/v_1, where V is the collision volume $V = \frac{4\pi}{3} r_0^3 (1 + A_T^{1/3})^3$ and A_T is the mass number of the target nucleus. Numerical values for $\frac{d\sigma}{d\omega}$ for the π^+ emission as NMCR in ^{208}Pb are given in Fig.5 and Table 2 for both, absorptive and non-absorptive, cases. The absorption was taken into account in the standard way using the relation

$$\left(\frac{d\sigma}{d\omega}\right)_{abs} = \left(\frac{d\sigma}{d\omega}\right)_{nonabs} \cdot < F_{abs} > \tag{22}$$

where

$$< F_{abs} > = \frac{1}{2R} \int_0^{2R} exp[-Bx] dx = \frac{1 - exp[-2BR]}{2BR} \tag{23}$$

and the absorption coefficient

$$B \equiv 2Imk(\omega) = 2qImn(\omega) = (\omega^2 - m_\pi^2)^{1/2} \cdot Imn^2(\omega)/Ren(\omega).$$

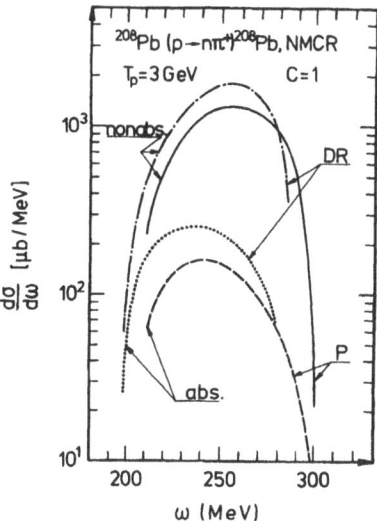

Figure 5. Differential cross sections $d\sigma/d\omega$ for $\pi^+ NMCR$ in ^{208}Pb.

Table 2. The total pion energy ω, Rek, $dRek/d\omega$, $v_{ph}(\omega)$, $\frac{d\sigma}{d\omega}$ and $<F_{abs}>$ at $T_p = 500\ GeV$ on ^{208}Pb for the two bands CB1 and CB2.

Band	ω (MeV)	Rek (MeV)	$\dfrac{d(Rek)}{d\omega}$	$v_{ph}(\omega)$	$\left(\dfrac{d\sigma}{d\omega}\right)_{nonabs}$ ($\mu b/MeV$) $T_p = 500\ GeV$	$<F_{abs}>$
$CB1$	197.7	200.6	2.015	0.98527	$3.37\ 10^{-3}$	0.448
	212.3	232.5	2.163	0.91306	$2.88\ 10^{-2}$	0.328
	227.8	264.7	2.161	0.86046	$6.00\ 10^{-2}$	0.226
	243.9	298.5	1.972	0.81709	$9.01\ 10^{-2}$	0.152
	260.5	328.8	1.561	0.79237	$9.75\ 10^{-2}$	0.102
	277.6	350.6	0.897	0.79189	$6.43\ 10^{-2}$	0.070
	295.1	357.0	-0.051	0.82653	-	0.052
	312.9	346.2	-1.313	0.90366	-	0.043
$CB2$	910.8	911.0	1.137	0.999776	$1.12\ 10^{-3}$	0.133
	930.5	932.2	1.016	0.998240	$8.71\ 10^{-3}$	0.121
	950.3	951.1	0.894	0.999187	$3.69\ 10^{-3}$	0.111

$R = r_0 A_T^{1/3}$ is the radius of the target nucleus. The values for the derivative $d(Rek)/d\omega$ are obtained from a polynomial fit and are given in Table 2, together with the values of $\left(\dfrac{d\sigma}{d\omega}\right)_{nonabs}$ and $<F_{abs}>$, for π^+ emission in $^{208}Pb(p,p\pi^0)^{208}Pb$ at the proton kinetic energy 500 GeV, for both pionic CB1 and CB2 bands.

5. NGCR as a Possible Mechanism for Production of VHEGR/UHEGR from Cosmic Sources

As already suggested in refs. [8] and [11] the NGCR effect can be considered as one of the competitive mechanisms for the production of very high energy gamma radiation (VHEGR) or/and of ultra high energy gamma radiation (UHEGR) from pulsars and X-ray binary systems.

The interest in VHEGR/UHEGR from astrophysical point sources (see e. g. ref.[13]) is stimulated by the existence of very powerful γ-ray sources in stellar systems. These are identified as neutron stars (e. g. pulsars such as Crab and Vela) or neutron stars with companion stars forming so-called binary systems (e. g. CygnusX-3, VelaX-1 etc.).

Table 3. *High-energy cosmic ray characteristics $<EGF>$ fitted with data from relevant literature.*

γ-Source	Distance (kps)	Periodicity	$<EGF>$ $eV.cm^{-2}.s^{-1}$
Crab Pulsar	2	33ms	6.45 ± 0.7
Vela Pulsar	0.5	89.2 ms	4.46 ± 1.2
PSR0355+54	1.5	156 ms	10.3 ± 2.6
Vela X-1	1.4	8.96 d	24.2 ± 8.4
LMC X-4	55	1.41 d	36.8 ± 13.6
Geminga	0.1	595	50 ± 30
Sco X-1		0.8 d	48 ± 10
Cygnus X-3	> 11.4	4.8 h	141.0 ± 27.3
Crab Nebula	2	variable	106 ± 25
M31 Andromeda	670		220 ± 70
Centaurus X-3	10	2.09 d	247 ± 171

VHEGR/UHEGR are produced by ultrarelativistic charged particles (e. g. electrons. muons, mesons, protons etc.) in interaction with the ambient medium by different mechanisms. Because of their coherence properties the nuclear (γ, π, etc.)-Cherenkov like effects must be considered among the favourite candidates for the production of VHEGR/UHEGR. Our results, presented in Fig. 3, show that the NGCR mechanism can become important at least in the spectral regions of CB-I and CB-II where the gamma coherence condition $v_{ph}(\omega) \leq v$ is fulfilled. The main advantage of the NGCR mechanism is that it can be well integrated into the pulsar or binary system scenarios and theories. The characteristics of the NGCR spectrum is essentially independent of the spectrum of the radiating particles, and the γ-Cherenkov photons are perfectly linearly polarized. The NGCR spectrum for VHEGR/UHEGR depends essentially on $[Re\, n^2(\omega) - 1]$. If the refractive index is continued into the region beyond $\omega = 500\, GeV$, as described by eq.(9) in Sect.3, then the differential flux $\frac{dF}{d\omega}$ and the integral flux $F(> E)$ predicted by the NGCR mechanism are given by

$$\frac{dF}{d\omega} \sim \omega^{-2} \quad and \quad F(> E) \sim E^{-1}. \tag{24}$$

These predictions are in surprising agreement with the experimental data from some important VHEGR/UHEGR sources, such as Crab and Vela pulsars as well as CygnusX-3 and VelaX-1 binary systems. The characteristic constants $< EGF >$, defined for each source by the NGCR fit to the data for the integral quantity $EGF \equiv E * F(\geq E)$, are presented in Table 3. However, for the confirmation of a NGCR nature of observed γ-ray fluxes from these sources more experimental data and, in addition, polarization measurements are needed.

6. Summary and Conclusions

In this paper a classical description of electromagnetic and scalar mesonic Cherenkov mechanisms in nuclear media was presented. The implications of high energy Cherenkov photons for the production processes of high energy cosmic gamma radiation was also discussed. Then, a quantum theoretical approach to Cherenkov pion production in nuclei was given. It should be noted that π^+-NMCR turned out to be an effect competitive with the "Δ-hole" excitation mode in ref.[14], even in light nuclei such as ^{12}C. These two processes, however, are of totally different character. The Cherenkov pions are radiated spontaneously as real pions from the projectiles while the $\Delta-$hole pions come from the excitation of the medium, i. e. from a doorway state [14]. The two kinds of pions belong to different branches of the pion spectrum in the nuclear medium. The Cherenkov pions must be coplanar with the incoming and outgoing projectile, furthermore they must fulfil the coherence condition correlating emission angle and energy.

We plan to extend our investigations to other mesons, including vector and strange mesons.

Acknowledgement

This work was partly supported by COSY (Jülich). We thank J. Speth for helpful discussions.

References

[1] P.A. Cerenkov, *Compt. Rend.* **2**, 451 (1934); I. Frank and I. Tamm, *C. R. Acad. Sci. USSR* **14**, 109 (1937); J. V. Jelley, *Cerenkov Radiation and its Applications*, (Pergamon Press, London-New York-Los Angeles, 1958); V. P. Zrelov, *Cerenkov Radiation in High-Energy Physics*, (Atomizdat, Moskwa, 1968), translated from Russian (Israel Program for Scientific Translations, Jerusalem, 1970); V. A. Bazylev et al., *Sov.Phys. JETP* **54**(5), 884 (1981); V. A. Bazylev and N. K. Zhevago, *Sov. Phys. Usp.* **25**, 565 (1982)

[2] D. B. Ion and W. Stocker, *Phys. Lett. B* **258**, 262 (1991); *Phys. Lett. B* **262**, 491 (Err.) (1991); *Ann. of Phys. (N.Y.)* **213**, 355 (1992)

[3] W. Wada, *Phys. Rev.* **75**, 981 (1949); D. Ivanenko and Gurgenidze, *Dokl. Acad. Nauk* **67**, 555 (1949); D. I. Blochintzev and V. L. Indenbom, *JETP* **20**, 1123 (1950); W. Czyz, T. Ericson and S. L. Glashow, *Nucl. Phys.* **13**, 516 (1959); W. Czyz and S. L. Glashow, *Nucl. Phys.* **20**, 309 (1960); G. Yekutieli, *Nuovo Cimento* **13**, 446 (1959); P. Smrž, *Nucl. Phys.* **35**, 165 (1962); D. B. Ion, *Studii Cercet. Fiz.* **22**, 125 (1970); D. B. Ion, *Doctoral Thesis, University of Bucharest*, 1971; D. B. Ion and F. G. Nichitiu, *Nucl. Phys.* **29**, 547 (1971); R. F. Sawyer, *Nucl. Phys. A* **271**, 235 (1976); D. F. Zaretskii and V. V. Lomonosov, *JETP Lett.* **23**, 614 (1976); R. F. Sawyer and A. Soni, *Phys. Rev. Lett.* **38**, 1383 (1977); I. M. Dremin, *JETP Lett.* **30**, 140 (1979); G. Bertsch, G. E. Brown, V. Koch and B. A. Li, *Nucl. Phys. A* **490**, 745 (1988); G. E. Brown, E. Oset, M. V. Vacas and W. Weise, *Nucl. Phys. A* **505**, 823 (1989); G. E. Brown, *Comments Nucl. Part. Phys.* **19**, 185 (1990); D. B. Ion, *Rev. Roum. Phys.* **36**, 595 (1991)

[4] D. B. Ion and W. Stocker, *Phys. Lett. B* **273**, 20 (1991); *Phys. Rev. C* **48**, 1172 (1993); *Rom. J. Phys.* **38**, in print (1993)

[5] L. L. Foldy, *Phys. Rev.* **67**, 107 (1945); M. Lax, *Rev. Mod. Phys.* **23**, 287 (1951)
[6] J. A. Armstrong et al., *Phys. Rev. D* **5**, 1640 (1972); *Nucl. Phys. B* **41**, 445 (1972)
[7] O. Dumbrajs, *Nucl. Phys. B* **149**, 264 (1979)
[8] D. B. Ion and W. Stocker, *Phys. Lett. B* **311**, 339 (1993)
[9] D. B. Ion and F. G. Nichitiu, *Nucl. Phys. B* **29**, 547 (1971); *Roum. J. Phys.* **37**, 763 (1992); D. B. Ion, A. Rosca and C. Petrascu, *Roum. J. Phys.* **37**, 991 (1992)
[10] E. Pedroni et al., *Nucl. Phys. A* **300**, 321 (1978)
[11] W. Stocker and D. B. Ion, *Proceedings of NATO Adv. Study Inst., Topics in Atomic and Nuclear Collisions*, Predeal 1992
[12] G. Höhler et al., *Handbook of Pion-Nucleon Scattering*, Karlsruhe Report Nr. **12-1** (1979)
[13] P. W. Ramana Murthy and A. W. Wolfendale, *Gamma-Ray Astronomy*, and references therein (Cambridge University Press, 1993); T. C. Weck, *Phys. Rep.* **160**, 1 (1988) and refs. quoted therein
[14] P. Oltmanns, F. Osterfeld and T. Udagawa, *Phys. Lett. B* **299**, 194 (1993)

INSTANTONS IN QCD

Calin Alexa

Department of High Energy Physics, Institute of Atomic Physics, Bucharest, P.O. Box MG-6, Romania

I will briefly review some of the basic instanton concepts. Then I will present some applications of instantons in QCD.

Because only at high energy QCD becomes perturbative, we need some special solutions of the Euler-Lagrange equations at low energy that give us the main contribution. The equations are obtained from the minimal value of the action S, because we always have $\exp(-S)$ in our formulas. To be more specific, in the field theory we always have relations like this: $\langle O \rangle \sim \int d\mu O \exp(-S)$, where $\langle O \rangle$ is the mean value. We define instantons as solutions of the Euler-Lagrange equations that give the minimal value of the action S.

In 1976 Belavin[1] et. al. found instantons using a pure Yang-Mills Lagrangian: $L = -\frac{1}{4}F^2$ for $SU(2)$ gauge fields. After this a lot of work has been done to find $SU(3)$ instantons and the general $SU(N)$ instantons. It is well-known that the proper geometrical approach of gauge fields is the modern theory of fibre bundles. We know from the book of Atiyah[2] that in principle we can compute any $SU(N)$ instanton, but until now nobody has obtained this solution.

If we work with a $SU(2)$ gauge theory we can compute at low energy a lot of interesting things[3]. It is possible to find $SU(N)$ instantons more readily but not in a very clean approach[4]. Here one uses the field strength formulation of non-Abelian YM-theories that yields the so-called field strength approach (FSA) of YM-theories. With the isotropic ansatz $F^a_{\mu\nu} = G^a_{\mu\nu}\Psi(x^2)$ ($G^a_{\mu\nu}$ is a constant) and with the $SU(3)$ embedding of the $SU(2)$ instanton we can get the $SU(3)$ instanton configuration. Then we can compute the quark condensation in the gluonic background. It seems that everything works, but for realistic applications of instantons in QCD the Lagrangian is not simply F^2. If we have a different Lagrangian one is tempted to interpret these solutions as a conglomerate of condensed instantons.

It is very interesting to mention that it is possible to observe QCD instantons[5]. If at high energy the QCD-instanton-induced cross section reaches its unitarity limit, this process should be considered as the main source of the large transverse energy events. If the emission of a large number of gluons compensates no more than one half of the exponential suppression of the instanton amplitude, the observation will be not so easy, but at $E_T \sim 30 GeV$ it can be hoped to select the events induced by the QCD instantons.

It is very interesting to use the instantons in supersymmetric gauge theories[6]. This SUSY approach has the possibility of performing explicit and finite calculations of condensates that are in contrast with one's inability of doing so in ordinary gauge theories such as QCD, where numerical methods are instead necessary. Unfortunately it is not yet known how supersymmetry will eventually be incorporated in the body of particle physics, but is very interesting to see the effects of SUSY non-perturbative approach in the properties of confinement and chiral symmetry breaking.

I wish to thank Laurentiu Lazarovici for helpful discussions during my work in this domain, Wolfgang Sakuler for his stimulating discussions about possible applications.

References

[1] A. A. Belavin et al., *Phys. Lett. B* **59**, 85 (1975)
[2] M. F. Atiyah, *Geometry of Yang-Mills Fields*, Bologna (1979)
[3] E. Reya, *The Spin Structure of the Nucleon*, Workshop on *QCD - 20 Years Later*, Aachen, June 1992
[4] K. Langfeld, H. Reinhardt, Preprint UNITU-THEP-18/1992
[5] I. I. Balitsky, M. G. Ryskin, *Phys. Lett. B* **296**, 185-190 (1992)
[6] D. Amati et al., *Phys. Rep.* **162**, No.4, 169-248 (1988)

WIGNER DISTRIBUTION FOR THE HARMONIC OSCILLATOR WITHIN THE THEORY OF OPEN QUANTUM SYSTEMS

Aurelian Isar

Department of Theoretical Physics, Institute of Atomic Physics, PO Box MG-6, RO-76900 Bucharest, Romania

Because dissipative processes imply irreversibility and, therefore, a preferred direction in time, it is generally thought that quantum dynamical semigroups are the basic tools to introduce rigorously the dissipation in quantum mechanics. The most general form of the generators of such semigroups was given by Lindblad[1]. This formalism has been studied for the case of damped harmonic oscillators[2,3] and applied to various physical phenomena, for instance, to the damping of collective modes in deep inelastic collisions in nuclear physics[4]. According to the axiomatic theory of Lindblad[1], the usual von Neumann-Liouville equation ruling the time evolution of closed quantum systems is replaced in the case of open systems by the following time-homogeneous quantum mechanical Markovian master equation for the density operator ρ:

$$\frac{d\rho(t)}{dt} = -\frac{i}{\hbar}[H,\rho(t)] + \frac{1}{2\hbar}\sum_j([V_j\rho(t),V_j^+] + [V_j,\rho(t)V_j^+]). \qquad (1)$$

Here H is the Hamiltonian of the system and V_j, V_j^+ are operators on the Hilbert space of the Hamiltonian. To obtain an exactly solvable model for the damped quantum harmonic oscillator, the Hamiltonian H is chosen of the form $H = H_0 + (\mu/2)(pq+qp)$, $H_0 = p^2/(2m) + m\omega^2 q^2/2$ and the operators V_j are taken as polynomials of first degree of the basic observables q and p of the one-dimensional quantum mechanical system. With these choices and introducing the annihilation and creation operators, the Markovian master equation (1) can be written[5]:

$$\frac{d\rho}{dt} = \frac{1}{2}(D_1+\mu)(a^+a^+\rho - a^+\rho a^+) + \frac{1}{2}(D_1-\mu)(\rho a^+a^+ - a^+\rho a^+)$$
$$+\frac{1}{2}(D_2+\lambda+i\omega)(a\rho a^+ - a^+a\rho) + \frac{1}{2}(D_2-\lambda-i\omega)(a^+\rho a - \rho aa^+) + H.c., \qquad (2)$$

where $D_1 = (m\omega D_{qq} - D_{pp}/(m\omega) + 2iD_{pq})/\hbar$, $D_2 = (m\omega D_{qq} + D_{pp}/(m\omega))/\hbar$; D_{pp}, D_{qq} and D_{pq} are the diffusion coefficients and λ the friction constant. They satisfy the following fundamental constraints[3]: $D_{pp} > 0$, $D_{qq} > 0$, $D_{pp}D_{qq} - D_{pq}^2 \geq \lambda^2\hbar^2/4$.

Generally the master equation for the density operator gains considerably in clarity if it is represented in terms of the Wigner distribution function. The Fokker-Planck equation, which is the c-number equivalent equation to the Lindblad master equation (2) and which is satisfied by the Wigner distribution function $W(x_1, x_2, t)$, where x_1 and x_2 are real coordinates (the position and momentum of the harmonic oscillator), has the form[5]:

$$\frac{\partial W}{\partial t} = \sum_{i,j=1,2} A_{ij} \frac{\partial}{\partial x_i}(x_j W) + \frac{1}{2} \sum_{i,j=1,2} Q_{ij} \frac{\partial^2}{\partial x_i \partial x_j} W, \qquad (3)$$

where $A = \begin{pmatrix} \lambda - \mu & -\omega \\ \omega & \lambda + \mu \end{pmatrix}$, $Q = \frac{1}{\hbar} \begin{pmatrix} m\omega D_{qq} & D_{pq} \\ D_{pq} & D_{pp}/(m\omega) \end{pmatrix}$. Since the drift coefficients are linear in the variables x_1 and x_2 and the diffusion coefficients are constant with respect to x_1 and x_2, eq. (3) describes an Ornstein-Uhlenbeck process[6]. The resulting differential equations of the Fokker-Planck type for the distribution functions can be solved by standard methods[6] and observables directly calculated as correlations of these distribution functions.

1) When the Fokker-Planck equation (3) is subject to a Gaussian (wave packet) type of the initial condition $W_w(x_1, x_2, 0) = (2/\pi) \exp\{-2[(x_1 - x_{10})^2 + (x_2 - x_{20})^2]\}$, the solution is found to be (x_{10} and x_{20} are initial values of x_1 and x_2, respectively):

$$W_w(x_1, x_2, t) = \frac{1}{\pi\sqrt{|B_w|}} \exp\{-\frac{1}{2B_w}[\phi_w(x_1-\bar{x}_1)^2 + \psi_w(x_2-\bar{x}_2)^2 + \chi_w(x_1-\bar{x}_1)(x_2-\bar{x}_2)]\},$$

where $B_w = g_1 g_2 - g_3^2/4$, $g_1 = g_2^* = q^* e^{2\nu_1 t} + D_{11}(e^{2\nu_1 t} - 1)/(2\nu_1)$, $g_3 = e^{-2\lambda t} + D_{12}(1 - e^{-2\lambda t})/\lambda$, $\phi_w = g_1 a^{*2} + g_2 a^2 - g_3$, $\psi_w = g_1 + g_2 - g_3$, $\chi_w = 2(g_1 a^* + g_2 a) - g_3(a + a^*)$ with $\nu_1 = \nu_2^* = -\lambda - i\Omega$, $q = \mu(\mu + i\Omega)/(2\omega^2)$, $a = (\mu - i\Omega)/\omega$ and $\Omega^2 = \omega^2 - \mu^2$. The functions $\bar{x}_1$ and $\bar{x}_2$, which are also oscillating functions, are given by $\bar{x}_1 = e^{-\lambda t}[x_{10}(\cos \Omega t + (\mu/\Omega)\sin \Omega t) + x_{20}(\omega/\Omega)\sin \Omega t]$, $\bar{x}_2 = e^{-\lambda t}[x_{20}(\cos \Omega t - (\mu/\Omega)\sin \Omega t) - x_{10}(\omega/\Omega)\sin \Omega t]$.

2) If the Fokker-Planck equation (3) is subject to the δ-function type of initial condition, the Wigner distribution function is given by

$$W(x_1, x_2, t) = \frac{1}{2\pi\sqrt{|B|}} \exp\{-\frac{1}{2B}[\phi_d(x_1 - \bar{x}_1)^2 + \psi_d(x_2 - \bar{x}_2)^2 + \chi_d(x_1 - \bar{x}_1)(x_2 - \bar{x}_2)]\},$$

where $B = f_1 f_2 - f_3^2$, $f_1 = f_2^* = D_{11}(e^{2\nu_1 t} - 1)/(2\nu_1)$, $f_3 = D_{12}(1 - e^{-2\lambda t})/(2\lambda)$, $\phi_d = f_1 a^{*2} + f_2 a^2 - f_3$, $\psi_d = f_1 + f_2 - f_3$ and $\chi_d = 2(f_1 a^* + f_2 a) - f_3(a + a^*)$. So, the Wigner functions are two-dimensional Gaussian distributions with the average values $\bar{x}_1$ and $\bar{x}_2$ and different widths. When $t \to \infty$, $\bar{x}_1$ and $\bar{x}_2$ vanish and we obtain the steady state solution: $W(x_1, x_2) = \exp[-\sum_{i,j=1,2}(\sigma)_{ij}^{-1}(\infty)x_i x_j/2]/(2\pi\sqrt{\det\sigma(\infty)})$. The stationary covariance matrix $\sigma(\infty)$ can be determined from the equation $A\sigma(\infty) + \sigma(\infty)A^T = Q$.

References

[1] G. Lindblad, *On the generators of quantum dynamical semigroups*, Commun. Math. Phys. **48**, 119 (1976)
[2] H. Dekker, *Classical and quantum mechanics of the damped harmonic oscillator*, Phys. Rep. **80**, 1 (1981)
[3] A. Sandulescu and H. Scutaru, *Open quantum systems and the damping of collective modes in deep inelastic collisions*, Ann. Phys. (N.Y.) **173**, 277 (1987)
[4] A. Isar, A. Sandulescu and W. Scheid, *Use of a characteristic function in open quantum systems and charge equilibration in deep inelastic reactions*, J. Phys. G **17**, 385 (1991)
[5] A. Isar, W. Scheid and A. Sandulescu, *Quasiprobability distributions for open quantum systems within the Lindblad theory*, J. Math. Phys. **32**, 2128 (1991)
[6] C. W. Gardiner, *Handbook of Stochastic Methods*, (Springer, Berlin, 1982)

THE QUANTUM DEFORMATION OF su(2) INTO su(1,1) AND THE POTENTIAL PICTURE

A. Ludu

Department of Theoretical Physics, University of Bucharest, Bucharest-Magurele PO Box MG-5211, Romania

The quantum group $su_q(2)$ generated by the operators $J_\pm$ and J_3 has its deformed Casimir in the form [1]:

$$C = [J_3]^2 + \frac{1}{2}(J_+J_- + J_-J_+). \tag{1}$$

The states of the representation are uniquely specified by the relations:

$$\begin{aligned} C|A,m> &= A|A,m>, \quad J_3|A,m> = m|A,m>, \\ J_\pm|A,m> &= ([j \mp m][j \pm m + 1])^{1/2}|A, m \pm 1>. \end{aligned} \tag{2}$$

In order to obtain the continuous part of the spectrum, starting from the discrete spectrum and using the commutator relations and the hermiticity conditions [2] we get

$$J_\pm J_\mp |A,m> = \left(C - [J_3 \mp \tfrac{1}{2}]^2\right)|A,m> \tag{3}$$

and consequently

$$A \geq \left[m \pm \frac{1}{2}\right]^2 \tag{4}$$

with $m \in Z$ or $\tfrac{1}{2}Z$. We choose $A \geq \frac{1}{\sin^2 s}$, and all integer and half-integer values of m are allowed. Thus we obtain for $su_q(2)$ the continuous series (C_x^o and $C_x^{1/2}$) of the representations of $su(1,1)$, taking also $A = [j][j+1]$ with $j = \tfrac{1}{2} + i\sigma$. For $s = \pi/2$ the imaginary part of A is zero and we have $A = \mathcal{R}e([j][j+1]) = [j][j+1] = 2\cosh(\sigma\pi) \geq 2$. Since in this case $[m + \tfrac{1}{2}]^2$ has only the values $\pm 1/\sqrt{2}$, 0 and ± 1 (depending whether m is odd, even or half-integer), the inequality given in eq.(4) is fulfiled for any integer or half-integer value of m.

We investigate the structure of the eigenvalues of J_+, C and $[J_3]^2$ for some limiting values of the parameter s (0, $\pi/2$ and π) for j and m integers or half-integers. The limit $s = 0$ ($s = 2k\pi$) gives the $su(2)$ properties. In the case $s \to \pi$, $J_\pm$ behaves exactly like in the $su(1,1)$ case :

$$J_\pm\Big|_{su_{s=\pi}(2)} |jm> = iJ_\pm\Big|_{su(2)} |jm>. \tag{5}$$

Then the commutator relations for $su_q(2)$ are the same as those of $su(1,1)$ when one uses the analytical prolongation of the operators $J_\pm$. For certain values of s we get a degeneracy of these levels, like quadruplets, triplets, doublets and even full degeneracy (one single level, for $s = 3\pi/2$).

By using a further extension for the q-deformed algebra $su_q(2)$ a four-dimensional Lie algebra can be obtained as a limiting symmetry of the $su_q(2)$ algebra in the form:

$$[J_3, J_\pm] = \pm J_\pm, \quad [J_+, J_-] = a[J_3] + b[I], \quad [J_\pm, I] = [J_3, I] = 0 \tag{6}$$

with the Casimir operator given by:

$$C_{a,b} = J_+ J_- + \frac{a}{2}[J_3]^2 + (b - \frac{a}{2})[J_3]. \tag{7}$$

We analyse eqs.(6) using the general Hopf algebra deformation in order to re-obtain all the limiting symmetries. Here the deformation operator is given by the expression:

$$\phi(J_3, q) = 2\frac{f^\alpha(J_3, q) - f^{-\alpha}(J_3, q)}{\alpha h(q)}. \tag{8}$$

If we take $\alpha = 1$ and consider that the function f is holomorphic in a neighbourhood of zero, we can use its Taylor expansion in a formal variable x and the coefficients $C_k(q)$. We obtain a recurrence relation for the coefficients and get:

$$f(J_3, q) = 1 + \frac{a(q)h(q)}{2}J_3 + \sum_{k \geq 0} C_{2k}(q) J_3^{2k}. \tag{9}$$

A coordinate representation of the $su_q(2)$ algebra and the corresponding potential approach in terms of the polar coordinates $\phi \in [0, 2\pi)$, $\rho \geq 0$ is given by:

$$J_\pm = g_q^\pm(\phi)\left(\pm\partial_\rho + f_1(\rho)\left(i[\partial_\phi]_{Defk} \mp \frac{1}{2}\right) + f_2(\rho)\right), \quad J_3 = -i\partial_\phi. \tag{10}$$

When we apply C on the wave function $\Psi_{jm}(\rho, \phi)$ we get the differential equation for the "radial" part of the wave function

$$\left(-2\partial_\rho^2 - 4f_1^2[im]^2 + 2f_1(1-\eta)\partial_\rho + i[im](2f_1 f_2 + 1)\right.$$
$$\left. + 2f_2^2 + f_1' + \frac{1}{2}f_1^2(2\eta - 1)\right)R_{jm} = [j][j+1]R_{jm} \tag{11}$$

This equation clearly represents a generalisation of the corresponding Schrödinger equation for $su(2)$ and $su(1,1)$ cases and shows a large variety of possibilities for potential shapes.

References

[1] J. Cseh, R. K. Gupta, A. Ludu, W. Greiner and W. Scheid, *J. Phys. G: Nucl. Part. Phys.* **18**, L73 (1992)

[2] N. A. Gromov and V. I. Mankov, *J. Math. Phys* **33**, 1374 (1992)

POTENTIALS IN THE ALGEBRAIC SCATTERING THEORY

A. Zielke and W. Scheid

Institut für Theoretische Physik, Justus-Liebig-Universität Gießen, Heinrich-Buff-Ring 16, D-35392 Gießen, Germany

The algebraic scattering theory (AST) allows a completely algebraic determination of the S–matrix for scattering systems with a given dynamical symmetry [1, 2]. Of particular interest for practical applications is the AST with $SO(1,3)$ or $SO(2,3)$ dynamical symmetry, since the S–matrix for Coulomb scattering appears as a special case of the more general, algebraically derived S–matrices. In an earlier contribution [3] we showed that the algebraic S–matrix is not unique, but comprises two different classes of S–matrices with algebraically undetermined phase factors. Since in the algebraic formalism the scattering potentials do not appear explicitly, the question naturally arises, how potentials corresponding to a given symmetry can be constructed. In the following, this problem is solved for the case of an $SO(1,3)$ and an $SO(2,3)$ dynamical symmetry.

The starting point for the derivation of scattering potentials is the observation that for each group the Hamiltonian can be written as a function of the quadratic Casimir invariant:

$$H = -\frac{C^{SO(1,3)} + 1}{\eta^2} E ,$$

$$H = -\frac{C^{SO(2,3)} + 9/4}{\eta^2} E .$$

Here $\eta = (Z_1 Z_2 e^2)/k$ denotes the Sommerfeld parameter and $E = k^2/2$ the scattering energy ($\hbar = m = 1$).

The Casimir invariants are quadratic functions in the generators of the group algebra:

$$C^{SO(1,3)} = \vec{L}^2 - \vec{K}^2 \qquad \text{6 generators,}$$
$$C^{SO(2,3)} = \vec{L}^2 + V^2 - \vec{A}^2 - \vec{B}^2 \qquad \text{10 generators.}$$

By constructing realizations of the generators in terms of differential operators one obtains realizations of the Casimir invariant and thus of the Hamiltonian. In general, the operator obtained in this way does not correspond to a Schrödinger type Hamiltonian. However, if one admits only realizations which are up to fourth order in the momentum operators and up to second order in the position operators (or vice versa), it is possible to transform the resulting equation into a Schrödinger equation by means of symmetry preserving transformations. The construction of appropriate realizations followed by a combination of several transformations leading finally to a Schrödinger equation is a highly non-trivial task. We found two different methods solving the problem.

The results are the following [4]:

- The potentials corresponding to an $SO(1,3)$ dynamical symmetry are special cases of the potentials corresponding to an $SO(2,3)$ dynamical symmetry.
- We only get two different classes of potentials. By the first method we obtain potentials of Pöschl–Teller type

$$V(r) = -\frac{E}{\eta^2}\left[l(l+1)\left(\frac{1}{(\frac{kr}{\eta})^2} - \frac{1}{\sinh^2(\frac{kr}{\eta})}\right) + \frac{v^2 - 1/4}{\cosh^2(\frac{kr}{\eta})}\right], \quad (1)$$

and by the second method those of Coulomb type

$$V(r,\vec{p}) = \frac{Z_1 Z_2 e^2}{r} + (v^2 - 1/4)\frac{2E}{r}\left(Z_1 Z_2 e^2 + r(E - \frac{\vec{p}^2}{2})\right)^{-1}\left(\frac{E - \vec{p}^2/2}{E + \vec{p}^2/2}\right)^2. \quad (2)$$

- In the case of the group $SO(2,3)$ the potentials contain the free parameter v. It allows the variation of the shape and the depth of the potential. For $v^2 = 1/4$ the potentials with $SO(2,3)$ dynamical symmetry reduce to the potentials with $SO(1,3)$ dynamical symmetry.
- The potentials are independent of the realization we start from.
- The modifications of the pure Coulomb potential in the $SO(2,3)$–AST are non–local. As seen from eq. (2) they are given by a complicated operator expression involving position and momentum operators.
- The S–matrices corresponding to the potentials (1) and (2) can be calculated analytically in the traditional scattering theory. The expressions agree with the form of the algebraic S–matrices.

We also considered generalizations of the $SO(1,3)$–AST to the situation when one has coupled reaction channels. As proposed by Alhassid, Iachello et al. [5] the algebraic S–matrix in the multi–channel case is obtained from the algebraic one–channel S–matrix by letting the c–number matrix elements S_l become matrices. Although it was attempted to justify this phenomenological procedure on algebraic grounds, a close look reveals that the approach amounts to leaving the algebraic foundations of the theory. Moreover, if one sticks to the ansatz of the one–channel theory that the Hamiltonian is a function of only the Casimir invariant of the chosen group, it is easy to see that there is no coupling of the channels. Therefore, we propose a different approach where the coupling of channels is achieved in an algebraic way by writing the Hamiltonian as a function of the Casimir invariant plus additional step–up and step–down operators. Work on this problem is in progress.

References

[1] Y. Alhassid, F. Gürsey and F. Iachello, *Ann.Phys.* **167**, 181 (1986)
[2] J. Wu, F. Iachello and Y. Alhassid, *Ann.Phys.* **173**, 68 (1987)
[3] A. Zielke and W. Scheid, *J. Phys. A: Math. Gen.* **25**, 1383 (1992)
[4] A. Zielke and W. Scheid, *J. Phys. A: Math. Gen.* **26**, 2047 (1993)
[5] Y. Alhassid and F. Iachello, *Nucl. Physics A* **501**, 585 (1989)

JAHN-TELLER DISTORTED EXCITED STATES OF THE C_{60} CLUSTER

Péter Surján[1,2], László Udvardi[2] and Károly Németh[1]

[1] *Dept. of Theoretical Chemistry, Eötvös University, H–1518 Budapest 112, POB 32, Hungary*
[2] *Quantum Theory Group, Institute of Physics, TU Budapest H–1111, Budafoki út 8, Budapest, Hungary*

Introduction: The geometry and the electronic structure of the ground state of the isolated cluster have been the subject of several theoretical investigations[1–4]. Similarly, the ionized structures C_{60}^-, C_{60}^+ have recently been studied in detail[5–8]. While the ground state of this cluster is known to have an icosahedral nuclear configuration, the geometry will be distorted in the degenerate ionized and excited states, according to the Jahn–Teller theorem. The present paper aims to discuss the energy and geometry of excited states C_{60}. The low–lying triplet and singlet states have been evaluated in some distorted states.

Within the π–electron approximation we built up the following extended–Hubbard Hamiltonian:

$$H = \sum_\mu \alpha_\mu \sum_\sigma a^+_{\mu\sigma} a_{\mu\sigma} + \sum_i \beta(r_i) \sum_\sigma \left(a^+_{i_1\sigma} a_{i_2\sigma} + h.c. \right) + \sum_i \gamma(r_i) n_{i_1} n_{i_2} + \sum_i f(r_i) \quad (1)$$

The index μ runs over all atomic sites, i labels bonds, i_1 and i_2 are the two sides of bond i, σ is the spin label. The particle number operator for the site μ is defined as $n_\mu = \sum_\sigma a^+_{\mu\sigma} a_{\mu\sigma}$. The $\beta(r)$, $\gamma(r)$, $f(r)$ are unique functions of the bond lengths denoted by r_i : $\beta(r) = -A exp(-\frac{r}{\zeta})$, $\gamma(r) = \frac{1}{\epsilon r}$. Since our Hamiltonian handles only the π–electrons, the $f(r)$ function has the role to describe the effect of the σ and core electrons. For the determination of the bond distances the $r_i = r_0 - P_{i_1,i_2}$ Coulson relation between the bond length and the P_{i_1,i_2} element of the first order density matrix was applied. With the requirement $\frac{\partial E}{\partial r_i} = 0$ the form of $f(r)$ can be determined[9].

The parameters of the model were obtained by fitting the bond lengths and the gap of the polyacetylene and the first triplet and singlet excitation energies of ethylene to experiments. As we checked, this set of parameters gives reliable results for butadiene, benzene and naphtalene.

The model was solved at Hartree–Fock level, the excited state wave function were determined by means of the Tamm–Dancoff approximation. Monitoring the Jahn-Teller distortion is possible by reducing the degenerate subspace of excited-state wave functions according to the irreducible representations of the relevant subgroups by appropriate projections. We considered the subgroups D_{5d}, D_{3d} and D_{2h} which have one-dimensional (i.e. Jahn-Teller inactive) irreducible representations contained in the degenerate icosahedral excited states. Depending on the selected particular subgroup (and irreducible representation), one will arrive at different excited state wave functions, excited densities and thus different excited state geometries.

Table 1. Total energies and Jahn–Teller distortions of the optimized excited states of C_{60}

Subgroup	Irred. Repr.	Total energy of the excited state (eV)		Jahn–Teller energy (meV)	
		S = 1	S = 3	S = 1	S = 3
D_{5d}	A_{1g}	-1371.920	-1372.075	-31	-28
	A_{2g}	-1371.959	-1372.352	-22	-185
	A_{1u}	-1371.393	-1371.461	-2	-3
	A_{2u}	-1371.400	-1371.573	-34	-20
D_{3d}	A_{1g}	-1372.025	-1372.057	-61	-10
	A_{2g}	-1371.986	-1372.226	-22	-59
	A_{1u}	-1371.443	-1371.617	-52	-11
	A_{2u}	-1371.420	-1371.649	-54	-43
D_{2h}	A_g	-1371.511	-1372.059	-120	-12
	A_u	-1372.088	-1371.649	-124	-183
	B_{1g}	-1371.463	-1372.287	-72	-120
	B_{1u}		-1371.668		-62

Results and Discussion: The optimized bond lengths of various distorted C_{60} structures with electronic wave functions belonging to different one-dimensional irreducible representations of the subgroups of I_h group have been calculated. There are only two different bond lengths in the icosahedral ground state. As the symmetry of the cluster is getting lower the number of different bond lengths is increasing. There are 8 different bond lengths in D_{5d}, 10 in D_{3d} and 15 in the D_{2h} symmetry. The largest change in bond lengths compared to the ground state is less than 0.02 Å.

The energies of the Jahn–Teller distortion for the investigated systems are summarized in Table 1. It can be seen that the lowest energy triplet is of D_{5d} symmetry. This can agree with the experimental results only in solution. At low temperatures in solid phase, since the cubic space group does not contain five-fold axes, the cluster will have the next lowest energy D_{2h} symmetry.

The energies of the systems with different symmetries are quite close to each other. If the temperature is increasing, transitions can occur between the different distorted geometries. Such a dynamic Jahn–Teller effect was also detected in EPR experiments[10].

References

[1] A. D. J. Haymet, *J. Am. Chem. Soc.* **108**, 319 (1986); M. D. Newton, R. E. Stanton, *ibid.* **108**, 2469 (1986)
[2] I. László and L. Udvardi, *Chem. Phys. Lett.* **136**, 418 (1987)
[3] I. László and L. Udvardi, *Journal of Molecular Structure (Theochem)* **183**, 271 (1989)
[4] H. P. Lüthi, J. Almlöf, *Chem. Phys. Letters* **135**, 357 (1986)
[5] K. Tanaka, M. Okada, K. Okahara and T. Yamabe, *Chem. Phys. Letters* **193**, 101 (1992)
[6] R. D. Bendale, J. F. Stanton, M. C. Zerner, *Chem. Phys. Letters* **194**, 467 (1992)
[7] F. Negri, G. Orlandi, F. Zerbetto, *Chem. Phys. Letters* **144**, 31 (1988)
[8] N. Koga, K. Morokuma, *Chem. Phys. Letters* **196**, 191 (1992)
[9] P. R. Surján, L. Udvardi, K. Németh, unpublished
[10] C. A. Steren, P. R. Levstein, H. Willigen, *Chem. Phys. Letters* **204**, 23 (1993)

NEW CLASSES OF POLES AND RESONANT STATES

Cornelia Grama, N. Grama and I. Zamfirescu

Institute of Atomic Physics, PO Box MG-6, RO-76900 Bucharest, Romania

The global method for all S-matrix poles identification [1] is used for a short range complex potential with a real barrier [2]. The pole function $k = k_l(g)$, where g ($g \in C$) is the strength of the nuclear potential, is analyzed by constructing the Riemann surface $R_g^{(l)}$ on which the pole function is single-valued and analytic. Because g covers all the complex plane each Riemann sheet contains simultaneously the well and the barrier, the absorptive and the emissive potentials. In order to construct the Riemann surface $R_g^{(l)}$ the branch-points of the function $k = k_l(g)$ have to be found. The division of the Riemann surface into sheets is done by taking cuts in the g-plane suitably joining the branch-points by rectilinear segments. By keeping the sheets of $R_g^{(l)}$ apart all the S-matrix poles laying on each sheet image over the k-plane can be identified. In this way by analyzing the k-plane image of each Riemann sheet new-class poles are identified and their properties are studied.

The new-class poles are located in finite regions of the k-plane in a neighborhood of some special points called "stable points" on the images of certain Riemann sheets, i.e. these poles do not become bound or virtual state poles as the depth of the potential well increases to infinity. The new-class poles approach to the stable points for a sufficiently deep potential well, the stable points acting like "attractors". For a given potential strength g there is a new-class pole on the image of a certain Riemann sheet and old-class poles on the other images of the Riemann sheets.

The existence of the new-class resonant state poles is related to the branch-points of the function $k = k_l(g)$. The cuts in the g-plane taken in order to separate a given Riemann sheet whose corresponding k-plane image contains new-class resonant state poles determine some windows of the absorptive part of the potential strength. If $Im\ g$ belongs to an absorption window then the new-class resonant state poles exist. In contrast to the imaginary part, the real part of the potential strength for which the new-class resonant state poles exist may take any value larger than a threshold value determined by the head of the cut.

The new-class resonant states that correspond to the new-class poles situated in the neighbourhood of the stable points have wave functions almost completely localized outside the potential well. The small amplitude of the wave function of such a resonant state inside the well leads to a small overlap with the adjacent resonant states of the continuum that are mostly confined to the region $r < R_{CN}$, where the compound nucleus radius R_{CN} is smaller than the potential well radius. This small overlap prevents the decay of the new-class resonant states into the complex neighbouring compound nuclear states. The new-class resonant states are doorway states whose stability against dissolution is a consequence of the localization of the wave function rather than of a symmetry. As an effect of the localization of the wave function outside the potential well radius the new-class resonant states are almost insensitive to the behaviour of the potential inside the well and almost completely determined by the potential in the barrier region. This is exceedingly important due to the fact that the potential is much better known in the barrier region than in the nuclear well region.

The number and position in the k-plane of the bounded regions where the new-class resonant state poles are located depends on the shape of the potential barrier. This is due to the fact that the wave functions of the new-class resonant states are mostly confined to the region outside the potential well. In order to determine the influence of the shape of the potential barrier the following potentials have been studied: Woods-Saxon (square) nuclear potential with a square or centrifugal barrier and Woods-Saxon (square) nuclear potential with Coulomb and centrifugal barrier. The stable points for a square nuclear potential are little shifted by introducing a small diffuseness. On the contrary the absoption windows that determine the occurrence of the new-class resonant state poles change as the diffuseness is introduced.

The existence of the barrier was shown to be essential for the occurrence of the new-class resonant state poles. There are no new-class resonant state poles in the absence of the barrier.

The same potential can support both types of resonant states: those localized in the potential well and those localized outside the potential well. This is because the same value of the potential strength g could provide a new-class resonant state pole or an old-class resonant state pole, depending on the Riemann sheet to which the given g belongs and on the value of $Im\ g$ with respect to the thresholds determined by the cuts.

The special points of the Riemann surface get a remarkable physical meaning: the branch-points are transition points of the quantum system from the old-class resonant states, localized in the region of the potential well, to the new-class resonant states, localized in the region outside the potential well. The stable points are points where the system is almost insensitive to the variations of the potential strength.

As one branch and only one branch of the pole function is associated with a sheet of the Riemann surface we use the label n of this sheet as a quantum number for the poles (both resonant and bound state poles) belonging to the corresponding sheet image in the k-plane and for the associated resonant (bound) states.

References

[1] C. Grama, N. Grama and I. Zamfirescu, *Global method for all S-matrix poles identification. New classes of poles and resonant states*, Ann. Phys. (N.Y.) **218**, 346 (1992)

[2] C. Grama, N. Grama and I. Zamfirescu, *New class of resonant states for potentials with Coulomb barrier. Quasimolecular states*, Ann. Phys. (N.Y.), in press

CLASSICAL PHASE SPACE STRUCTURE INDUCED BY SPONTANEOUS SYMMETRY BREAKING

M. Grigorescu

Department of Theoretical Physics
Institute of Atomic Physics, PO Box MG-6, RO-76900 Bucharest, Romania

The spontaneous symmetry breaking and the structure of the physical vacuum are known as outstanding problems of the present field theory[1], but in a similar form these problems appear also in nuclear physics. The close analogy[2] between the algebraic structure of the Poincaré transformation group for the space-time coordinates and of the group CM(3) for the nuclear collective coordinates suggests that in both cases we face the same basic phenomenon consisting in the occurrence of classical structures at quantum level, but at a different energy scale.

A localized or deformed nuclear mean-field indicates the breaking of translational or rotational symmetry in the many-body ground state and Goldstone bosons appear as the "spurious modes" of the RPA approximation. These modes are related to large amplitude collective motions of the system and their treatment requires to define at least locally canonical coordinates and momenta. Here[3] this problem is solved by considering the collective dynamics as a special case of low-energy quantum dynamics occurring when the quantum ground state $|g>$ is non-invariant to a continuous symmetry group $\mathcal{G}$ of the Hamiltonian. In the many-body case this situation appears often for the approximate ground states $|g^M>$ given by HF or HFB calculations which are critical points of minimum energy for the classical system obtained by constraining the quantum dynamics from the Hilbert space $\mathcal{H}$ to some finite-dimensional trial manifold M. This is assumed to be a $\mathcal{G}$-invariant symplectic manifold (M, ω^M) with ω^M defined naturally[4] by restricting to M the symplectic form on $\mathcal{H}$, $\omega_\psi^\mathcal{H}(X,Y) = 2Im < X|Y >$, $X, Y \in T_\psi \mathcal{H}$.

If $|g^M>$ is a symmetry breaking ground state, then a whole critical submanifold $Q \subset M$, $Q = \mathcal{G} \cdot |g^M>$ may be generated by the action of $\mathcal{G}$. Suppose that ω^M vanishes on the tangent space TQ of Q, if Q is an isotropic submanifold of M. When the algebra g of $\mathcal{G}$ is semi-simple, then the representation operators for g have vanishing expectation values on $|g^M>$. In particular for the deformed ground states the average of the angular momentum operators should be zero, and Q appears as the coordinate space for the rotational collective degrees of freedom.

If Q is isotropic, then at any $q \in Q$ the tangent space $T_q Q$ has a coisotropic ω^M-orthogonal complement F_q and the quotient $E_q = F_q/T_q Q$ is a symplectic vector space[4]. Consider P_q to be the complement of F_q, that is $T_q M = P_q + F_q$. Then P_q is isotropic, with the same dimension as $T_q Q$, and such that ω^M restricted to $P_q \times T_q Q$ is non-degenerate. Thus locally at every point $q \in Q$ one has a classical phase space structure with P_q representing the space of the momenta canonically conjugate to the collective coordinates. The remaining "intrinsic" variables are represented within E_q.

A simple and instructive application of this geometrical construction concerns the nontrivial case of the canonical momentum associated to a single angle coordinate ϕ. Let $\mathcal{G}_x$ be the group of rotations around the X axis generated by the orbital angular momentum operator L_x, $\mid g^M >$ a deformed ground state and $J : M \to R$ the "momentum mapping" $J(\mid Z >) = < Z \mid L_x \mid Z >$. Then for any regular value I of J, $\mathcal{F}_I = J^{-1}(I)$ is invariant to $\mathcal{G}_x$ and $Q = \mathcal{G}_x \cdot \mid g^M > \subset \mathcal{F}_0$. Moreover $F_q = T_q \mathcal{F}_0$, and a 1-dimensional complement P_q for F_q in M is provided by the tangent to any curve transversal to $\mathcal{F}_0$ at q. This ambiguity may be further solved by using dynamical arguments and selecting the "yrast" transversal joining continuously the minimum energy points from each $\mathcal{F}_I$ near $\mathcal{F}_0$. Such constrained minima of the Hamiltonian H can be generated using a standard cranking calculation[5], obtaining first a solution $\mid Z >_\omega$ of the variational equation $\delta < Z \mid H - \omega L_x \mid Z >= 0$, and then fixing ω at ω_I given by $J(\mid Z >_{\omega_I}) = I$. The result is a symplectic manifold $\mathcal{S} = \{\mid Z >_{(\phi,I)} = e^{-i\phi L_x} \mid Z >_{\omega_I}\}$ parameterized by the canonical variables ϕ and I.

If H is a nuclear Hamiltonian consisting of a single-particle spherical oscillator term and a quadrupole-quadrupole (QQ) interaction, and M is the manifold of the HF states, then $\mid Z_\omega >$ is a Slater determinant constructed with the cranked anisotropic oscillator eigenstates. These eigenstates are connected with the spherical harmonic oscillator eigenfunctions by the unitary operator $U = exp(-i\lambda c_x)exp(-i\sum_{k=1}^{3} \theta_k s_k)$ with $c_x = b_2^+ b_3 + b_3^+ b_2$, $s_k = i(b_k^{+2} - b_k^2)/2$, $b_k^+ = \sqrt{m\omega_0/2}(x_k - \frac{i}{m\omega_0}p_k)$, $\omega_0^2 = (\omega_2^2 + \omega_3^2)/2$, $tan2\lambda = 2\omega/\omega_0\eta$, $sinh2\theta_k = \omega_0(1 - \omega_k^2/\omega_0^2)/2\Omega_k$. Here ω_k are the anisotropic oscillator frequencies, $\eta = (\omega_2^2 - \omega_3^2)/2\omega_0^2$, $\Omega_1 = \omega_1$, $\Omega_{2,3}^2 = (\omega_0 + \epsilon_{2,3})^2 - (\omega_0\eta/2)^2$, $\epsilon_2 = -\epsilon_3 = \omega_0\eta/2cos2\lambda$. The operators $s_{1,2,3}$ generate the transition from a "spherical" to a "deformed" basis, while c_x appears as an "angle" operator because it generates the shift of the expectation value in the angular momentum. A direct application of this angle operator is provided by the treatment of the nuclear low-lying isovector "scissors" vibrations in deformed nuclei considered in the study of nuclear magnetism. Constructing manifolds $\mathcal{S}_p$ and $\mathcal{S}_n$ separately for protons and neutrons, and using then their direct product as trial function in a time-dependent variational calculation for a microscopic Hamiltonian including both isovector and isoscalar QQ interaction terms, it can be shown[3,6] that the excitation operator for the scissors modes is $B^+ = [a^p L_x^p - a^n L_x^n - \frac{i\Omega}{\omega_2-\omega_3}(a^p C_x^p - a^n C_x^n)]/2$, where Ω is the scissors frequency, $C_x^{p,n} = \sum_{i=1}^{Z,N} c_x^i$, and $a^{p,n}$ are the quantized angular amplitudes.

References

[1] T.D. Lee, *Missing symmetries, unseen quarks and the physical vacuum*, Nucl. Phys. A **538**, 3c (1992)

[2] G. Rosensteel, E. Ihrig, *Geometric quantization of the CM(3) model*, Ann. of Phys. **121**, 113 (1979)

[3] M. Grigorescu, *Classical phase space structure induced by spontaneous symmetry breaking*, Preprint MSUCL-860-September (1992)

[4] R. Abraham, J. Marsden, *Foundation of mechanics*, (Benjamin/Cummings Publishing Company, 1978)

[5] P. Ring, P. Schuck, *The nuclear many-body problem*, (Springer, Berlin, 1980)

[6] M. Grigorescu, *Cranking model for M1 states*, Rev. Roum. Phys. **34**, 1147 (1989)

Author Index

Abbondanno, U., 421
Ackermann, D., 145
d'Agostino, M., 421
Alexa, C., 479
Anagnostatos, G. S., 285
Apagyi, B., 419
Ardisson, G., 263
Aryaeinejad, R., 101
Avrigeanu, M., 99
Avrigeanu, V., 99, 403

Babu, B. R. S., 101
Bahri, C., 189
Barci, V., 263
Basrak, Z., 421
Bass, St. A., 439
Batra, J. S., 307
Beghini, S., 145
Berger, J. F., 181
Berinde, A., 257
Bettiolo, M., 421
Bitaud, L., 181
Blomqvist, J., 283
Börner, H. G., 261
Bozdog, H., 395
Brâncuş, I.M., 395
Braunss, G., 217
Brentano, P. von, 221, 245, 253, 255
Brillard, L., 263
Brown, B. A., 355
Bruno, M., 421
Butler-Moore, K., 101

Cata-Danil, G., 253
Cerruti, C., 435
Cindro, N., 421
Clausnitzer, G., 353, 355
Cole, J. D., 101
Constantinescu, O., 263
Corradi, L., 145
Covello, A., 207
Cseh, J., 295

Daniel, A. V., 101
Decharge, J., 181
Dingfelder, M., 243
Draayer, J. P., 189
Duchêne, G., 259
Duma, M., 395
Dumitrescu, O., 73

Faessler, A., 243
Florescu, A., 79
Freeman, E. S., 369

Garcia, J. B., 435
Gargano, A., 207

Gazaix, S., 435
Gelberg, A., 255
Gherghescu, R., 147
Girod, M., 181
Gönnenwein, F., 113
Grama, C., 433, 489
Grama, N., 433, 489
Graw, G., 253
Greenwood, R. C., 101
Greiner, M., 217, 447
Greiner, W., 3, 45, 57, 407, 439
Grigorescu, M., 491
Grün, N., 453
Gupta, R. K., 129, 295, 307

Hamilton, J. H., 101
Harangoza, A., 403
Hartnack, C., 439
Herzberg, R.-D., 221
Hess, P. O., 217, 307
Heumann, D., 217
Hodgson, P. E., 99
Hofer, D., 253
Hoffstadt, J., 453
Horoi, M., 355
Hussonnois, M., 263

Iacob, V. E., 257
Insolia, A., 79, 283
Ion, D. B., 467
Isar, A., 481
Itaco, N., 207

Johnson, N. R., 101
Jolie, J., 261
Jungclaus, A., 261

Kirch, K., 245
Klapdor-Kleingrothaus, H. V., 311
Kliman, J., 101
Kneissl, U., 221
Kormicki, J., 101
Kretschmer, M., 395
Kubo, K., 379

LeDu, J. F., 263
Lee, I. Y., 101
Lévai, G., 295
Lieb, K. P., 261
Lind, P., 281
Liotta, R. J., 81, 87, 281, 283
LoIudice, N., 229
Lovas, I., 441
Lovas, R. G., 87
Lu, Q., 101
Ludu, A., 57, 483

Ma, W.-C., 101

493

Magda, M. T., 169
Maglione, E., 281
Malik, S. S., 307
Mathes, H.J., 395
McGowan, F. K., 101
Milazzo, P. M., 421
Mirea, M., 141
Mişicu, Ş., 267
Montagnoli, G., 145
Morhac, M., 101
Mueller, L., 145
Müller-Zanotti, E., 253
Münzenberg, G., 157

Napoli, D. R., 145
Németh, K., 487
Neuneyer, G., 245
Nica, N., 257
Nojarov, R., 243

Oganessian, Y. T., 101, 263
Oros, A. M., 253

Park, J. Y., 407
Parlog, M., 257
Pascovici, G., 395
Petcu, M., 395
Petrache, C., 145
Petrovici, A., 279
Pitz, H. H., 221
Poenaru, D. N., 45
Polhorsky, V., 101
Pollarolo, G., 145
Popa, G., 71
Popeko, G. S., 101
Preiß, M., 353
Price, P. B., 39, 369

Raduta, A. A., 331
Ramayya, A. V., 101
Rebel, H., 359, 395
Ricci, R. A., 421
Rowley, N., 145

Sandulescu, A., 57, 71, 143
Sandulescu, N., 283
Scarlassara, F., 145
Scheid, W., 71, 217, 295, 307, 407, 419, 421, 453, 485
Schmid, K. W., 269, 279
Schmidt, J., 421
Scintei, N., 257
Segato, G. F., 145
Shi, D., 101
Siems, G., 245
Signorini, C., 145
Silisteanu, I., 71
Skalski, J., 151
Smolańczuk, R., 151

Snowden-Ifft, D. P., 369
Sobiczewski, A., 151
Soff, G., 447
Soramel, F., 145
Spolaore, P., 145
Stan-Sion, C., 257
Stefanescu, E., 143
Stefanini, A. M., 145
Stocker, W., 467
Stöcker, H., 439
Stoica, S., 343
Surján, P., 487
Szeglowski, Z., 263

Ter-Akopian, G. M., 101
Thiel, A., 407
Thiel, J., 453
Trache, L., 253, 257
Troltenier, D., 189
Trubert, D., 263

Udvardi, L., 487
Ulbig, S., 261

Vannini, G., 421
Vannucci, L., 421
Varga, K., 87, 295
Vertse, T., 281
Vidović, M., 447
Vogel, O., 255

Wang, M. G., 101
Warburton, E. K., 355
Wentz, J., 395
Wiedenhöver, I., 245

Zamfirescu, I., 433, 489
Zell, K. O., 253
Zhu, S., 101
Zielke, A., 485
Zilges, A., 221
Zimmer, K.W., 395

Subject Index

abundance of ^{9}Be, 379
algebraic nuclear models, 189, 295
algebraic scattering theory, 485
alpha decay, 45, 57, 71, 73, 79, 81, 87, 99, 151, 157, 181
^{242}Am(n,f), 113
anharmonic and deformation effects, 331
atomic force microscopy, 369

Ba isotopes, 245
back-bending effect, 307
barrier penetrability, 45
barrier shape, 147
BCS approximation, 207
BCS shell model configuration mixing, 79
Berggren representation, 81
beta stability, 181
beta transition strength, 331
big bang models, 379
binary fission modes of shell nuclei, 295
boson expansion, 331, 343
boson models, 189
bosonized formalism, 453
branching ratios, 39
burst counter, 395

C_{60} clusters, 487
Ce-isotopes, 101
^{252}Cf SF data, 101
chemistry of transactinides, 157
Cherenkov mechanisms, 467
classical fragmentation, 435
cluster decay, 3, 87, 129, 435
cluster emission, 45, 141
cluster formation, 435
cluster models, 285, 295
cluster preformation probability, 129
cluster radioactivity, 3, 39, 57, 71, 101, 129
cluster states, 295
cluster structure of the quark gluon plasma, 3
clustering of nucleons, 81
cold fission, 3, 45, 101, 113, 129, 143, 147
cold fusion, 3, 147
cold mass asymmetric fission, 113
cold reaction valleys, 129
cold scission, 113
cold shape asymmetric fission, 113
colour-singlets, 3
combined shell and cluster model, 87
complete fusion, 157
continuum random phase approximation, 281
cosmic gamma radiation, 467
cosmic rays, 395
Coulomb dissociation, 359
coupled channel calculations, 145, 269, 407, 453
CWKB calculations, 145

damped harmonic oscillator, 481
dark matter, 369
decay of ^{76}Ge, 311
decay of neutral to charged vacuum, 3
decay rates and width, 39, 87
deformation and angular momentum, 307
deformation energy, 147
deformation of fragments, 3
deformed nuclei, 207
Dirac particle, 311
dissipative processes, 143, 481
Doppler technique, 261
double beta decay, 311, 331, 343
double magic superheavies, 157
DWBA cross sections, 243
dynamical O(6) symmetry, 245
dynamical symmetry, 485

electromagnetic particle production, 447
electron scattering, 267, 269
electron-positron pair creation, 453
electron-positron-coincidences, 3
element Z=110, 157
energy spectra and electric transitions, 295
energy surfaces, 189
^{163}Er, 207
escape width of giant resonances, 81
Euler-Lagrange equations, 141
even-odd effects in fission, 113
excitation function, 421
exotic decays, 79

fast radiochemical separation, 263
fermion dynamical symmetries, 189
finite difference method, 453
fission, 157
fission barriers, 181
fission dynamics, 45
fission trajectories, 141
Fokker-Planck equation, 481
folded Yukawa potential, 181
fragment deformation, 3
fullerenes, 487
fusion cross sections for ^{58}Ni + ^{58}Ni, 145
fusion hindrance, 157

gamma and meson production, 467
gamma transition in Xe and Cs, 255
gamma-gamma coincidences, 257
Gamow penetrability, 39
Gamow theory, 81
Geiger-Nuttall law, 39, 129
giant monopole resonance in ^{208}Pb, 281
giant pairing resonance, 81
ground state yrast band, 307
GUT model, 311

495

hadron production, 379
hadronic weak interaction, 353
hafnium isotopes, 263
half-lives, 39
harmonic oscillator, 285
Hassium, 157
heavy ion reactions, 157, 169, 407, 419, 421, 435, 439
heavy ion resonances, 407, 419, 421, 433
heavy-particle decay, 73
Heidelberg-Moscow experiment, 311
hexadecapole stabilized nuclei, 157
hidden flavor, 441
high spin states, 101, 283
higher order QRPA, 343
hindrance factors, 45
hot nuclear matter, 439
hydrodynamical model, 267
hydrodynamical model for solitons, 57

in-flight separation, 157
inelastic electron scattering, 243
instantons in QCD, 479
interacting boson model, 189, 245
intranuclear transition rates, 403
inverse scattering problem, 419, 485
isomeric beta-decay, 245

Jahn-Teller distorted states, 487

K-hole creation probability, 3

Landau-Zener effect, 407
^{11}Li, 285
Lie algebra, 483
linearization of Schrödinger equation, 217
liquid drop models, 285
low lying collective states in Ba, 245

M1 transitions, 229, 243
magic deformed clusters, 3
Majorana particle, 311
master equation, 143, 481
Meitnerium, 157
Memos, 3
mica crystals, 369
molecular particle core model, 407
moment of inertia, 259, 307
multi-Λ-hypernuclei, 3
multi-Ξ-hypernuclei, 3
multi-pionic atoms, 3
multinucleon transfer, 145, 169

N-Z sum rule, 331
neutral vector mesons, 441
neutrino mass, 311
neutron decay of giant resonances, 81, 281
neutron multiplicities, 101
neutron-gated angular distributions, 257

^{58}Ni + ^{58}Ni, 421
^{58}Ni + ^{62}Ni, 421
Nielsbohrium, 157
NN amplitudes, 379
non Yrast states, 245
non-compact groups SO(2,3), SO(3,1), 485
nuclear astrophysics, 359, 369, 379, 395, 467
nuclear level density, 403
nuclear media, 467
nuclear molecules, 407, 419, 421
nuclear pair conversion, 3
nuclear radii, 285
nuclear reactions with ^{138}Ba, ^{139}La, ^{140}Ce, ^{141}Pr, ^{142}Nd, ^{143}Nd, 221
nuclear resonance fluoresence, 221
nuclear shapes, 307
nuclear shock waves, 3
nuclear tracks in solids, 369
nuclei far-off stability, 157
nucleon bags, 285
nucleosynthesis, 359, 379

octupole deformation in Ba nuclei, 101
octupole phonon, 221
one photon approximation, 359
open quantum systems, 143, 481
optical model potentials, 99, 403
Ornstein-Uhlenbeck process, 481
OSIRIS cube detector array, 245
overcritical QED-vacuum, 3

pairing correlations, 207
parity mixing, 353
parity non-conservation in A=18-21, 355
particle modes in ^{145}Sm, 253
penetration factor, 39, 81
phonon modes in ^{145}Sm, 253
photodisintegration, 359
photon scattering, 221
pion emission, 439
PNC matrix element, 355
potentials for ^{12}C + ^{12}C, 419
pre-compound particle emission, 157
pre-equilibrium emission, 99, 403
preformation factor, 57
preformation probability, 45
preformed cluster model, 129
production of antibaryons and antinuclei, 3
projection techniques, 269, 279
prompt gamma rays in spontaneous fission, 101
properties of heaviest nuclei, 151
proton rich Ba isotopes, 45
pseudo-spin symmetries, 189
pseudo-SU(3)-model, 189

QRPA calculations, 243, 311, 343
quadrupole phonon, 221
quantum deformation of SU(2), 483
quantum tunneling, 143

quark-gluon plasma, 3, 441
quasimolecular states, 407, 419, 421, 433, 489
quasiparticle excitations, 283

R-matrix formalism, 79
RACHEL, 263
radiative capture, 359
radioactive nuclear beams, 169
reaction (n,α), 99, 403
relativistic atomic heavy ion collisions, 453
relativistic heavy ion collisions, 3, 439
relativistic mean field theory, 3
relativistic quantum molecular dynamics, 3
resonance matter, 3
resonance scattering, 71
resonance structures, 421
resonant states, 281, 489
RICH detectors, 441
Riemann surface, 433
ring image Cherenkov detectors, 441
rotational band coexisting with a spherical band, 257
rotational motion, 189
rotational symmetry, 259
rotor model, 207
RPA formalism, 81

S-matrix poles, 433, 489
saddle-point fission model, 129
sampling calorimeter, 395
scissors modes, 189, 229, 243, 491
secondary beams, 157
semimicroscopic algebraic approach to clusterization, 295
shape coexistence in ^{111}Sb, 257
shell model, 279, 285
shell stabilization effects, 129, 157
SHIP, 157
short lifetimes, 261
Siegert theorem, 267
softening of rigid rotator, 307
solitons on sphere, 57
SOSY approach, 479
specialization energy, 45
spin polarization, 379
spontaneous fission, 101
spontaneous fission half-lives, 151
spontaneous symmetry breaking, 491
statistical model, 99, 403
SU(3)/Sp(3,R) models, 189
superasymmetric fission model, 3
superdeformation, 229, 259
superdeformed states, 295
superfluid transition, 71
superheavy elements, 151, 157, 181
symplectic model, 189
synthesis of heavy and superheavy elements, 3, 151, 157, 169

ternary fission, 113
TeV muons, 395
^{229}Th(n,f), 113
themal neutron capture, 261
toroidal multipoles, 267
track-recorder, 39
tracks of recoil atoms, 369
trans-Francium nuclei, 45
transactinides, 157, 181, 263
transfer channels, 145
translational invariance, 269
triaxial rotor plus particle model, 255
two neutrino double beta decay, 331
two phonon and particle states, 221
two-center shell model, 307, 407
two-rotor model, 229

U^{92+} (10 GeV/nucleon) + U^{92+}, 453
U-U giant nuclear system, 3
^{235}U(n,f), 113
ultra high resolution crystal spectrometer, 261
ultrarelativistic heavy ion collisions, 3, 441, 447

variational methods, 279
vibrational motion, 189

weakly interacting massive particles, 369
Wigner distribution function, 481
Wigner-Thomas theory, 81
WKB approximation, 79

Yang-Mills theory, 479

The manufacturer's authorised representative in the EU is Springer Nature Customer Service Centre GmbH, Europaplatz 3, 69115 Heidelberg, Germany. If you have any concerns regarding our products, please contact ProductSafety@springernature.com

Printed and bound by CPI Group (UK) Ltd, Croydon, CR0 4YY
02/01/2026
02028262-0010